Concise Medical Textbooks

Concise Medical Textbooks

Biochemistry

J. H. Ottaway

BSc, PhD, DSc
Senior Lecturer in Biochemistry
University of Edinburgh
Medical School

D. K. Apps

MA, PhD
Lecturer in Biochemistry
University of Edinburgh
Medical School

Fourth Edition

Baillière Tindall

London

Published by Baillière Tindall
1 St Anne's Road, Eastbourne
East Sussex BN21 3UN

First published 1965
Second edition 1969
Third edition 1976
Reprinted 1979
Fourth edition 1984

Printed and bound in Great Britain
at the University Press, Cambridge

British Library Cataloguing in Publication Data

Ottaway, J. H.
Biochemistry — 4th ed.
1. Biological chemistry
I. Title   II. Apps, D. K.
III. Datta, S. P.
574.19'2.   QP514.2

ISBN 0-7020-0903-2

Contents

Preface to the Fourth Edition

Dr Datta and I completed the first edition of *Aids to Biochemistry* some twenty-five years ago, never doubting that it would be both our first and last foray into book-writing. Since 1960 the text has, however, passed in one or another guise through six editions and the present volume will be the fourth in the Concise Medical Textbook series. The fact that it continues to supply a need among students has enabled the publishers to use a more generous and attractive format with more illustrations, and two-colour printing.

The first edition was unashamedly a 'cram book' for examinations, designed to include the maximum amount of information within the minimum compass. This approach has been somewhat relaxed in subsequent editions, but our aim is still to present, as concisely as is consistent with lucidity, a core of biochemical knowledge and understanding that is desirable for all medical students. The choice of this core material is in the last instance a matter of judgement, and it is our judgement that has dictated the content of this book.

The principles of relevance and conciseness have helped us to keep the book as small as possible; nevertheless biochemistry has grown enormously in the last twenty-five years. For each new edition a choice has been made of new material to be brought in, and older material which must perforce be left out. In this edition the emphasis has rightly swung towards nucleic acid and protein synthesis, and the structure and function of membranes. The section on the acid–base balance of blood has been greatly reduced. (Baillière Tindall have published a book devoted entirely to this topic, entitled *Medical Acid–Base Balance: The Basic Principles*, by Dr M. L. G. Gardner.)

The chapter on hormones has been completely deleted because of the immense growth in knowledge, which has unfortunately far outstripped our capacity to codify or even to explain it. To an extent, this is due to the recent realization of the very large number of compounds, secreted by one cell type and influencing another, which are nevertheless not products of the classical endo-crine system. Prostaglandins, thromboxanes, endorphins and somatostatin are all examples of extracellular effector molecules that do not readily fit into the present conceptual framework. When one finds that even the biologically active derivative of vitamin D has been described in scientifically respectable circles as a 'hormone', one comes to feel that the subject is both too large, and too incoherent, to be satisfactorily treated in a concise text such as this. It is our hope that some new author, endowed with both ruthlessness and ingenuity, will soon come forward to assist biochemical endocrinology.

We have retained a chapter on control mechanisms, dealing with such topics as receptor binding, covalent modification of proteins, feedback and cascade mechanisms, and the balance between enzyme synthesis and destruction, because we believe that it is through these mechanisms that a majority of extra-cellular effectors operate.

Finally, on a more personal note, the appearance of this volume signals the end of a long co-operation with a friend and colleague of many years, Prakash Datta. He is replaced by a colleague from this department, David Apps, whose talents and outlook will undoubtedly maintain the personality that this book has achieved for itself over the years.

J. H. Ottaway

Edinburgh
June 1983

Definitions

Before starting the book proper it will be useful to define some units which frequently occur in biochemistry.

Mole. One mole of a substance is the mass of that substance in grams that is numerically equal to its molecular weight.

Molar solution. A molar solution (abbreviation: M) of a substance is one which contains one mole of that substance in one litre of solution.

The terms mole and molar solution are extended to ions; thus a one molar solution of disodium hydrogen phosphate (Na_2HPO_4) is one molar with respect to phosphate ions and two molar with respect to sodium ions.

Equivalent. One equivalent (abbreviation: equiv.) of an ion is the weight of one mole of that ion divided by the number of its electric charge. Thus one equivalent is equal to one mole for singly charged ions (Na^+, Cl^-), to $\frac{1}{2} = 0.5$ moles for doubly charged ions (Ca^{2+}, HPO_4^{2-}), to $\frac{1}{3} = 0.33$ moles for triply charged ions (Fe^{3+}, citrate^{3-}), etc. Or, conversely, one mole of a singly charged ion contains one equivalent, one mole of a doubly charged ion contains two equivalents, one mole of a triply charged ion contains three equivalents, etc.

Multiples and submultiples of units are shown by prefixes as follows:

Multiplier	Prefix	Symbol	Example
10^9	giga	G	gigabecquerel (GBq)
$1\,000\,000 = 10^6$	mega	M	megaunit (Mu)
$1\,000 = 10^3$	kilo	k	kilogram (kg)
$0.1 = 10^{-1}$	deci	d	decimetre (dm)
$0.01 = 10^{-2}$	centi	c	centimetre (cm)
$0.001 = 10^{-3}$	milli	m	millilitre (ml)
$0.000001 = 10^{-6}$	micro	μ	microlitre (μl)
10^{-9}	nano	n	nanometre (nm)
10^{-12}	pico	p	picogram (pg)
10^{-15}	femto	f	femtomole (fmol)

Only one multiplying prefix is used at one time to a given unit. Thus one thousandth of a millimole is not called a millimillimole (1 mmmol) but is known as one micromole (1 μmol) and instead of 1 millimicrometre (0.001 μm) one writes 1 nanometre (nm).

Examples. The following examples will be found useful in the laboratory.

A 1 molar (M) solution contains 1 mole per litre (1 mol/l), 1 millimole per millilitre (1 mmol/ml), and 1 micromole per microlitre (1 μmol/μl).

A 1 millimolar (1 mM) solution contains 1 millimole per litre (1 mmol/l), 1 micromole per millilitre (1 μmol/ml), and one nanomole per microlitre (1 nmol/μl).

Each millilitre of a solution which is 1 millimolar with respect to *both* KH_2PO_4 and Na_2HPO_4 contains 1 μmole K^+, 2 μmoles Na^+ and 2 μmoles total phosphate ($H_2PO_4^- + HPO_4^{2-}$). *Each litre* of such a solution contains 1 meq. K^+, 2 meq. Na^+, and 3 meq. total phosphate, made up of 1 meq. $H_2PO_4^-$ and 2 meq. HPO_4^{2-} since each mole of HPO_4^{2-} is equal to 2 equivalents.

Joule (J). The joule is a unit of energy and is numerically equivalent to 0.239 calorie. One calorie is equal to 4.184 joules (J) and one kilocalorie is equal to 4.184 kilojoules (kJ).

Ångström (Å). This is a unit of length.

$$1 \text{ Å} = 10^{-8} \text{ cm} = 10^{-10} \text{ m} = 0.1 \text{ nm}.$$

Pascal (Pa). This is a unit of pressure, equal to 1 newton per square metre. The newton (N), a unit of force, is given by

$$1 \text{ N} = 1 \text{ kg m s}^{-2}.$$

Hence $1 \text{ Pa} = 1 \text{ N m}^{-2} = 1 \text{ kg m}^{-1} \text{ s}^{-2}.$

$$1 \text{ kPa} \equiv 75 \text{ mmHg}.$$

Becquerel (Bq). The S.I. unit of radioactivity. 1 Bq is the quantity of a radioisotope which disintegrates in one second. An obsolescent unit of radioactivity, the curie (Ci), is defined as the quantity of a radioisotope undergoing 3.7×10^{10} disintegrations per second, i.e. $1 \text{ Ci} = 3.7 \times 10^{10}$ Bq, or $1 \text{ mCi} = 37 \text{ MBq}$.

1

Hydrogen Ion Concentration and Buffers

Acids and Bases

An acid is a molecular species tending to lose a hydrogen ion (proton), whereas a base is a species tending to add on a hydrogen ion. The dissociation of a hydrogen ion from an acid may be represented by the equilibrium

$$HA \rightleftharpoons B + H^+ \qquad (1)$$

Since the dissociation is reversible the species B formed when A loses a hydrogen ion is in fact a base. Such a pair of species is known as a conjugate acid–base pair. Since an acid loses a hydrogen ion to form its conjugate base it follows that the acid must always have a charge which is one unit more positive than its conjugate base. These points are illustrated in the following equilibria:

$$HCl \rightleftharpoons H^+ + Cl^- \qquad (2)$$

$$CH_3 \cdot COOH \rightleftharpoons H^+ + CH_3 \cdot COO^- \qquad (3)$$

$$H_2PO_4^- \rightleftharpoons H^+ + HPO_4^{2-} \qquad (4)$$

$$NH_4^+ \rightleftharpoons H^+ + NH_3 \qquad (5)$$

In all biological systems the solvent is water, which can itself act as an acid or as a base, as is shown by the following equilibria:

$$CH_3 \cdot COOH(A_1) + H_2O(B_2) \rightleftharpoons H_3O^+(A_2)$$
$$+ CH_3 \cdot COO^-(B_1) \qquad (6)$$

and

$$R \cdot NH_2(B_1) + H_2O(A_2) \rightleftharpoons OH^-(B_2) + R \cdot NH_3^+(A_1) \qquad (7)$$

The ion product of water

Because of the acidic and basic properties of water it follows that interactions between water molecules themselves will give rise to H_3O^+ and OH^- ions, thus

$$H_2O(A_1) + H_2O(B_2) \rightleftharpoons H_3O^+(A_2) + OH^-(B_1) \qquad (8)$$

and the ion product of water, K_w, is given by

$$K_w = [H_3O^+][OH^-] \qquad (9)$$

The constant, K_w, has a value of about 10^{-14} moles^2/litre^2 at ordinary temperatures, and it follows from (9) that there is a reciprocal relation between $[H_3O^+]$ and $[OH^-]$. When $[H_3O^+] = [OH^-]$, the concentration of

$H_3O^+ = \sqrt{K_w} = 10^{-7}$ moles/litre. This is the concentration of H_3O^+ ions at *neutrality*.

When acids dissociate in water, as in (6), they give rise to the hydronium ion H_3O^+ and not to the hydrogen ion H^+, though for simplicity we shall always refer to the hydrogen ion and write H^+.

Strong acids

When strong mineral acids are dissolved in water the dissociation of the hydrogen ion may be considered to be complete. Thus HCl, $HClO_4$, HNO_3 and the first hydrogen of H_2SO_4 are completely ionized in dilute solution. In other words equilibrium (1) is completely over to the right.

Weak acids

When weak acids such as $CH_3 \cdot COOH$, H_3PO_4, $H_2PO_4^-$, HPO_4^{2-}, HSO_4^- and $CH_3 \cdot NH_3^+$ are dissolved in water they are incompletely dissociated, that is to say both the acids and their conjugate bases are present in the solution in similar concentrations. All these dissociations may be represented by the general equilibrium

$$HA \rightleftharpoons A^- + H^+ \qquad (10)$$

where the charge on the conjugate base, A^-, is one unit less positive than on the conjugate acid, HA.

Acid dissociation constants

The Law of Mass Action may be applied to these equilibria giving (from 10)

$$K_{HA} = \frac{[A^-][H^+]}{[HA]} \qquad (11)$$

where K_{HA} is the equilibrium or acid dissociation constant of the acid HA. The constant K_{HA} has the dimensions of concentration and is a measure of the 'strength' of the acid; the larger the value of K_{HA} the 'stronger' the acid. The following acids are arranged in order of their 'strengths' at 25°C: H_3PO_4, $K = 8.91 \times 10^{-3}$ M; $CH_3 \cdot COOH$, $K = 2.24 \times 10^{-5}$ M;

3

$H_2PO_4^-$, $K = 1.58 \times 10^{-7}$ M; $CH_3 \cdot NH_3^+$, $K = 2.40 \times 10^{-11}$ M and $K_w = 10^{-14}$ M.

pH and pK

The numerical values of $[H^+]$ and K with which we have to deal are very small, such as the values of K listed above; $[H^+]$ at neutrality $= 10^{-7}$ moles (100 nanomoles)/litre. To simplify calculations the pH and pK scales are used; these are defined as the negative logarithms to the base 10 of the hydrogen ion concentration and the acid dissociation constant respectively.

$$pH = -\log[H^+] = \log\frac{1}{[H^+]} \qquad (12)$$

$$pK = -\log K = \log\frac{1}{K} \qquad (13)$$

The hydrogen ion concentration of the blood, which is kept fairly constant, can also conveniently be expressed as nanomoles of hydrogen ion per litre. Normally blood has a pH = 7.4 or $[H^+] = 40$ nmol/l (see chapter 15).

It follows from these definitions that the functions pH and pK have the following important properties:

(a) The higher the hydrogen ion concentration $[H^+]$, in moles/litre, the lower the pH and vice versa, e.g. if $[H^+] = 3 \times 10^{-7}$ M, pH = 6.523 and if $[H^+] = 2 \times 10^{-4}$ M, pH = 3.699. Similarly the lower the pK the greater K and the 'stronger' the acid; thus at 25°C for H_3PO_4, pK = 2.05; $CH_3 \cdot COOH$, pK = 4.65; $H_2PO_4^-$, pK = 6.8; $CH_3 \cdot NH_3^+$, pK = 10.62, and $pK_w = 14$.

(b) A tenfold change in $[H^+]$ or K corresponds to a change of one unit in pH or pK; e.g.

$[H^+] = 10^{-6}$ mol/l or 1 μmol/l, pH = 6

$[H^+] = 10^{-7}$ mol/l or 100 nmol/l, pH = 7

$[H^+] = 10^{-8}$ mol/l or 10 nmol/l, pH = 8

The relation between pH and pK

We can now rewrite equation (11) in terms of pH and pK to give the very important equations:

$$pK = pH - \log\frac{[A^-]}{[HA]} \qquad (14)$$

and

$$pH = pK + \log\frac{[A^-]}{[HA]} = pK + \log\frac{[\text{conjugate base}]}{[\text{conjugate acid}]} \qquad (15)$$

From (15), the Henderson–Hasselbalch equation, it follows that pH = pK when $[A^-] = [HA]$, i.e. when the acid is half neutralized.

The pH of a solution of a weak acid

A weak acid dissociates in solution as shown in equations (10) and (11), and further it is necessary for the solution to remain electrically neutral, i.e. there must be the same *total* number of positive charges on the ions as there are negative charges. Then ignoring the small $[OH^-]$ in an acid solution and assuming HA is uncharged, the electroneutrality condition is:

$$[A^-] = [H^+] \qquad (16)$$

Let the total concentration of acid be A_T, then from (10)

$$A_T = [A^-] + [HA] \qquad (17)$$

and from (11)

$$[H^+][A^-] = K[HA] \qquad (18)$$

Combining (16), (17) and (18) we have:

$$[H^+]^2 = K(A_T - [H^+]) \qquad (19)$$

In a dilute solution of a weak acid $[H^+]$ may be assumed to be small compared with A_T, so (19) becomes

$$[H^+] \approx \sqrt{(KA_T)} \qquad (20)$$

or

$$pH = \tfrac{1}{2}pK - \tfrac{1}{2}\log A_T \qquad (21)$$

For example, if we have solutions of acetic acid (pK = 4.65) at concentrations of 0.1, 0.01 and 0.001 M, their pH's will be given by:

$$pH = \tfrac{1}{2}(4.65) - \tfrac{1}{2}\log(0.1, 0.01 \text{ and } 0.001)$$

$$= 2.325 + 0.5, 1.0 \text{ and } 1.5$$

$$= 2.825, 3.325 \text{ and } 3.825 \text{ respectively.}$$

Buffers

These are solutions of weak acids, HA, and their salts, MA; such systems resist changes in the pH when acid or alkali is added to the solution.

The acid HA is by definition a weak acid so we may assume that it is only very slightly dissociated and the concentration [HA] is equal to the *total concentration of acid added*. Further we may assume that the salt MA, if it is an alkali metal salt, is completely dissociated into

M^+ and A^- and hence the concentration of the conjugate base, $[A^-]$, is equal to the *total concentration of salt added*. Equation (15) then becomes:

$$pH = pK + \log\frac{[\text{salt}]}{[\text{acid}]} \qquad (22)$$

Both monobasic and polybasic acids form buffers. Typical examples of the first type are the acetic acid–acetate buffers and of the second type, the phosphate buffers. When the pK's of the various groups of a polybasic acid are near each other (e.g. citric acid) the analysis of the buffer system is more complex and will not be considered.

Acetate buffers

The buffering action of mixtures of acetic acid (pK 4.65) and Na acetate is illustrated in Fig. 1.1, which shows the changes in pH when 100 ml 0.2 M $CH_3\cdot COOH$ is titrated with 2 M NaOH.

On each addition of NaOH, the pH rises sharply from A to B and then less rapidly past C, the point of half neutralization, to D. Then the pH rises very sharply until the equivalence point is reached at E.

Fig. 1.1 shows that buffers most strongly resist changes in pH near the point of half neutralization (C), that is when pH = pK. Further it is seen that the range over which buffers are effective is about 1 pH unit on either side of the pK, i.e. from [salt]/[acid] = 1/10 to 10/1 (equation (22)).

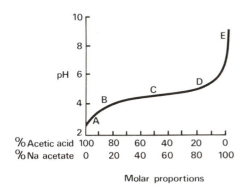

Fig. 1.1   pH titration curve of $CH_3\cdot COOH$.

Putting this in another way, we can say that the rate of change of pH with titre is minimal when pH = pK, but it should be noted that the rate of change of hydrogen ion concentration, $[H^+]$, with titre is least at the lowest values of $[H^+]$ (highest pH's). On the other hand, the rate of change of hydroxide ion concentration,

$[OH^-]$, with titre is large at low $[H^+]$ (high pH) and low at high $[H^+]$ (low pH).

Phosphate buffers

In the titration curve of phosphoric acid shown in Fig. 1.2, three distinct regions of buffering can be distinguished; these correspond to the three dissociations of phosphoric acid. As phosphoric acid, H_3PO_4, can lose 3 hydrogen ions per mole, 3 equivalents of alkali are required to neutralize it completely. There are, therefore, 3 sodium (or other metal) phosphates, NaH_2PO_4, Na_2HPO_4, and Na_3PO_4. The main features of the phosphate system are indicated below and in Fig. 1.2.

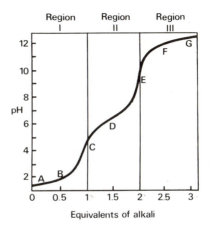

Fig. 1.2   pH titration curve of H_3PO_4.

The pH of any solution of phosphoric acid and/or its alkali metal salts can be determined from equations (23), (24) or (25) below. The addition of alkali to any phosphate solution will cause the pH to rise along the curve in Fig. 1.2, while the addition of a strong acid will cause the pH to fall along the curve.

Region I

Equilibrium $H_3PO_4 \rightleftharpoons H_2PO_4^- + H^+$

Curve   A   to   C

Mid-point of equilibrium, B

$$pK_1 = 2.0$$

$$pH = pK_1 + \log\frac{[H_2PO_4^-]}{[H_3PO_4]} \qquad (23)$$

Region II

Equilibrium $\quad H_2PO_4^- \rightleftharpoons HPO_4^{2-} + H^+$

Curve $\qquad\qquad$ C to E

Mid-point of equilibrium and point of maximum buffering, D

$$pK_2 = 6.8$$

$$pH = pK_2 + \log\frac{[HPO_4^{2-}]}{(H_2PO_4^-]} \qquad (24)$$

Region III

Equilibrium $\quad HPO_4^{2-} \rightleftharpoons PO_4^{3-} + H^+$

Curve $\qquad\qquad$ E to G

Mid-point of equilibrium, F

$$pK_3 = 11.7$$

$$pH = pK_3 + \log\frac{[PO_4^{3-}]}{[HPO_4^{2-}]} \qquad (25)$$

The Measurement of pH

The fundamental instrument for the measurement of pH is the hydrogen electrode, though for routine use the glass electrode is more convenient.

The glass electrode

If a thin bulb of a special glass is placed in a solution, it acquires a potential which depends on the pH in the same way as does that of a hydrogen electrode. In order to measure the potential of the glass membrane it is necessary to have a reference electrode (generally $Ag \cdot AgCl \cdot HCl$) *inside* the glass bulb as well as a reference electrode connected to the test solution by a salt bridge. Then the potential difference between the two reference electrodes is given by the equation:

$$E = E' + \frac{2.303RT}{F} \times pH \qquad (26)$$

where R is the gas constant, F the Faraday constant, T the temperature in kelvins, and E' a constant for the system.

In practice it is always first necessary to measure the potential of the glass electrode system in a standard buffer of known pH and then in the test solution. If E_S is the potential of the electrode system in a standard buffer pH_S, then the pH of the test solution, pH_X, is given by:

$$pH_X = pH_S + \frac{(E_X - E_S)F}{2.303RT} \qquad (27)$$

The potential of the hydrogen-saturated calomel electrode system can be measured with an ordinary potentiometer. The glass electrode system, on the other hand, has so high a resistance that the potential has to be measured with a high input impedance voltmeter usually arranged as a pH meter, that is to say the potentiometer is divided to read directly in pH units. Even though the scale is calibrated in pH units, it must be emphasized that a calibration measurement in a buffer of known pH must always be made before measuring the pH of the test solution.

Indicators

Indicators are weak organic acids which change colour on ionization. Thus the acid form of methyl red is red while the conjugate base is yellow. Similarly with phenolphthalein, the acid is colourless while the base is pink. The dissociation of the indicator, HI, may be represented thus

$$HI \rightleftharpoons H^+ + I^- \qquad (28)$$

with a dissociation constant K_I. Then equation (15) becomes:

$$pH = pK_I + \log\frac{[I^-]}{[HI]} \qquad (29)$$

Since the species HI and I^- are of different colours (red and yellow for methyl red, colourless and pink for phenolphthalein) the colour of the solution depends on the ratio $[I^-]/[HI]$, i.e. it depends on the second term on the right of equation (29) and therefore on the pH of the solution.

Because of the difficulty in discriminating between small changes in one colour in a large excess of another

Table 1.1 Table of indicators

Indicator	Useful range of pH and colour change
Thymol blue (acid range)	1.2 red–2.8 yellow
Tropaeolin–thymol blue	1.0 red–3.5 yellow
Methyl orange	3.0 red–4.4 yellow
Bromophenol blue	2.8 yellow–4.6 blue
Methyl red	4.2 red–6.3 yellow
Chlorophenol red	5.0 yellow–6.6 red
Bromothymol blue	6.0 yellow–7.6 blue
Phenol red	6.8 yellow–8.4 red
Phenolphthalein	8.3 colourless–10.0 violet-red
Thymol blue (alkali range)	8.0 yellow–9.6 blue
Thymol violet	9.0 yellow–13.0 violet

colour, the useful range of an indicator is only about 1 pH unit, i.e. over the range $pK_I \pm 0.5$ pH (see Table 1.1). The measurement of pH then resolves itself into the choice of an indicator with a suitable pK_I, near the pH to be measured, and the determination of the concentration ratio of the two colours. This is most easily done by comparing the colour of the unknown solution containing a little indicator, with the colour of standard buffers of known pH containing the same concentration of indicator. Indicators are not so reliable as glass or hydrogen electrodes for the measurement of pH, as pK_I is often affected by the presence of salts, proteins, etc. Indeed the 'protein error' of a paper strip soaked in a suitable indicator can be used as a test for the presence of protein, e.g. in urine.

Physiological Buffers

The buffers important in vivo are those which are effective around pH 7.4, the pH of blood. The pH of urine, however, can vary between 4 and 9. The chief systems are listed below.

Bicarbonate

The apparent pK_1 of carbonic acid is 6.1 (see below), so the base/acid ratio is 20 at pH 7.4; bicarbonate is therefore a good buffer when blood is being acidified, but very poor if it is being made alkaline. The concentration of HCO_3^- ions in plasma is about 0.03 M. Bicarbonate is also useful in buffering urine.

Phosphate

The pK of the equilibrium $H_2PO_4^- \rightleftharpoons HPO_4^{2-}$ is 6.8, i.e. the ratio $[HPO_4^{2-}]/[H_2PO_4^-]$ in plasma is 4/1. This makes phosphate a more efficient buffer than bicarbonate at physiological pH's, but its concentration in plasma is only 0.002 M. In cells, the various phosphate esters, which have very roughly the same pK as inorganic phosphate, come to about 0.08 M, and are therefore important buffers. Inorganic phosphate is the chief buffer in urine.

Amino acids

Most of these compounds are dibasic, i.e. in going from pH 1 to pH 10 they lose two protons. The pK's of the COOH and NH_3^+ groups are, however, far removed from 7.4 and they are not important, except in buffering the HCl released in the gastric juice. The free amino acid concentration is also small.

Proteins

Many of the amino acids in peptide chains have acidic or basic groups not forming part of a peptide bond (e.g. glutamic acid, lysine). These groups can buffer solutions but, as with the amino acids, the pK's are far removed from 7.4, with one exception: *histidine* (pK 6.0). As haemoglobin is so concentrated in blood (14 g/100 ml $\equiv$ 0.008 M) and it contains a good deal of histidine, haemoglobin accounts for 60% of the buffer capacity of whole blood. On the same principle, the plasma proteins account for another 20%.

Carbon Dioxide Transport and the Carbonic Acid Buffering System

This system is of special and central importance in living organisms because it removes one of the chief products of the oxidation of food, the acid anhydride CO_2. About 25 000 mEq H^+ ions are formed every day as the result of CO_2 production. At the same time the system is able to buffer effectively against acids and to a lesser extent against alkalis. Numerical values for the constants defined below are given in Table 1.2 for 38°C in a solution like plasma.

Hydration of dissolved CO_2

Carbon dioxide is formed in tissue cells. As such it is dissolved in water where some of it is hydrated to carbonic acid according to the following reaction:

$$CO_2 + H_2O \rightleftharpoons H_2CO_3 \qquad (30)$$

This reaction is slow but in the red cells there is an enzyme, *carbonic anhydrase*, which greatly speeds up the attainment of equilibrium. Because of this enzyme and the ability of CO_2 and H_2CO_3 to penetrate the red cell

Table 1.2 Values of constants relating to the carbonic acid buffering system appropriate to blood composition and temperature

$[H_2CO_3]/[CO_2] = K'_h$	≈ 0.00296	$pK'_h \approx 2.53$
$[H^+][HCO_3^-]/[H_2CO_3] = K'_{H_2CO_3}$	$= 1.66 \times 10^{-4}$	$pK'_{H_2CO_3} = 3.78$
$[H^+][HCO_3^-]/[CO_2 + H_2CO_3] = K'_1$	$= 7.94 \times 10^{-7}$	$pK'_1 = 6.1$
$[H^+][CO_3^{2-}]/[HCO_3^-] = K'_2$	$= 1.58 \times 10^{-10}$	$pK'_2 = 9.8$
$[H^+][OH^-] = K_w$	$= 2.51 \times 10^{-14}$	$pK_w = 13.6$
$[CO_2 + H_2CO_3]/P_{CO_2} = q^\star$	$= 3.16 \times 10^{-2}$	$-\log q = 1.5$

$\star$ For P_{CO_2} in mm Hg and $[CO_2 + H_2CO_3]$ in mmol/l.

membranes, in cells and plasma dissolved CO_2 is in equilibrium with its hydrated form H_2CO_3. The concentration equilibrium constant for the hydration equation (30), K'_h, is defined by:

$$K'_h = \frac{[H_2CO_3]}{[CO_2]} \qquad (31)$$

Since it is difficult to determine H_2CO_3 separately from dissolved CO_2, it is usual in blood analysis to determine the sum of the concentrations of dissolved CO_2 and H_2CO_3.

Solubility of CO_2

Carbon dioxide obeys Henry's law and its concentration in solution is proportional to its partial pressure, P_{CO_2}, in equilibrium with the solution. Remembering that CO_2 is partially hydrated, we have:

$$[CO_2] + [H_2CO_3] = qP_{CO_2} \qquad (32)$$

where q is the Henry's law coefficient. An important consequence of equation (32) is that the concentration of *both* CO_2 *and* H_2CO_3 will vary, in plasma, with the partial pressure of CO_2 in equilibrium with the plasma (in the alveolar air).

Ionization of H_2CO_3

The 'true' first ionization of carbonic acid is:

$$H_2CO_3 \rightleftharpoons H^+ + HCO_3^-$$

for which the acid dissociation constant, $K'_{H_2CO_3}$, is given by:

$$K'_{H_2CO_3} = \frac{[H^+][HCO_3^-]}{[H_2CO_3]} \qquad (33)$$

The 'overall' first ionization constant

Because, in blood, H_2CO_3 is in equilibrium with dissolved CO_2, we have the overall equilibrium:

$$CO_2 + H_2O \rightleftharpoons H_2CO_3 \rightleftharpoons H^+ + HCO_3^- \qquad (34)$$

for which the equilibrium constant, K'_1, is given by:

$$K'_1 = \frac{[H^+][HCO_3^-]}{[CO_2] + [H_2CO_3]} = \frac{[H^+][HCO_3^-]}{qP_{CO_2}} \qquad (35)$$

Combining (31), (33) and (35) we obtain:

$$K'_1 = K'_{H_2CO_3}\left(\frac{K'_h}{K'_h + 1}\right) \qquad (36)$$

From Table 1.2 we see that carbonic acid is a fairly strong acid with $pK'_{H_2CO_3} = 3.78$. However, because of the dehydration of the acid and the volatility of its anhydride, for the 'overall' or apparent first dissociation in plasma $pK'_1 = 6.1$.

The second ionization constant

The bicarbonate ion, HCO_3^-, can lose another proton

$$HCO_3^- \rightleftharpoons H^+ + CO_3^{2-} \qquad (37)$$

to form the carbonate ion, CO_3^{2-}. The acid dissociation constant for equilibrium (37), K'_2, is given by

$$K'_2 = \frac{[H^+][CO_3^{2-}]}{[HCO_3^-]} \qquad (38)$$

Since $pK'_2 = 9.8$ this dissociation is unimportant at physiological pH values.

Buffering by the carbonic acid system

From equations (32) and (35) we have:

$$[HCO_3^-] = \frac{K'_1}{[H^+]}qP_{CO_2} \qquad (39)$$

and from (32), (38), and (39) we have:

$$[CO_3^{2-}] = \frac{K'_1 K'_2}{[H^+]^2}qP_{CO_2} \qquad (40)$$

Assuming respiratory control is adequate to keep the alveolar air composition constant with $P_{CO_2} = 40$ mm Hg, we can, from (39) and (40), and the values in Table 1.2, calculate the concentrations of HCO_3^- and CO_3^{2-} ions which would occur in plasma at different pH's. These concentrations are shown in Fig. 1.3. From it one can see that for $P_{CO_2} = 40$ mmHg, $[HCO_3^-] = 25.2$ mEq/litre at pH 7.4; this value rises to 31.8 mEq/litre at pH 7.5 and falls to 20.0 mEq/litre at pH 7.3. The shape of the curve is not the same as that of the usual

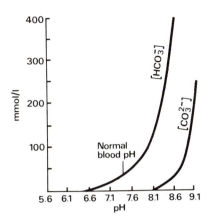

Fig. 1.3 Concentration of bicarbonate and carbonate ions in a solution in equilibrium with gas containing a fixed partial pressure of carbon dioxide (P_{CO_2} 40 mm Hg), as a function of pH. Small scale, over a wide pH range.

titration curve (see Fig. 1.1) because it is plotted on a different basis; here the acid concentration is held fixed (P_{CO_2}) and that of the conjugate base (HCO_3^-) varies with pH. At pH 7.4 and $P_{CO_2} = 40$ mm Hg, the carbonate ion concentration $[CO_3^{2-}] = 0.105$ mmol/l or 0.21 mEq/litre, less than 1 % of the bicarbonate ion concentration. The curve in Fig. 1.3 shows that $[CO_3^{2-}]$ only begins to become significant above pH 8.

Near the pH of blood (7.4) we have therefore only to consider the 'overall' first ionization of carbonic acid and we have the most important relationship:

$$pH = 6.1 + \log \frac{[HCO_3^-]}{qP_{CO_2}} \qquad (41)$$

which may be compared to the *Henderson–Hasselbalch equation* (15).

2

Amino Acids, Peptides and Proteins

Amino Acids

Amino acids are carboxylic acids which also contain an amino (—NH$_2$) group. Those occurring in proteins are almost exclusively α-amino acids, in which the carboxyl and amino groups are attached to the same carbon atom (the α-carbon). Their general structure is

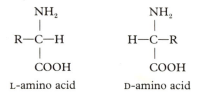

A few amino acids, not usually found in proteins, have the amino group attached to the β- or γ-carbon; proline, which occurs in proteins, is an imino acid, having a secondary amino group which is part of a heterocyclic ring.

Optical activity. In all α-amino acids except glycine the α-carbon is asymmetric, so there are two optically active isomers. Those which occur in proteins have the L configuration. In the systematic nomenclature used for stereoisomers, many, but not all, L-amino acids have the S configuration. D-Amino acids are not commonly found in proteins, but occur in some antibiotics and in bacterial cell walls.

Classification

Amino acids are classified according to the side chain, R; the most useful way to do this is according to their hydrophobicity, which determines their function in proteins. The most hydrophobic are those with large apolar side chains, the most hydrophilic those with charged side chains. The structures of the 20 amino acids commonly occurring in proteins are given in Table 2.1, together with their pK_a values, and ΔG_T, the free energy of transfer of the amino acid (when it is part of a protein) from an aqueous to a hydrophobic medium. This gives an idea of the polarity of the amino acid; apolar amino acids have large negative free energies of

transfer, while transfer of polar amino acids to a hydrophobic environment is energetically unfavourable, so the free energy change is positive.

Abbreviations. Amino acids are often denoted by abbreviations of three letters (in most cases the first three letters of the name) or one letter. These are useful in writing out the long sequences of amino acids that occur in peptides and proteins.

Neutral aliphatic amino acids. These include glycine, alanine, the branched-chain amino acids valine, leucine and isoleucine, two hydroxyamino acids, serine and threonine, and two sulphur-containing amino acids, cysteine and methionine. Isoleucine and threonine (and hydroxylysine and hydroxyproline, discussed below) contain a second asymmetric carbon, and therefore have four isomers, only one of which occurs in proteins.

Cysteine can be oxidized to its disulphide, cystine:

$$2 \; \underset{\text{Cysteine}}{\begin{array}{c} CH_2SH \\ | \\ CHNH_2 \\ | \\ COOH \end{array}} \; \underset{\text{reduction}}{\overset{\text{oxidation}}{\rightleftharpoons}} \; \underset{\text{Cystine}}{\begin{array}{c} CH_2-S-S-CH_2 \\ | \qquad\qquad | \\ CHNH_2 \qquad CHNH_2 \\ | \qquad\qquad | \\ COOH \qquad COOH \end{array}}$$

Both cysteine and cystine occur in proteins, and disulphide links are important in maintaining the structures of some proteins.

Aromatic amino acids. These are phenylalanine, tyrosine and tryptophan. They absorb light in the near-ultraviolet (260–300 nm) and are responsible for the absorbance in this region that is shown by most proteins. They are also fluorescent, with maximum emission at 350 nm.

Basic amino acids. Arginine and lysine are strongly basic, as they have two amino groups and only one carboxyl. Histidine, with a heterocyclic imidazole side chain, is more weakly basic.

Table 2.1 Structures and properties of amino acids

Amino acid*	Structure	pK_a values α-carboxyl	α-amino	side chain	$\Delta G_T^{\ddagger}$ (kJ/mol)
Alanine (Ala, A)	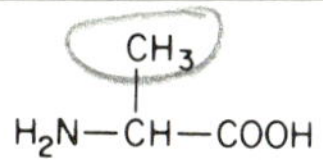 CH₃—CH(NH₂)—COOH	2.4	9.7	—	−12
Arginine (Arg, R)	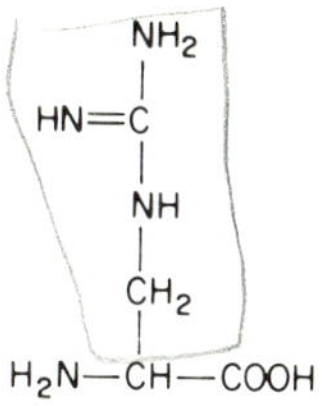 H₂N—C(=NH)—NH—CH₂—CH(NH₂)—COOH	2.2	9.1	12.5	+39
Asparagine (Asn, N)	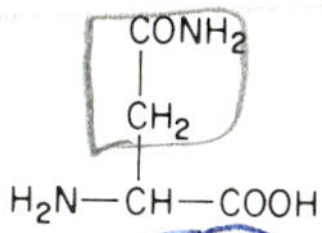 CONH₂—CH₂—CH(NH₂)—COOH	2.2	8.8	—	+4
Aspartic acid (Asp, D)	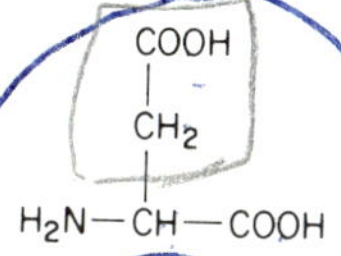 COOH—CH₂—CH(NH₂)—COOH	2.1	9.8	3.9	+23
Cysteine (Cys, C)	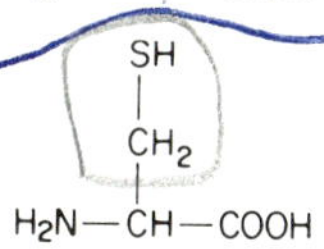 SH—CH₂—CH(NH₂)—COOH	1.7	10.8	8.3	+4
Glutamine (Gln, Q)	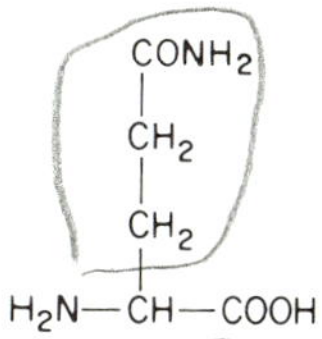 CONH₂—CH₂—CH₂—CH(NH₂)—COOH	2.2	9.1	—	+6
Glutamic acid (Glu, E)	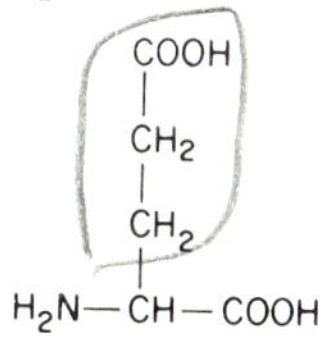 COOH—CH₂—CH₂—CH(NH₂)—COOH	2.2	9.7	4.3	+8
Glycine (Gly, G)	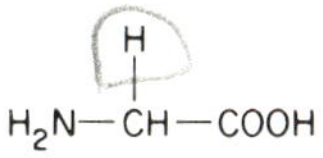 H—CH(NH₂)—COOH	2.3	9.6	—	−8
Histidine (His, H)	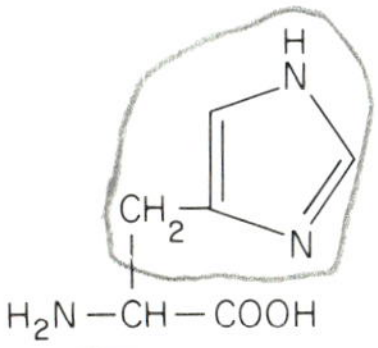 imidazole—CH₂—CH(NH₂)—COOH	1.8	9.2	6.0	+6
Isoleucine (Ile, I)	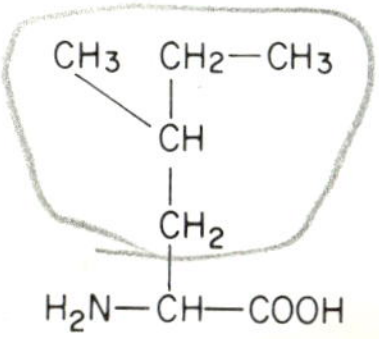 CH₃ CH₂—CH₃ / CH—CH₂—CH(NH₂)—COOH	2.4	9.7	—	−18

Base   Acdic

11

Table 2.1   Structures and properties of amino acids (*cont.*)

Amino acid*	Structure	α-carboxyl	pKₐ values α-amino	side chain	$\Delta G_T^\dagger$ (kJ/mol)
Leucine (Leu, L)	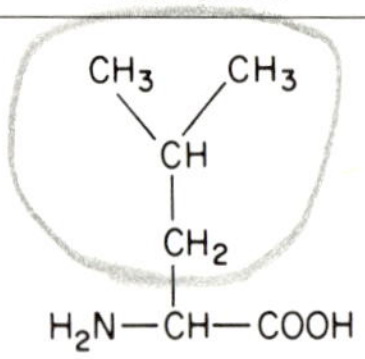	2.4	9.6	—	−18
Lysine (Lys, K)	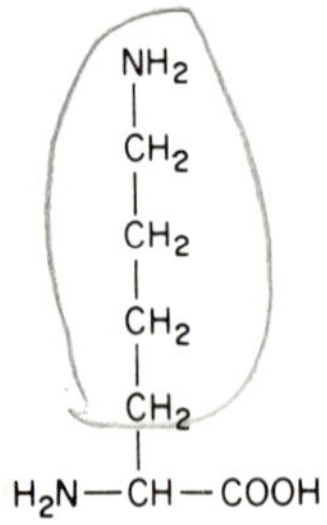	2.2	9.0	10.5	+10
Methionine (Met, M)	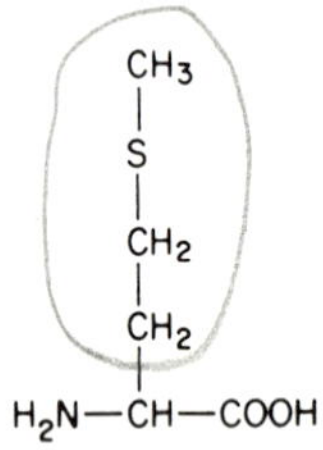	2.3	9.2	—	−9
Phenylalanine (Phe, F)	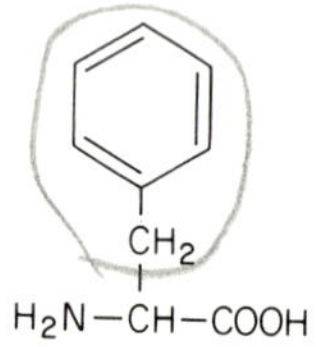	1.8	9.1	—	−22
Proline (Pro, P)	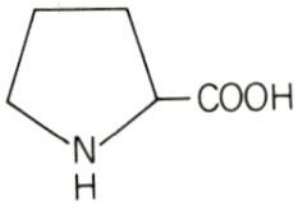	2.0	10.6	—	+6
Serine (Ser, S)	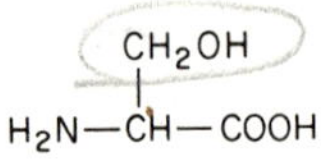	2.2	9.2	—	−2
Threonine (Thr, T)	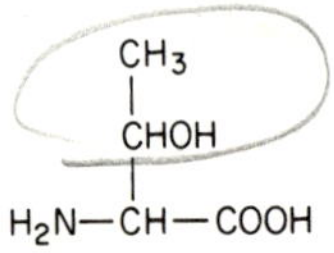	2.6	10.4	—	−4
Tryptophan (Trp, W)	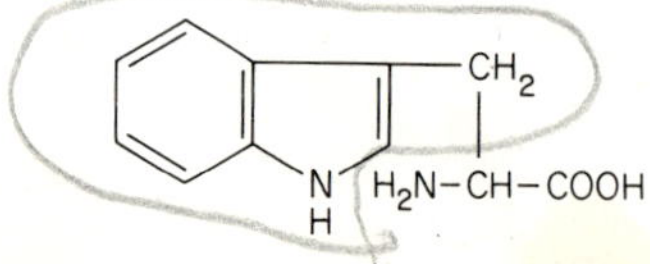	2.4	9.4	—	−16

Table 2.1 Structures and properties of amino acids (*cont.*)

Amino acid*	Structure	pK_a values α-carboxyl	α-amino	side chain	ΔG_T† (kJ/mol)
Tyrosine (Tyr, Y)		2.2	10.1	9.1	−2
Valine (Val, V)		2.3	9.6	—	−16

* The names of amino acids may be abbreviated by one or three letters. Asx or B indicates *either* asparagine *or* aspartic acid; Glx or Z indicates *either* glutamine *or* glutamic acid.

† ΔG_T is the free energy of transfer of the amino acid (in peptide linkage) from random coil in aqueous solution to helical conformation in a hydrophobic environment.

Dicarboxylic amino acids. Aspartic and glutamic acids contain two carboxyls and one amino group; they are therefore acidic. Their amides, asparagine and glutamine, are neutral.

Imino acids. Proline contains no primary amino group, but this imino acid forms peptide bonds in the same way as α-amino acids.

Unusual amino acids. Most proteins contain only the 20 amino acids listed in Table 2.1, but a few contain unusual amino acids such as 4-hydroxyproline and 3-hydroxylysine, which occur in collagen; ε-trimethyllysine, found in several Ca^{2+}-binding proteins; γ-carboxyglutamic acid, 3-methyl histidine, pyroglutamic acid, and others. They are formed by modification of the parent amino acid after incorporation into protein.

A few naturally-occurring amino acids are not found in proteins: these include L-ornithine and L-citrulline, constituents of the urea cycle (p. 131) and β-alanine, which occurs in coenzyme A, the vitamin pantothenic acid, and the muscle peptides carnosine and anserine.

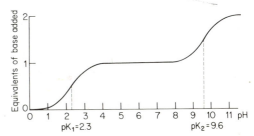

Pyroglutamic acid
(pyrrolidone
carboxylic acid)

β-alanine

3-methyl histidine

Ionization of amino acids

Free amino acids have at least two ionizing groups, and exist in several ionic states which, for a neutral amino acid, are:

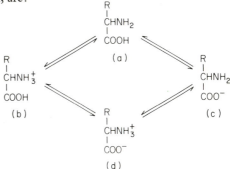

In neutral solution and in crystals the predominant form is the zwitterion (d), rather than the unionized form (a). The titration curve of glycine, a typical neutral amino acid, is shown in Fig. 2.1; there are two inflections,

Fig. 2.1 The titration curve of glycine, a neutral amino acid.

corresponding to ionization of the carboxyl group $(b \rightleftharpoons d)$ and amino group $(d \rightleftharpoons c)$, the pK_a values of which are respectively 2.3 and 9.6. These groups have similar pK_a values in other α-amino acids (Table 2.1).

Amino acids with ionizable side chains show a third inflection of the titration curve; for aspartic acid, for example, a simplified ionization scheme is

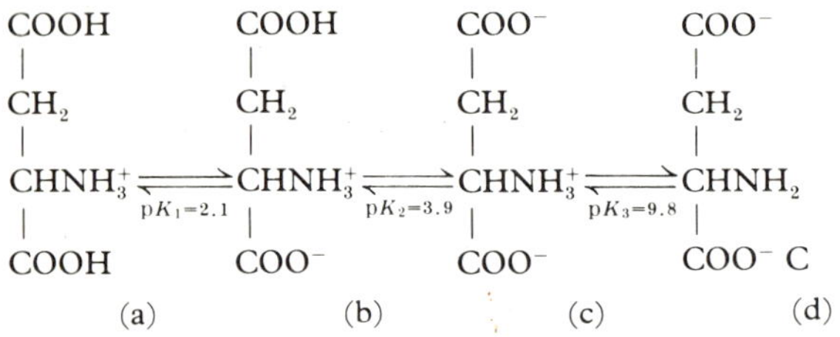

$$
\begin{array}{cccc}
\text{COOH} & \text{COOH} & \text{COO}^- & \text{COO}^- \\
| & | & | & | \\
\text{CH}_2 & \text{CH}_2 & \text{CH}_2 & \text{CH}_2 \\
| & | & | & | \\
\text{CHNH}_3^+ \underset{pK_1=2.1}{\rightleftharpoons} & \text{CHNH}_3^+ \underset{pK_2=3.9}{\rightleftharpoons} & \text{CHNH}_3^+ \underset{pK_3=9.8}{\rightleftharpoons} & \text{CHNH}_2 \\
| & | & | & | \\
\text{COOH} & \text{COO}^- & \text{COO}^- & \text{COO}^- \text{ C} \\
(a) & (b) & (c) & (d)
\end{array}
$$

At low pH the predominant species is (a); as the pH is raised, (a) (net charge $+1$) is successively converted to (b), (c), and (d), with charges of 0, -1 and -2. There are two other possible ionization states, of charge 0 and -1, but these are present in low concentrations and are not shown.

Some other amino acids also have ionizing side chains. These are glutamic acid, which ionizes in the same way as aspartic acid; lysine, arginine and histidine, which are basic and therefore positively charged at low pH;

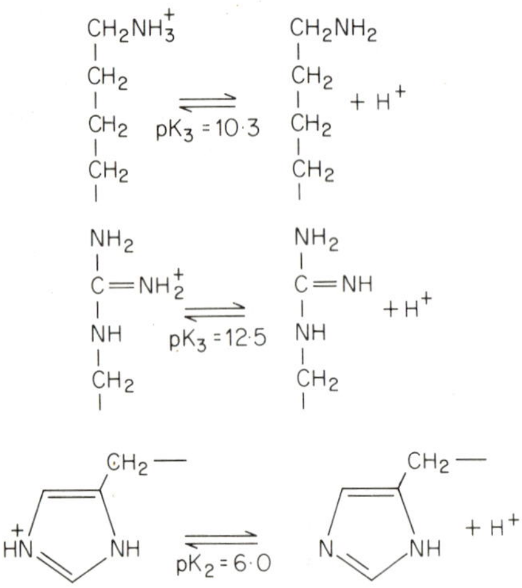

$$
\begin{array}{cc}
\text{CH}_2\text{NH}_3^+ & \text{CH}_2\text{NH}_2 \\
| & | \\
\text{CH}_2 & \text{CH}_2 \\
| & | \quad + \text{H}^+ \\
\text{CH}_2 \underset{pK_3 = 10.3}{\rightleftharpoons} & \text{CH}_2 \\
| & | \\
\text{CH}_2 & \text{CH}_2 \\
| & |
\end{array}
$$

$$
\begin{array}{cc}
\text{NH}_2 & \text{NH}_2 \\
| & | \\
\text{C}=\text{NH}_2^+ & \text{C}=\text{NH} \\
| \quad \rightleftharpoons & | \quad + \text{H}^+ \\
\text{NH} \quad pK_3=12.5 & \text{NH} \\
| & | \\
\text{CH}_2 & \text{CH}_2 \\
| & |
\end{array}
$$

$$
\text{imidazole} \underset{pK_2=6.0}{\rightleftharpoons} \text{imidazole} + \text{H}^+
$$

and cysteine and tyrosine, which can lose a proton to become negatively charged:

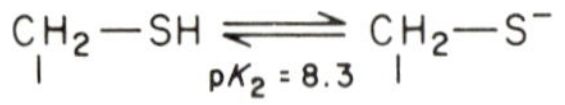

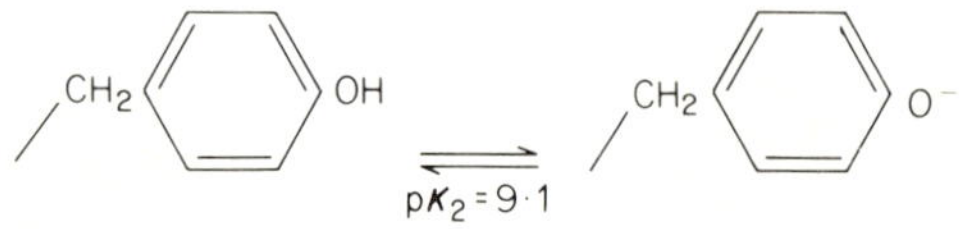

$$
\text{CH}_2-\text{SH} \underset{pK_2=8.3}{\rightleftharpoons} \text{CH}_2-\text{S}^-
$$

$$
\text{CH}_2-\text{C}_6\text{H}_4-\text{OH} \underset{pK_2=9.1}{\rightleftharpoons} \text{CH}_2-\text{C}_6\text{H}_4-\text{O}^-
$$

No other amino acid ionizes in the physiological range.

Chemical reactions of amino acids

The ninhydrin reaction

This is a useful colour reaction of amino acids, and it is used to detect and measure them in various types of separation.

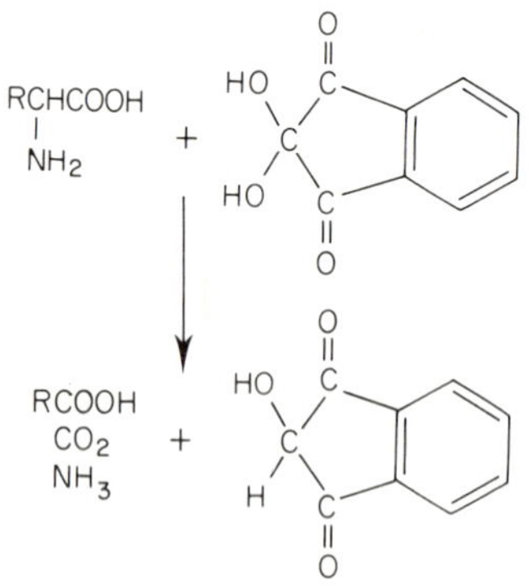

Ninhydrin oxidizes α-amino acids, the reduced ninhydrin product reacting with ammonia and a molecule of oxidized ninhydrin to give a product which is blue or purple in most cases; that from proline is yellow, and that from asparagine brown. In certain circumstances primary amines may react, but amino acids in peptide linkage do not give a positive reaction.

The fluorescamine reaction

This reagent reacts with free amino groups to give a product which is highly fluorescent, permitting the sensitive detection of free amino acids:

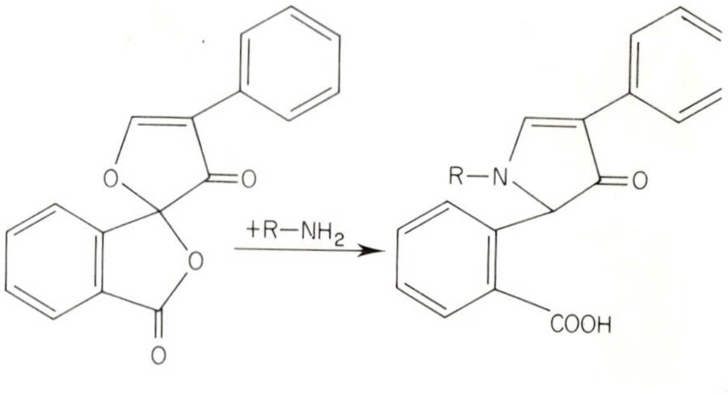

Other colour reactions

Some specific reactions give colours with particular amino acids, e.g. the Sakaguchi reaction (α-naphthol and sodium hypochlorite giving a red colour with arginine), and the Pauly reaction (diazotized sulphanilic acid giving a red colour with histidine or tyrosine).

Peptide bond formation

An amide link can be formed between the —COOH of one amino acid and the —NH$_2$ of another, with the elimination of water:

$$H_2N-\underset{\underset{R_1}{|}}{CH}-COOH + H_2N-\underset{\underset{R_2}{|}}{CH}-COOH \longrightarrow$$

$$H_2N-\underset{\underset{R_1}{|}}{CH}-CO-NH-\underset{\underset{R_2}{|}}{CH}-COOH + H_2O$$

Such links are called peptide bonds, and polymers of amino acids formed in this way are peptides. Bonds involving the β- or γ-carboxyls of aspartate and glutamate, or the ϵ-amino of lysine, do not occur in most proteins, but glutamate–lysine links are important in blood clotting (p. 240).

Peptide bond formation does not occur spontaneously: it can be induced in vitro by certain chemical reagents, and in vivo by the elaborate cellular machinery which synthesizes proteins (chapter 12).

Separation of amino acids

Paper chromatography

Amino acids may be separated by chromatography, the usual medium being paper, or a thin layer of silica spread on a glass plate. The amino acid mixture is applied as a spot to the dry paper or plate, and the chromatogram developed by allowing a solvent (usually a mixture of organic solvents and water) to permeate it from one end.

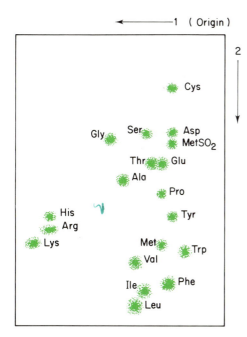

Fig. 2.2 Two-dimensional separation of amino acids on paper, by (1) electrophoresis at pH 2.2, followed by (2) chromatography in butanol/butyl acetate/acetic acid/water at pH 6.5.

The amino acids travel at different rates, because each one has a different partition coefficient between the mobile (liquid) and stationary (solid) phases: they are therefore separated as the solvent advances. When development is complete the chromatogram is dried, and sprayed with ninhydrin to locate the amino acid spots. Resolution may be improved by two-dimensional separation, in which the chromatogram is developed first in one solvent, then in a different one, run at right angles to the first. Two-dimensional separation may also be performed using chromatography and electrophoresis (see below, and Fig. 2.2).

Ion-exchange chromatography

The charge on an amino acid is determined by the pH: at the isoelectric point (pI) the molecule is electrically neutral, having an equal number of positive and negative charges. For a monocarboxylic amino acid,

$$pI = \frac{pK_1 + pK_2}{2}$$

Here pK_1 and pK_2 are the pK_a values of the α-NH_2 and α-COOH; if there are two basic or two acidic groups, the pK_a values of the groups of *like* charge are substituted in this formula.

Ion-exchange resins are insoluble polymers (polyacrylamide, cellulose, polystyrene, etc.) containing either acidic (such as $-SO_3H$ or $-COOH$) or basic (such as quaternary amino) groups. The former behave as *cation exchangers*, as they can exchange protons for Na^+, K^+, Ca^{2+} or other cations; the latter as *anion exchangers*, which bind OH^-, Cl^-, HPO_4^{2-} or other anions. If the components of a mixture bear different charges, they can be separated by chromatography on a column of the appropriate ion exchange resin, as they have different strengths of ionic interaction with the resin, and therefore move at different speeds.

Amino acids can be separated on cation exchange resins at low pH; the binding of any amino acid to the resin depends on pH, ionic strength and temperature. The automatic amino acid analyser employs a strongly acidic, sulphonated polystyrene resin; amino acids are sequentially eluted from the column by a buffer of changing composition, the most basic amino acids emerging last. The eluate is automatically mixed with ninhydrin, and the absorbance of the product measured by a spectrophotometer. The analyser produces a continuous trace of absorbance; each absorbance peak represents an amino acid which is identified by the position of the peak, the amount of each being calculated from the area of the peak.

Electrophoresis

Amino acids can be separated by their different rates of migration in an electric field. The mobility depends on

15

the charge on the amino acid: at pH values above pI migration is toward the anode, below pI toward the cathode. Since different amino acids have different values of pK_a and therefore of pI, they can be separated by electrophoresis (usually on paper) at a carefully chosen pH.

Peptides

There is no sharp distinction between peptides and proteins. Usually the term protein is applied to polypeptides of molecular weight more than a few thousand: this is large enough to prevent passage through a dialysis membrane, and to confer a defined, folded structure.

Linear polymers of amino acids have one α-NH$_2$ and one α-COOH which are not involved in peptide-bond formation; the amino acids bearing these are known as

the N-terminus and C-terminus, respectively. Amino acid sequences are conventionally written from N- to C-terminus, using one- or three-letter abbreviations, e.g.

$$\text{Asp–Arg–Val–Tyr–Ile–His–Pro–Leu–Phe}$$
$$\text{D —R —V —Y —I —H —P —L —F}$$

in which the N-terminus is aspartic acid, the C-terminus phenylalanine. Peptides with biological function range from di- and tripeptides to polymers of molecular weight several thousand. A few of the smallest are listed in Table 2.2; larger peptides include hormones secreted by the pancreas, pituitary and parathyroid.

In some cases the terminal amino or carboxyl may be modified, for example by acetylation of α-NH$_2$, amidation of α-COOH or cyclization of N-terminal glutamate to pyroglutamate. A few cyclic peptides are also known.

Table 2.2   Some peptides of biological importance

Posterior pituitary	GlyLeuProCysAsnGlnIleTyrCys
	     | |
	     S —————— S
		Oxytocin

	GlyArgProCysAsnGlnPheTyrCys
	     | |
	     S —————— S
		Vasopressin

Hypothalamus	AlaGlyCysLysAsnPhePheTrpLysThrPheThrSerCys
	     | |
	     S ——————————————————— S
		Somatostatin
		(growth-hormone release inhibiting factor)

	PGAHisTrpSerTyrGlyLeuArgProGlyNH$_2$	Luteotrophin releasing factor
	PGAHisProNH$_2$	Thyrotrophin releasing factor
Intestine	GluGluProTrpLeuGluGluGluGluGluAlaTyrGlyTrpMetAspPheNH$_2$	Gastrin
Kidney	AspArgValTyrIleHisProPheHisLeu	Hypertensin I
	AspArgValTyrIleHisProPhe	Hypertensin II
Central nervous system	TyrGlyGlyPheMet	Met-enkephalin
	TyrGlyGlyPheLeu	Leu-enkephalin
Muscle	β-AlaHis	Carnosine
	β-Ala 3-MeHis	Anserine
Ubiquitous	γ-GluCysGly	Glutathione

PGA = pyroglutamic acid; β-Ala = β-alanine; γ-Glu = glutamic acid, linked through its γ-carboxyl; 3-MeHis = 3-methyl histidine.

Proteins

The molecular weights of proteins range from a few thousand to several millions. Most proteins are folded into defined structures which are maintained by large numbers of relatively weak, non-covalent bonds, and the biological activities of proteins usually depend critically on the maintenance of the correctly folded conformation.

Ionization

Most proteins have large numbers of ionizable groups: amino acid side chains, the N- and C-termini (providing these are not blocked) and sometimes other attached groups. Whether or not any particular group carries a charge depends on its pK_a in relation to the pH of the solution. At low pH there are more positively-charged groups than negatively charged; the protein is cationic, and migrates toward the negative electrode (cathode) in an electric field. At high pH, negatively-charged groups predominate, and the protein is anionic. Only at the isoelectric point, at which there are equal numbers of positive and negative charges, is a protein immobile in an electric field.

The pK_a values of amino acid side chains in proteins are not always the same as in the free amino acids, because ionizations are dependent on the environment, and affected by neighbouring charged groups, and by the exclusion of water in the interior of the folded protein. Because any group can occur in many different environments, the pK_a within a protein cannot be stated exactly, but usually falls within a certain range:

γ-carboxyl (glutamic acid) $\big\rbrace$	3.0–4.8
β-carboxyl (aspartic acid) $\big\rbrace$	
imidazole (histidine)	5.6–7.0
sulphydryl (cysteine)	8.2–8.6
phenolic (tyrosine)	9.8–10.4
ϵ-amino (lysine)	9.4–10.6
guanidino (arginine)	11.6–12.6

Exceptionally, a group may have a pK_a outside this range, particularly if it is involved in catalysis, at the active centre of an enzyme. Thus aspartate and histidine in the 'charge relay' of chymotrypsin (p. 40) have pK_a values of 6.8 and 3.3, and the essential lysine at the active centre of the bacterial enzyme acetoacetate decarboxylase has a pK_a of about 6.

Solubility

Proteins vary greatly in their solubility: most globular proteins are soluble in water or dilute salt solution, and are precipitated by concentrated salt, strong acid or organic solvents. Many structural proteins are highly insoluble, because they are highly cross-linked. Proteins are least soluble at their isoelectric points, when ionic forces separating the molecules are minimized.

Osmotic pressure

Because of their large size, protein molecules do not pass through semipermeable membranes such as cellophane or the walls of capillary blood vessels. Proteins contained within such membranes exert an osmotic pressure.

Hydrolysis

Proteins may be hydrolysed by heating with acids or alkalis, or by the action of proteolytic enzymes. The products of hydrolysis are peptides and amino acids.

Protein structure

Because the folding of proteins is very complex, different elements of structure are considered separately as primary, secondary, tertiary or quaternary structure.

Primary structure

This means the covalent structure of the protein: the sequence of amino acid residues, and the position of any interchain links (such as the disulphide cystine). The primary structure is unique to each polypeptide, and determines which higher structure it assumes. Semi-automatic methods of sequence determination are now highly developed, and the primary structures of hundreds of proteins are known.

Secondary structure

This is the regular folding of the polypeptide 'backbone', without reference to the side chains. It may extend over hundreds of residues, as occurs in fibrous proteins, or may change several times within a short stretch of polypeptide, as in most globular proteins. Only a few types of secondary structure are possible, and these are dictated by the properties of the peptide bond (Fig. 2.3), and the geometry of the 'peptide unit'. The

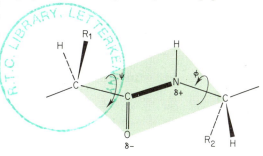

Fig. 2.3 Configuration of a 'peptide unit', showing the six atoms that lie in the plane of the peptide bond. The dispositions of neighbouring peptide units are defined by the rotational angles ψ and ϕ.

sp^2 electrons of the amino group are delocalized onto the peptide bond itself, giving it partial double-bond character. This has the following consequences:

i. The peptide bond is shorter (0.133 nm) than a normal C—N bond (0.150 nm) although longer than a C=N double bond (0.127 nm).

ii. Rotation about the peptide bond does not occur.

iii. Six atoms (the α-carbons of two neighbouring amino acids, with the CO and NH of the peptide bond) all lie in a plane (Fig. 2.3).

iv. There are two configurations of the 'peptide unit', *cis* and *trans*. In proteins, the *cis* arrangement is hardly ever found.

v. The nitrogen is somewhat positively charged, and the oxygen negatively charged. This makes hydrogen bonds between different peptide units particularly strong.

The polypeptide chain therefore consists of a series of linked 'peptide units', of fixed geometry. Rotation is possible about the C—C and N—C bonds (marked ψ and ϕ in Fig. 2.3), but because of steric constraints, only a limited number of configurations is possible. The most stable structures are those that maximize the amount of H-bonding, but this demands accurate angular disposition of the groups involved. When ψ and ϕ have constant values through a number of successive peptide units, there is a regular folding of the chain.

Two particularly stable arrangements, the α-helix and the β-pleated sheet, are shown in Fig. 2.4.

In the α-helix ($\psi = 130°$, $\phi = 120°$), the peptide chain is in a right-handed coil, maintained by H-bonds between the C=O and N—H of every fifth amino acid within the chain. These bonds are parallel to the axis of the helix, and are particularly strong because water is excluded from the interior of the helix. There are 3.6 amino acids per turn; the α-helix has a pitch of 0.54 nm, and a rise of 0.15 nm per amino acid. The amino acid side chains protrude externally from the helix: with the naturally-occurring L-amino acids steric contact is minimized in a *right*-handed helix, which is therefore more stable than a left-handed helix.

In the β-pleated sheet ($\psi = 315°$, $\phi = 40°$) the peptide chain is almost fully extended: H-bonding is perpendicular to the axis of the chain and may be between chains running in the same direction (parallel pleated sheet) or opposite directions (antiparallel pleated sheet). The side chains are perpendicular to the sheet, above and below it.

There are other possible structures, such as the 3_{10} and π-helices, which are respectively more and less tightly wound than the α-helix, and the triple helix which is peculiar to collagen and is discussed below. Globular proteins often contain mixtures of different secondary structures, and regions where no *regular* structure is recognizable: this is because of interactions between side chains, such as attractions or repulsions between ionic groups. The imino acid proline has bond angles different from α-amino acids, and does not fit into an α-helix, although it is important in the collagen structure. Proline is often found at the ends of α-helical sequences in globular proteins.

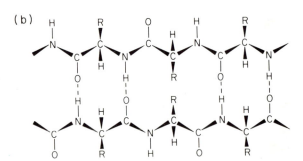

(a)

(b)

Fig. 2.4 Two common protein secondary structures: (a) the α-helix, and (b) the antiparallel β-pleated sheet. In a variant of this structure, the parallel sheet, H-bonding occurs between polypeptide chains aligned in the same direction.

Tertiary structure

This is the *overall* folding of the polypeptide chain. The backbone may have regular folding (secondary structure) which is maintained by H-bonding, while interactions between different regions of the chain fold it to produce tertiary structure. A number of different types of bonding are involved:

i. *Hydrophobic bonding.* In most globular proteins at least half of the amino acids have hydrophobic side chains, which are internally clustered in the folded protein, while hydrophilic, polar side chains are external. Water is excluded from the interior of the protein molecule, and the interaction between apolar groups is called hydrophobic bonding. It is primarily *entropic* in character. Water molecules in contact with apolar groups form an ordered structure, with loss of entropy: a more stable structure is formed if apolar groups are removed from contact with water, with a gain in entropy of the solvent.

ii. Hydrogen bonding. Several amino acid side chains can form hydrogen bonds, and such bonds contribute to protein folding. Since many polar groups are on the surface of folded proteins, they may bond to water rather than to each other, but H-bonds can be particularly strong in the hydrophobic interior of the protein.

iii. Ionic bonding. Side chains of opposite charge attract each other. Interactive forces of this type are called ionic or electrostatic bonds. The strength of such bonds decreases with increase in the dielectric constant of the intervening medium; on the surface of the protein, where most ionic groups are situated, the dielectric constant is high, and bonding rather weak. In the interior of the protein, charged groups are rather rare, but interactions between ion pairs can be very strong.

iv. Covalent bonds. The only common type of interchain bond is the disulphide bond of cystine, although covalent links of other types occur in a few proteins, such as collagen. Disulphide bonds are common in structural proteins and in extracellular enzymes, which they help to stabilize against denaturation, but are rather rare in intracellular, globular proteins.

Quaternary structure

This is the association of separate polypeptide chains into oligomers. Many globular proteins are oligomers of similar or identical chains; for example many glycolytic enzymes, such as aldolase and lactate dehydrogenase, are tetramers. Others have several different types of chain, with different structures and functions; pyruvate dehydrogenase has four different types of subunit. In each case the subunits are symmetrically arranged in the oligomeric protein. Quaternary structure occurs in most (though not all) proteins which show cooperative and allosteric effects (chapter 13).

In most cases quaternary structure is maintained by non-covalent interchain bonds, of the same type that maintain tertiary structure.

Conjugated proteins

Conjugated proteins have covalently-attached substituents, which may be of various kinds; low molecular weight substituents include sulphate, single sugar residues and short oligosaccharides, (and phosphate, which is found in pepsin (p. 123) and the milk protein casein. Some enzymes have covalently-attached prosthetic groups, which are required for activity; many carboxylases contain *biotin* (p. 38), which is covalently

linked to an ϵ-NH_2 group (lysine), while *lipoic acid* (p. 38) is similarly attached to 2-oxoacid dehydrogenases. Glycoproteins contain long, often branched oligosaccharide chains, which are attached to serine, threonine or asparagine (p. 176).

Denaturation

Denaturation is an imprecise term meaning conversion of a protein to an unfolded or altered state. Fully denatured proteins are in a 'random coil' configuration, with no recognizable, regular structure, but denaturation may be progressive, mild treatments producing only partial unfolding. The consequences of denaturation are loss of biological activity, solubility and the ability to form crystals. Occasionally denaturation may involve the making or breaking of covalent bonds and is therefore irreversible. Normally it can be reversed by some procedure—even proteins denatured to the point of insolubility, like boiled egg white, can be redissolved in a strong solution of urea, which can then be removed by dialysis.

Denaturation occurs at extremes of pH and at high temperatures, and is also brought about by agents which interfere with the bonds maintaining higher structure: chaotropic ions (I^-, SCN^-), large anions derived from strong acids (ClO_4^-, CCl_3COO^-), concentrated urea or guanidine hydrochloride, ionic detergents (sodium dodecyl sulphate, p. 178), and organic solvents.

If the extended random coils of denatured globular proteins can be arranged parallel to each other, as by extrusion through a fine orifice, the resulting bundles can be given the elasticity and texture of muscle fragments. This is the basis of spun vegetable protein as a meat substitute.

Protein folding

The sequence of amino acids in a protein (primary structure) contains enough information to define all higher structure. This is shown by the fact that many denatured proteins can be induced to refold in vitro, with full return of biological activity. In vivo, folding occurs spontaneously as the protein is synthesized, although some post-translational modification, such as limited proteolytic cleavage or the attachment of a prosthetic group, may be necessary.

The mechanism of folding is poorly understood. Since it forms spontaneously, the native structure is obviously an energy minimum, but it is only one of very many possible structures, so it is surprising that proteins fold so rapidly. Probably folding starts by *nucleation*—the formation of limited regions of order, probably secondary structure, in different parts of the polypeptide chain; these develop into folded *domains*, which together form the native protein.

Structures of individual protein types

Globular proteins

Most soluble proteins in the body, such as enzymes and blood proteins, are *globular*—ovoid molecules in which the polypeptide chain is compactly folded, with different types of secondary structure in different regions. Each molecule of a given protein has the same structure, and many proteins can be crystallized. The detailed structure of the protein can be derived from the X-ray diffraction pattern of crystals.

Fig. 2.5 shows the folding of myoglobin, a protein which is unusual in having a high content of α-helix, but no β-sheet. The nine α-helical regions contain almost 80% of the amino acids; between these regions are bends, many of which contain proline. The planar haem group, which binds oxygen, is buried in a cleft, with only one edge exposed. It is held by non-covalent interactions, the iron atom being coordinated to a histidine side chain; in this way the protein prevents oxidation of the iron to the ferric state, which does not bind oxygen.

The four subunits of haemoglobin are structurally similar to myoglobin, the slight differences in folding permitting inter-subunit contacts. These interactions are largely hydrophobic, although there are some H-bonds and ionic interactions.

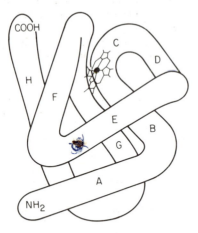

Fig. 2.5 Folding of the polypeptide chain in the oxygen-storage protein myoglobin, showing the eight helical regions (labelled A–H).

Fibrous proteins

Fibrous protein molecules are long and thin, and when assembled in bundles they form fibrils. The structural material of the body is largely made up of this group of insoluble proteins. The structural protein of wool and hair is almost completely α-helical, bundles of keratin molecules being aligned in parallel, and wound together

(probably in bundles of 3) to form a 'profibril'; these are packed together, forming fibres, which are embedded in an amorphous protein matrix within the cell. The α-helices are elastic; on stretching, the extended protein chains assume the β-configuration, breaking being prevented by numerous interchain disulphide bonds.

Fibroin, the structural protein of silk, forms sheets of antiparallel chains in the β-configuration, in which side chains are alternately above and below the plane of the sheet. Since these sheets are stacked one upon another, large side chains disrupt the structure, and silk is comprised largely of glycine, alanine and serine (3:2:1), with small amounts of other amino acids.

Myosin, the contractile protein of muscle, has elements of globular and fibrous structure, each polypeptide chain having a globular head and an extended α-helical tail (Fig. 2.6). Two such molecules associate, with the tails wound in a supercoil. The globular heads are each associated with two light chains, and have ATPase activity; these make and break contacts with actin as the filaments slide past each other, without change in conformation from the α-helical form.

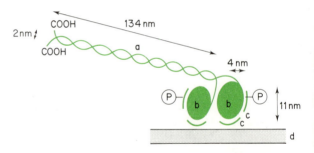

Fig. 2.6 The contractile protein myosin. Two 'heavy chains' are associated by supercoiling of α-helical tails (a); the globular heads (b) are associated with 'light chains' (c) which can become phosphorylated, and which make sliding contacts with the filamentous protein actin (d).

Collagen occurs in skin and connective tissue, and in the cornea; it makes up about a third of the protein in the body. Three parallel, helical collagen chains are wound in a triple helix, maintained by interchain H-bonds between the carboxyl and amino groups of the peptide bonds (Fig. 2.7). This structure is only possible when most amino acid side chains are small; collagen contains 33% glycine, 20% proline and hydroxyproline, and 10% alanine. These triple helices form a supercoil of tropocollagen; collagen fibrils contain staggered tropocollagen molecules (see p. 221).

Collagen and elastin, a related protein which occurs in lungs, arteries and some ligaments, contain 4-hydroxyproline and 3-hydroxylysine, formed after completion of the polypeptide chain; they also contain

Fig. 2.7   The triple-helical structure of collagen.

interchain cross-links, formed from lysine (see chapter 12).

Separation of proteins

Because proteins are susceptible to denaturation, separation procedures must always employ mild conditions. *Fractional precipitation* can be used on a large scale, but usually gives only limited purification. Proteins are separated according to their different solubilities in solutions of salt (ammonium sulphate often being used), aqueous acetone or ethanol, or polyethylene glycol. Careful addition of one of these selectively precipitates some proteins, with others remaining in solution. This is the basis of an obsolescent classification of proteins into albumins (which are soluble in half-saturated ammonium sulphate) and globulins (which are not).

Various forms of chromatography are applicable to proteins: of those described on p. 15, paper chromatography is unsuitable, as it causes denaturation, but *ion-exchange chromatography* on columns of chemically-modified cellulose or other polymers is widely used. *Exclusion chromatography* is performed on columns packed with polymers of controlled pore-size. Proteins too bulky to enter the pores are excluded from the gel, and pass through the column at the same rate as the eluting buffer, being eluted in the 'void volume'. Smaller proteins are retarded, by diffusing into the pores. Proteins can therefore be separated according to their molecular size, and calibrated exclusion chromatography columns can be used for approximate

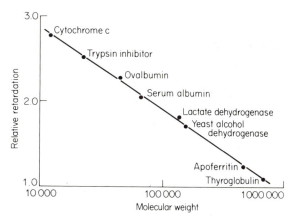

Fig. 2.8   Determination of the molecular weight of a protein in its native state, by exclusion chromatography.

determinations of the molecular weights of proteins (Fig. 2.8). Various types of exclusion chromatography gel are available, and they are made of polysaccharides (dextran or agarose), polyacrylamide or glass. Different grades of these are suitable for separating particles in different molecular-weight ranges, from a few hundreds (peptides) up to many millions (such as viruses).

Affinity chromatography is a very powerful separation technique which is applicable if the protein has specific affinity for some substance, such as a substrate or inhibitor. This substance is covalently linked to a solid support such as agarose, and the protein mixture passed through a column (an 'affinity column') of this material. Proteins with affinity for the immobilized ligand are retained by the column, and can be eluted by washing the column with the free ligand, or by changing the conditions to produce desorption. Very large purifications can be achieved in one step. Affinity chromatography can be used to purify enzymes (e.g. kinases on agarose-bound ATP, or cytochrome oxidase (p. 189) on immobilized cytochrome c) and is also applicable to nucleic acids. Affinity columns of lectins (p. 178) or specific antibodies are also useful in the purification of certain proteins.

Electrophoresis is difficult to apply on a large scale, but it is used analytically and for small-scale preparations. Proteins separate according to their mobility in an electric field; this depends on their charge (and therefore on the isoelectric point, relative to the pH) and molecular size. Hydrated gels of polyacrylamide, agarose or starch are usually used for protein separation.

In *electrofocussing* (isoelectric focussing) a pH gradient is set up in a column or gel by the use of *ampholytes*, a mixture of low molecular weight substances (often amino acid derivatives) of different p*I*. In an electric field proteins migrate to a point where the pH equals their isoelectric points; here they have no net charge, and remain.

Molecular weight determination

The most accurate estimates of protein molecular weights are obtained by ultracentrifugation, either from measurements of sedimentation velocity (which depends on the weight, shape and density of the molecule) or of sedimentation equilibrium, in which a concentration gradient is established, with diffusion opposing sedimentation; this depends only on the molecular weight of the protein. Such methods require large amounts of purified protein.

Exclusion chromatography (Fig. 2.8) gives a less accurate estimate, but is applicable to crude mixtures if the protein under study can be detected, for example by its biological activity. Gel electrophoresis in the presence of the detergent SDS (p. 178) can be used for very small amounts of protein. This anionic detergent denatures the protein and binds to the extended polypeptide chain, giving the molecule a uniform charge density; the rate of electrophoresis then depends only on molecular size, and the mobility is compared with that of proteins of known molecular weight (Fig. 2.9).

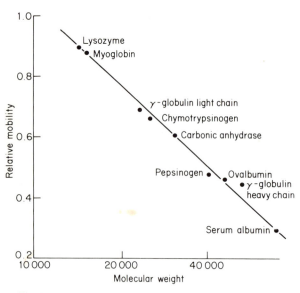

Fig. 2.9 Determination of the subunit molecular weight of a protein, by polyacrylamide gel electrophoresis in the presence of sodium dodecyl sulphate (SDS).

If the protein has quaternary structure this procedure usually gives the molecular weights of the subunits, which are dissociated by the detergent. The molecular weight of the undissociated protein can be determined electrophoretically after chemical cross-linking of the subunits, or under non-denaturing conditions by exclusion chromatography or ultracentrifugation.

Determination of amino acid sequences

The primary structure of a protein is a unique sequence of amino acids, which may be 1000 or more residues long. The sequence is usually determined by cleavage of the molecule into short peptides, which are then separated and analysed separately.

Cleavage of a protein at specific points, to produce a defined series of peptides, is performed using enzymes or chemical reagents. Some proteases are highly specific: trypsin cleaves peptide bonds on the carboxyl side of lysine or arginine, and a protease from *Staphylococcus aureus* cleaves next to aspartate or glutamate. Less specific proteases make more cuts and therefore generate more peptides. Certain reagents cleave peptide bonds at particular residues: for example cyanogen bromide (CNBr) breaks methionyl bonds.

Sequences of amino acids in peptides are determined by sequential removal and identification of amino acids from the N-terminal or C-terminal end. Phenylisothiocyanate (Edman's reagent) reacts with the N-terminus; the resultant thiohydantoin is identified by chromatography, or amino acid analysis of the residual peptide can be used to determine which residue has been removed.

Carboxypeptidases sequentially remove single amino acids from the C-terminus, and the sequence can be inferred from the kinetics of their liberation.

The order in which the peptides occur in the protein is found by generating a different set of peptides, overlapping with the first, and determining their amino acid sequences. Automated procedures using Edman's reagent can sequence peptides of 50 amino acids or more. Sequencing procedures for DNA are also very refined (p. 203) and if a gene can be isolated, this is the most rapid way of determining the amino acid sequence of the protein it specifies.

Detection of proteins

The concentration of proteins in solution can be measured spectrophotometrically in various ways, such as the biuret reaction, which forms a coloured complex between Cu^{2+} ions and peptide bonds; the Folin–Lowry procedure, which depends on this and other colour-forming reactions; measurement of the absorbance of ultraviolet light (280 nm) by aromatic amino acids in the protein; or measurement of an absorbance change occurring when the dye Coomassie Blue G binds to proteins. Each of these methods may be subject to interference by contaminants such as salts, nucleic acids, metabolites or detergents.

In electrophoretograms proteins are usually detected by staining with dyes, such as Coomassie Blue or Amido Black. Specific proteins can be localized in tissues by microscopy after staining with specific antibodies covalently linked to dyes.

3

Enzymes

Most of the chemical reactions which occur in the body will not proceed in vitro, except under extreme conditions of reactant concentration, pH or temperature, unless they are catalysed. Almost all physiological reactions are catalysed by enzymes, which are the most efficient catalysts known, accelerating the rates of reactions by factors of 10^6–10^{12} and allowing the rapid attainment of equilibrium, without side reactions, in dilute solution, at low temperature and under neutral conditions.

All enzymes are proteins, and their catalytic activities depend on all the factors which affect protein conformation: activity is invariably lost on denaturation of the enzyme, and may even be abolished by chemical modification of a single amino acid side chain. Enzymes exhibit two further catalytic features which are of great biological importance: high specificity (i.e. they are selective for a single substrate or group of substrates), and the ability to be controlled, catalytic activity often being regulated by the concentrations of effectors which do not themselves take part in the reaction.

Catalysis

An essential property of all catalysts is that, while they enter into chemical reactions, they remain unchanged at the end of the reaction. Many enzymes are rather labile and, in vitro at any rate, lose activity during the course of the reaction they catalyse: this is because of denaturation, and has nothing to do with the catalysis itself. Although it is not unknown for enzymes to catalyse 'suicide' reactions which result in their own inactivation, examples are very rare.

In common with other catalysts, enzymes accelerate both forward and reverse reactions; in the reaction

$$A + B \underset{k_2}{\overset{k_1}{\rightleftharpoons}} C + D \qquad (1)$$

where k_1 and k_2 are the forward and reverse rate constants, the forward and reverse rates of the reaction must be equal at equilibrium:

$$k_1[A][B] = k_2[C][D] \qquad (2)$$

and K, the equilibrium constant, is given by

$$K = \frac{[C][D]}{[A][B]} = \frac{k_1}{k_2} \qquad (3)$$

Since K depends only on thermodynamic factors, it cannot be altered by a catalyst, so if k_1 is increased by catalysis, k_2 must be increased by the same factor.

Catalysis is explained by the transition-state theory. Fig. 3.1 shows the energy profile of a reaction, in which the Gibbs free energy of the system is plotted against

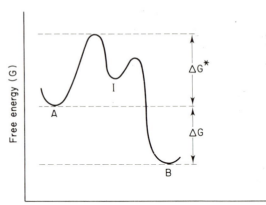

Fig. 3.1 Energy profile of a reaction. A, reactants; B, products; C, an intermediate; ΔG, free energy change of the reaction; $\Delta G^\star$, activation energy.

the 'reaction coordinate', an indication of distance along the reaction pathway. The equilibrium constant is determined by the overall difference in free energy between reactants and products:

$$\Delta G_0 = -RT \ln K \qquad (4)$$

However, the rate of attainment of equilibrium depends on the activation energies ($\Delta G^\star$) of the various steps in the reaction. In Fig. 3.1, the pathway from reactants (A) to products (B) is shown proceeding via an intermediate state (I) which may be an enzyme–substrate complex. In the uncatalysed reaction there is a large energy barrier between A and B, whereas catalysis greatly reduces the activation energy, and therefore accelerates the reaction. In general, the rate constant (k_n) for a particular step is given by

$$k_n = \frac{kT}{h} e^{-\frac{\Delta G^\star}{RT}} \qquad (5)$$

where R is the gas constant, k is Boltzmann's constant, and h is Planck's constant. The factor kT/h is of course independent of the nature of the reactants, so k_n is determined by the activation energy, $\Delta G^\star$, and by the absolute temperature, T.

Enzyme–substrate complexes

Enzymic catalysis involves the formation of transient, highly specific complexes between enzyme and substrate. Usually the enzyme is much larger than the substrate: typically, enzymes have molecular weights in the range 10^4–10^6, substrates in the range 10^2–10^3. In the case of enzymes acting on polymers such as proteins or nucleic acids, the enzyme binds to only a limited region of the substrate. The enzyme usually contains a cleft in which the substrate is bound. The high reactivity of enzyme–substrate complexes is due to a number of factors:

(1) *Proximity and orientation*. In reactions involving more than one substrate, which includes most cases, the substrates are bound in close proximity, and at the correct angular disposition for reaction between them to occur. The reaction within the complex proceeds as an intramolecular, rather than intermolecular, process: accelerations of up to 10^8-fold are predicted from this effect alone.

(2) *Chemical catalysis*. The substrate-binding cleft contains catalytic groups provided by the enzyme, and these participate in the reaction. The catalytic groups may donate protons to, or accept protons from, the reactants (general acid or base catalysis), or may act as nucleophiles, becoming transiently combined with the substrate and thus providing a reaction pathway of low activation energy. The catalytic groups may be amino acid side chains, such as —COOH, —NH$_2$, —OH, —SH or imidazole, or may be provided by non-protein

components of the enzyme, called prosthetic groups (see p. 36). It may be difficult to distinguish experimentally between those groups which actually participate in the reaction, and those which are simply required for binding the substrate.

(3) *Strain*. The activation energy ($\Delta G^\star$) between successive steps of a reaction can be decreased if the enzyme has a high affinity for the intermediate or *transition state* of the reactants, thus straining the substrates into the correct conformation for reaction to occur. This is kinetically more favourable than simply binding the substrate tightly. In some cases it has been possible to prove this by guessing at the nature of the transition state for a particular reaction, synthesizing a stable analogue of similar geometry, and showing that this binds to the enzyme far more tightly than the substrate itself. In the substrate of lysozyme, for example, the sugar next to the bond which is hydrolysed is normally in the 'chair' configuration (Fig. 3.2a); the transition state is thought to be in the 'sofa' configuration (b). The compound (c), which is structurally similar to (b), binds to the enzyme some 100-fold more tightly than the substrate: it is known as a 'transition state analogue'.

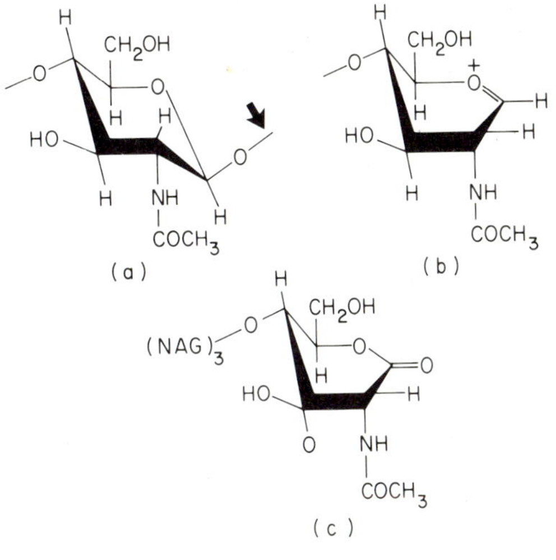

Fig. 3.2 Conformations of the substrate in lysozyme. (a) *N*-acetyl glucosamine in the 'chair' form: the scissile bond is shown by an arrow; (b) the transition state, in the 'sofa' form; (c) a transition-state analogue, which binds very tightly to the enzyme.

Factors Affecting the Rates of Enzyme-Catalysed Reactions

1. *Time course of a reaction*

The progress of an enzyme-catalysed reaction is shown by a plot of the concentration of any of the products (or substrates) against time; the reaction rate (moles/litre/second) at any particular time is then given by the tangent to the curve at that time (Fig. 3.3). During in vitro studies of enzymes it is usual to measure the *initial velocity* of the reaction (v_0), which is the slope of the progress curve at the origin: at this time the substrate concentration is not depleted by the reaction, and the product concentration is zero. As the reaction proceeds, the reaction velocity falls, through substrate depletion and product inhibition (p. 32). Furthermore, as the product concentration increases, the velocity of the back reaction rises, until at equilibrium the forward and back reactions are occurring at the same rates, and the net reaction rate is zero.

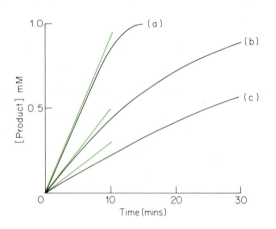

Fig. 3.3 Time course of enzyme-catalysed reactions. In each case the original substrate concentration is 1 mM, and $V_{max} = 1$ μmol/ml/min. (a) $K_m = 0.1$ mM; (b) $K_m = 1.0$ mM; (c) $K_m = 3.0$ mM.

The initial velocity may be quite difficult to measure if the time-course is sharply curved, and for this reason it is preferable to use an enzyme assay which produces a continuous record of the progress of the reaction, rather than measuring a single concentration of product after a fixed time.

2. *Kinetics of enzyme-catalysed reactions*

A simple equation, known as the Michaelis equation, describes the dependence of the initial reaction velocity (v_0) on the concentrations of enzyme, [e], and substrate,

[S]. We assume that enzyme and substrate form a complex which breaks down to give enzyme and product: at the start of the reaction the product concentration is zero, so combination of enzyme and product does not occur.

$$E + S \underset{k_2}{\overset{k_1}{\rightleftharpoons}} ES \overset{k_3}{\rightarrow} E + P \qquad (6)$$

Having assigned rate constants to the various steps of the reaction, we can state the reaction velocity, which is equal to the rate of appearance of product:

$$v_0 = \frac{d[P]}{dt} = k_3[ES] \qquad (7)$$

We now need to know [ES], the concentration of the enzyme–substrate complex. To obtain it we make two further assumptions:

(i) That the concentration of substrate is very much greater than the concentration of enzyme:

$$[S] \gg [e] \qquad (8)$$

This is almost always true in vitro: enzymes are such efficient catalysts that the concentration needed to produce a measurable reaction rate is usually in the range 10^{-10}–10^{-7} M, whereas the substrate concentrations used are dictated by the affinity of the enzyme for the substrates, and are usually 10^{-6}–10^{-2} M. The concentration of *free* substrate is not significantly diminished by formation of the enzyme–substrate complex, and is therefore approximately equal to the *total* substrate concentration, [S].

(ii) That while the reaction velocity is measured, a *steady state* applies: although there is a flux of substrate through the reaction pathway, the concentrations of the various forms of the enzyme (just E and ES, in our example) are unchanged. We can now write

$$[e] = [E] + [ES] \qquad (9)$$

$$\frac{d[E]}{dt} = 0 = k_3[ES] + k_2[ES] - k_1[E][S] \qquad (10)$$

$$\frac{d[ES]}{dt} = 0 = k_1[E][S] - k_2[ES] - k_3[ES] \qquad (11)$$

From equations (9) and (10)

$$[e] = [ES]\left(1 + \frac{k_2 + k_3}{k_1[S]}\right) \qquad (12)$$

and substituting this in equation (7),

$$v_0 = \frac{k_3[e]}{1 + \dfrac{k_2 + k_3}{k_1[S]}} \qquad (13)$$

The primary rate-constants in this equation (k_1, k_2 and k_3) may not be known, and cannot be measured in a steady-state experiment; even the enzyme concentration ([e]) may not be known, unless the enzyme used in the experiment is pure and of known molecular weight. It is therefore often useful to write the equation in a form containing simpler constants:

$$v_0 = \frac{V_{max}}{1 + \dfrac{K_m}{[S]}} \qquad (14)$$

This is the Michaelis equation, which is valid not only for the simple case shown in equation (6), but for all reactions involving a single substrate (providing there is only a single, linear reaction pathway). K_m is the Michaelis constant and V_{max} the maximum velocity. If the reaction scheme contains more intermediates than are shown in equation (6), then K_m and V_{max} have more complicated identities than those discussed below.

Dependence of reaction rate on enzyme concentration

From equation (13) it can be seen that the reaction velocity v_0 is directly proportional to enzyme concentration, providing the substrate concentration is fixed. This assumes that in the absence of enzyme, the reaction rate is negligible, as is usually the case. In the Michaelis equation the enzyme concentration is contained in the term V_{max}, which is the maximum velocity of the reaction *for a given value of* [e]. Obviously V_{max} is directly proportional to [e]: in the case derived, the rate-limiting step is the breakdown of ES, and the proportionality constant is k_3. In more complex cases there may be no single rate-limiting step, and V_{max} may contain several constants. In general, we can write

$$V_{max} = k_{cat}[e] \qquad (15)$$

where k_{cat} is the *catalytic rate constant*, defined as the number of substrate molecules formed by each enzyme molecule in unit time. k_{cat} may also be called the turnover number or molecular activity of the enzyme. It has the dimensions mols product/mol enzyme/time, in other words reciprocal time, and values of k_{cat} for different enzymes range from 0.1 s^{-1} to 10^7 s^{-1} or more. It is a more useful constant than V_{max}, but to calculate it one must know the concentration, purity and molecular weight of the enzyme. If the number of active sites per molecule of enzyme is known, k_{cat} can be expressed as the turnover number at each site (catalytic centre activity).

Dependence of reaction rate on substrate concentration

A plot of v_0 against [S] is called a Michaelis plot, and is shown in Fig. 3.4; it is a rectangular hyperbola, asymptotic to V_{max} on the v_0 axis. The Michaelis constant, K_m, is the concentration of substrate which

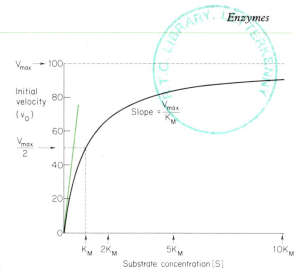

Fig. 3.4 The Michaelis plot for an enzyme-catalysed reaction.

gives *half* the maximum velocity: it is easily shown from equation (14) that if [S] = K_m, then $v_0 = \dfrac{V_{max}}{2}$.

K_m is a practical and very useful parameter, as it defines the substrate concentration range that an enzyme requires in order to work efficiently. It has the dimensions of concentration, and K_m values are usually in the range 10^{-6}–10^{-2} M. In the sample case shown in equation (6) it has the identity $\dfrac{k_2 + k_3}{k_1}$, while in more complicated cases it is a more complex combination of constants. Whatever its identity, the *lower* the K_m, the *higher* will be the reaction rate with a given non-saturating substrate concentration. K_m therefore *appears* to be inversely related to the affinity of the enzyme for its substrate, but the true affinity is defined by K_s, the equilibrium dissociation constant of the enzyme–substrate complex.

$$K_s = \frac{[E][S]}{[ES]} \qquad (16)$$

From equation (6) it can be shown that $K_s = \dfrac{k_2}{k_1}$, so that K_m and K_s are approximately equal *only* if $k_2 \gg k_3$, in other words as long as the catalytic reaction is not fast enough to disturb the binding equilibrium. This is by no means always so.

From equation (14) we see that at very low substrate concentrations $v_0 = \dfrac{V_{max}}{K_m}[S]$: the reaction velocity is directly proportional to substrate concentration near the origin of the Michaelis plot. At very high substrate concentrations $v_0 \simeq V_{max}$, so reaction velocity becomes independent of substrate concentration: the enzyme is then said to be saturated with substrate. V_{max} is actually achieved only at infinite substrate concentration: even if [S] = $10 K_m$, the reaction velocity is only $0.91 V_{max}$.

Determination of K_m *and* V_{max}

In principle, these constants might be obtained from a series of measurements of initial reaction velocity at different substrate concentrations, by use of the Michaelis plot (Fig. 3.4); in practice this is impossible to do with accuracy since V_{max} is only approached at very high substrate concentrations, which may be experimentally difficult to achieve, so there are difficulties in defining the asymptote, and in deciding whether the best fit of the data is really the rectangular hyperbola predicted by equation (14). It is much more satisfactory to use a form of the Michaelis equation which produces linear plots, by rearranging equation (14) to give

$$\frac{[S]}{v_0} = \frac{[S]}{V_{max}} + \frac{K_m}{V_{max}} \qquad (17)$$

A plot of $\dfrac{[S]}{v_0}$ against [S] is a straight line, with a slope

of $1/V_{max}$: the intercept on the ordinate is $\dfrac{K_m}{V_{max}}$, and on the abscissa $-K_m$ (the Hanes plot, Fig. 3.5). This plot is satisfactory from a statistical point of view, in that any errors in the determination of v_0 carry similar weights over a wide range of [S], and deviations of the enzyme from Michaelis behaviour (such as cooperative substrate binding, chapter 13) are easy to detect.

Other linear plots (of $1/v_0$ versus $1/[S]$, the Lineweaver–Burk plot, or $v_0/[S]$ against v_0, the Hofstee–Eadie plot) are less easily used and are statistically unsatisfactory, so they are not recommended.

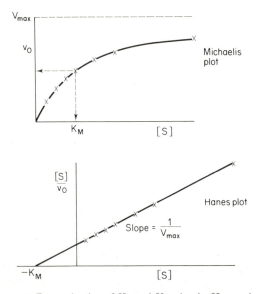

Fig. 3.5 Determination of K_m and V_{max} by the Hanes plot. Note the difficulty in extrapolating to V_{max} in the Michaelis plot of the same data points.

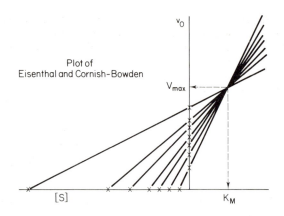

Fig. 3.6 Determination of K_m and V_{max} by the plot of Eisenthal and Cornish-Bowden. The data points are the same as those plotted in Fig. 3.5.

However, an excellent method for determining K_m and V_{max} is provided by the *direct linear plot* of Eisenthal and Cornish-Bowden, in which each observation is represented by a line, drawn from the value of [S], on the horizontal axis, through the corresponding value of v_0, on the vertical axis (Fig. 3.6). A series of such lines meet at a point, the horizontal and vertical coordinates of which are K_m and V_{max}. Usually there is some scatter in the experimental data and the lines do not meet at a single point; the best estimates of K_m and V_{max} are each given by the *median* value (the one in the middle) of the different intercepts.

The Michaelis equation is a differential equation, describing the reaction rate, $\dfrac{d[P]}{dt}$ or $-\dfrac{d[S]}{dt}$, in terms of [S] and [e]; it can be integrated to a form which describes the entire time-course of the reaction (Fig. 3.3), relating [P] to [e], the initial substrate concentration and the time. This integrated rate equation contains the constants K_m and V_{max}, but is not often used to measure them.

Significance of K_m *and* V_{max}

Knowledge of K_m and V_{max} is essential if one is attempting to calculate enzyme activities in vivo, although there are many difficulties in extrapolation from in vitro conditions, such as the possibility of various types of inhibition, and the fact that, in the cell, the assumption (equation 8) that concentrations of enzyme are much lower than those of substrate may not be valid. Most enzymes have more than one substrate, and their behaviour is described by more complex rate equations than equation (14); consequently the *apparent* K_m for one substrate may depend upon the concentration of another, and to predict v_0, more kinetic constants have to be measured—four for a two-substrate reaction, and eight when there are three substrates!

Michaelis constants are of course independent of enzyme concentration. V_{max} has the same units as v_0 (e.g. μmol/min/ml) but is proportional to enzyme concentration and can be used to calculate the *specific activity* of the enzyme, usually expressed as μmol substrate/mg protein/min. The International Unit (IU) of enzyme activity is the amount of enzyme which will produce 1 μmol of product per minute (under particular conditions, defined for each enzyme), and the specific activity of commercial enzyme preparations is usually given in IU/mg: this can be used to calculate k_{cat} (p. 27), if the molecular weight and purity are known.

The S.I. unit of enzyme activity is the katal (kat), which is the amount of enzyme which will form one mole of product per second. Thus 1 I.U. = 16.7 nkat.

Table 3.1 shows the kinetic parameters of some glycolytic enzymes.

Table 3.1 Kinetic parameters of some enzymes of glycolysis in muscle

Enzyme	k_{cat} (s^{-1})	*Substrate*	K_m (μM)
Phosphoglucomutase	200	Glucose-1-phosphate	8
Glucose-phosphate isomerase	800	Glucose-6-phosphate	700
Phosphofructokinase	215	ATP	45
		Fructose-6-phosphate	30
Aldolase	11	Fructose-1,6-bis-phosphate	100
Triose phosphate isomerase	4400	Dihydroxyacetone phosphate	750
Glyceraldehyde-3-phosphate de-hydrogenase	50	Glyceraldehyde-3-phosphate	155
		NAD$^+$	46
		Phosphate	290
Phosphoglycerate kinase	1600	1,3-diphospho-glycerate	2.2
		ADP	350
Phosphoglycerate mutase	225	3-phosphoglycerate	5000
Enolase	60	2-phosphoglycerate	70
Pyruvate kinase	300	Phosphoenolpyruvate	70
		ADP	300
Lactate dehydrogenase	190	Pyruvate	170
		NADH	10

Inhibitors

The activity of an enzyme can often be reduced by addition of a particular substance to the reaction mixture; excluding non-specific effects which might affect its activity, such as changes in pH or ionic strength, this is known as *inhibition*. Inhibitors are divided into two major classes, reversible and irreversible, and there are further subdivisions within these classes.

Irreversible inhibitors

These are substances, usually not of biological origin, which react covalently with an enzyme, preventing substrate binding or catalysis. For example, many enzymes are irreversibly inactivated when certain cysteine side chains are alkylated by reaction with iodoacetamide:

$$Enzyme—CH_2SH + ICH_2CONH_2 \rightarrow$$
$$Enzyme—CH_2S—CH_2CONH_2 + HI$$

This does not necessarily imply that the reactive thiol participates in catalysis: it may be distant from the substrate binding site, but important in maintaining the correct conformation of the enzyme, or it may be near the substrate binding site, and its alkylation may cause steric hindrance of substrate binding. On the other hand a particular group may be susceptible to attack by irreversible inhibitors precisely because it is catalytically important, and its environment makes it especially reactive: examples are the essential cysteine at the active site of glyceraldehyde 3-phosphate dehydrogenase (p. 72), and the essential serine at the active site of chymotrypsin (p. 40), which is highly nucleophilic and blocked by the nerve gas di-isopropyl fluorophosphate (DFP):

$$Enzyme—CH_2OH + iPr—\overset{\overset{\displaystyle O}{\|}}{\underset{\underset{\displaystyle iPr}{|}}{P}}—F \rightarrow$$

$$Enzyme—CH_2—O—\overset{\overset{\displaystyle O}{\|}}{\underset{\underset{\displaystyle iPr}{|}}{P}}—iPr + HF$$

Other serine residues in chymotrypsin are not modified by this reagent. The specificity of irreversible inhibitors is increased if they bear a structural resemblance to the substrate of an enzyme; they may then be bound at the active site, and react covalently with a nearby amino acid, permitting the identification of at least part of the active site. Two examples, which are structurally similar to ATP and also contain covalently reactive groups, are 8-azido ATP and *p*-fluorosulphonyl benzoyl adenosine (FSBA). These compounds irreversibly inhibit many enzymes, such as phosphotransferases (kinases), which have ATP as a substrate. They are known as 'active site directed irreversible inhibitors'.

8 - azido ATP

FSBA

Irreversible inhibition is usually total, although sometimes a modification reduces, rather than abolishes, enzyme activity. It is time-dependent, the rate of loss of activity depending on the inhibitor concentration (Fig. 3.7) and is not amenable to the type of analysis applied to reversible inhibitors (see below) since after addition of an irreversible inhibitor the enzyme is a mixture of unreacted, fully active protein, and reacted, inactive protein: the effect of an irreversible inhibitor is therefore to reduce the concentration of active enzyme.

Fig. 3.7 Inactivation of an enzyme by two concentrations of an irreversible inhibitor, one (●) twice the other (▲). In (a) the activity is plotted on a linear scale, and in (b) on a logarithmic scale.

Reversible inhibitors

Reversible inhibitors do not react covalently with enzymes, but undergo rapid, equilibrium binding. The velocity of the enzyme-catalysed reaction is reduced by the formation of enzyme–inhibitor or enzyme–substrate–

inhibitor complexes. Reversible inhibitors are classified as competitive, uncompetitive or non-competitive, according to their effects on the steady-state kinetic parameters K_m and V_{max}: these effects depend on which form of the enzyme binds the inhibitor.

Competitive inhibitors

These *increase* the apparent K_m for the substrate, without effect on V_{max}: this is caused by the inhibitor binding to the free enzyme only:

$$I + S + E \underset{k_2}{\overset{k_1}{\rightleftharpoons}} ES \overset{k_3}{\longrightarrow} E + P \qquad (18)$$

$$k_4 \Big\updownarrow k_5$$

$$EI$$

The concentration of enzyme–inhibitor complex is given by

$$\frac{[E][I]}{[EI]} = \frac{k_5}{k_4} = K_i \qquad (19)$$

where K_i is the inhibition constant for this particular inhibitor. The rate equation becomes

$$v_0 = \frac{V_{max}}{1 + \dfrac{K_m}{[S]}\left(1 + \dfrac{[I]}{K_i}\right)} \qquad (20)$$

where [I] is the concentration of inhibitor. Inhibition of this type is usually caused by the inhibitor binding to the substrate-binding site, so that occupation by substrate and inhibitor are mutually exclusive. For a given inhibitor concentration, inhibition is most severe at low substrate concentrations, when formation of EI predominates over ES; as [S] increases, so does the steady-state concentration of the productive complex ES, until at very high substrate concentrations inhibition is slight. From equation (20) it is apparent that V_{max} is unaffected by the inhibitor, whereas the K_m is raised by the factor $\left(1 + \dfrac{[I]}{K_i}\right)$. A well-known example of competitive inhibition is the inhibition of succinate dehydrogenase by malonate, a dicarboxylic acid which is an analogue of the substrate, succinate, but which cannot be dehydrogenated to fumarate. Oxaloacetate is also a competitive inhibitor, for the same reason.

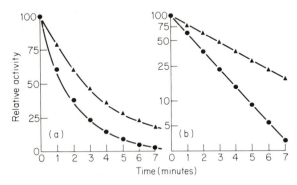

Uncompetitive inhibition

Uncompetitive inhibitors *decrease* both K_m and V_{max}: this is caused by the inhibitor binding to the enzyme–substrate complex only:

$$I+S+E \underset{k_2}{\overset{k_1}{\rightleftharpoons}} ES \overset{k_3}{\longrightarrow} E+P \qquad (21)$$
$$\underset{EIS}{\overset{\quad k_6 \Big\| \uparrow k_7}{}}$$

The rate equation for this type of inhibition is

$$v_0 = \frac{V_{max}}{1 + \dfrac{K_m}{[S]} + \dfrac{[I]}{K_i}} \qquad (22)$$

where K_i is defined in a similar way as earlier:

$$K_i = \frac{[ES][I]}{[ESI]} = \frac{k_7}{k_6} \qquad (23)$$

Equation (22) may be rearranged to give

$$v_0 = \frac{V_{max} \Big/ \left(1 + \dfrac{[I]}{K_i}\right)}{1 + K_m \Big/ [S]\left(1 + \dfrac{[I]}{K_i}\right)} \qquad (24)$$

from which it is apparent that both K_m and V_{max} are decreased, being divided by the same factor $\left(1 + \dfrac{[I]}{K_i}\right)$.

At high substrate concentrations this type of inhibition is not alleviated, but becomes relatively worse, as increased steady-state concentrations of ES favour the binding of the inhibitor, to give the 'dead-end' complex EIS.

Examples of uncompetitive inhibition of single-substrate enzymes are rare, since such enzymes do not usually contain more than one type of binding site; but in multisubstrate reactions there is often a fixed order of addition of substrates:

$$E \underset{}{\overset{S_1}{\rightleftharpoons}} ES_1 \underset{}{\overset{S_2}{\rightleftharpoons}} ES_1 S_2 \longrightarrow EP_1 P_2 \overset{P_1}{\underset{}{\nearrow}} EP_2 \overset{P_2}{\underset{}{\nearrow}} E \qquad (25)$$
$$\underset{ES_1 I}{\overset{\Big\| \uparrow}{}}$$

An inhibitor which binds to ES_1, forming the dead-end complex $ES_1 I$ by occupying the binding site for S_2, is competitive with S_2, and uncompetitive with S_1.

Non-competitive inhibition

It may happen that an inhibitor binds to both free enzyme and enzyme–substrate complex:

$$I+S+E \underset{k_2}{\overset{k_1}{\rightleftharpoons}} ES \overset{k_3}{\longrightarrow} E+P \qquad (26)$$
$$\underset{EI \rightleftharpoons ESI}{\overset{k_4 \Big\| \uparrow k_5 \quad k_6 \Big\| \uparrow k_7}{}}$$

The rate equation derived from this scheme is

$$v_0 = \frac{V_{max} \Big/ \left(1 + \dfrac{[I]}{K_i'}\right)}{1 + \dfrac{K_m}{[S]}\left(1 + \dfrac{[I]}{K_i}\right) \Big/ \left(1 + \dfrac{[I]}{K_i'}\right)} \qquad (27)$$

where

$$K_i = \frac{k_5}{k_4}, \quad K_i' = \frac{k_7}{k_6} \qquad (28)$$

This is called non-competitive inhibition; both K_m and V_{max} are affected, the relative changes depending on K_i and K_i'. Non-competitive inhibition used to be defined as causing a reduction in V_{max} but no change in K_m: obviously, from equation (27), this only happens if K_i and K_i' are fortuitously equal, and there is no reason to limit the definition to this special case.

Non-competitive inhibition is common in multisubstrate enzymes, but rare when there is only one substrate; but one could consider a hydrogen ion to meet the definition, when an essential ionizing group binds protons with different affinities (i.e. has different pK_a values) in E and ES.

Diagnosis of type of inhibition

The different types of inhibitor are distinguished by their different effects on K_m and V_{max}, which can be seen in Fig. 3.8. In this type of plot the 'pure' forms of inhibition, competitive and uncompetitive, result in a

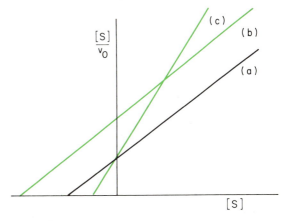

Fig. 3.8 The effects of a reversible inhibitor on an enzyme: (a) uninhibited; (b) competitive inhibition; (c) uncompetitive inhibition.

change in intercept and slope, respectively; non-competitive inhibition affects both and, depending on the relative size of K_i and K_i', either slope or intercept effects may dominate. These plots show the effect of varying the substrate in the absence of inhibitor, and with a single inhibitor concentration; to characterize the inhibition properly, several different inhibitor concentrations should be used, producing a family of lines. The value of K_i can be calculated from the increase in slope or intercept over the uninhibited case.

Product inhibition

An enzyme–product complex is part of the reaction sequence of enzyme-catalysed reactions; in the presence of large concentrations of product, this complex is dominant, and product inhibition is observed. This is one of the reasons why the velocity of the reaction falls as equilibrium is approached, and it is of physiological importance in controlling pathways by negative feedback (p. 227). In single-product enzymes (equation 6) addition of the product causes reversal of the reaction, but if there are two or more products, both are needed for reversal to occur, so inhibition of the forward reaction by each can be studied separately. Usually each product is competitive with the substrate from which it is derived; for example in the hexokinase reaction (p. 72), ADP is competitive with ATP, and glucose-6-phosphate is competitive with glucose.

Substrate inhibition

It is sometimes found that as the substrate concentration is increased, the reaction velocity increases, but then falls at high substrate concentrations. This is known as substrate inhibition, and it is most often found in enzymes which have an ordered sequence of substrate binding—if a substrate binds out of turn it may form an unreactive 'dead-end' complex. For example, in equation (25), at high concentration S_2 might bind to E, forming the 'dead-end' complex ES_2. Liver alcohol dehydrogenase, which has a mechanism in which NAD^+ binds first, is inhibited by high concentrations of the second substrate, alcohol.

The effect of pH on reaction velocity

The velocity of an enzyme-catalysed reaction is usually found to be fairly sharply pH-dependent (Fig. 3.9) and to be maximal at a given pH value (the optimal pH) which may, however, depend on other conditions of assay, such as temperature, substrate concentration, ionic strength, and nature of the buffer.

This effect is due to ionizations of the enzyme, substrate or enzyme–substrate complex. Usually, only one ionic form of the free enzyme and substrate will

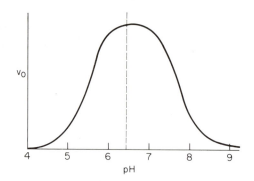

Fig. 3.9 The effect of pH on the initial velocity, v_0, of an enzyme-catalysed reaction.

combine to form the enzyme–substrate complex, and only one ionic form of this complex undergoes the catalytic reaction to form products. The pH dependence of v_0 (Fig. 3.9) is therefore a result of changes in K_m and/or V_{max}, above or below the optimal pH. From the pH dependences of these parameters one can determine the pK_a values of groups involved in substrate binding and catalysis. Having determined these pK_a values it is not always possible to identify the amino acids involved, as the pK_a of a group depends very much on its environment, and may fall within quite a wide range (p. 17).

Over a narrow range of pH these effects are usually reversible, but exposure to strongly acidic or alkaline conditions may bring about irreversible denaturation of the enzyme.

Effects of temperature

In general, the rate of enzyme-catalysed reactions increases with temperature; at high temperatures, however, the increased turnover is offset by an increased rate of denaturation of the enzyme. For this reason a plot of reaction velocity against temperature may *appear* to pass through a maximum, suggesting an 'optimal temperature'. This concept has little value because, as shown in Fig. 3.10, the 'optimal temperature' is an artefact of the assay. As the temperature is increased the *initial* velocity (v_0) always increases, but at high temperatures the reaction rate quickly falls through denaturation. A plot of v_0 against temperature does not go through a maximum: the 'optimal temperature' is apparent only if the reaction rate is derived from measurements of product-concentration after some fixed time interval, when denaturation is significant.

The transition-state theory predicts the variation of k_{cat} with temperature (equation 5); from this expression it can be seen that a plot of $\log k_n$ (or v_0) against $1/T$ is a straight line of slope $-\dfrac{2.3\Delta G^\star}{R}$. This is known as an

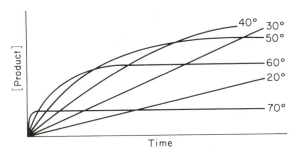

Fig. 3.10 The effect of temperature on the time course of an enzyme-catalysed reaction. Both the initial reaction rate and the rate of enzyme inactivation increase with temperature.

Arrhenius plot, and is used to measure the activation energy of a reaction: this is always lower for a catalysed than for an uncatalysed reaction. An alternative and less precise way of stating the activation energy is to quote the Q_{10} for a particular reaction, which is the factor by which the reaction velocity is increased when the temperature is raised by 10°C, in other words the ratio of reaction rates (v_0) at, say, 25°C and 35°C. Q_{10} values are often about 2 for enzyme-catalysed reactions.

Arrhenius plots sometimes show discontinuities, probably indicating conformation changes in the enzyme which occur at a particular temperature. This is particularly common for membrane-bound enzymes, because their activity is critically dependent on the surrounding phospholipid bilayer, which undergoes structural transitions in the range 25–45°C.

Enzyme activators

Enzymes are often found to require for their activity the presence of some substance, other than the substrate, which is loosely termed an 'activator'. Activators may be components of the enzyme which have been lost during purification, such as loosely-bound prosthetic groups (p. 36), or they may be metabolites which act as control signals in vivo, binding to specific sites on an enzyme and inducing it to form an active conformation—

these are called allosteric effectors, and are discussed in chapter 13.

Some enzymes show specific ion requirements which do not fall into either of these categories; for example most enzymes acting on phosphorylated substrates require a divalent cation, such as Mg^{2+}, Mn^{2+} or Ca^{2+}, for activity. This is because the enzyme acts on a complex formed by substrate and cation:

$$ATP^{4-} + Mg^{2+} \rightleftharpoons ATPMg^{2-}$$

Nucleoside triphosphates bind cations very tightly (dissociation constant 10^{-5}–10^{-4} M) and in the cell, the Mg^{2+} concentration is high enough to ensure that nucleoside triphosphates are predominantly in this form.

Zymogens

Some enzymes are synthesized as inactive precursors, which are converted to active forms by proteolysis. These precursors are called zymogens. The proteases of the exocrine pancreas are stored and secreted as zymogens, which are activated on reaching the digestive tract. The formation of active chymotrypsin from its zymogen, chymotrypsinogen, is described on p. 40; trypsin and elastase are derived from trypsinogen and proelastase, the former by removal of an N-terminal peptide of 6 amino acids. In the digestive tract, trypsin is responsible for the activation of all the proteolytic zymogens; small amounts of trypsin are generated from trypsinogen by the action of enterokinase, a protease occurring only in the duodenum. Self-activation of zymogens in the pancreas is prevented by a trypsin inhibitor, a small protein which binds very tightly to the active site of trypsin.

Pepsin, the protease of gastric juice, is secreted as pepsinogen, which is stable at pH 5 or above but spontaneously activated at pH 2. Activation is probably autocatalytic, and occurs with a loss of a 44-amino-acid peptide from the N-terminus (see p. 123).

The blood-clotting cascade, which involves the proteolytic activation of a series of zymogens, is described on p. 240.

Coenzymes and Prosthetic Groups

These two classes of substances include non-protein compounds, often complex molecules, which participate in enzyme-catalysed reactions, but there is a fundamental distinction between them: coenzymes do *not* remain permanently bound to the enzyme—having been acted upon by one enzyme (for example reduced, acylated or phosphorylated), a coenzyme then becomes the substrate

for a different enzyme, which reverses the process. Coenzymes therefore act as the links between metabolic pathways; they exist in quite high concentrations in the cell, are found in two or more forms, and can be considered as special intracellular enzyme substrates. A prosthetic group, on the other hand, is a non-protein component of an enzyme which is a necessary part of its

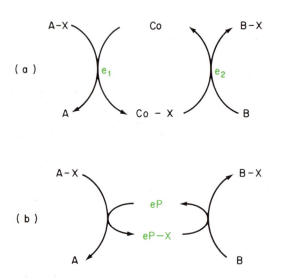

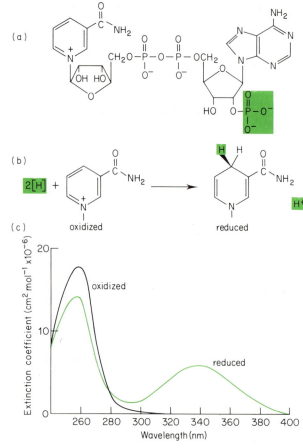

Fig. 3.11 Distinction between a coenzyme and a prosthetic group. In (a) transfer of group X from A to B proceeds by transfer to a coenzyme, which is released by enzyme-1 (e_1) and then reacts with enzyme-2 (e_2). In (b), a single enzyme with a prosthetic group (e_p) catalyses the transfer.

structure, usually in the active site. Some prosthetic groups are covalently linked to proteins; others are less tightly bound and may be removed from the enzyme by dialysis or charcoal treatment to give the *apoenzyme*. Unlike coenzymes, they remain associated with the enzyme for a complete reaction cycle, and are not regenerated by reaction with different enzymes.

Fig. 3.11 illustrates the difference between coenzymes and prosthetic groups: individual compounds are discussed below.

1. *Nicotinamide nucleotides*

Two coenzymes, nicotinamide adenine dinucleotide (NAD) and nicotinamide adenine dinucleotide phosphate (NADP) are together known as nicotinamide nucleotides. They were previously called di- and triphosphopyridine nucleotides (DPN and TPN). The structures are shown in Fig. 3.12; they are dinucleotides in that two mononucleotides (of adenine and nicotinamide) are linked by a pyrophosphate bond, and the coenzymes differ in the 2′-phosphate on the adenosine of NADP. They are coenzymes for a large number of dehydrogenases, most of which are specific for one or the other, and although both are electron carriers, they have quite different metabolic functions: NAD is involved in oxidative degradation (glycolysis, the TCA cycle and degradation of acyl CoA) and NADP participates in reductive biosynthesis, such as the synthesis of fatty acids and sterols. For this reason NAD is maintained mainly in the oxidized form ([NAD$^+$]/[NADH] is about 100 in liver cytosol) while NADP is

Fig. 3.12 Nicotinamide nucleotide coenzymes. (a) NADP$^+$; NAD$^+$ lacks the 2′-phosphate shown in green; (b) reduction of the nicotinamide ring; (c) the absorption spectra of the oxidized and reduced forms of the coenzymes.

predominantly in the reduced form ([NADP$^+$]/[NADPH] is about 0.01).

Reduction of the coenzymes occurs on the nicotinamide ring: dehydrogenases transfer two electrons and one proton from their substrates to the coenzymes, a second proton being liberated:

$$NAD(P)^+ + 2H \rightleftharpoons NAD(P)H + H^+$$

The oxidized coenzymes have an absorbance peak at 260 nm, due to the adenine and nicotinamide rings; in the reduced coenzymes dihydronicotinamide absorbs maximally at 340 nm, so that interconversion involves a change in the spectrum (Fig. 3.12) and can be followed by measuring absorbance at 340 nm. This is the usual way to assay dehydrogenases. The reduced coenzymes are also fluorescent, with an emission maximum at 450 nm, whereas the oxidized coenzymes are not, so fluorescence measurements can also be used in the assay of nicotinamide-nucleotide dependent dehydrogenases.

The presence of a nicotinamide ring in these coenzymes is responsible for the dietary requirement for nicotinic acid (niacin).

2. Adenine nucleotides

Adenosine triphosphate (ATP) functions as an energy store within the cell, linking exergonic reactions, to which its synthesis is coupled, and endergonic reactions, which are driven by the energy released during its hydrolysis. It is termed a 'high energy compound', because of its large free energy of hydrolysis ($\Delta G_0 = -30$ kJ/mol at pH 7)

$$ATP^{4-} + H_2O \rightarrow ADP^{3-} + HPO_4^{2-} + H^+$$

The value for simple phosphate esters is much lower: $\Delta G_0 = -13.8$ kJ/mol for the hydrolysis of glucose-6-phosphate, for example. The large free energy of hydrolysis derives from the electronic structure of the triphosphate group, and is similar for other nucleoside triphosphates. ATP is synthesized in degradative reactions such as glycolysis, and the oxidation of reduced coenzymes by the mitochondrial electron-transport chain (p. 186). The energy available from these reactions is used in energy-consuming processes such as muscle contraction (p. 20), active transport across membranes (p. 182) and biosynthetic reactions. The free energy stored in ATP may be released by its hydrolysis either to ADP or to AMP and pyrophosphate:

$$ATP^{4-} + H_2O \rightarrow AMP^{2-} + HP_2O_7^{3-} + H^+$$

For this reaction, $\Delta G_0 = -32$ kJ/mol; since pyrophosphate is further hydrolysed to phosphate in vivo, the overall free energy of the process is even larger (p. 156).

ATP may also act as a donor of phosphoryl, pyrophosphoryl, adenylyl or adenosyl groups: the points of cleavage of the molecule in these reactions are shown in Fig. 3.13. A derivative of ATP, *S*-adenosyl

methionine, functions as a coenzyme in methyl transfer, and is discussed in chapter 8.

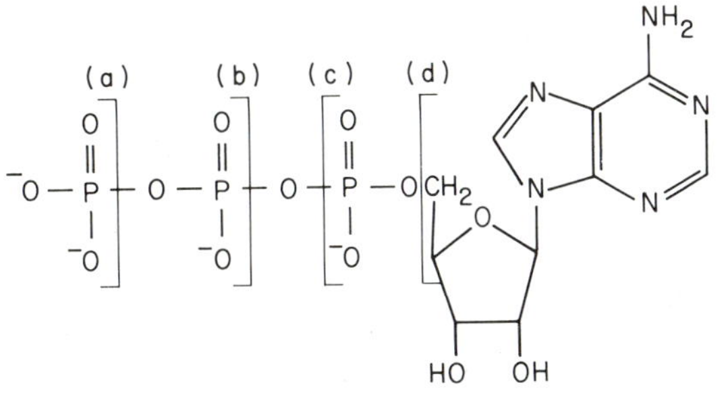

3. Other nucleoside triphosphates

The biosynthesis of complex lipids and polysaccharides utilizes nucleotide-linked intermediates, which are generated from nucleoside triphosphates. Thus CDP-choline, CDP-diglyceride and other intermediates occur in the synthesis of phospholipids (p. 108), and UDP- and GDP-sugars in polysaccharide synthesis.

Hydrolysis of GTP provides the energy required by ribosomes for protein synthesis (p. 216) and nucleoside triphosphates are the substrates for all nucleic acid synthesis (chapter 11).

4. Tetrahydrofolic acid

This is the reduced form of the vitamin folic acid: it acts as a carrier of 'one-carbon' fragments, which are attached at N^5, N^{10} or both together. After transfer to the coenzyme from a donor such as serine, the one-carbon residue can be oxidized or reduced before transfer to an acceptor (p. 141). The coenzyme (Fig. 3.14) contains a reduced pteridine ring, coupled to glutamate through 4-aminobenzoate: its synthesis in bacteria is inhibited by sulphonamides, which are structural analogues of 4-aminobenzoate. These simple compounds were among the first man-made antibiotics, and are still in use (see p. 143).

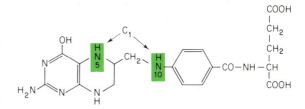

Fig. 3.14 Tetrahydrofolic acid. 1-C groups react with the two nitrogen atoms shown in green.

5. Pteridine coenzymes

Pteridines can act as electron-transferring coenzymes, in the same way as NAD; their use is limited to a very few enzymes, of which phenylalanine hydroxylase (p. 149) is one.

6. Ubiquinone (coenzyme Q)

Ubiquinone (Fig. 3.15) acts as a hydrogen carrier in mitochondrial electron transport, being reduced by a number of complex dehydrogenases and reoxidized by Complex III, in which the immediate electron acceptor is cytochrome b (see chapter 9).

Fig. 3.13 Sites of cleavage of adenosine triphosphate by (a) phosphoryltransferases (kinases); (b) pyrophosphoryltransferases; (c) adenylyltransferases and ligases; (d) adenosyl transferases.

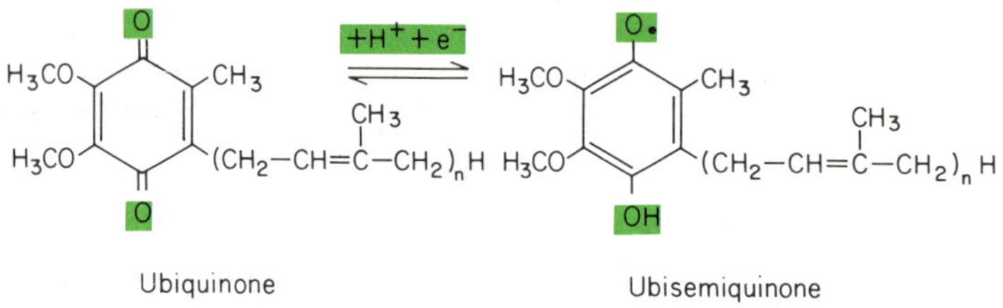

Fig. 3.15 Oxidized and reduced forms of ubiquinone (UQ, coenzyme Q). The predominant form of ubiquinone has $n = 10$.

7. Coenzyme A

Coenzyme A (Fig. 3.16) is a carrier of acyl groups, which form a thiolester bond with the free sulphydryl of the coenzyme. Free fatty acids are lipid-soluble, and dissolve freely in membranes: esterification to CoA prevents this, and provides a nucleotide 'handle' which

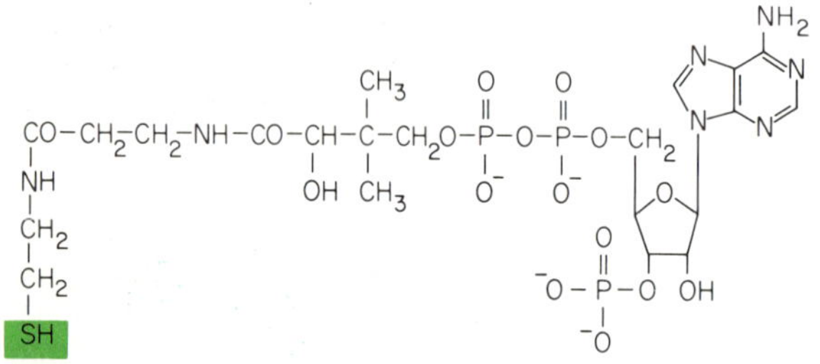

Fig. 3.16 Coenzyme A. Acyl groups become esterified to the free sulphydryl shown in green.

is recognized by enzymes of fatty-acid metabolism. The thiolester bond has a large free energy of hydrolysis (about 33.5 kJ/mol) and its formation by acyl-CoA synthetases requires the hydrolysis of ATP

$$RCOOH + CoA{-}SH + ATP \rightarrow RCO{-}S{-}CoA + AMP + P\text{-}P_i$$

Hydrolysis of succinyl-CoA is used to drive synthesis of nucleoside triphosphate (p. 158).

The pantothenate residue of CoA cannot be synthesized in man, so there is a dietary requirement for pantothenic acid.

The coenzymes discussed above occur as substrates in a number of (sometimes very many) enzymic reactions. We now consider prosthetic groups, which, although they have essential functions in catalysis, never become separated from enzymes; from a kinetic point of view they can be treated as part of an enzyme (Fig. 3.11).

Flavins

There are two flavin derivatives: flavin mononucleotide (FMN), in which flavin is linked to ribitol phosphate,

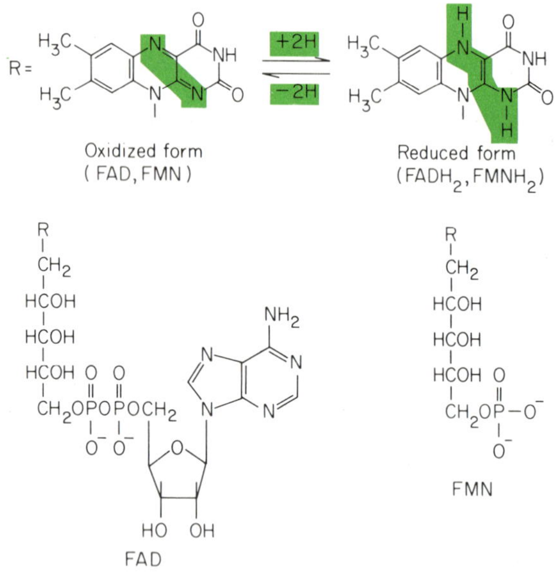

Fig. 3.17 Structures of the flavin prosthetic groups.

and flavin adenine dinucleotide (FAD, Fig. 3.17). These are the prosthetic groups of a number of complex dehydrogenases, including those in mitochondria which transfer electrons to ubiquinone from NADH, succinate, acyl-CoA and glycerol-1-phosphate. They act as carriers of two hydrogen atoms:

$$\text{substrate}{-}H_2 \diagdown \text{enzyme}{-}FAD \diagdown \text{acceptor}{-}H_2$$
$$\text{product} \diagup \text{enzyme}{-}FADH_2 \diagup \text{acceptor}$$

Flavoproteins frequently contain other prosthetic groups, such as iron, copper or molybdenum; the electron acceptors include haemoproteins, ubiquinone and oxygen. Synthesis of flavoproteins requires a dietary supply of riboflavin (vitamin B_2). The prosthetic group is not usually covalently bound, and can often be removed by denaturation of the protein, or in some cases even by dialysis. An exception to this is succinate dehydrogenase, which contains FAD covalently bound to histidine.

Haem

Haem is an iron-containing tetrapyrrole compound, found in haemoglobin and myoglobin, where its function is to bind oxygen (p. 249), in cytochromes, where it acts as an electron carrier (p. 164), and in some enzymes, such as tryptophan oxygenase (p. 146), where it functions as a prosthetic group in the reduction of oxygen. Only in the c-type cytochromes is haem covalently linked to proteins (see p. 48).

Non-haem iron

Some proteins contain iron linked to sulphur (either inorganic sulphur, or the side chains of cysteine): they are known as iron–sulphur proteins, or non-haem-iron proteins, and their structures are described in chapter 9. They are usually involved in redox reactions, although aconitase (p. 156) also contains one iron–sulphur cluster, the function of which is unknown.

Metal ions

Ions such as Mg^{2+} often act as dissociable activators of enzymes (p. 33); other ions may be tightly bound to proteins, fulfilling diverse catalytic roles. Enzymes with metal ions bound through amino acid side chains are called metalloproteins: Mn, Zn, Co, Cu, Mo and Ni have been found in various enzymes, although their function is not always clear. These involvements in enzyme activity account for the dietary requirement for these trace metals.

Pyridoxal phosphate

This is a derivative of pyridoxine (vitamin B_6), and is the prosthetic group of a number of enzymes of amino acid metabolism, including transaminases, racemases and decarboxylases. The amino acid reacts directly with the prosthetic group to form an imine ('Schiff's base') intermediate (p. 127). Decarboxylation occurs by

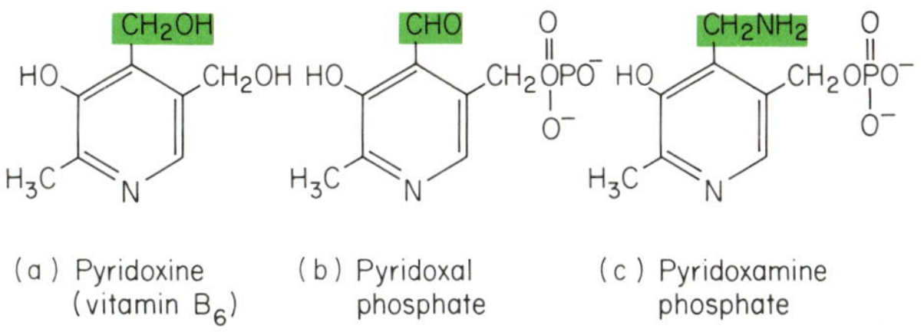

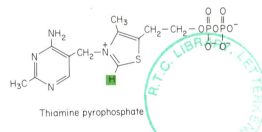

(a) Pyridoxine (vitamin B_6) (b) Pyridoxal phosphate (c) Pyridoxamine phosphate

Fig. 3.18 Pyridoxine and its derivatives.

cleavage of a C—C bond, and racemization (alteration of the stereochemical configuration of the amino acid, for example D → L) by breakage and reformation of a C—H bond. Transamination occurs by hydrolysis of a C—N bond, releasing the oxoacid and leaving the prosthetic group as pyridoxamine phosphate. This can react with a different oxoacid in a reversal of the sequence, thereby converting it to the corresponding amino acid:

amino acid$_1$ ⟩ enzyme—pyridoxal-P ⟨ amino acid$_2$
oxoacid$_1$ ⟨ enzyme—pyridoxamine-P ⟩ oxoacid$_2$

In transaminases the aldehyde group of pyridoxal phosphate is linked to the ϵ-NH_2 of a lysine, and it is this conjugate that reacts with the incoming amino acid.

Thiamine pyrophosphate

This derivative of thiamine (vitamin B_1) is involved in the oxidative decarboxylation of pyruvate, 2-oxoglutarate and the branched-chain amino acids (p. 143), and in the transketolase reaction (p. 78). In these reactions the proton attached to C-2 of the thiazole ring dissociates, and the resulting anion attacks the substrate (Fig. 3.19). In the metabolism of oxoacids,

Thiamine pyrophosphate

Fig. 3.19 Thiamine pyrophosphate. The ionizable hydrogen atom shown in green can be replaced by the substrate during the reaction cycle of the prosthetic group (see p. 74).

decarboxylation then occurs and the resulting group ($R \cdot CH(OH)$—) is either released as the aldehyde, or transferred to lipoic acid for oxidation (see below). In the transketolase reaction the group —$CO \cdot CH_2OH$ is transferred, via thiamine pyrophosphate, in a similar way.

Lipoic acid

This disulphide derivative of octanoic acid is a prosthetic group in oxoacid dehydrogenases (p. 74), where it is involved in the transfer of the substrate from thiamine pyrophosphate to coenzyme A. During transfer the substrate is oxidized, with reduction of lipoic acid to a dithiol (Fig. 3.20)—the prosthetic group is recycled, being oxidized back to a disulphide by FAD. Lipoic acid is covalently linked to protein, by an amide link to the ϵ-NH_2 of lysine.

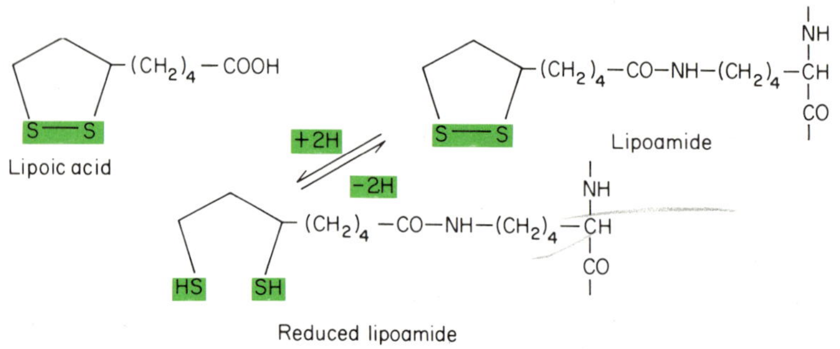

Fig. 3.20 Lipoic acid and its derivatives.

Biotin

Biotin is the prosthetic group in a number of carboxylation reactions, and functions as a carrier of CO_2. It is carboxylated in an ATP-dependent reaction (Fig. 3.21), and the carboxyl group is then transferred to the acceptor substrate:

HCO_3^-
ATP ↘ enzyme—biotin ↗ substrate—CO_2^-
$ADP + P_i$ ↙ enzyme—biotin—CO_2^- ↖ substrate

Typical biotin-dependent carboxylases are those for acetyl-CoA, propionyl-CoA and pyruvate. Biotin is covalently linked through its carboxyl to the ϵ-NH_2 of a protein lysine.

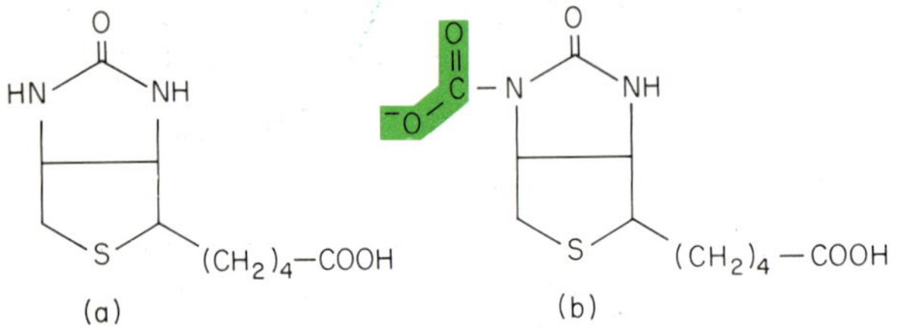

Fig. 3.21 Biotin (a) and its carboxylated form (b).

Cobalamin

The cobalamin prosthetic groups are derived from vitamin B_{12}, containing Co^{2+} surrounded by a tetrapyrrole corrin ring (Fig. 3.22). The fifth ligand to the metal ion is dimethylbenzimidazole, and the sixth is 5′-deoxyadenosine. In bacteria it is involved in a number of important reactions, including the reduction of ribonucleotides to deoxyribonucleotides (p. 56) and synthesis of methionine. In higher animals one reaction in which vitamin B_{12} is definitely known to be involved is methylmalonyl-CoA mutase (p. 106); the bond between Co and the 5′-C of deoxyadenosine is replaced by another Co—C bond, to —CO—SCoA as it migrates. There also appears to be a link between B_{12} and folate deficiencies, which suggests that B_{12} could be involved in the metabolism of methionine.

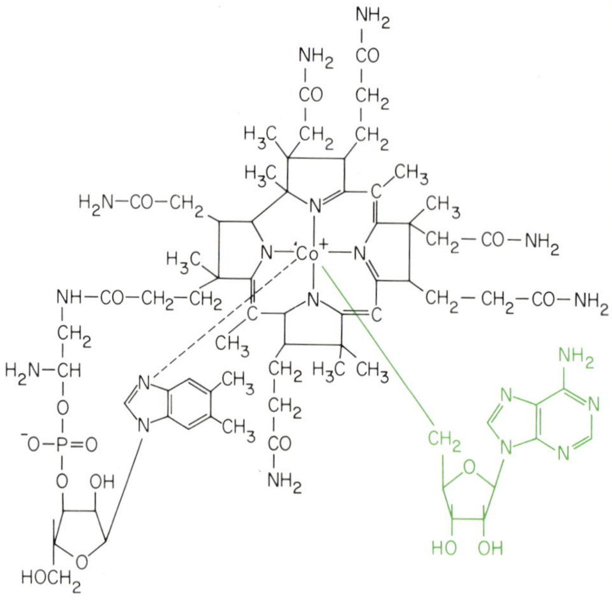

Fig. 3.22 The cobalamin prosthetic group derived from vitamin B_{12}. The substrate can replace the 5′-deoxyadenosine shown in green.

The Mechanism of Enzyme Action

Enzymic catalysis can be investigated at different levels, a total description of the process necessarily involving experiments of many different types. Only in a few favourable cases has it been possible to account completely for the rate enhancement observed, to identify the intermediates and to calculate the rates of their interconversions. The types of experiments used are discussed below: in most cases, large amounts of purified enzyme are required.

1. *Study of the chemical reaction*

Obviously, the stoichiometry of the reaction must be known; more detailed information, such as the stereochemistry of the reactants and products, and the identification of those atoms which exchange with ions in the solvent, may suggest the chemical pathway of catalysis.

2. *Kinetic experiments*

Steady-state kinetics, including the use of reversible inhibitors, can elucidate the 'formal' mechanism of the enzyme: the order in which the substrates bind and products are released, and which steps are rate-limiting. These experiments give no idea of the chemistry of the steps involved, and some intermediates may not be detected; the constants that one can calculate, such as K_m and V_{max}, are composites of many rate constants. Pre-steady-state studies, in which the very fast rates of individual steps are measured in special apparatus, give a more complete description of the reaction kinetics, but still no chemical information.

3. *Studies of enzyme structure*

For even simple data to be interpreted fully, it is necessary to know something of the structure of an enzyme—its molecular weight, subunit composition, prosthetic group content and so on. A detailed 3-dimensional map of the structure is provided by studies of the X-ray diffraction patterns of enzyme crystals: this reveals the configuration of the polypeptide chain, and even the positions of individual amino acid side chains.

4. *Identification of the active site*

The participation of particular residues in catalysis can be inferred from the pH-dependence of reactions, and from the use of irreversible inhibitors (p. 30), if the loss of enzymic activity can be correlated with the modification of a particular amino acid in the enzyme. In particular, active-site directed inhibitors identify amino acids near the substrate-binding site, and this identification can be confirmed by X-ray crystallography.

5. *Model studies*

It may be possible to mimic the process occurring at the active site, using simple compounds in which the chemistry is easily studied. Reaction rates are usually much lower than in enzymes, but the processes involved may be similar.

6. *Study of the enzyme–substrate complex*

The enzyme may be crystallized together with a bound substrate or competitive inhibitor, in which case X-ray crystallography shows the disposition of amino acids in the active site, and the areas of contact between enzyme and substrate. Such studies are not confined to stable complexes: transient intermediates can be studied at low temperatures, after diffusion of the substrate into the enzyme crystals.

A combination of these experimental approaches has given a fairly complete picture of the reaction mechanism of a few enzymes, notably extracellular, degradative enzymes such as lysozyme, ribonuclease, and various proteases, which are small (mol. wt 15–30000), stable and easily isolated; and a few glycolytic enzymes, which are more complex but are at least available in large quantities. Two examples are discussed below.

Chymotrypsin

This is one of the serine proteases, so called because of the essential serine at the active site: others are trypsin, elastase and thrombin, which is involved in blood-clotting (p. 240). Chymotrypsin hydrolyses not only peptides, but also esters and amides, the greatest rates of hydrolysis occurring when the acyl part of the substrate is hydrophobic. In the hydrolysis of proteins there is a preference (rather than an absolute specificity) for bonds on the carboxyl side of tryptophan, tyrosine and phenylalanine. The reaction occurs in two stages: *acylation* of the enzyme, when serine-195 attacks the carbonyl of the substrate, and *deacylation*, as the new bond is hydrolysed.

Serine-195 is a particularly good nucleophile, as it is H-bonded to histidine-37, itself bonded to aspartate-102, this whole 'charge-relay' system being buried in the hydrophobic interior of the protein (Fig. 3.23). Chymotrypsin is secreted as an inactive zymogen, chymotrypsinogen, which is activated by proteolysis.

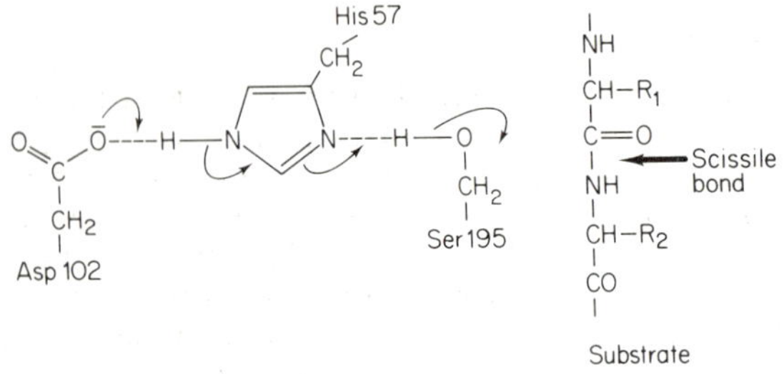

Fig. 3.23 The 'charge relay' system in chymotrypsin. Other serine proteases have similar mechanisms.

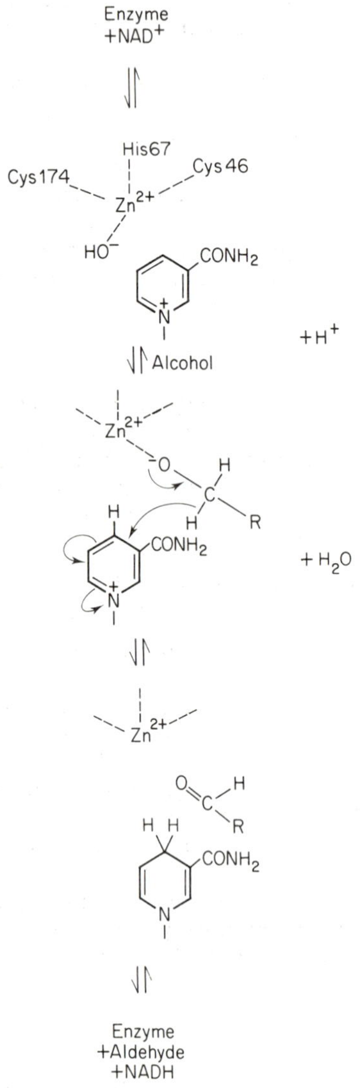

The essential 'charge-relay' is intact in the zymogen, but the substrate cannot bind because an H-bond, between its carbonyl and glycine-193, is not formed. On activation of chymotrypsinogen, by tryptic cleavage of the peptide bond between arginine-15 and isoleucine-16, a conformation change rotates glycine-193 into the correct position for substrate binding.

Liver alcohol dehydrogenase

This enzyme contains two polypeptide chains of molecular weight 40 000, each containing two atoms of Zn, one in the active site, the other some distance away and apparently having a purely structural role. Kinetic studies showed that NAD^+ is the first substrate to bind: this produces a conformation change in the enzyme, releases a proton and permits binding of the second substrate, alcohol. Probably the proton comes from a Zn-bound water molecule; hydride transfer occurs directly from alcohol to the nicotinamide ring of NAD^+ (Fig. 3.24).

Liver alcohol dehydrogenase will oxidize many aliphatic alcohols, which are bound in a hydrophobic pocket next to the nucleotide-binding crevice: one form of the enzyme will even oxidize steroid alcohols.

Enzyme specificity

Unlike most inorganic catalysts, enzymes are specific for the substrates of the reactions they catalyse. Some are absolutely specific for a single substrate, for example urease hydrolyses only urea; others act on a number of structurally related compounds, often at different rates, because different substrates have different values of K_m or V_{max}. For the esterification of coenzyme A there is a series of acyl-CoA synthetases of broad and overlapping specificity, each having maximal activity with fatty acids of a particular chain length (Fig. 3.25).

A wide variation in the degree of specificity is found

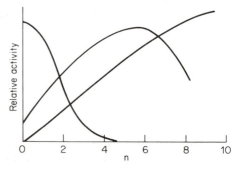

Fig. 3.25 Effect of chain length on the activity of fatty-acyl CoA synthetases, which catalyse the reaction

$$R(CH_2)_n COOH + ATP + CoA \rightarrow$$
$$R(CH_2)_n COSCoA + AMP + P\text{-}P_i.$$

Fig. 3.24 A possible catalytic mechanism for liver alcohol dehydrogenase.

in proteolytic enzymes. Trypsin hydrolyses only peptide bonds from the basic amino acids lysine and arginine, chymotrypsin shows a preference for aromatic amino acids but also attacks many other bonds, and carboxypeptidase A, which removes C-terminal amino acids, also prefers aromatic amino acids but will attack all except lysine, arginine and proline.

Most enzymes show stereospecificity: for example most enzymes of mammalian amino acid metabolism act on the naturally-occurring L-forms, and are inactive with the D-isomers, exceptions being amino acid racemases and D-amino acid oxidases. In some enzyme-catalysed reactions a substrate which is structurally symmetrical behaves as though it were unsymmetrical: in the aconitase reaction (p. 156) for example, for which the substrate is the symmetrical molecule citrate, the hydroxyl group migrates to the carbon atom derived from oxaloacetate, and not to that from acetyl-CoA (Fig. 3.26). This is because the active site is asymmetric: if it binds the substrate at three points (or more) the transition state is asymmetric, and the isomerization proceeds asymmetrically.

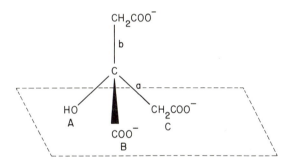

Fig. 3.26 Binding of citrate at the active site of aconitase. Dehydration and rehydration occur across the bond (a), but not (b).

An interesting example of control acting at the level of specificity occurs with β-lactoglobulin, a widely-distributed protein which catalyses the synthesis of *N*-acetyllactosamine, which is used in the synthesis of mucopolysaccharide:

$$UDP—Gal + NAcGlc \rightarrow Gal—GlcNAc + UDP$$

where   Gal = galactose and Glc = glucose

α-Lactalbumin, a protein secreted only in mammary gland, forms a complex with this enzyme, modifying its specificity so that it synthesizes lactose:

$$UDP—Gal + Glc \rightarrow Gal—Glc + UDP$$

Competing substrates

Discrimination between different possible substrates, which is a most important function of biological specificity, depends on the interactions between enzyme and transition state, which occur in the enzyme–substrate complex. Obviously there is no difficulty in discriminating against a competing substrate which is too large for the active site: the provision of a substrate-binding cleft of limited size excludes larger molecules. This cannot occur if the competing substrate is the same size as the natural substrate, or smaller: discrimination then depends on differences in binding strengths between the two. If, for example, they differ in size by a methyl group, the free energies of binding differ by about 12 kJ/mol, resulting in only a 100–200 fold difference in K_m or k_{cat}.

This is a particular problem for those enzymes involved in protein synthesis, since some amino acids are structurally very similar, but incorrect insertion of an amino acid results in the synthesis of an altered protein which may itself generate further errors: such positive feedback could lead to the rapid breakdown of order in the cell. The aminoacyl-tRNA synthetases (p. 212) have evolved a special 'editing' mechanism which reduces errors of this kind.

Induced fit

Water may be a competing substrate in some reactions: for example in a phosphotransferase reaction, such as that catalysed by hexokinase, it should be able to enter the active site and accept a phosphate group from ATP—in other words, the phosphotransferase should act as an ATPase:

$$Glucose + ATP \rightarrow Glucose\text{-}6\text{-}P$$
$$+ ADP \text{ (normal reaction)}$$

$$H_2O + ATP \rightarrow P_i + ADP \text{ (competing reaction)}$$

Despite the high (55 M) concentration of water in the cell, such side reactions are of little importance, as binding of the real substrate produces a conformation change which correctly aligns those groups in the active site involved in phosphate transfer. The entire structure of the glucose molecule is necessary for this conformation change, which is not induced by water: hexokinase therefore does not remove the terminal phosphate of ATP until the correct acceptor is present in the active site. This is known as 'induced fit'.

Quaternary structure of enzymes

Extracellular enzymes, such as lysozyme, ribonuclease and the proteases, tend to be small and robust, stability being very important in the extracellular environment;

they are usually composed of one polypeptide chain, with intramolecular disulphide bonds. Intracellular enzymes are often much larger, with molecular weights in the range 10^5–10^6, and are composed of subunits, which are non-covalently linked. Subunits may be identical or non-identical: usually there is only one active site on each polypeptide chain, although there are exceptions to this—DNA polymerase I, for example, has a single chain with three different activities (p. 199). Even when enzymes are composed of several subunits, there may be no apparent interaction between them: aldolase, for example, has four identical, non-interacting subunits, which retain their activity when separated. However, interaction between identical or similar subunits can lead to cooperativity in substrate binding: this alters the shape of the substrate binding curve, and is an important feature in the regulation of some enzymes (chapter 13).

Other enzymes are composed of non-identical subunits, each having a different function. Protein kinase (p. 229) has two types, one containing the active site, the other the binding site for the regulator, cAMP. The 2-oxoacid dehydrogenases (p. 74) are really multi-enzyme complexes, as they contain five types of subunit, catalysing different reactions in the sequence: transfer of the substrate between associated subunits is more rapid than diffusion between separate enzyme molecules, and unstable intermediates can be transferred directly from one prosthetic group to another. Other examples of multi-enzyme complexes are found in the synthesis of pyrimidines (p. 54) and of fatty acids. The fatty acid synthetase complex catalyses seven sequential reactions: in *E. coli* this enzyme has seven different subunits, non-covalently associated, but in higher animals it has only two very large polypeptides, one of which has three different activities, the other four. Such enzymes appear to have evolved by fusion of the genes coding for separate proteins, to produce a single gene coding for a protein with several different active sites. The protein is folded into a series of *domains*, each with a different function; sometimes these can be separated, without loss of activity, after proteolysis to break the peptide chain which connects them.

Isoenzymes

Sometimes multiple molecular forms of an enzyme are found: these may differ in physical and kinetic properties, and are called isoenzymes. Different isoenzymes may have a different location within the cell: both malate dehydrogenase and glutamate-oxaloacetate transaminase exist in cytoplasmic and mitochondrial forms, which catalyse the same reaction but have no close structural relationship. In other cases different isoenzymes are found in different tissues, and these are the products of related genes.

Lactate dehydrogenase is a tetramer, composed of subunits (mol. wt 35 000) which may be of two types, called H and M. These combine to produce five different isoenzymes: H_4, H_3M, H_2M_2, HM_3 and M_4 (also known as LDH_1, $LDH_2 \ldots LDH_5$). The H_4 form predominates in heart, M_4 in skeletal muscle and liver: other tissues contain various proportions of the five isoenzymes. The H and M subunits have different kinetic properties, and those of the isoenzymes vary according to the subunit composition. They may be separated by electrophoresis in starch or agarose gels, and detected by a stain for LDH activity, in which an artificial electron carrier phenazine methosulphate (PMS) reoxidizes NADH generated by enzyme activity, reducing a tetrazolium dye and so staining the gel:

$$\text{lactate} \diagdown \text{NAD}^+ \diagup \text{PMSH}_2 \diagdown \text{dye} \quad \text{(colourless)}$$
$$\text{LDH}$$
$$\text{pyruvate} \diagup \begin{array}{c}\text{NADH}\\+\text{H}^+\end{array} \diagdown \text{PMS} \diagup \begin{array}{c}\text{reduced (blue-black)}\\\text{dye}\end{array}$$

H_4 (LDH_1) is the most acidic of the five isoenzymes, and runs closest to the anode: M_4 (LDH_5) runs closest to the cathode (Fig. 3.27). The tissue damage that occurs in certain disease states results in selective increases in particular isoenzymes in the serum. Thus myocardial infarction (increase in LDH_1, and to a lesser extent LDH_2), liver disease (LDH_5), Duchenne's muscular dystrophy (LDH_1 and LDH_2 equally) and diseases of the lungs, leukaemia, pericarditis and viral infections (LDH_3 and LDH_4) can all be diagnosed by examination of serum isoenzymes.

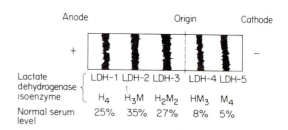

Fig. 3.27 Electrophoretic separation of lactate dehydrogenase isoenzymes.

Creatine kinase (p. 73) has two subunits, which may be of two types, B and M, generating three isoenzymes— BB (CK_1, found in brain, lung and bowel), BM (CK_2, found in myocardium) and MM (CK_3, found in skeletal muscle and myocardium). Electrophoretic measurement of creatine kinase isoenzymes is also used in the diagnosis of disease states; other isoenzymes which may be measured include those of glutamate-oxaloacetate transaminase, and acid- and alkaline-phosphatases.

Classification of enzymes

Enzymes are classified according to the reactions they catalyse: in addition to its trivial name, each has a systematic name, which is a description of the reaction catalysed, with the suffix *-ase*, and an Enzyme Commission number, the first figure of which indicates the class to which the enzyme belongs, the second and third the type of reaction within that class, in terms of the chemical groups involved. The fourth figure is the individual number of the enzyme: in general, all enzymes catalysing the same reaction have the same EC number. Six classes of enzyme are defined:

1. *Oxidoreductases*. This class contains enzymes catalysing several different types of reaction, including dehydrogenases, oxidases, mono-oxygenases (hydroxylases) and di-oxygenases.

2. *Transferases*. Enzymes in this class catalyse the transfer of acyl-, phosphoryl-, glyceryl-, amino- or other groups, according to the general scheme

$$A—X + B \rightarrow B—X + A$$

3. *Hydrolases*. This class contains enzymes which hydrolyse esters, amides, peptides, etc., and includes all of the digestive enzymes: the reaction catalysed is of the type

$$A—B + H_2O \rightarrow AOH + BH$$

4. *Lyases*. These enzymes catalyse the non-hydrolytic removal of a group from the substrate, in reactions of several types, including dehydration, and the scission of C—C bonds (as occurs in decarboxylation).

5. *Isomerases*. These catalyse the isomerization of substrates.

6. *Ligases*. These catalyse the formation of C—C, C—N, C—O or C—S bonds, in reactions requiring energy from the hydrolysis of nucleoside triphosphates.

Appendix to chapter 3: definitions

Maximum velocity (V_{max}). The maximal initial rate of an enzyme-catalysed reaction, which is only achieved at infinite concentration of all substrates: it is directly proportional to the enzyme concentration.

Molecular activity, or turnover number (k_{cat}). The rate-constant of an enzyme-catalysed reaction, such that $V_{max} = k_{cat}[e]$. It is equal to the number of substrate molecules transformed by each molecule of enzyme in unit time.

Michaelis constant (K_m). The substrate concentration at which the initial rate of an enzyme-catalysed reaction is equal to half the maximum velocity. It is independent of enzyme concentration, provided this is much lower than the substrate concentration. If there is more than one substrate, the K_m for each is the concentration which gives half-maximal velocity, with infinite concentrations of other substrates.

Substrate dissociation constant (K_s). The dissociation constant of the enzyme–substrate complex, equal to the substrate concentration which produces half-maximal saturation.

$$K_s = \frac{[E][S]}{[ES]} = \frac{k_2}{k_1} \text{ (see equation 6)}$$

The association constant is the reciprocal of the dissociation constant:

$$K_{ass} = \frac{1}{K_s} = \frac{k_1}{k_2}$$

Inhibition constant (K_i). The dissociation constant of an enzyme–inhibitor complex, equal to the inhibitor concentration which produces half-maximal inhibition.

Activation constant (K_a). The dissociation constant of an enzyme–activator complex, equal to the concentration of an activator which produces half-maximal activation.

Specific activity. The activity of an enzyme preparation relative to the weight of protein. It is typically expressed in units/mg (μmol product/mg protein/min) or katals/g (moles product/g protein/s).

4

Haem Pigments and Bile Pigments

The important haem-containing compounds in the body are haemoglobin, myoglobin, the cytochromes, and some enzymes such as catalase, peroxidase and tryptophan oxygenase. They are conjugated proteins in which the prosthetic group (haem) is a porphyrin ring containing an iron atom.

Porphyrins are all derivatives of *porphin*. They are large, flat, heterocyclic ring structures made up of four *pyrrole* rings linked together by methine (—CH=) bridges. In all the naturally occurring porphyrins there are eight side chains on the ring in the positions marked 1–8. This allows a large number of different isomers to exist, since all the side chains are not the same. The relationships between these isomers will be clearer after the biosynthesis of the porphyrins has been described; they are dealt with below.

When the side chains contain carboxyl radicals, the porphyrins are amphoteric with iso-electric points in the range pH 3–4.5. They have absorption bands in the visible spectrum and show ultraviolet fluorescence in

Pyrrole

Porphin

solution in mineral acids. When the tetrapyrrole ring is broken the resulting compounds are the bile pigments, which are described later in this chapter.

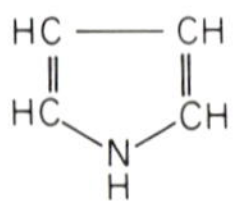

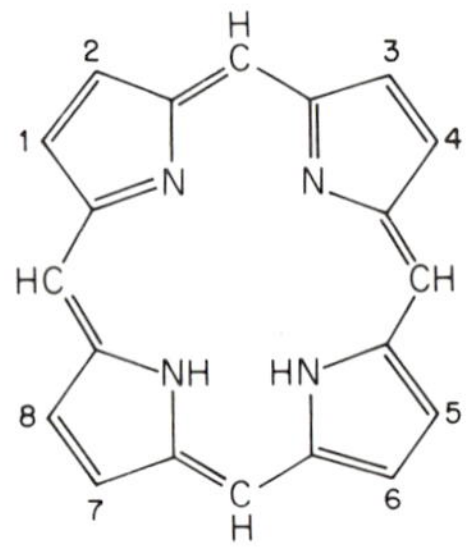

Porphyrinogens are compounds similar to the porphyrins but the pyrrole rings are linked together by —CH_2— (methylene) bridges; they contain six more hydrogen atoms than do the corresponding porphyrins.

Biosynthesis of Porphyrins

The pathways by which the porphyrins are made in vivo are shown in Figs 4.1 and 4.2. These figures also show the way in which the nitrogen atom and the methylene carbon atom of glycine are incorporated into the porphyrin structure; all the other carbon atoms come from succinyl coenzyme A.

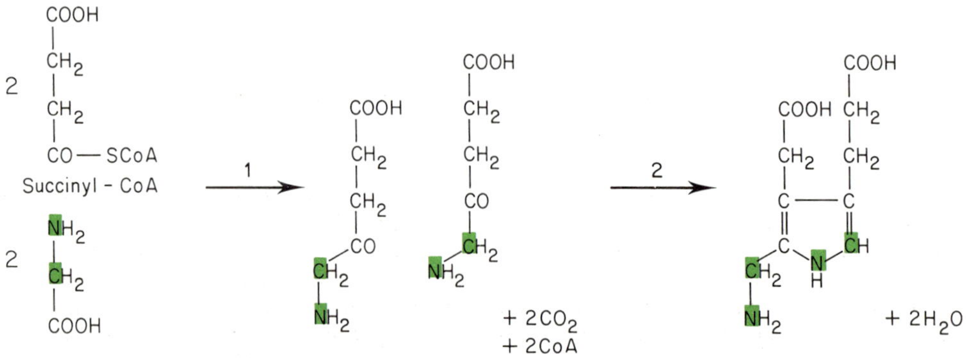

Fig. 4.1 The biosynthesis of porphobilinogen. The atoms shown in green arise from the atoms of glycine similarly marked.

44

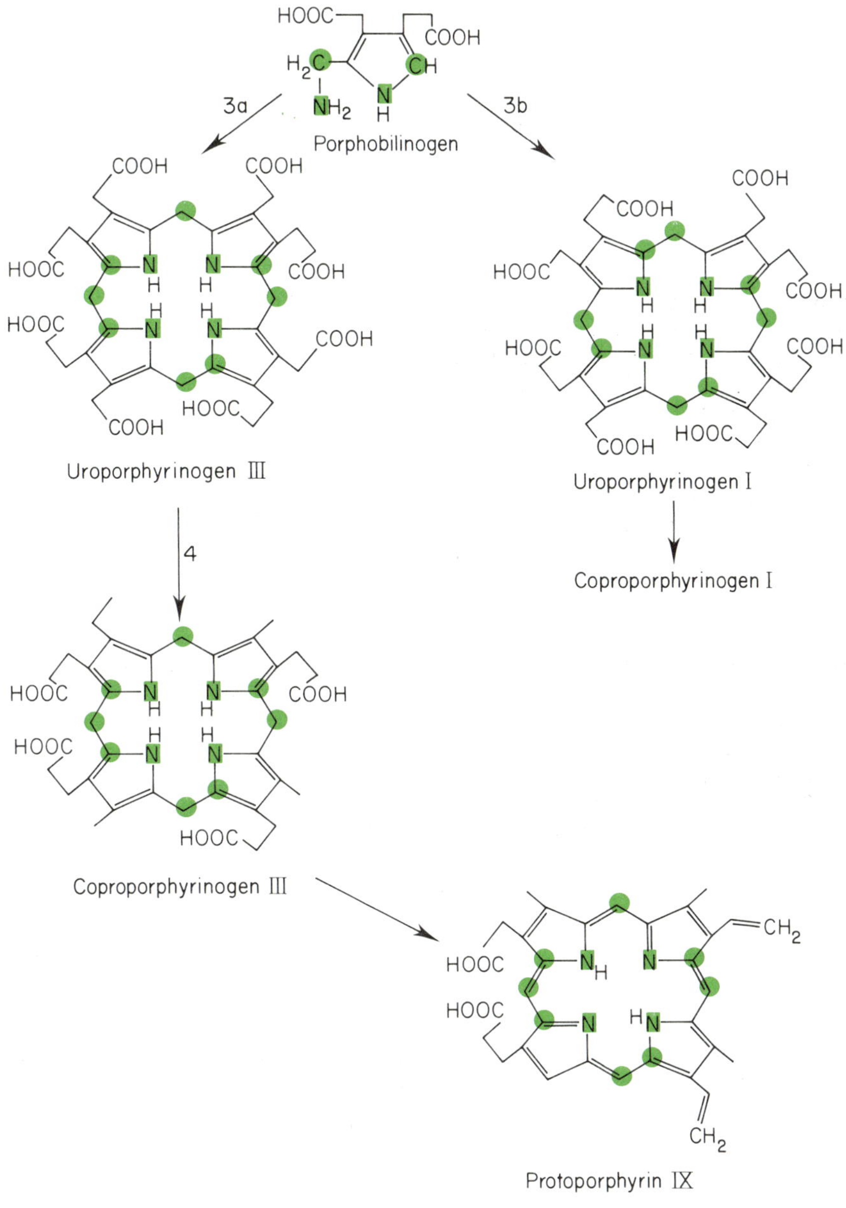

Fig. 4.2 The biosynthesis of porphyrins. Atoms shown in green arise from glycine (see Fig. 4.1), the remainder from acetate.

Reaction 1 (Fig. 4.1), the condensation of succinyl coenzyme A and glycine, requires pyridoxal phosphate. The hypothetical reaction product, α-amino-β-oxo-adipic acid, decarboxylates spontaneously at neutral pH to form δ-aminolaevulinic acid, thereby eliminating the glycine carboxyl carbon atom. The enzyme that catalyses the reaction, δ-*aminolaevulinic acid synthetase*, is found to be present in very elevated amounts in the liver of patients suffering from two forms of congenital porphyria. Acute attacks of the disease are frequently precipitated by barbiturates or by other compounds which induce synthesis of δ-aminolaevulinic acid synthetase, and are accompanied by excretion of large amounts of δ-aminolaevulinic acid and of porphobilinogen. The increased concentration of the synthetase is of particular interest because inborn errors of metabolism are usually associated with a *deficiency* of an enzyme.

45

Reaction 2 (Fig. 4.1) involves the condensation of two molecules of δ-aminolaevulinic acid, with the elimination of water, to form the pyrrole derivative porphobilinogen. The enzyme catalysing this reaction is *δ-aminolaevulinic acid dehydrase*. Both this and the preceding enzyme are normally inhibited by end products of the pathway, such as protohaemin IX. Mutations that lead to insensitivity to these feedback inhibitors also produce porphyria.

The next step, the deamination and polymerization of four molecules of porphobilinogen to form a cyclic tetrapyrrole, is catalysed by uroporphyrinogen synthetase. Because the two side chains of porphobilinogen are different, there are different ways of putting the molecules together, leading to four isomeric porphyrinogens. Only porphyrins and porphyrinogens derived from uroporphyrinogens I and III occur in the body. To form uroporphyrinogen III (reaction 3a, Fig. 4.2), uroporphyrinogen synthetase requires a protein modifier, or co-synthetase; if this co-synthetase is absent or inactive, uroporphyrinogen I is formed (reaction 3b, Fig. 4.2). The products formed from this isomer do not inhibit δ-aminolaevulinic acid synthetase, and large amounts of porphyrins are produced.

The four acetic acid side chains of the uroporphyrinogens can be decarboxylated enzymically (reaction 4, Fig. 4.2) to the corresponding coproporphyrinogens.

Coproporphyrinogen III is converted into a protoporphyrin (reaction 5, Fig. 4.2) by an enzyme which removes two hydrogen atoms and a carboxyl group from each of two of the four propionic acid side chains, turning them into vinyl groups. At the same time the four methylene bridges are dehydrogenated to methine bridges, the porphyrinogen thereby being converted into a porphyrin. Protoporphyrin, therefore, has four methyl, two vinyl, and two propionic acid side chains. There are 15 possible isomers of such a porphyrin but only one, type IX, is formed.

The arrangement of the side chains in the naturally occurring porphyrins is shown in Table 4.1.

Table 4.1 The arrangement of the side chains of the porphyrins

Compound	Side chains at the various positions of porphin							
	1	2	3	4	5	6	7	8
Uroporphyrin I	A	P	A	P	A	P	A	P
Coproporphyrin I	Me	P	Me	P	Me	P	Me	P
Uroporphyrin III	A	P	A	P	A	P	P	A
Coproporphyrin III	Me	P	Me	P	Me	P	P	Me
Protoporphyrin IX	Me	V	Me	V	Me	P	P	Me
Deuteroporphyrin IX	Me	H	Me	H	Me	P	P	Me
Mesoporphyrin IX	Me	Et	Me	Et	Me	P	P	Me
Haematoporphyrin IX	Me	HE	Me	HE	Me	P	P	Me

A = —CH$_2$·COOH Et = —CH$_2$·CH$_3$
H = hydrogen HE = —CH(OH)·CH$_3$
Me = —CH$_3$ P = —CH$_2$·CH$_2$·COOH
V = —CH=CH$_2$

Occurrence of Porphyrins in the Body

Porphyrins are present in the excreta, bile, bone marrow and blood. In *blood* most of the porphyrins are in the red cells but their concentration is very small: about 20 μg protoporphyrin and less than 1 μg coproporphyrin per 100 ml blood. The *bile* contains 40–60 μg porphyrin per 100 ml and a considerable amount of this undergoes enterohepatic circulation. The *faeces* contain porphyrins of both endogenous and exogenous origin. The former enter the gut in the bile and are mainly protoporphyrin and coproporphyrin. The protoporphyrin can be altered by the bacteria in the gut to mesoporphyrin and deuteroporphyrin. Between 150 and 400 μg total porphyrin per day is normally excreted in the faeces, more than half being protoporphyrin and most of the rest coproporphyrin.

In normal human males the *urine* contains 166 ± 45 μg and in females 134 ± 42 μg total coproporphyrin per day. It is believed that it is excreted by the kidney as coproporphyrinogen and is oxidized spontaneously to the porphyrin. Normal urine also contains small amounts of uroporphyrin. The coproporphyrin occurs as a mixture of both the I and III isomers. Isotope tracer experiments have shown that urinary porphyrins are by-products of haem or porphyrin synthesis and are not derived from degradation of haemoglobin.

Haematin Compounds

The porphyrins can form coordination complexes with heavy metals and the iron complexes of the porphyrins are the haematin compounds. The insertion of an iron atom into protoporphyrin IX, is catalysed by a specific ferrochelatase. This enzyme is very susceptible to inhibition by heavy metal ions, such as Pb^{2+}; many of the characteristic symptoms of lead poisoning, such as anaemia and excretion of large amounts of δ-aminolaevulinic acid and coproporphyrin III, arise from inhibition of ferrochelatase.

Haem

Table 4.2 lists the principal haematin compounds.

Haem is the prosthetic group of haemoglobin. It contains ferrous iron and can easily be oxidized to haemin or haematin in which the iron is ferric. The iron atom is quite strongly held in the porphyrin. Haem alone cannot combine reversibly with oxygen as can haemoglobin.

Haemin and haematin. These are both protoporphyrin IX containing a ferric iron atom, coordinated only to N atoms in the planar tetrapyrrole ring, i.e., the iron is not coordinated to N atoms in protein nor to chromogenic nitrogenous compounds. The name haemin is reserved for the compound with a net positive charge (Table 4.2).

Haemochromes (these contain ferrous iron). Haem is able to combine with a large number of nitrogenous bases to form haemochromes. Among these bases are denatured protein, ammonia, amino acids, pyridine, etc.; combination occurs in the proportion of two molecules of base per ferrous iron atom. It is important when making haemochromes to add a reducing agent to keep the iron in the ferrous state. For example, when haemoglobin is treated with alkali the globin is denatured but remains attached to the haem to give *denatured globin haemochrome.* Pyridine haemochrome is formed by treating haemoglobin with pyridine, sodium hydroxide and a little dithionite as a reducing agent. The

Table 4.2 Structure of haematin compounds

Compound	Structure	Charge	Iron atom Valency	Iron atom Ferrous or ferric
Haem	Fe coordinated to 4 N	0	2	Ferrous
Haemochrome	Fe coordinated to 4 N and 2 N'	0	2	Ferrous
Haematin (in alkaline solution)	Fe coordinated to 4 N, HOH and OH⁻	0	3	Ferric
Haemin (chloride, etc.) (in acid solution)	Fe coordinated to 4 N	+1	3	Ferric
Haemichrome (chloride, etc.)	Fe coordinated to 4 N and 2 N'	+1	3	Ferric

N = porphyrin nitrogens which contribute 2 negative charges to the complex. N' = non-porphyrin nitrogens.

haemochromes have absorption spectra with a very sharp band near 550 nm; they do *not* form compounds reversibly with molecular oxygen.

Haemichromes (these contain ferric iron). They correspond to the haemochromes apart from the oxidation state of the iron. However, negatively charged ligands (e.g. CN^-, OH^-) bind more strongly to the haemichromes, which have a net positive charge, than they do to the haemochromes (cf. cyanmethaemoglobin, p. 48). The association constant for the hydroxyl ion is in fact so large that compounds with neutral ligands such as pyridine can only be formed in acid conditions.

A very sensitive test for haem is the oxidation by hydrogen peroxide of guaiacol, benzidine or *o*-toluidine to coloured (blue-green) products. This reaction is catalysed by haem but is non-enzymic, and therefore survives boiling or other treatments that denature proteins. This test is used to detect the presence of blood, for example in urine.

Haem-Containing Proteins

Haemoglobin, the protein responsible for the transport of oxygen from the lungs to tissues by red cells, is discussed elsewhere (chapter 14), as is the related protein myoglobin, which stores oxygen in muscle cells. The prosthetic group of these proteins is iron-protoporphyrin IX (known as protohaem or simply haem); it is bound to globin by non-covalent forces, and is released on denaturation of the protein. Each haem combines reversibly with one molecule of oxygen, the iron atom remaining in the ferrous state. Haemoglobin has an affinity for carbon monoxide that is 200-fold greater than for oxygen, so CO will displace O_2 from oxyhaemoglobin, forming cherry-red carboxyhaemoglobin. Haemoglobin also combines with other neutral ligands, such as nitric oxide.

Methaemoglobin is made by conversion of the haem of haemoglobin to the ferric state, by autoxidation of oxyhaemoglobin or by the action of oxidants such as ferricyanide. It is normally present in blood in small quantities, but in some diseases appears in abnormally large amounts (methaemoglobinaemia). Some variant haemoglobins, such as the M types, have amino acid substitutions which stabilize the oxidized form of the prosthetic group. Methaemoglobin does not bind oxygen, but combines with cyanide: the 'cyanmet method' of estimating haemoglobin involves its oxidation to methaemoglobin, formation of the cyanmethaemoglobin complex, and measurement of its absorbance at 542 nm.

Cytochromes (chapter 9) have an essential function in electron transport. The prosthetic group can assume the oxidation states $+2$ (ferrous) or $+3$ (ferric), the midpoint potential for oxidation/reduction depending on the environment provided by the protein. In b-type cytochromes (including cytochrome P450) the prosthetic group is haem, as in haemoglobin, and is not covalently attached to the protein. The c-type cytochromes have the same prosthetic group, but the two vinyl side chains are reduced and linked by thioether bonds to cysteine residues of the protein.

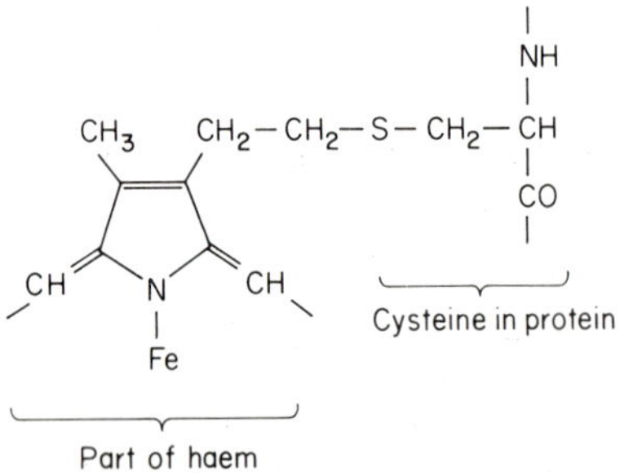

Part of haem

The cytochrome oxidase complex (p. 189) contains cytochromes a and a_3; in each case the prosthetic group is haem A (see below), which has a formyl group at position 8 and a fifteen-carbon farnesyl group at position 2. Haem A is bound to the protein by non-covalent forces.

Enzymes with haem prosthetic groups include catalase (p. 57), peroxidase (p. 167) and tryptophan oxygenase (tryptophan pyrrolase, p. 146). In each case the prosthetic group is related to protoporphyrin IX.

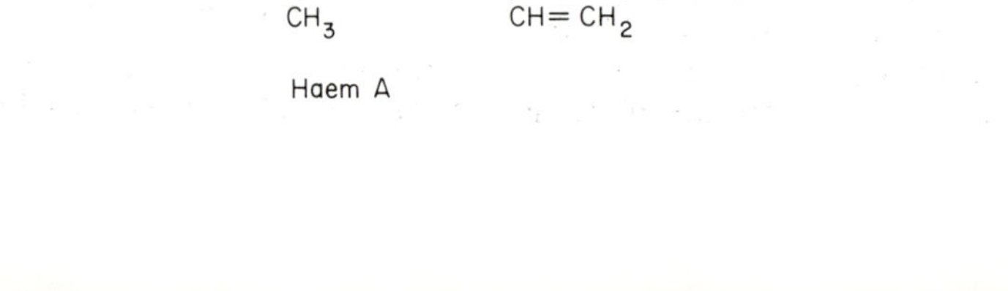

Haem A

Bile Pigments

These compounds have an open chain of four pyrrole rings, joined by methylene (—CH$_2$—) or methine (=CH—) groups. They are derived from proto-porphyrin IX by oxidative scission of the α-methine link to form biliverdin, which is then reduced to other members of the group. They are classified into bilanes, bilenes, biladienes and bilatrienes, which contain 0, 1, 2 or 3 methine bridges, respectively. The bilenes, such as urobilinogen and stercobilinogen, are colourless, the other compounds highly coloured.

Degradation of haemoglobin

After a life-span of about 120 days, red cells are taken up by the reticuloendothelial system (e.g. spleen) and destroyed. Some 6–7 g of haemoglobin are degraded daily in adults, by the pathway shown in Fig. 4.3. First, oxidative scission of the α-methine bridge liberates carbon monoxide and forms a green biliverdin–iron–globin complex (verdoglobin). Iron is removed and attached to transferrin for transport to storage sites, globin is degraded to amino acids, and the released biliverdin is reduced at the γ-methine bridge to bilirubin.

Bilirubin is relatively insoluble in water, and is transferred to the liver as a complex with plasma proteins (albumins and α-globulins); in this state it is known as 'indirect bilirubin', and is not filtered into the urine. In the liver the two propionic acid side chains of bilirubin are esterified to the hydroxyl at C-1 of glucuronic acid:

Bilirubin + 2 UDP—glucuronic acid →
    Bilirubin diglucuronide + 2 UDP

The water-soluble bilirubin diglucuronide is secreted into the bile, and passes from the gall bladder to the intestine. There, probably largely by the action of bacteria, bilirubin is deconjugated and reduced to various bilanes, collectively called urobilinogen; some of these are dehydrogenated to bilenes, known as urobilin. Some bilenes are reabsorbed into the blood, from which they are removed by the liver and kidneys, which secrete them into bile and urine respectively. These pathways are summarized in Fig. 4.4.

The ability to conjugate bilirubin is almost entirely lacking in the neonatal liver, and some degree of jaundice is common; this and other forms of hepatic jaundice, such as a failure of bilirubin absorption, lead to high plasma levels of unconjugated bilirubin. Pre-hepatic jaundice is usually caused by excessive erythrocyte breakdown, for example because of incom-

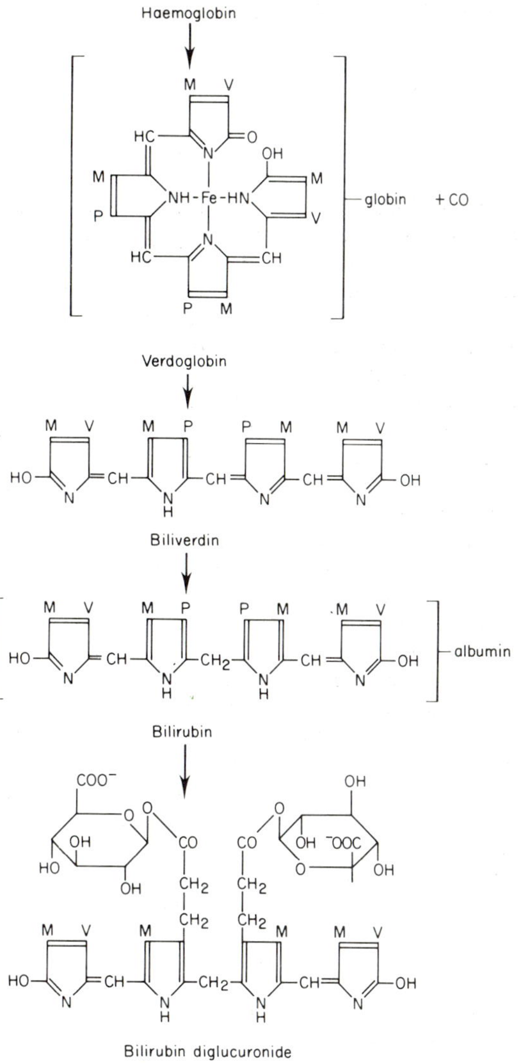

Fig. 4.3 The degradation of haem. M = methyl, V = vinyl, P = propionic acid.

patible maternal antibodies; the plasma level of free bilirubin is elevated, and the pigment is selectively absorbed by brain tissues, causing irreversible damage (*kernicterus*). In post-hepatic jaundice, caused for example by blockage of the bile duct, conjugated bilirubin appears in the plasma.

The van den Bergh reaction

This is used as a test for plasma bilirubin, which reacts with diazotized sulphanilic acid to give a red colour. In

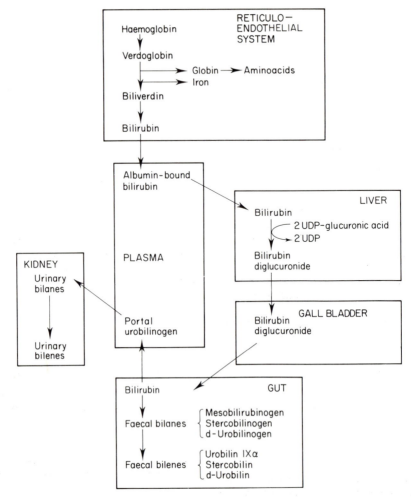

Fig. 4.4 The formation and metabolism of bile pigments.

normal serum, which contains protein-bound, uncon-jugated bilirubin, this reaction does not occur without the addition of ethanol (the 'indirect reaction'). A 'direct reaction', occurring without the addition of ethanol, reveals the presence of mono- and di-glucuronides of bilirubin.

5

Nucleotides and Nucleotide Metabolism

Nucleotides

Nucleotides are the monomeric units from which nucleic acids are synthesized; nucleotides of many types also occur free within the cell, and some of them function as coenzymes.

A nucleotide has three structural components: a heterocyclic compound (sometimes loosely termed a 'base'), sugar, and phosphate. In nucleic acids, the bases are purines and pyrimidines; the parent compounds (which do not occur in nature) have the following structures, with the atoms numbered as shown:

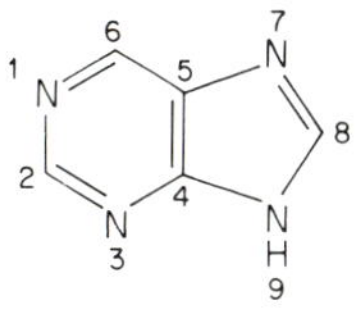

Purine 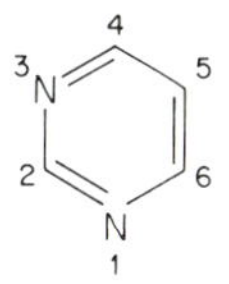   Pyrimidine

The sugar is either D-ribose or D-2-deoxyribose; each occurs in the furanose form, the configuration at C-1 being β.

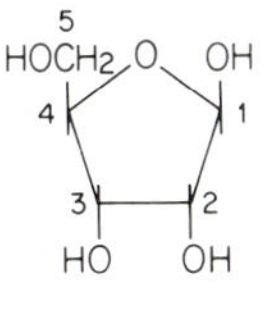

Ribose 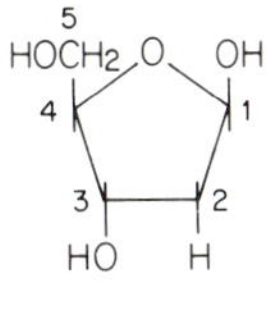   2-deoxyribose

The heterocyclic compound is attached by an N-glycosidic link to C-1. A compound of 'base' and sugar is called a nucleoside; when the sugar is phosphorylated it becomes a nucleoside phosphate, or nucleotide. Phosphate may be attached at C-5, C-3 or (in ribose only) C-2.

Nucleic acids contain nucleoside-5'-phosphates (5'-nucleotides), joined by phosphodiester links to C-3 of the next nucleotide unit (see Fig. 11.3, p. 195).

Pyrimidines

Nucleic acids contain cytosine, uracil (not found in DNA) and thymine (not found in most types of RNA). Orotic acid and dihydroorotic acid are also important metabolically (p. 54).

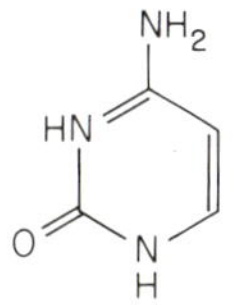

Cytosine   Uracil   Thymine
(2-oxy, 4-amino-   (2,4-dioxy-   (5-methyluracil)
pyrimidine)   pyrimidine)

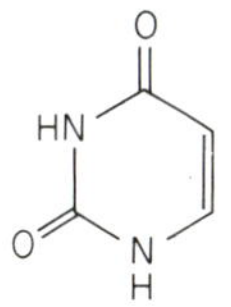

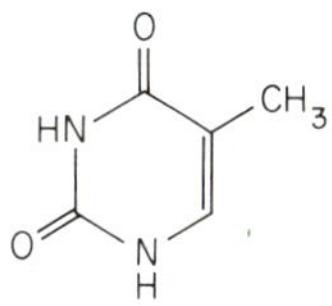

Because hydroxy- or amino-substituents occur next to ring nitrogens, there is keto–enol tautomerism:

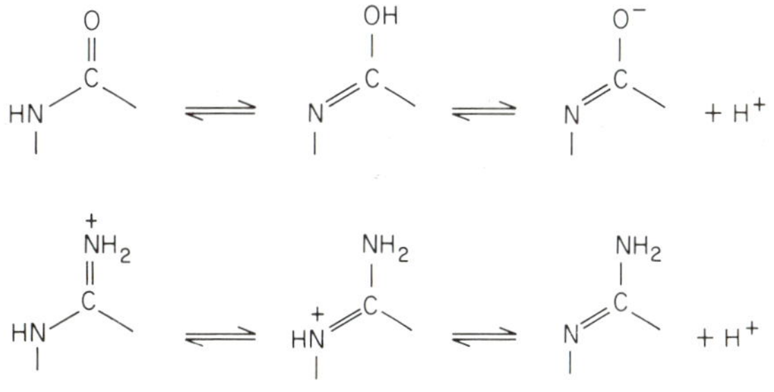

Oxypyrimidines are weakly acidic, and aminopyrimidines weakly basic; the pK_a values are shown in Table 5.1. The keto–enol equilibrium is therefore pH-dependent. The conjugated double-bond system absorbs light in the near-ultraviolet, with maximal absorption at 260–270 nm, the absorption spectra also being somewhat pH-dependent. Pyrimidines are moderately soluble in water.

51

Table 5.1 Derivatives of purines and pyrimidines

	Nucleoside	Nucleotide	pKa of free base Amino	pKa of free base Enol
Purines				
Adenine	Adenosine	Adenylic acid (adenosine monophosphate, AMP)	4.2	—
Guanine	Guanosine	Guanylic acid (guanosine monophosphate, GMP)	3.3	9.2
Hypoxanthine	Inosine	Inosinic acid (inosine monophosphate, IMP)	—	8.9
Uric acid	—	—	—	5.4, 10.3
Xanthine	Xanthosine	Xanthylic acid (xanthosine monophosphate, XMP)	—	7.4, 11.1
Pyrimidines				
Cytosine	Cytidine	Cytidylic acid (cytidine monophosphate, CMP)	4.6	12.2
Orotic acid	Orotidine	Orotidylic acid (orotidine monophosphate, OMP)	—	9.4 (2.4 carboxyl)
Thymine	Thymidine	Thymidylic acid (thymidine monophosphate, dTMP)	—	9.8
Uracil	Uridine	Uridylic acid (uridine monophosphate, UMP)	—	5.4

Purines

Adenine and guanine are the only purines which commonly occur in nucleic acids, although they may be modified by methylation.

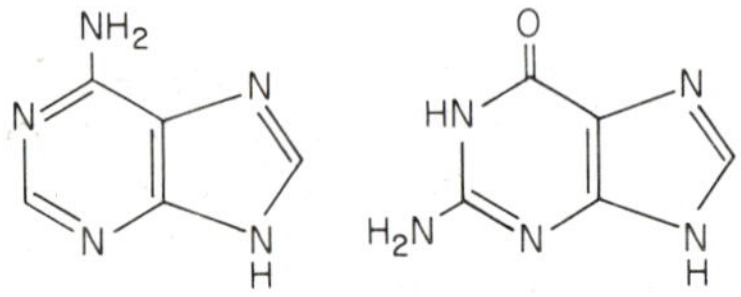

Adenine
(6-aminopurine)

Guanine
(2-amino,6-oxypurine)

Others which are important in metabolism are hypoxanthine, xanthine and uric acid (p. 57). As occurs with pyrimidines, several tautomeric structures are in pH-dependent equilibrium, and there is strong absorbance of ultraviolet light around 250–260 nm. Purines, particularly xanthine and uric acid, are rather insoluble in water, at pH 7. Uric acid has a relatively low pK_a (Table 5.1), and is predominantly anionic in neutral solution.

Nomenclature of nucleosides and nucleotides

The *N*-glycosidic link of purine and pyrimidine nucleotides is between C-1 of ribose (or deoxyribose) and N-1 of pyrimidines, or N-9 of purines. The nomenclature of these compounds is somewhat confusing, and is given in Table 5.1. Nucleosides are named as derivatives of the 'bases', the prefix *deoxy* -indicating that the sugar is 2-deoxyribose. Because thymine occurs mainly in DNA, thymidine is assumed to contain deoxyribose; however, ribothymidine, in which the sugar is ribose, occurs in some transfer RNAs (p. 210). An invariant minor component of tRNA is pseudouridine, in which ribose is attached to C-5 of uracil:

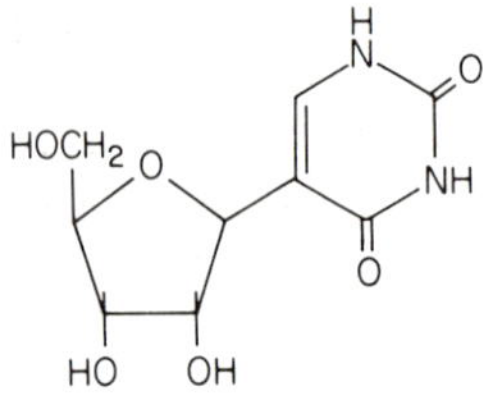

Pseudouridine

Nucleotides are named as the phosphates of nucleosides, the primed number indicating the position of esterification on ribose. Esterification on C-5 may be with di- or triphosphates: the phosphorus atoms are labelled α, β and γ, starting from that nearest to C-5.

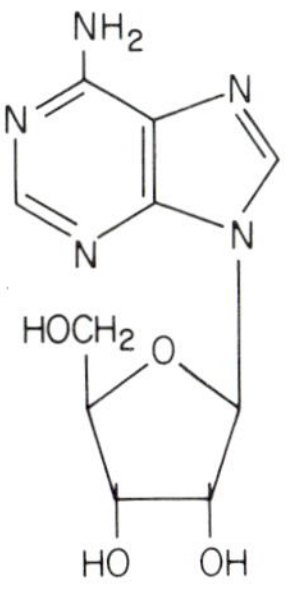

Adenosine

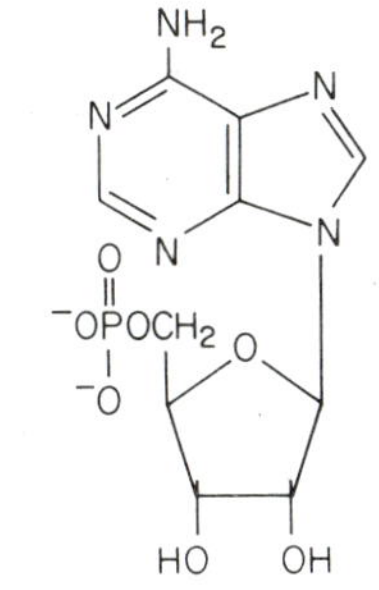

Adenosine 5'-monophosphate
(5' Adenylic acid, 5'–AMP)

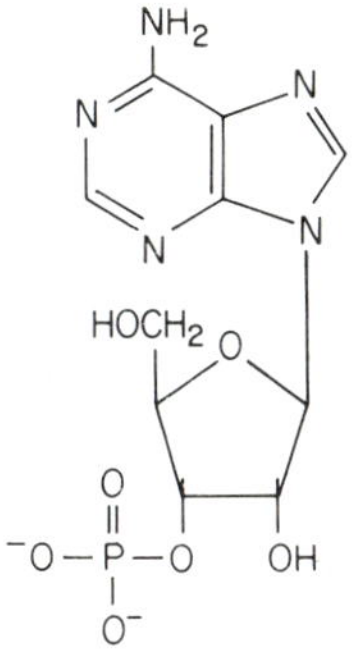

Adenosine 3'-monophosphate
(3' Adenylic acid, 3'–AMP)

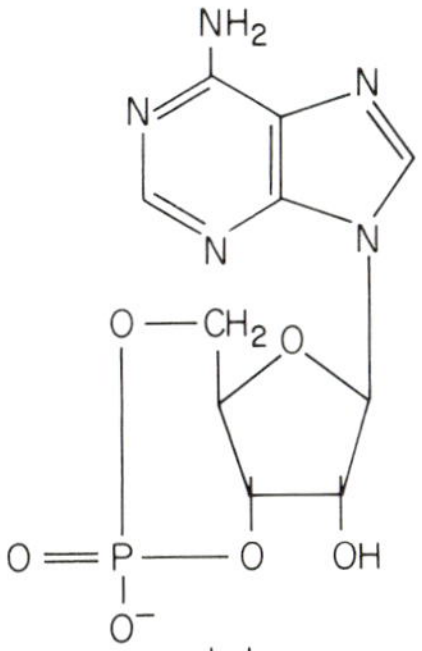

Adenosine 3',5'-monophosphate
(cAMP)

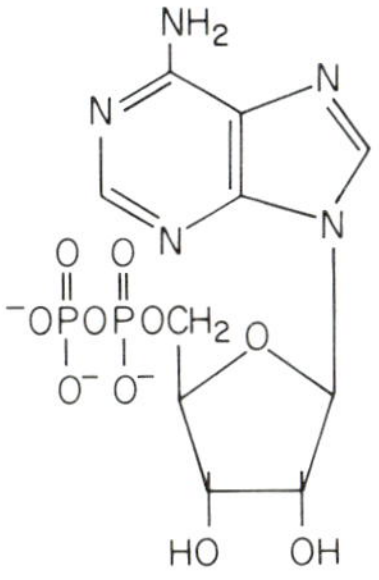

Adenosine 5'- diphosphate
(ADP)

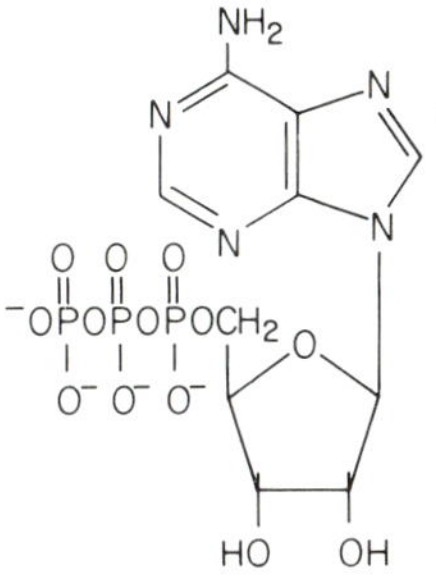

Adenosine 5'- triphosphate
(ATP)

Phosphate monoesters have a primary ionization with pK_a 1–2, and secondary ionization with pK_a 6.5–7.0:

Nucleotides are therefore very acidic and water-soluble. All the phosphates of nucleoside di- and triphosphates are ionized, but only the terminal phosphate has the second ionization (pK_a 6.5–7.0). These nucleotides have strong affinity for divalent cations such as Mg^{2+} and Ca^{2+}, which bind to the β and γ or α and β phosphates.

Nucleotide coenzymes

Adenine nucleotides have a variety of metabolic roles, such as energy conservation, and the transfer of phosphate and pyrophosphate groups. Nucleotides of uracil and guanine are involved in polysaccharide synthesis, sugars being transferred as glycosides of UDP or GDP (p. 180), and CDP has a similar role in lipid synthesis.

The nicotinamide nucleotide coenzymes (NAD and NADP, p. 34) contain nicotinamide mononucleotide (in which N-1 or nicotinamide is linked to ribose), joined by a phosphodiester link to AMP; these compounds are therefore dinucleotides.

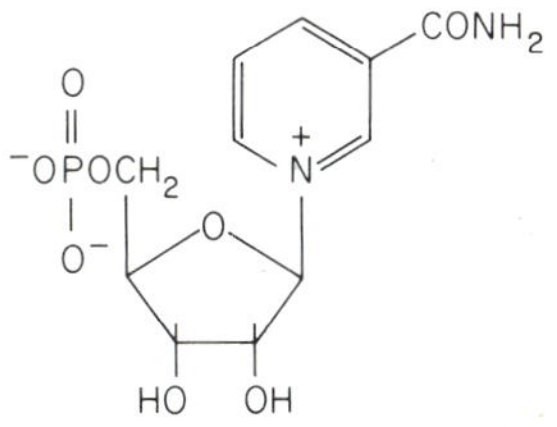

Nicotinamide mononucleotide (NMN)

In the flavin prosthetic groups (FMN and FAD, p. 36), flavin is linked to ribitol phosphate which, in FAD, is attached to AMP. Coenzyme A (CoA, p. 36) is also a nucleotide, structurally related to AMP.

Biosynthesis of Purines and Pyrimidines

Purine and pyrimidine nucleotides are used in large quantities in the biosynthesis of nucleic acids, so despite their complex structures they are synthesized *de novo* by most organisms. In higher animals the biosynthetic pathways are present in many types of cell, although others depend on scavenger pathways which synthesize nucleotides from purines and pyrimidines salvaged from the degradation of nucleic acids.

The origins of the atoms in the rings are shown in Fig. 5.1. The biosynthetic pathways appear to be basically similar in all organisms, although regulatory mechanisms differ.

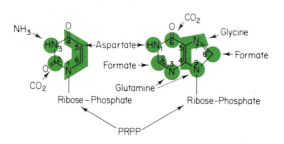

Fig. 5.1 The origins of the atoms in purine and pyrimidine nucleotides.

Pyrimidine biosynthesis

The six enzymic steps in the formation of UMP are shown in Fig. 5.2. In the first, carbamyl phosphate is synthesized from glutamine and bicarbonate, with the consumption of two molecules of ATP:

$$glutamine + 2ATP + HCO_3^- \rightarrow NH_2 \cdot COOPO_3^{2-}$$
$$+ glutamate + 2ADP + HPO_4^{2-}$$

This reaction occurs in the cytoplasm; a second carbamyl phosphate synthetase, which uses ammonia rather than glutamine and requires *N*-acetyl glutamate for activity, occurs in mitochondria, where it supplies carbamyl phosphate for the urea cycle (chapter 8).

The condensation of carbamyl phosphate with aspartate is catalysed by aspartate transcarbamylase; the product, carbamyl aspartate, is cyclized to dihydroorotic acid, which is then oxidized to orotic acid. This is converted to its nucleotide, orotidine monophosphate, by reaction with 5'-phosphoribose-1-pyrophosphate (PRPP), with elimination of pyrophosphate. Decarboxylation of orotidine monophosphate completes the synthesis of UMP.

Cytidine nucleotides are formed in a glutamine-

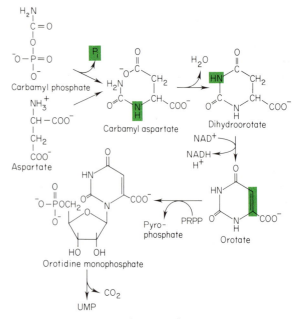

Fig. 5.2 *De novo* biosynthesis of uridine monophosphate.

dependent amination, the substrate of which is UTP, formed from UMP by two successive phosphorylations.

$$UMP + ATP \rightarrow UDP + ADP$$

$$UDP + ATP \rightarrow UTP + ADP$$

$$UTP + glutamine + ATP \rightarrow CTP + glutamate$$
$$+ ADP + P_i$$

In bacteria the six reactions of UMP synthesis are catalysed by separate enzymes, encoded by separate genes. In higher animals, only three proteins are involved:

i) a complex which is probably a trimer of identical chains; each chain (mol. wt 200000) has carbamyl phosphate synthetase, aspartate transcarbamylase and dihydroorotase activities;
ii) dihydroorotate dehydrogenase, an iron-containing flavoprotein which is located in the inner mitochondrial membrane, with its active site facing the cytoplasm;
iii) a protein with orotate phosphoribosyltransferase and orotidylate decarboxylase activities.

The intermediates in the first three steps are not released from the multi-enzyme complex, but are passed from one active site to another.

PRPP is formed by pyrophosphorylation of ribose-5-phosphate, which is derived from the pentose-phosphate pathway.

$$\text{Ribose-5-phosphate} + \text{ATP} \rightarrow \text{PRPP} + \text{AMP}$$

This reaction, catalysed by PRPP synthetase, is an important control point in nucleotide biosynthesis. PRPP is the starting point for the synthesis of purines; it is also used in the biosynthesis of NMN (p. 53), which then reacts with ATP to form NAD^+.

Purine biosynthesis

The *de novo* purine biosynthesis pathway forms IMP, the nucleotide of hypoxathanine; there are 10 enzymic steps and, in contrast to the biosynthesis of pyrimidines, all of the intermediates are nucleotides, the heterocyclic ring system being built up while attached to ribose-5-phosphate.

In the first step, PRPP is converted to 5-phosphoribosylamine, in a glutamine-dependent reaction which introduces N-9 of the purine ring.

$$\text{PRPP} + \text{glutamine} \rightarrow \text{5-phosphoribosylamine} + \text{P-P}_i + \text{glutamate}$$

This intermediate reacts with glycine, which is incorporated to provide C-4, C-5 and N-7. The remaining atoms of the purine are added one at a time (Fig. 5.3).

IMP may be converted to GMP, by oxidation and amination at C-2, or to AMP by amination at C-6 (Fig. 5.4). In the latter reaction the amino donor is aspartate, which is deaminated to fumarate; a comparable reaction occurs in the introduction of N-1 of the purine ring (Fig. 5.3) and in the urea cycle (p. 131).

Purine biosynthesis requires a supply of glutamine and of one-carbon residues (carried by the coenzyme tetrahydrofolate). Glutamine-dependent aminations are inhibited by such glutamine analogues as 6-diazo,

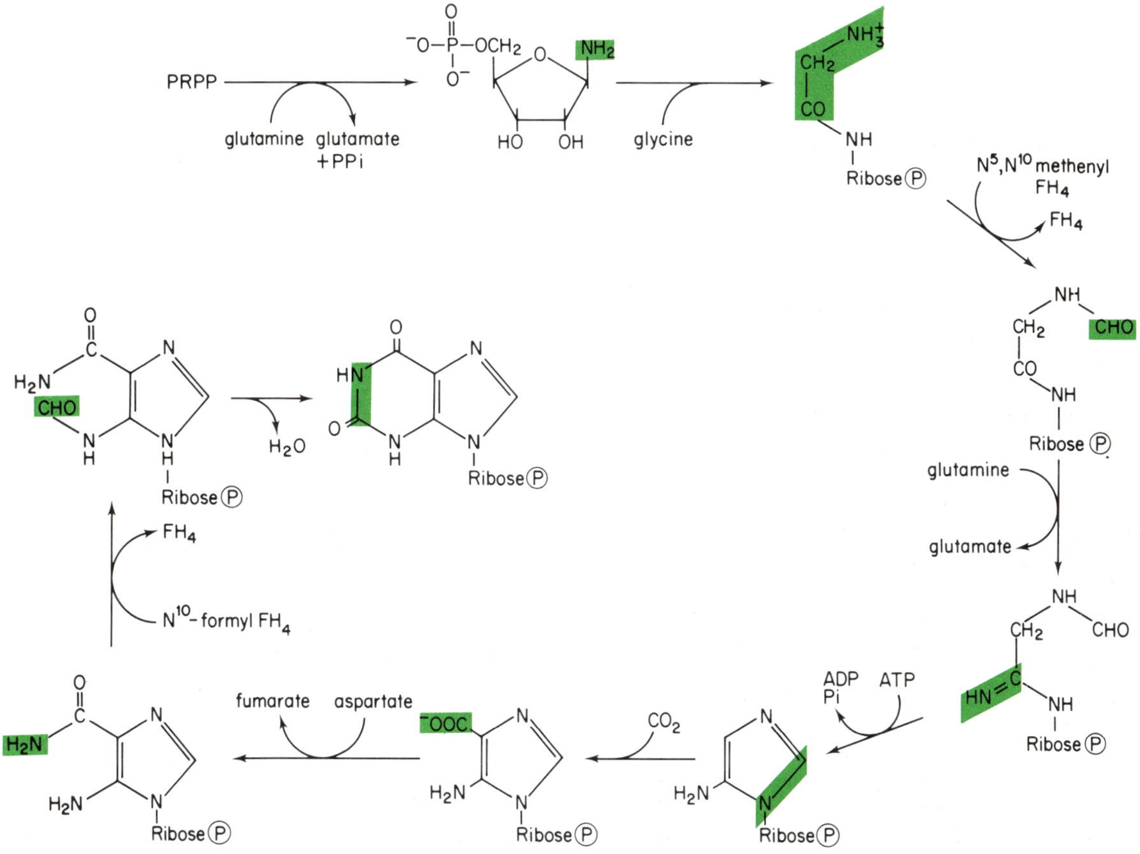

Fig. 5.3 De novo biosynthesis of inosine monophosphate.

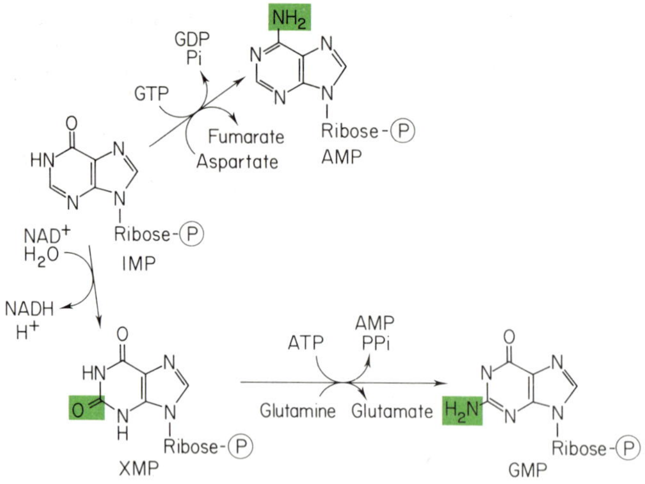

Fig. 5.4 Interconversions of purine nucleotides.

5-oxo-norleucine (DON) and azaserine, antibiotics produced by *Streptomyces*.

$$H_2N—CH—CH_2CH_2CONH_2 \qquad \text{Glutamine}$$
$$\qquad |$$
$$\qquad COOH$$

$$H_2N—CH—CH_2CH_2OCO·CH=\overset{+}{N}=\overset{-}{N} \qquad \text{DON}$$
$$\qquad |$$
$$\qquad COOH$$

$$H_2N—CH—CH_2OCO·CH=\overset{+}{N}=\overset{-}{N} \qquad \text{(Azaserine)}$$
$$\qquad |$$
$$\qquad COOH$$

The sulphonamide antibiotics (p. 143), which block the synthesis of folic acid in bacteria, indirectly inhibit purine synthesis by limiting the supply of methenyl- and formyl-tetrahydrofolate: they have no effect on purine synthesis in man, since folic acid is a vitamin, and not synthesized. Note that in both folate-dependent steps, the one-carbon group is introduced at the oxidation state of formate.

Synthesis of deoxyribonucleotides

Deoxyribonucleotides are formed directly from ribonucleotides, by reduction at the 2'-carbon; in animals the substrates of these reactions are nucleoside diphosphates. The immediate reductant is thioredoxin, a protein of molecular weight 12 000 which can donate two electrons, with the oxidation of two cysteines to cystine; oxidized thioredoxin is reduced by NADPH. Thioredoxin thus functions as a coenzyme in the reduction of ribonucleotides by NADPH (Fig. 5.5).

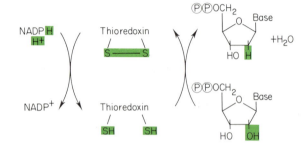

Fig. 5.5 Conversion of ribonucleotides to deoxyribonucleotides.

Some bacteria possess a nucleoside *tri*phosphate-specific ribonucleotide reductase with a vitamin B_{12} prosthetic group, but this enzyme seems not to be present in man. In some cases glutathione can also act as the electron donor for deoxyribonucleotide synthesis, but it is not clear whether this also occurs in man.

Synthesis of thymine nucleotides

Following the reduction of ribose to 2-deoxyribose in UDP, the pyrimidine may be methylated to thymine (Fig. 5.6). The substrate for this reaction is deoxyuridylic acid (dUMP), and the one-carbon donor N^5,N^{10}-methylenetetrahydrofolate, in which the transferred atom is at the oxidation state of formaldehyde. During transfer it is reduced to a methyl group, the electron donor being tetrahydrofolate, which becomes oxidized to dihydrofolate. It must be reduced before it can be reused; inhibitors of dihydrofolate reductase, such as methotrexate (amethopterin) and aminopterin, therefore inhibit pyrimidine synthesis, and block cell

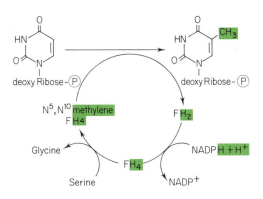

Fig. 5.6 Formation of dTMP by methylation of dUMP.

division. Because these drugs are most active against rapidly dividing cells they are effective anti-tumour agents and immunosuppressants; methotrexate, for example, is used in the treatment of leukaemia. 5-Fluorouracil is converted in vivo to fluorodeoxy-thymidylate, which in turn inhibits thymidylate syn-thetase, and fluorouracil has also been used as an anti-cancer drug.

Catabolism of Nucleotides

The principal products of nucleic acid breakdown are nucleoside-5′-monophosphates, which are degraded as shown in Fig. 5.7. The most important enzymes involved are:

i)   *5′-nucleotidase*, which hydrolytically removes the 5′-phosphate:

   e.g. $AMP + H_2O \rightarrow adenosine + P_i$

ii) *nucleoside phosphorylases*, which break the glycosidic bond:

   e.g. $adenosine + P_i \rightarrow adenine + ribose\text{-}1\text{-}phosphate$

iii) *deaminases*, which remove the 6-NH_2 of adenine; separate enzymes act on adenosine and AMP:

$$adenosine + H_2O \rightarrow inosine + NH_3$$
$$AMP + H_2O \rightarrow IMP + NH_3$$

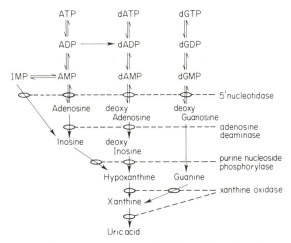

Fig. 5.7 Enzymes involved in the degradation of purine nucleotides.

Free purines and pyrimidines may be degraded, or reincorporated into nucleotides by salvage pathways.

Degradation of purines

Purines are oxidized by molecular oxygen, through the action of xanthine oxidase (Fig. 5.8), a complex enzyme

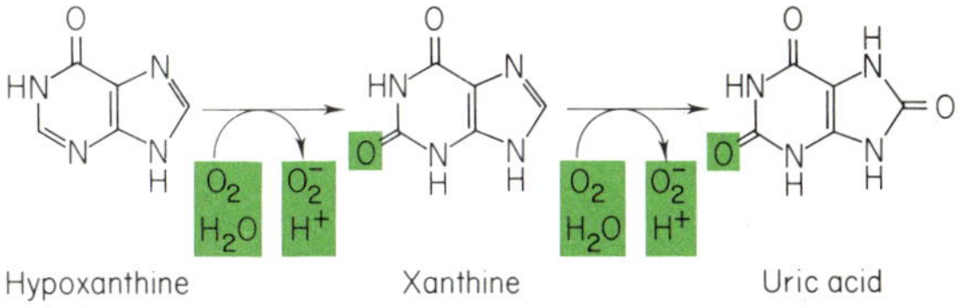

Fig. 5.8 Conversion of hypoxanthine to uric acid, by xanthine oxidase.

which oxidizes hypoxanthine to xanthine, and xanthine to uric acid. Its prosthetic groups are FAD, molybdenum and non-haem iron, and it reduces oxygen to superoxide (O_2^-). This dangerously reactive radical is removed by superoxide dismutase and catalase:

$$2\,O_2^- + 2\,H^+ \xrightarrow{\underset{\text{dismutase}}{\text{superoxide}}} H_2O_2 + O_2$$

$$2\,H_2O_2 \xrightarrow{\text{catalase}} 2\,H_2O + O_2$$

Uric acid is excreted by man and other primates, and is the principal nitrogen-containing excretory product of birds and reptiles. In other mammals purine catabolism continues to allantoin, which in fish and lower animals is degraded to urea and glyoxylate.

Degradation of pyrimidines

Pyrimidines are usually completely broken down (Fig. 5.9), and do not appear in the urine unless large quantities are ingested. One of the products, β-alanine, can be incorporated into coenzyme A, or such peptides as carnosine or anserine (p. 16).

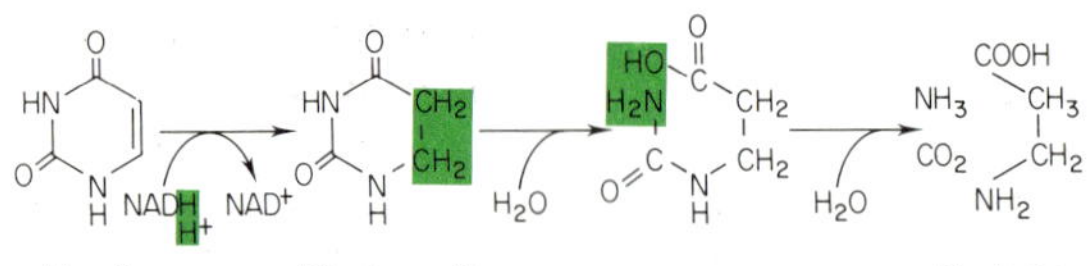

Fig. 5.9 Degradation of uracil. Thymine is degraded by a similar pathway, to give β-amino isobutyric acid.

Scavenger pathways

Purines may be reconverted to nucleotides by the action of phosphoribosyl transferases

$$\text{e.g.} \qquad \text{adenine} + \text{PRPP} \rightarrow \text{AMP} + \text{P-P}_i$$

There are two enzymes of this type, one (APRT) acting on adenine, the other (HGPRT) on hypoxanthine or guanine. Although the reactions are reversible, the concentration of pyrophosphate within cells is invariably low, so these enzymes probably do not function in nucleotide degradation. They are important in the acquisition of purines from the blood, by erythrocytes and peripheral tissues.

There are also scavenger pathways for pyrimidines. These are used experimentally in the labelling of DNA: radioactive thymidine is rapidly converted to its nucleotides, which are then incorporated into DNA.

Control of Nucleotide Synthesis

Pyrimidines

In bacteria, pyrimidine synthesis is controlled by feedback inhibition of aspartate transcarbamylase, the inhibitor being CTP in *E. coli* (see p. 227). In animals this enzyme is part of a complex which catalyses the first three steps of pyrimidine biosynthesis, and it does not appear to be under allosteric control; instead, synthesis of carbamyl phosphate, catalysed by another enzyme in the complex, is inhibited by many pyrimidine nucleotides, such as UDP, UTP, dUDP, dUTP, and CTP: it is also activated by PRPP. Carbamyl phosphate is channelled to aspartate transcarbamylase without leaving the complex, and the rate of its synthesis controls the pathway.

Purines

Synthesis of purine nucleotides is controlled by feedback inhibition at several points:

1. Synthesis of PRPP is inhibited by AMP, GMP and IMP.

2. The next step, amination of PRPP to phosphoribosylamine, is synergistically inhibited by adenine and guanine nucleotides.
3. AMP inhibits its own formation from IMP, but this step requires a guanine nucleotide (GTP); similarly, GMP inhibits oxidation of IMP to xanthosine monophosphate, the precursor of GMP; amination of this to GMP requires ATP.

These reciprocal positive and negative effects balance the synthesis of adenine and guanine nucleotides.

Deoxyribonucleotides

Ribonucleotide reductase is under allosteric control by a complex set of feedback mechanisms which ensure that deoxyribonucleotides are synthesized in the proportions required for DNA synthesis. The reduction of all ribonucleoside diphosphates is inhibited by dATP, while ATP *stimulates* reduction of UDP and GDP, and reduction of GDP and ADP is stimulated by dCTP and dTTP.

Defects of Purine and Pyrimidine Metabolism

Orotic aciduria

Orotic acid is sometimes excreted in the urine; this may be a result of a defect in orotate phosphoribosyltrans-ferase, but can also occur if there is a defect in one of the urea cycle enzymes, which leads to an excess of carbamyl phosphate in the mitochondria of liver and kidney. This enters the cytoplasm, stimulating pyrimidine biosynthesis.

Gout

Raised uptake or endogenous overproduction of purines leads to hyperuricacidaemia, with the precipitation of uric acid and sodium urate in the joints (gout). This painful condition is alleviated by allopurinol, a purine analogue which cannot be oxidized by xanthine oxidase, but acts as a potent inhibitor.

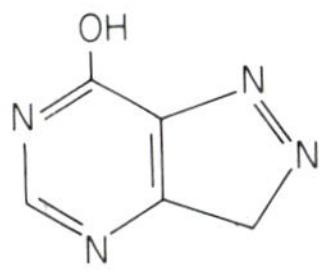

Allopurinol

This drug stops purine catabolism at hypoxanthine, which is more soluble than uric acid. It is converted in vivo to its nucleotide, by reaction with PRPP; this inhibits the synthesis of phosphoribosylamine, and reduces the rate of purine production.

The Lesch–Nyhan syndrome

This disease is associated with the congenital absence of HGPRT; the symptoms appear in young children, and include neurological abnormalities, self-mutilation and uric acid production at up to 20 times the normal rate. The activity of HGPRT is normally highest in brain, especially in the basal ganglia, but the connection between the lesion and the symptoms is obscure. Normally, at least 90% of purines released from nucleic acids are reused; with the defect in the salvage pathway, there is inadequate feedback control of purine biosynthesis. The overproduction cannot be controlled by allopurinol, since individuals with this defect cannot convert the drug to its nucleotide.

Severe combined immune deficiency

The Lesch–Nyhan syndrome demonstrates that a single enzyme defect, in what appears to be a minor pathway, can have very serious consequences. Severe combined immune deficiency disease, in which there is a defect in both B- and T-lymphocytes, is often associated with a deficiency of one of three enzymes of purine catabolism: adenosine deaminase, purine nucleoside phosphorylase or 5′-nucleotidase. In the absence of adenosine deaminase there is excretion of deoxyadenosine, and increased intracellular dATP; purine nucleoside phosphorylase deficiency increases the concentration of dGTP in cells. These nucleotides inhibit ribonucleotide reductase, blocking DNA synthesis and cell division, so the rapid proliferation of lymphocytes is impossible.

This suggests that inhibitors of adenosine deaminase might be used as immunosuppressants. Some other immunosuppressant drugs act on purine synthesis: these include 6-mercaptopurine, 6-thioguanine and azathioprine, all of which are converted in vivo to their ribonucleotides, which act as inhibitors at control points in purine biosynthesis.

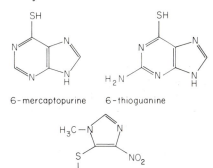

6-mercaptopurine   6-thioguanine

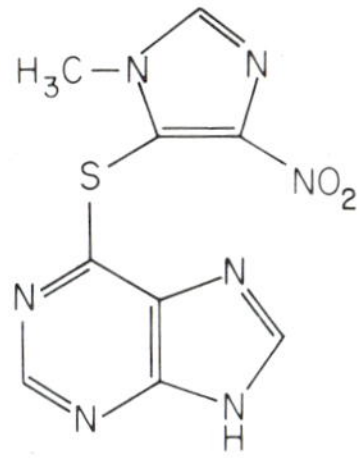

Azathioprine

6

Carbohydrate Structure and Metabolism

A. Structure

The structures of simple sugars

Carbohydrates, unlike lipids, have a precise empirical formula which enables them to be easily classified. This formula is $(CH_2O)_n$, and it is obeyed from the simplest carbohydrate, *glyceraldehyde* ($n = 3$) to the largest polysaccharides, whose molecular weights are numbered in millions of daltons. These polymers are made up of repeated units of the monosaccharides described in this section. Many polysaccharides, however, especially those which have a structural role, do not conform in all respects to the strict empirical formula because their component monomers include amino sugars, deoxy sugars or sugar acids. These derivatives are always formed from true monosaccharides, and they and their products are classed with the carbohydrates.

Fig. 6.1 shows two structures for glyceraldehyde in

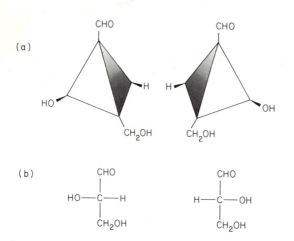

Fig. 6.1 (a) The two forms of glyceraldehyde have identical compositions, but their 3-D structures are not superimposable on one another. They are mirror images, or *enantiomers*. (b) These are the conventional (Fischer) projections of the same two molecules. The atoms that project forwards from the asymmetric C atom are placed horizontally. When the —OH group lies to the right, and the —CHO group is upwards, this is taken to represent D-glyceraldehyde. The L-isomer is on the left.

a conventional two-dimensional representation. In a three-dimensional model, the two structures cannot be superimposed; they are mirror images (*enantiomers*) of one another. They do not differ in physical properties, but each rotates the plane of polarized light in the opposite direction to the other. (*N.B.* The direction of rotation cannot be inferred from the absolute configuration symbols D- and L-, and is recorded by the lower-case symbols *d*- and *l*-.)

The Prelog convention

This is a convention for deciding the *chirality*, or 'handedness' of asymmetric molecules. Because many very complex organic molecules exist, the convention has many complicated rules, but the principles can be explained fairly simply. One takes the imaginary tetrahedron that bounds the asymmetric atom, and arranges it in space with the lowest rank substituent (often —H) away from one. It is helpful to think of fixing the tetrahedron in the centre of the steering wheel of a car, with the lowest rank substituent down in the steering column. One then looks at the ranks of the other three constituents, those that lie in the plane of the wheel, and decides whether they are arranged in clockwise (R) or anti-clockwise (S) order.

The complexity lies in deciding the ranks of the substituents, and no attempt will be made to treat this thoroughly. Briefly, atoms are given a rank corresponding to their atomic number, thus $O > N > C > H$. If the atoms are attached directly to the asymmetric carbon (as with —OH or —NH$_2$), this is easy. If they are attached through a C atom, the latter gets a priority in virtue of its substituent(s), e.g., $CH_2OH > CH_2NH_2$. In the case of glyceraldehyde, the groups (other than —H) attached to C-2 are ranked $OH > CHO > CH_2OH$. Inspection of Fig. 6.1(a) shows that in D-glyceraldehyde, one moves from the highest to the lowest-ranking substituent in an anti-clockwise direction; therefore D-glyceraldehyde is (2*S*)-glyceraldehyde.

Aldose sugars

In a linear molecule, the only way of satisfying the formula $(CH_2O)_n$ is for one C atom to carry a carbonyl group (aldehyde or ketone), while all the others carry a hydroxyl. The two kinds of group, $=CO$ and $-CHOH$, have different chemical reactivities which are important in understanding carbohydrate biochemistry, and are discussed later in this section. The structure is exemplified in D-glucose (Fig. 6.2).

Each of the C atoms added to glyceraldehyde to

(a)

(b)

(c)

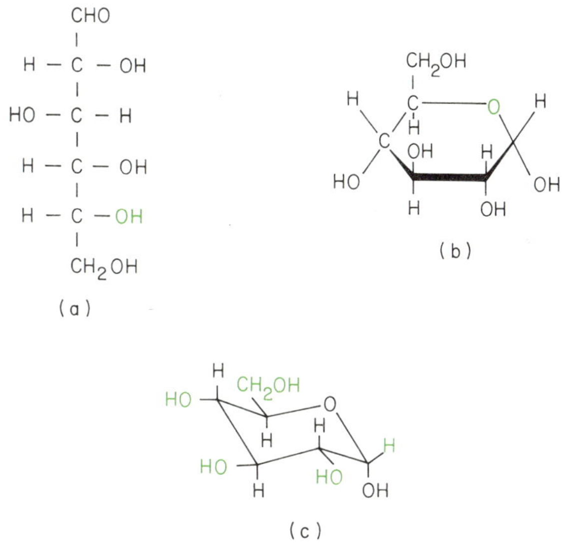

Fig. 6.2 Two dimensional representations of the D-glucose molecule. (a) Fischer conventional projection. (b) Conventional planar representation of the pyranose ring structure of (α-1)-glucose. The O atom that takes part in the formation of the ring is coloured green. (c) Chair form of the puckered structure of α-D-glucopyranose. All the substituents coloured green are *equatorial*, the others are *axial*.

construct this hexose sugar is optically active, so that the aldohexose D-glucose has 15 isomers ($2^4 = 16$); these isomers are not mirror images of D-glucose, and have slightly different physical properties, such as melting point, solubility, sweetness, etc. Most of the isomers have no biological importance; only galactose and mannose are considered here.

The formulae above have been drawn in a ring notation, and it is not immediately apparent that this and the straight-chain representation (cf. Fig. 6.2a) are equivalent. The pyranose (6-membered) ring is formed as the result of hydration of the carbonyl group in aqueous solution, to form a group $>CH(OH)_2$, followed by condensation between this and one of the $-CHOH$ groups elsewhere in the molecule to form an internal *hemi-acetal* (Fig. 6.3).

Such a condensation reaction also occurs in forming polysaccharides, as discussed in a later section. With simple sugars, both forward and reverse reactions occur rapidly and spontaneously in aqueous solution, although they can be speeded up by a *mutarotase* enzyme, but in compounds with other molecules in which the carbonyl group takes part, the configuration, α or β, is 'frozen'. The configuration found in nature will depend on the specificity of the enzyme catalysing the synthesis.

Since in organic compounds 6- and 5-membered rings are most stable, the ring structures to be found are usually pyranose or *furanose* (5-membered). Furanose rings are planar, but pyranose rings are puckered, so that the depiction above is over-simplified. The extended, or 'chair', formation of pyranose rings is the more common.

Ketose sugars

It is not essential that the $=CO$ group should be at the end of a molecule for the empirical formula to be satisfied. The diagram on p. 62 shows another way in which this can be arranged. It is theoretically possible for the carbonyl group to be anywhere in the centre of

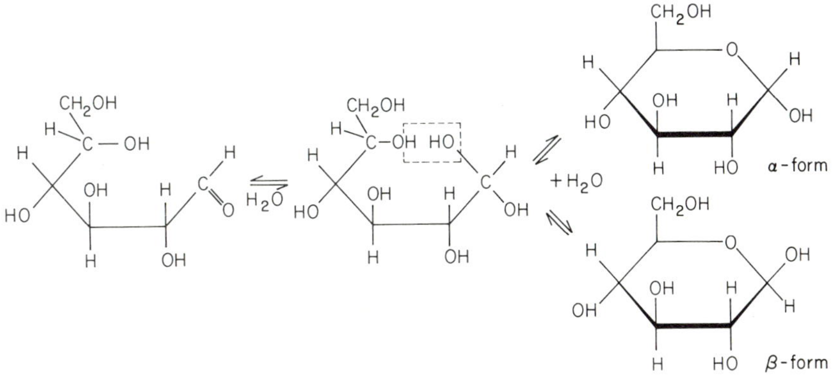

Fig. 6.3 Formation of a hemi-acetal bond and the cyclization of glucose.

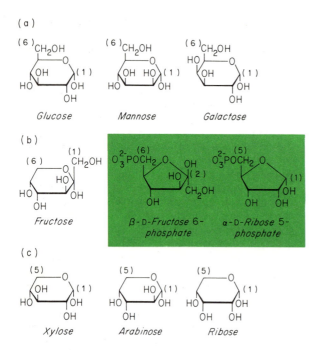

(a)

Glucose Mannose Galactose

(b)

Fructose β-D-Fructose 6-phosphate α-D-Ribose 5-phosphate

(c)

Xylose Arabinose Ribose

Fig. 6.4 Structures of some common monosaccharides. The forms shown for the free sugars are all pyranose; this is almost universal, even for pentoses (row (c)). In row (b), however, two furanose forms are shown (green shading). These are both esters, one of a hexose (fructose) and one of a pentose (ribose). The furanose form is the more usual for these two sugars when in combination.

the sugar molecule, but in practice only 2-keto sugars, such as fructose, shown here, are commonly found. Note that because neither the terminal —CH₂OH nor the =CO groups are optically active, ketose sugars have one less centre of symmetry than their aldose counterparts; thus there are 7 isomers of D-fructose. The physical properties of ketoses are very similar to those of aldoses, but the chemical properties differ somewhat. In particular, the terminal group cannot be readily oxidized to give a sugar acid, so that when ketoses are oxidized, the molecule usually splits in two. In addition, it is more common for the hemi-acetal ring structure to have the furanose than the pyranose form. All the known derivatives of fructose, for example, have a furanose ring.

In addition to glyceraldehyde (and its ketose isomer, dihydroxyacetone), monosaccharides with 4–9 C atoms are known, but those with 5 for 6 C atoms are the most common. *Erythrose*, which has 4, and *sedoheptulose*, which has 7, will be found in Fig. 6.19. The structures, in ring form, of some of the most important monosaccharides are shown in Fig. 6.4.

Sugar alcohols and sugar acids

Both aldoses and ketoses can be reduced, chemically or biologically, to polyhydric alcohols H·(CHOH)ₙ·H, and some of these compounds occur free in nature, mostly in plants. They are not metabolically very active (see however p. 64), but have some medical importance in that they are used as non-glucose-forming sweeteners in foodstuffs for diabetics. Sorbitol (C_6) and xylitol (C_5) are the most commonly used.

Both C-1 and C-6 of aldoses may be biologically oxidized, to give, e.g. from glucose, *gluconic acid* and *glucuronic acid*, respectively. With keto-sugars only the C-6 (or C-5) is oxidized. Although the 'uronic' acids are very important units in the structures of some structural polysaccharides (see p. 65), they do not occur free in nature to any great extent.

Chemical and physical properties of the monosaccharides

The physical properties of the simple sugars, and of the sugar alcohols, are largely dictated by the multiplicity of hydrophilic —OH groups. The compounds are generally crystalline solids with a high melting point, very soluble in water, and more or less sweet to the taste. These properties are also possessed by the 'unnatural' L-sugars, which may be used as non-metabolizable sweeteners.

The sugars have two sets of chemical properties—those that depend on the —OH groups, and those that depend on the =CO group. For the former, esterification with acids is the reaction of most importance; biologically speaking, esters of phosphoric and sulphuric acids are more important than all others. With free mono-saccharides, terminal esters, i.e. with the acyl groups attached to C-1 or C-6 (or C-5), are most important, but in polymers esterification at any position may be found, e.g. the 3′–5′ diesters of ribose which make up the backbone structure of RNA (chapter 11), or the complex esters of the connective tissue polysaccharides (p. 65). Note that because the —OH group at C-1 of aldoses is derived from a carbonyl group, esters formed from it are more unstable (easily hydrolysed) than other sugar esters. They provide suitable leaving groups for the transglycosylation reactions that lead to the synthesis of polysaccharides and other sugar derivatives (see p. 80).

We have seen already that the hydrated carbonyl group provides one partner in the easily formed internal hemi-acetals that underlie the ring structures of the monosaccharides. A second condensation gives a full acetal: normally the two reactants are different in carbohydrate chemistry. This may be summarized as follows

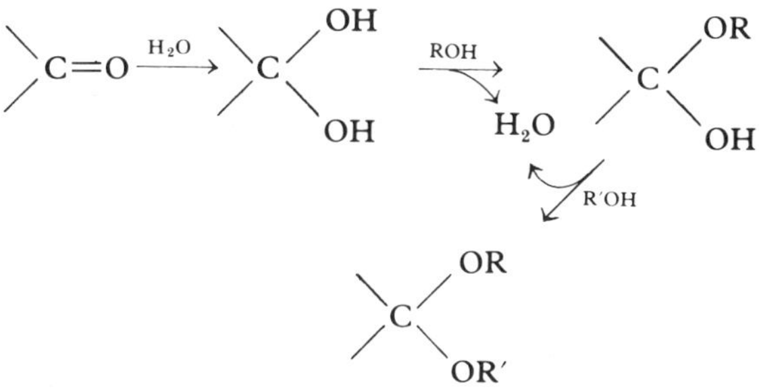

The chemical structures at top left show:

$$\overset{\diagdown}{\diagup}C=O \xrightarrow{H_2O} \overset{\diagdown}{\underset{\diagup}{}}C\overset{OH}{\underset{OH}{\diagup}} \xrightarrow[H_2O]{ROH} \overset{\diagdown}{\underset{\diagup}{}}C\overset{OR}{\underset{OH}{\diagup}}$$

$$\overset{\diagdown}{\underset{\diagup}{}}C\overset{OR}{\underset{OR'}{\diagup}} \quad \xleftarrow{R'OH}$$

Usually the first donor ROH is an —OH group of the monosaccharide itself, and the second donor R′OH is an —OH group of another sugar molecule. However, very similar bonds can be formed with an $\diagup$NH group:

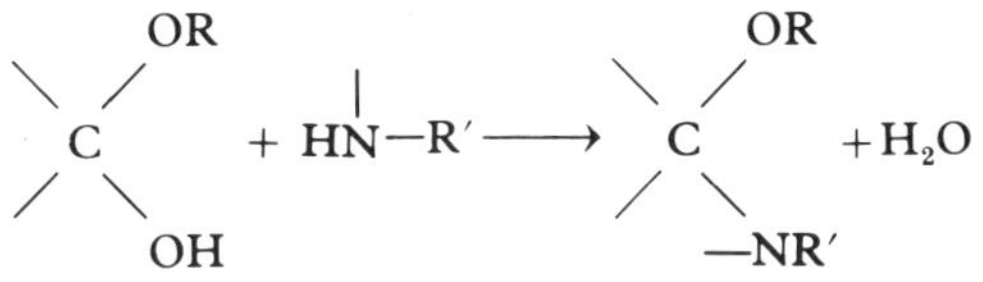

$$\overset{\diagdown}{\underset{\diagup}{}}C\overset{OR}{\underset{OH}{\diagup}} + HN{-}R' \longrightarrow \overset{\diagdown}{\underset{\diagup}{}}C\overset{OR}{\underset{-NR'}{\diagup}} + H_2O$$

This reaction provides the link between sugar and base in the nucleosides, and hence in the nucleic acids (see the reactions of phosphoribose pyrophosphate, chapter 5).

The aldehyde group of an aldose sugar can be oxidized to a carboxylic acid, and the existence of a hemi-acetal ring structure does not prevent this happening, although the immediate product is then a lactone and not the free acid (cf. the oxidation of glucose-6-phosphate, p. 78). Both chemically and biologically the oxidation can be accomplished with mild oxidizing agents. However, a full acetal is resistant to oxygenation; one may put this another way by saying that in a linear polysaccharide only one sugar residue is oxidizable—the terminal residue in which the aldehyde group is not involved in a bond between sugars. Ketose sugars are not readily oxidized.

Both aldoses and ketoses can be reduced to the corresponding sugar alcohol. In vivo, NADH is sufficient to bring this about.

The acetal reactions join molecules together by a —C—O—C—link. Of equal importance in biochemistry is a reaction involving aldehydes or ketones, giving rise to a —C—C— bond, i.e. it makes a more complex molecule from two simpler ones. Frequently the reaction is reversible in vivo, so that it may appear in catabolism as well as anabolism. It is called the *aldol condensation*. In its simplest form, the reaction involves an aldehyde and another group (not a true acid), from which a proton may be lost in the presence of a suitable base. This second group may be another aldehyde or even a reactive methyl group (as in acetyl-CoA). When reduced to the minimal requirements, the reaction may be represented as in Fig. 6.5.

The best known example of an aldolase reaction is that catalysed by fructose diphosphate aldolase (p. 72), but the transaldolase reaction of the hexose monophosphate shunt is a similar type of reaction. Note that the carbonyl group which is attacked by the enolate ion is invariably converted into a hydroxyl group. With this information, one can see that the condensation of the acetyl residue with oxaloacetate to form citrate (p. 156) is also an aldol condensation.

The hexose monophosphate shunt also includes two enzymes designated as *transketolases*. The reactions they catalyse also involve the formation of carbon–carbon bonds, but are rather different in character, in that one of the reactants does not occur free, but as an adduct attached to the prosthetic group thiamine pyrophosphate (p. 74).

Some important reactions of monosaccharides are summarized in Fig. 6.6.

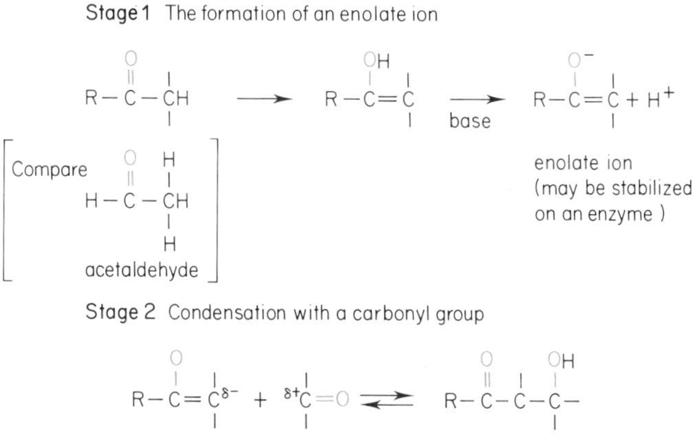

Stage 1   The formation of an enolate ion

$$R-\overset{O}{\overset{\|}{C}}-\overset{|}{\underset{|}{C}}H \longrightarrow R-\overset{OH}{\overset{|}{C}}=\overset{|}{\underset{|}{C}} \xrightarrow{base} R-\overset{O^-}{\overset{|}{C}}=\overset{|}{\underset{|}{C}} + H^+$$

$$\left[\text{Compare} \quad H-\overset{O}{\overset{\|}{C}}-\overset{H}{\underset{H}{\overset{|}{C}}}H \right. \\ \left. \text{acetaldehyde} \right]$$

enolate ion
(may be stabilized
on an enzyme)

Stage 2   Condensation with a carbonyl group

$$R-\overset{O}{\overset{|}{C}}=\overset{|}{\underset{|}{C}}{}^{\delta-} + {}^{\delta+}\overset{|}{\underset{|}{C}}=O \rightleftharpoons R-\overset{O}{\overset{\|}{C}}-\overset{|}{\underset{|}{C}}-\overset{OH}{\overset{|}{\underset{|}{C}}}-$$

Fig. 6.5 Aldol condensation.

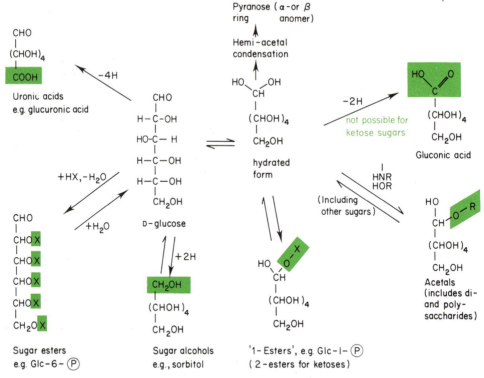

Fig. 6.6 Chemical transformations of importance in the metabolism of monosaccharides.

Disaccharides

Many compounds are known which have been formed by an acetal condensation between the carbonyl group of one monosaccharide and the —OH group of another. The linkage is usually head to tail (e.g. lactose, maltose) but it can be head to head, with no free reducing group remaining (e.g. sucrose, trehalose). Only three disaccharides are of importance in medical biochemistry (see Fig. 6.7).

Lactose, β-D-galactosyl-1,4-D-glucose, occurs to the extent of about 5% in milk. It is less soluble, and less sweet, than most other sugars, and is not fermented by most yeasts. It is a reducing sugar in virtue of the free aldehyde group on the glucose residue.

Sucrose, α-D-glucosyl-1,2-β-D-fructose, is the sugar of commerce. Its double acetal bond between the two sugars is very readily hydrolysed by acid. Trehalose, which occurs in fungi and in insect blood, is the non-reducing 1,1-disaccharide of glucose.

Maltose is α-D-glucosyl-1,4-D-glucose. It does not occur free in nature, but is an important end product of the enzymic digestion of starch.

Cells in the upper part of the human small intestine contain specific hydrolysing enzymes for each of the four disaccharides mentioned above (see p. 184).

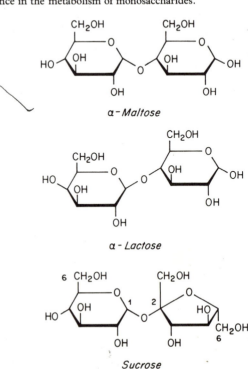

Fig. 6.7 The most important naturally-occurring disaccharides.

Polysaccharides

There is no sharp dividing line between disaccharides and polysaccharides, but few oligosaccharides (i.e. compounds containing 5–20 sugar residues) occur free except during the enzymic hydrolysis of starch. The so-called *blood-group oligosaccharides*, for example, are attached covalently to amino acid residues of proteins, and are considered below under glycoproteins.

The polysaccharides can be conveniently divided into those which contain only one kind of repeating unit, and those, called the heteropolysaccharides, which contain two or more repeating monomers. From the point of view of the organization of this chapter, it is more convenient to consider the latter group first.

Heteropolysaccharides

Plant polysaccharides

Roughage. The structural homopolysaccharide of plants is *cellulose*. This is a straight-chain polysaccharide consisting of β-1,4 linked glucose units. The β-configuration at C-1, which is apparently a trifling difference, is very important, as no vertebrate has a digestive enzyme able to attack it. Ruminants digest cellulose with the aid of symbiotic micro-organisms. Cellulose is very insoluble. Chemical modification gives carboxymethyl-cellulose, which is soluble but bulky, and is sometimes used in slimming products. There are some other homopolysaccharides, and a few have a storage, rather than a structural, function, e.g. *inulin*, a polymer of fructose found in some tubers. None of these compounds is of great importance in mammalian biochemistry.

It is rare for cellulose to occur pure in plant structures. It is usually accompanied by pentose polymers such as xylans (D-xylose) and arabinans (L-arabinose), and by hetero-polymers containing acidic groups. All these polymers are known collectively as hemicelluloses. None of them is significantly attacked by pancreatic amylase, and they are excreted from the gut unchanged, except insofar as they have been digested by gut bacteria. This bacterial metabolism is often associated with the production of uncomfortable volumes of gas, chiefly CH_4 and H_2. Nevertheless it has been recognized that Western diets containing much white flour and meat products are likely to lead to a decrease in faecal bulk, with the consequent possibility of serious constipation. There has been a trend towards the deliberate re-introduction of non-digested polysaccharides ('roughage') into the diet. Cereal husks (bran) provide an abundant source of hemicelluloses,

but other sources exist. Gum arabic contains arabinans; the seaweed product agar, containing agarose, a galactose polymer, has been used therapeutically for many years. Many heteropolymers, especially agarose and those containing sulphate ester or amino sugar residues, are associated in solution with large volumes of water, which aids in the maintenance of optimal faecal consistency. Cellulose does not possess this water-retaining property.

Glycosaminoglycans

This group of compounds used to be known as the mucopolysaccharides. They are characterized by the presence, in regular alternation with other residues, of amino sugars (glucosamine or galactosamine), which are not found in plant heteropolysaccharides.

It is now known that most of the glycosaminoglycans are attached by covalent bonds to a protein 'core', which typically accounts for some 10% of the dry weight. The size of the core, and the number of attached carbohydrate chains, are both very variable. This group of compounds is called the *proteoglycans*; each carbohydrate chain is usually unbranched, whereas in the glycoproteins (see p. 67) branching is common. Hyaluronic acid and heparin are the most important of the glycosaminoglycans that are not attached covalently to protein.

Taken as a group, the glycosaminoglycans are fairly acidic, both because 'uronic' acids (glucuronic and galacturonic) occur regularly in the chains, and because esterification by sulphate is common.

Hyaluronic acid

This is a large linear polymer made up of the disaccharide repeating unit shown in Fig. 6.8. Note that there is no sulphate ester in hyaluronic acid. Typically there are 400–4000 repeating units, giving a molecular weight range of 1.5×10^5–1.5×10^6. The largest molecules would be about 0.4 mm long if they were uncoiled,

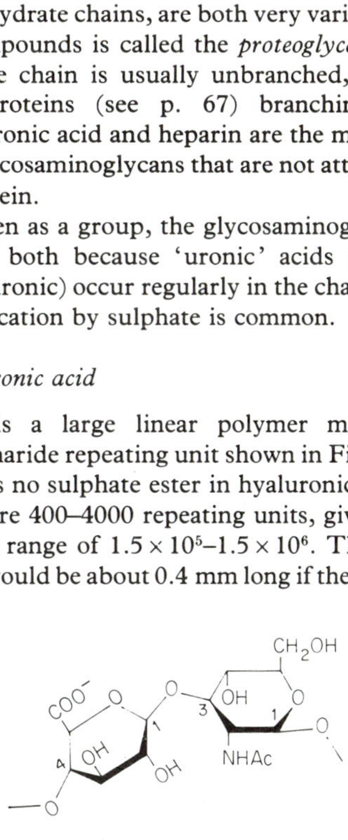

Glucuronic acid   N-Acetyl glucosamine

Fig. 6.8 Repeating disaccharide unit of hyaluronic acid. Note how the 1,3-link between the sugars tends to give the molecule a coiled structure.

but in fact the molecules coil and entwine to make a very firm gel at very low concentration ($\sim 0.1\%$). The gel excludes other large molecules and also micro-organisms, so that the rate of spread of bacterial infection is hindered. Many micro-organisms secrete a *hyaluronidase*, which by shortening the average chain length of the polymer, reduces dramatically the viscosity of the gel.

Gels made from hyaluronic acid have good resistance to compression, and appear as lubricants and shock-absorbing components in synovial fluid, in sub-cutaneous connective tissue (where it is progressively replaced in adult life by the arguably less elastic dermatan sulphate) and in many other tissues. In cartilage hyaluronic acid is found in very small quantities, but plays a rather special role (see p. 67).

Heparin

This is a small molecular weight polymer (mol. wt 15–20 000) derived from the mast cells lining the walls of blood vessels, especially in liver and lung; it does not occur in connective tissue. Unlike hyaluronic acid, heparin is very highly sulphated, and hence very acidic. The sulphate groups on the $-NH_2$ groups of the glucosamine residues are very readily hydrolysed by acid. When this happens, heparin loses its biological activity.

Heparin is a powerful inhibitor of blood clotting (see chapter 14). There is still some doubt about its structure, but the major repeating structure is thought to be as shown in Fig. 6.9.

Proteoglycans

Our knowledge of connective tissue polysaccharides has been very dependent on advances in the technique of extracting them. For many years they were always extracted with alkali, which happens to hydrolyse the bond between xylose, the terminal sugar, and the hydroxy-amino acid in the peptide chain. Thus until recently the major polymers—chondroitin sulphate (esterified either in the 4- or the 6-position), keratan sulphate, dermatan sulphate and heparan sulphate—were

N-Acetyl galactosamine
Glucuronic acid   -6-sulphate

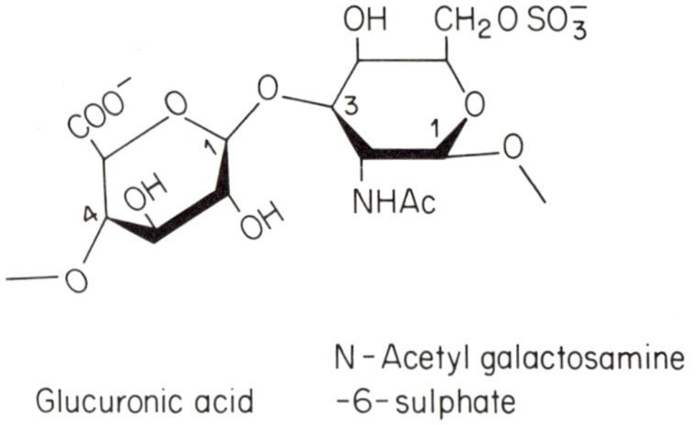

Fig. 6.10 Repeating disaccharide unit of chondroitin-6 sulphate.

not recognized as proteoglycans. The substances are fairly widely distributed in the extracellular matrix of tissues, although dermatan sulphate is more especially found in skin, and chondroitin-4-sulphate in cartilage, where it may form up to 40% of the dry weight. The structure of the repeating disaccharide unit of chondroitin is shown in Fig. 6.10. With the exception of heparan, which is primarily a less highly sulphated version of heparin, the other polymers mentioned above have a repeating disaccharide unit that is similar to chondroitin, but with different monomers, as indicated below:

Chondroitin β-glucuronic acid-β-N-acetyl-galactosamine
Dermatan α-iduronic acid-β-N-acetylgalactosamine
Keratan β-galactose-β-N-acetylglucosamine

In diaphysial bone and tendon there is very little proteoglycan, but in cartilage, which is resilient, there is a great deal, and it is rather highly organized. The bulk of the structure is provided by *proteoglycan subunits* (PGS), which consist of a core protein rich in serine and threonine, to which are attached chains of keratan sulphate (5–6 disaccharide units long) and chondroitin sulphate (40–50 units long), by a —Gal—Xyl— link. The protein content of a PGS is about 11%. One end of the protein core is free from carbohydrate, as shown in Fig. 6.11, and this associates non-covalently with hyaluronic acid. A small link protein helps to stabilize the arrangement, which extends over several disaccha-

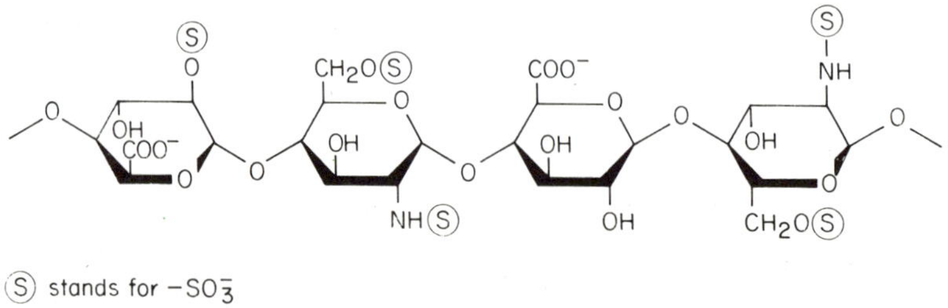

Ⓢ stands for $-SO_3^-$

Fig. 6.9 Repeating tetrasaccharide unit of heparin. Note the high density of acidic groups, including the labile sulphonic acid amide on the rightmost residue. The leftmost residue is L-iduronic acid, an isomer of D-glucuronic acid. It occurs in several heteropolysaccharides (cf. Hunter's syndrome, p. 111).

(a)

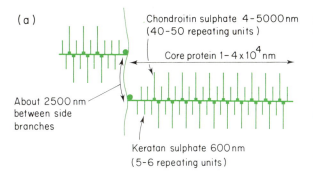

Chondroitin sulphate 4-5000 nm
(40-50 repeating units)

Core protein $1-4 \times 10^4$ nm

About 2500 nm between side branches

Keratan sulphate 600 nm
(5-6 repeating units)

(b)

Hyaluronic acid
400-4000 repeating units
(max. length 4×10^5 nm $\equiv 0.4$ mm)

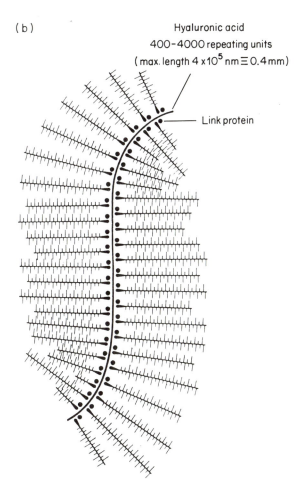

Link protein

Fig. 6.11 Structure of a proteoglycan unit of cartilage. (a) Detailed view of a single subunit, to show the sizes of the components. (b) Tentative model of a proteoglycan aggregate, which makes up the bulk of the amorphous ground substance. The overall length of the largest hyaluronic acid chains, 0.4 mm, would make the unit large enough to see with a hand lens if it had a regular structure. There is a good deal of hydrogen bonding between these carbohydrate groups and the oligosaccharide side chains attached to collagen (see p. 221).

ride units. There will be a PGS unit about every 25 repeating units of hyaluronic acid, which gives from 16 to 160 units per assembly. The total molecular weight of even the smallest assemblies is many millions of daltons. The assemblies are to some extent flexible, and are of various sizes, so that the ground substance does not have a pseudo-crystalline structure, but it is probable that the PGS carbohydrate associates with the short carbohydrate chains attached to collagen. This association may well be strengthened by divalent metal ions.

Glycoproteins

In this chapter, emphasis must be on the carbohydrate component of complex polymers, but it is worth noting that there is no absolute dividing line between proteoglycans and glycoproteins; the mucoproteins described below contain only an average of 15% protein, compared with the 11% of the proteoglycans discussed above. Most glycoproteins do indeed contain much less carbohydrate than this, often as branched, bushy structures. Some examples are given in chapter 10.

Mucins

These are the chief components of the viscous slimy secretions of the intestinal tract, and also of the bronchi (sputum); large quantities of the characteristic carbohydrates have been obtained from ovarian cysts. The overall structure resembles that of PGS, with a core protein rich in serine and threonine, to which the carbohydrate chains are attached, but the two classes of glycoprotein are quite distinct. In particular, the carbohydrate of mucins contains no uronic acids. Instead, galactose and glucosamine predominate. In addition, in about 80% of the population (secretors), some of the mucin polysaccharides have blood-group antigenicity. Connective tissue polysaccharides do not possess this property.

It is presumed that mucins have lubricating and protective properties, although direct evidence for this is sparse. In *cystic fibrosis*, overproduction of mucin occurs both in the digestive system and in the lungs, where it forms insoluble plugs obstructing the airways. It is not yet known whether the glycoprotein is also abnormal in structure.

Blood group substances

At least nine antigen systems can be detected in human erythrocytes, of which only ABO and rhesus are

(*Figure 6.11*(b) is taken from an illustration by L. Rosenberg in *Dynamics of Connective Tissue Macromolecules* by Buleigh & Poole, with kind permission of the publishers, Elsevier Biomedical Press.)

(a)

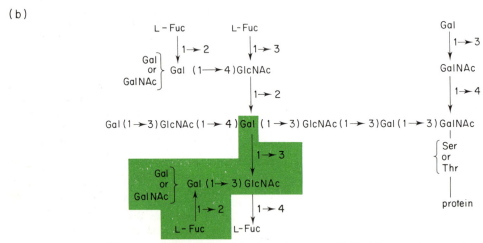

(b)

Fig. 6.12 (a) Space-filling representation of a minimal membrane glycolipid with antigenic properties. Note the coiling of the chain (cf. Figs 6.8 and 6.10). The terminal tetrasaccharide, as shown, gives Group O specificity. If Gal or N-Ac-Gal are attached to position 3 of the outer Gal residue (arrows), the specificity becomes Group B or Group A respectively. The presence of the fucose residue helps to determine the Lewis antigen system. (b) Condensed representation of the protein-linked oligosaccharide that will cause agglutination of red blood cells carrying it, with A or B antibodies. The subsection enclosed in green is identical with the minimal oligosaccharide in Fig. 6.12 (a).

clinically important. The antigen in all nine systems is an oligosaccharide. This is not inevitably the case; the histocompatibility antigens are glycoproteins, but the protein moiety is the antigen. The blood group antigens can be surprisingly small, and a change in a single monosaccharide residue can alter the antigenicity. An example is shown in Fig. 6.12a. Note how the $1 \rightarrow 3$ glycoside bonds promote coiling of the chain. Red blood cell agglutination depends on a rather larger oligosaccharide covalently bonded to a membrane protein (not glycophorin for the ABO system). However, the same minimal configuration (enclosed in green in Fig. 6.12b) occurs in this highly branched structure. A very similar oligosaccharide is found in the mucoprotein of secretors.

Homopolysaccharides

Starch

This is a mixture of two main components, one soluble in boiling water and making up to 10–20% of the total, called *amylose*; the other 80–90% is insoluble in boiling water and is called *amylopectin*. Both are made up of D-glucose units. Amylose is unbranched, containing 200–2000 glucose units linked α-1,4 in a straight line. Amylopectin, on the other hand, is highly branched. It has one end-group to 24–30 glucose units, which means that the outer chains are about 13–18 residues long. The molecule is very large, containing 250–5000 units. The main linkage is α-1,4, but at the branch points a third molecule of glucose is joined in the 6-position (see Fig. 6.13).

The mixture of amylose and amylopectin known as

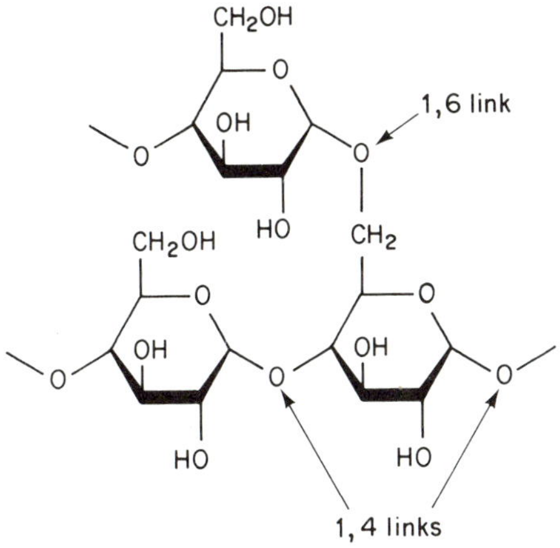

Fig. 6.13   Structure of starch (amylopectin).

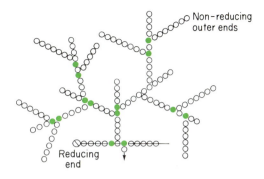

Fig. 6.14   Diagram of the structure of glycogen. The open rings represent glucose residues linked only at carbon atoms 1 and 4 while the green circles represent branch points where glucose residues are linked at carbon atoms 1, 4 and 6. The molecule is symmetrical about the original chain; the lower half of the molecule has been suppressed for the sake of clarity.

starch forms a gel when concentrated solutions cool. It gives a blue colour with iodine.

   Dextran. This is a polysaccharide, consisting of relatively unbranched 1,6 linked glucose molecules, produced by the bacterium, *Leuconostoc mesenteroides*, acting on sucrose. It is soluble, but colloidal, and can be used as a plasma substitute in the treatment of shock. A similar polymer, produced by bacteria from sucrose, forms the plaque that coats teeth, and precedes decay.

Glycogen

This polysaccharide is found only in animals. It has a structure very like that of amylopectin, except that it is even more highly branched. The average chain-length of the exterior chains is only 8 glucose units (13–18 in amylopectin), and in the main chains there is a branch point every 3 units on the average (every 5–6 units in amylopectin). The molecular weight is very high, about $5\,000\,000$ ($\equiv 25\,000$ units). It gives a red colour with iodine (amylopectin gives a red-violet). Its structure is shown in Fig. 6.14. Glycogen forms a colloidal solution, but in cells is often found as glycogen particles, mol. wt up to 2×10^7, and containing associated enzymes.

B. Metabolism

The Digestion of Carbohydrates

Polysaccharides

Starch is the only polysaccharide of any importance in food; post mortem glycolysis normally ensures that the glycogen content of meat products is negligible. Both saliva and pancreatic juice contain an *amylase* which attacks 1–4 glucosyl bonds at random, so long as they are not near chain ends or branch points. The designation α-amylase refers to the configuration at the free C-1 of the products. Both this and β-amylase have an obligatory requirement for Cl^-.

   The products of amylase action on amylose are linear fragments, mostly maltose and maltotriose. Similar fragments are formed from amylopectin, together with a mixture of small branched fragments with 5–9 residues, known collectively as α-dextrins. The 1–6 bonds are hydrolysed by an *oligo-1,6-hydrolase* (α-dextrinase), and the linear oligosaccharides by a group of glucosidases, of which *maltase* is the best known. Most of the hydrolysis of starch occurs in the small intestine.

Disaccharides

Sucrose and lactose are the only two disaccharides of any importance in foodstuffs, but sucrose may make up as much as 40% of the carbohydrate intake. In addition, maltose is formed during the hydrolysis of starch. The mucosal cells of the duodenum and jejunum contain several disaccharidases which hydrolyse specific disaccharides. The evidence is that the hydrolysis occurs within the cells. If the enzyme action is defective, significant traces of the sugars, especially lactose, may be found in blood and later in urine, but normally complete hydrolysis occurs, and the resulting monosaccharides are transported from the cells into the blood stream.

Adult *lactose intolerance* is so common that it would be truer to say that only Western Europeans (and North Americans) retain the ability to hydrolyse lactose in adult life. The persistence of the sugar in the lumen may cause water retention, with abdominal cramps and diarrhoea, which is made worse by bacterial fermentation. Other adults may however be able to tolerate milk without symptoms, although they have little mucosal lactase. It seems that the enzyme begins to disappear at about the time of weaning, since older children may show primary lactose intolerance. Secondary lactose (and also sucrose) intolerance arises if the short-lived mucosal cells are not replaced. This may happen in protein malnutrition (see chapter 16) and in coeliac disease.

Monosaccharides

The concentration of glucose in the lumen of the intestine during carbohydrate digestion is very variable, because it depends on other constituents of the diet, and the speed of gastric emptying, etc. In any event, a specific transporter removes glucose from the lumen, against a concentration gradient if need be. The transporter will also carry galactose and xylose, but glucose competitively inhibits carriage of both these sugars. The carrier requires Na^+ ions (normally present in the lumen) to provide co-transport. The Na^+ ions are probably removed from the serosal side of the mucosal cells by a Na^+/K^+ pump, but this has not been so precisely defined for intestine as it has been for kidney, where a similar transport system operates. Since the Na^+/K^+ pump requires ATP, this is active transport (see chapter 10). The system can also be visualized as a glucose-activated Na^+ transport, and the addition of glucose has been shown to improve electrolyte uptake by cholera patients.

Fructose is absorbed rapidly enough to suggest that a carrier is involved, but low rates of absorption when the fructose concentration in the lumen is low suggests that the carrier may not be energy-linked. Other sugars cross the mucosal barrier by passive diffusion.

The fate of absorbed glucose

Glucose catabolism

A test meal of glucose or sucrose is absorbed from the gut within an hour. It is easy to show that the rise in blood sugar during the absorptive period is far less than would be expected if the absorbed monosaccharide were

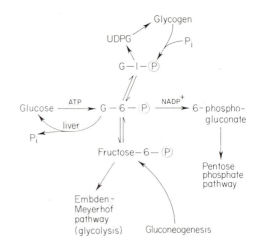

Fig. 6.15 The central role of glucose-6-phosphate in carbohydrate metabolism.

simply to be distributed in the body water. The liver is responsible for much of this, either storing the glucose residues as glycogen, or converting them to fatty acids, but a rise in the Respiratory Quotient (chapter 16) to a value around 1.0 shows that every tissue in the body is oxidizing the ingested glucose as a major nutrient. Whatever the detailed fate of the glucose, the initial reaction is the same—phosphorylation to form glucose-6-phosphate (G-6-P).

Except in liver (and, to a small extent, in kidney) the formation of G-6-P is irreversible, so that any glucose residue that is phosphorylated is committed to some kind of metabolic transformation, as summarized in Fig. 6.15. The formation of G-6-P is therefore—directly or indirectly—rigorously controlled. The sequence of reactions leading to the formation of pyruvate, usually called the Embden–Meyerhof pathway, is quantitatively the most important pathway of glucose metabolism, and it is the only one leading to the irreversible loss of carbohydrate structure, so we will begin with it.

The Embden–Meyerhof pathway

Fig. 6.16 shows the pathway in outline. It is not proposed to discuss every reaction; the following notes refer to some of the most important steps. It will be observed that the intermediates are all phosphate esters (which cannot leave the cell). Simple sugar derivatives do not occur in glucose metabolism (see, however, fructose, p. 82).

1. Apart from G-1-P (see p. 80), the phosphate esters are formed by phosphoryl transfer from ATP. *Hexokinase* catalyses the phosphorylation of glucose; *phosphofructokinase* catalyses the formation of fructose bisphosphate. There is a considerable free energy drop, and the reactions are essentially irreversible.

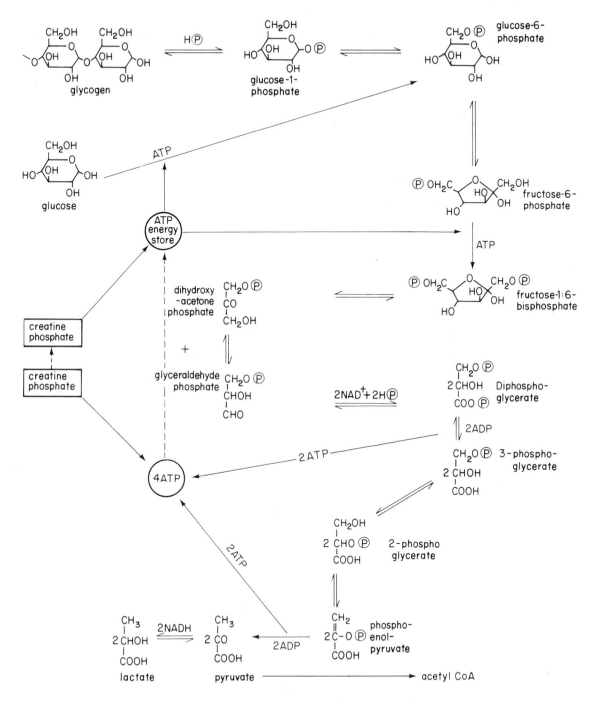

Fig. 6.16   The Embden–Meyerhof scheme of glucose catabolism.

All the kinases have a low K_m for MgATP^{2-}, so that they are saturated with this substrate in vivo. Most tissue hexokinases also have a low K_m for glucose, so that they would also be saturated with this substrate at the concentrations normally found in plasma. The overall control of G-6-P formation then lies with the rate of glucose transport into the cell. Intracellular (free) glucose may be very low. G-6-P may also act as a product inhibitor if it is not removed rapidly.

In liver, however, there is an enzyme with a much lower affinity for glucose, usually called *glucokinase*. The rate of hepatic glucose phosphorylation is then controlled by the plasma glucose concentration, since liver cell membranes are freely permeable to glucose. The long-term control is, however, exerted by regulation of the number of glucokinase molecules per cell (induction, see chapter 13). Glucokinase is not inhibited by G-6-P.

Phosphofructokinase is also highly regulated, but by a different set of mechanisms (allosteric control, see chapter 13).

2. The splitting of the hexose molecule into two trioses catalysed by *aldolase* is perfectly reversible, and except at very low product concentrations, the balance lies in the direction of synthesis. There are several aldolases with different substrate affinities, but this cannot affect the position of equilibrium.

Of the two products of the reaction, only glyceraldehyde phosphate (GAP) is metabolized by the Embden–Meyerhof pathway, but dihydroxyacetone phosphate (DHAP) does not accumulate because it is continuously converted to GAP by *triose phosphate isomerase*. DHAP can be reduced to glycerol phosphate (see chapter 7, p. 101 and chapter 9, pp. 162–163).

3. The oxidation of GAP to phosphoglyceric acid (PGA) is very important, because it is accompanied by synthesis of ATP from ADP. Thus an exergonic reaction (oxidation) is coupled to an endergonic one (phosphorylation). This type of reaction is called *substrate level phosphorylation* because definite phosphorylated intermediates can be isolated (cf. oxidative phosphorylation, chapter 10). Two enzymes are involved.

The first stage, catalysed by *glyceraldehyde phosphate dehydrogenase*, operates as follows:

$$-CH_2O\,\textcircled{P}\cdot CHOH\cdot CHO + HS\cdot Enz\text{---}NAD^+$$

$$\overset{\text{OH}}{CH_2O\,\textcircled{P}\cdot CHOH\cdot \overset{|}{C}H\cdot S\cdot Enz\cdot NAD^+} \quad (1)$$

$$CH_2O\,\textcircled{P}\cdot CHOH\cdot CO\cdot S\cdot Enz + NADH + H^+ \quad (2)$$

The reaction here is the oxidation of an aldehyde adduct to an acyl thioester, with concomitant reduction of NAD$^+$.

$$CH_2O\,\textcircled{P}\cdot CHOH\cdot CO\cdot S\cdot Enz + P_i \quad (3)$$

$$CH_2O\,\textcircled{P}\cdot CHOH\cdot COO\,P + HS\cdot Enz \quad (4)$$

The acyl thioester is split by phosphorolysis to yield the stable intermediate, 1,3-diphosphoglycerate (1,3-DPG). The acyl phosphate grouping is however easily hydrolysed, and is 'high-energy'.

The second stage, catalysed by a separate enzyme, *phosphoglycerate kinase*, leads to the formation of ATP.

$$CH_2O\,\textcircled{P}\cdot CHOH\cdot COO\,\textcircled{P}$$
$$+ADP \rightleftharpoons CH_2O\,\textcircled{P}\cdot CHOH\cdot COOH + ATP \quad (5)$$

For the straightforward oxidation of the aldehyde to the acid, $\Delta G_0'$ is very negative (-43 kJ/mol). For the formation of 1,3-DPG it is, on the other hand, slightly positive ($+6.3$ kJ/mol). The phosphorylation of ADP is exergonic (-18.8 kJ/mol), so that the sum of reactions (1)–(5) gives an overall energy change of -12.5 kJ/mol. In effect, about 30 kJ/mol have been conserved, and can be used elsewhere in the cell by reactions linked to ATP hydrolysis.

The two actions are reversible, and indeed all glucose residues formed by gluconeogenesis in animals, or by photosynthesis in plants, come from precursors which have passed through these two stages.

4. For the shunt at this stage in the Embden–Meyerhof pathway in red blood cells, see p. 239.

5. The phosphate ester of enol pyruvic acid (phospho-enol pyruvate, PEP) is another easily hydrolysable compound, so that $\Delta G_0'$ for the reaction catalysed by *pyruvate kinase*

$$ADP + PEP \rightleftharpoons ATP + Pyr \quad (6)$$

is about -22 kJ/mol. This means that the equilibrium of reaction (6) is so far to the right that the reaction is essentially irreversible. This has important consequences in gluconeogenesis (p. 78). There are two forms of pyruvate kinase, one found in muscle, and the other in liver. The liver (L-) form is reversibly inactivated by phosphorylation (see p. 79), thus restricting futile cycling between PEP and pyruvate.

6. For each molecule of glucose that is catabolized, two molecules of pyruvate are formed, and four molecules of ATP are synthesized (two in the oxidation of 2 molecules of GAP, and two in the transfer of $\sim$ P from 2 molecules of PEP). Two molecules of ATP were used in the phosphorylation of glucose and of F-6-P, so that the net gain of ATP in the overall reaction

$$\begin{array}{c} \text{glucose} \\ +2NAD^+ \end{array} \longrightarrow \begin{array}{c} \text{2 pyruvate} \\ +2NADH+4H^+ \end{array}$$

$$\Delta G_0' = -147 \text{ kJ/mole}$$

is 2 moles per mole of glucose converted. The efficiency of the energy conservation is about 40% in standard conditions, rather more at the concentrations prevailing in cells.

7. In muscle cells, but to only a very small extent in other tissues, a substance called creatine phosphate is found. This can act as an immediate source of useful energy in the first few seconds of muscular contraction, before the complete pathways of metabolism can be speeded up to produce the extra energy required. An active enzyme in muscle, *creatine phosphokinase* (CPK), catalyses the equilibrium

$$\text{creatine-} \textcircled{P} + ADP \rightleftharpoons \text{creatine} + ATP$$

the reaction going from left to right in the condition described above. In resting muscle, creatine phosphate is rapidly resynthesized by the reverse of this reaction (see p. 137 for formulae of creatine and creatine phosphate).

8. The NADH formed in the overall process (equation 6) is normally re-oxidized, although indirectly, by mitochondria, with a further gain in ATP trapped by oxidative phosphorylation (for the energetics of this, see p. 171). However, in cells which have no mitochondria (e.g. red blood cells), or in certain other specialized cells (e.g. retina, some tumours), re-oxidation of NADH by reduction of pyruvate, as indicated in Fig. 6.16, is normal. Almost all cells have the capacity to reduce pyruvate if the O_2 supply is cut off. This then makes the Embden–Meyerhof pathway self-sufficient in the oxidation–reduction sense. In bacteria and yeasts the equivalent process, with production of ethanol as reduced end product, may continue indefinitely, but in most animal cells, lactate production in *anaerobic glycolysis* is a temporary measure only. Brain cells, for example, can endure interruption of the oxygen supply for less than a minute.

Complete oxidation of glucose requires 6 molecules of oxygen for each molecule of glucose:

$$6\,O_2 + C_6H_{12}O_6 \rightarrow 6\,CO_2 + 6\,H_2O$$

The reactions by which this oxidation is completed start with decarboxylation of pyruvate. The complete conversion to CO_2 is described in a later chapter. After describing pyruvate oxidation we return to the other pathways that are initiated by an enzymic attack on glucose-6-phosphate (refer back to Fig. 6.15).

Pyruvic acid

This is a very important compound in carbohydrate metabolism because it can undergo many reactions, and thus the carbon atoms of glucose can be directed into one of several channels.

1. It can be *decarboxylated*. In animals this is always an oxidative reaction, whose overall equation is:

$$CH_3 \cdot CO \cdot COOH + HS \cdot CoA + NAD^+ \rightarrow$$
$$CH_3 \cdot CO \cdot S \cdot CoA + CO_2 + NADH$$

It is catalysed by a multi-enzyme system, the *pyruvate dehydrogenase complex*. The mechanism of the reaction is given in Fig. 6.17. The complete formula of thiamine pyrophosphate is given on p. 37.

The dehydrogenase complex resides on the inner side of the mitochondrial membrane, so that pyruvate has to be transported across that membrane before it can be oxidized. The complex has a molecular weight of about 5×10^6, and contains many units of each of the three enzymes E_1, E_2 and E_3 (pyruvate dehydrogenase, transacetylase and lipoamide dehydrogenase) depicted in Fig. 6.17.

Similar mitochondrial enzyme complexes oxidatively decarboxylate other 2-oxo-acids, e.g. oxoglutarate (p. 157), and the branched chain oxo-acids (p. 143). The chief differences are the specificities of the decarboxylases (E_1) and the transacylases (E_2).

In yeasts, by contrast, the adduct formed after decarboxylation breaks down to free enzyme and acetaldehyde:

$$CH_3 \cdot CO \cdot COOH \rightarrow CH_3 \cdot CHO + CO_2$$

which may be reduced to ethanol. Thiamine pyrophosphate is also the prosthetic group of the yeast enzyme.

The acetyl-coenzyme A formed in the oxidative reaction can, with oxaloacetic acid, form citric acid and so be oxidized in the tricarboxylic acid cycle (see chapter 9), or it can be synthesized into long-chain fatty acids (see chapter 7).

2. Pyruvic acid can be carboxylated:

$$\text{pyruvate} + ATP + CO_2 \rightarrow \text{oxaloacetate} + ADP + P_i.$$

This reaction, catalysed by *pyruvate carboxylase*, a biotin-containing enzyme, is essentially irreversible. The enzyme has an absolute requirement for acetyl-CoA as a cofactor.

The reaction catalysed by *malic enzyme*

$$\text{malate} + NADP^+ \rightarrow \text{pyruvate} + NADPH + CO_2$$

is reversible, and might apparently serve to carboxylate pyruvate. The evidence is, however, that it usually operates from left to right as written above. In addition, the enzyme is not very active, even in liver.

3. Pyruvate can be transaminated by glutamic acid to form alanine:

$$CH_3 \cdot CO \cdot COOH + HOOC \cdot (CH_2)_2 \cdot CH(NH_2) \cdot COOH$$
$$\rightleftharpoons CH_3 \cdot CH(NH_2) \cdot COOH + HOOC \cdot (CH_2)_2 \cdot CO \cdot COOH$$

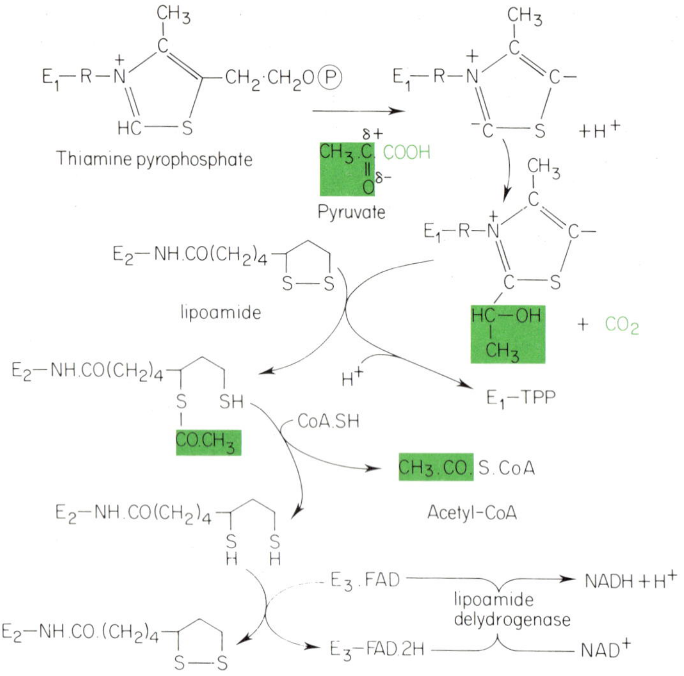

Fig. 6.17 The mechanism of pyruvate oxidation. E_1, E_2, and E_3 are the component enzymes of the mitochondrial pyruvate dehydrogenase complex.

In muscle, this reaction leads to a net production of alanine which leaves the tissue and is taken up by the liver (the *alanine cycle*, see chapter 8).

4. It can be reduced to *lactic acid*, by an NADH-requiring dehydrogenase. The cytoplasm of most cells contains much lactate dehydrogenase activity, and the [lactate]/[pyruvate] ratio in tissues usually reflects the cytoplasmic [NAD$^+$]/[NADH] ratio. However, as pyruvate is not freely permeable between many tissues and the blood, the blood [lactate]/[pyruvate] ratio does not necessarily reflect the tissue redox ratios.

Lactate metabolism

In theory, lactate is an end product of carbohydrate metabolism in any anaerobic tissue, but because continuously active tissues like brain, kidney and liver cannot support their function for more than a few seconds purely from glycolysis, in practice the lactate appearing in blood almost always reflects glucose catabolism in muscle. Thus it is impossible to discuss lactate production without briefly considering muscle activity. To be brief is difficult, because a great deal of research has been done on the fuels of muscular contraction, much of it with the aim of improving the speed or the endurance of competitive athletes, while the present text ought to be more concerned with the average healthy adult. The differences can be consider-

able: there is little doubt that the maximal capacity of the body is strongly related to the ability of the heart to supply oxygenated blood to the muscles. This can be significantly improved by training. Moreover, the concentration of mitochondria in muscle can be doubled by training.

There is also the question of muscle structure. It has been established that human muscles contain a mixture of 'fast twitch' fibres, which rely on anaerobic glycolysis for their ATP replenishment, and 'slow twitch' fibres relying mainly on oxidative phosphorylation. Within a given muscle, the large motor units are composed of fast fibres, the small units of slower ones. It is generally believed that the size of the motor unit brought into action increases as the force exerted by the muscle increases. It follows that the large and forceful motor units that are made up of fast twitch glycolytic fibres are used rather infrequently. Finally, because of differences in muscle size, capillary blood supply and other factors, muscles in different parts of the body respond differently to stimulation. For an equal intensity of work, the blood lactate level rises much more from arm exercise than from leg exercise.

With these limitations in mind, one may now consider the time relationships of ATP resynthesis in active muscle. These are summarized in Table 6.2.

(a) Heavy exercise leading to exhaustion within 10 sec does *not* give rise to lactate production. The ATP required is formed by phosphoryl transfer from phosphocreatine (P-Cr, see p. 73). Exercise at a rate below maximal leads after a few seconds to the establishment of a steady state concentration of P-Cr below the resting level. This reduces the maximal power attainable in a sudden spurt after a period of steady exercise (details not shown in Table 6.2.). P-Cr breakdown is sometimes called the alactic mechanism.

(b) Anaerobic glycolysis, sometimes called the lactic mechanism. This appears in two contexts: *early lactate* and *supramaximal exercise*.

Early lactate. It takes about 2 min for respiration to become established at a higher rate in continued exercise. During this early period the lactate concentration in blood may rise. Table 6.1 shows some illustrative figures.

Table 6.1

Extra O_2 consumption (ml/kg/min)	'Early' lactate from exercise in	
	arm	leg
10	1.6*	0
20	4.8	0.4
50	—	4.0

* Values in mmoles/litre blood.

When respiration has become established at a higher rate, the blood lactate may become steady at a higher level, or it may even decrease. In both cases lactate production is set off against reoxidation, which takes place chiefly in the liver. This is the 'Cori cycle'. It is calculated that after 20 min exercise at maximal respiration rate the *net* contribution of anaerobic glycolysis to extra energy production is about 1.4% of the total in athletes, and perhaps 3–4% in untrained persons. This clearly conceals a good deal of lactate production in fast twitch muscles, balanced by reoxidation in liver. The amount is difficult to estimate, but very few of us ever exercise at full respiratory capacity for anything like 20 min. At lower exercise rates, lactate production may be negligible, as Table 6.1 shows.

Supramaximal exercise. For short periods (say up to 2 min) it is possible to sustain a work load 2–3 times greater than the energy equivalent of the maximum respiratory capacity. As Fig. 6.18 shows, there is a linear relationship between this supramaximal work and the rise in blood lactate. It is interesting that if the energy equivalent is equal to or less than the maximal extra O_2 intake, the lactate accumulation is zero, as implied in the preceding paragraph.

The maximum anaerobic power output from muscles depends partly on the maximum rate of glycolysis, but much more on the maximum lactate concentration that can be tolerated in body water. This is about 17 mM in blood, although higher values have been reported for very brief exercise. The maximum which can be tolerated drops sharply with age, and is only 5 mM at 70 years. It is presumed that the limiting factor is actually the pH change in muscle and blood.

After the exercise is over, 'supramaximal' lactate disappears exponentially from the blood with a $t_{\frac{1}{2}}$ of about 15 min. It is doubtful whether it is oxidized at all in human muscle, other than heart muscle.

(c) Although the maximum power obtainable in muscle (as, for example, in a standing high jump) is only one-third as great from oxidative phosphorylation as

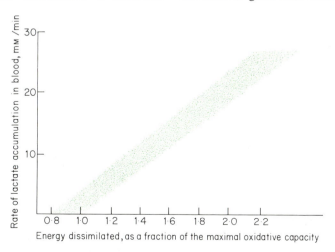

Fig. 6.18 'Supramaximal' lactate production. The maximal oxidative capacity is the rate at which energy is produced, by oxidizing fat or CHO, when the lungs are taking in the maximal volume of O_2 of which they are capable. For short periods of time (up to 2 min), the muscle can expend energy at a greater rate than this. The graph shows that in these circumstances lactate accumulates in the blood, but does not do so when the capacity of the lungs is greater than the rate of energy expenditure.

Table 6.2 Maximal power output and capacity of musculature. The data relate to average male subjects aged about 25 years, and are expressed in joules/kg body weight. The methods of calculation of the figures, and the restrictions on their interpretation, are discussed in the text.

	Maximum power $(J/kg/s)$	Maximum capacity (J/kg)	Absolute upper time limit for which maximum effort could be sustained
Phosphocreatine break-down ('alactic' mechanism)	56.5	750	7.5 s
Anaerobic glycolysis ('lactic' mechanism)	26	up to 1175 (muscle glycogen stores only equivalent to ~ 750)	45 s (probably much less if only a single muscle group involved)
Oxidation (both CHO and long-chain fatty acids)	17.5	$987t - 1.03t^2$ for t in minutes (valid up to 6 hr)	$940 - 20\,\dot{V}_{O_2}$* min

$\star$ $\dot{V}_{O_2}$ is the respiratory rate, above the resting level, in ml O_2/kg/min. The maximum rate for an average young man is 46 ml/kg/min, which is equivalent to about 20 minutes delivery of maximum oxidative power.

from P-Cr breakdown (Table 6.2), nevertheless this is the kind of energy which sustains most of our movements, because the *capacity* of the musculature to resynthesize ATP by this means is almost infinite. The appearance of t in column 3 of Table 6.2 reflects the fact that the energy stores of the body do eventually become exhausted.

A word of caution about the interpretation of Table 6.2 is in order. The experimental observations on which the figures are based were mainly of the oxygen consumption (above the resting level) during the *recovery* period, for the 'alactic' and 'lactic' mechanisms; for the oxidation mechanism they were based on the O_2 consumption during the exercise. The figures have been converted to joules, using a value of 20.9 J/ml O_2, which is equally valid, to a first approximation, for both fat and carbohydrate oxidation (see chapter 16). However, too many assumptions would have to be made for a further conversion, to the rate of ATP breakdown during the exercise, to be practicable. Nevertheless, the few direct measurements that have been made, by muscle biopsy, of phosphocreatine and glycogen breakdown, and of lactate accumulation, suggest that the values in each row and column of the table do relate to each other reasonably well.

It must also be borne in mind that the figures for the 'alactic' and 'lactic' mechanisms (especially the latter) may have been obtained with the blood supply to a group of muscles, e.g. in the arm, completely occluded during the exercise. Although such methods give a value for the time for which maximum power *could* be sustained, it is inevitably somewhat artificial. To use the figures to attempt to prove, for instance, that *all* muscular contraction lasting 45 s is powered *only* by anaerobic glycolysis would be completely to misinterpret the table.

Either carbohydrate or fat can be used as fuel for oxidative phosphorylation in muscle. The relative utilization of carbohydrate (estimated chiefly from R.Q. measurements) is about 25% at rest and about 80% for work intensity near the maximal aerobic capacity, but it does depend on the carbohydrate stores of the body, so that in prolonged exercise, the carbohydrate contribution decreases with time. Trained subjects have a 10% lower R.Q. than untrained, which may be related to the increased mitochondrial contents of their muscles. As is discussed later (p. 82), in phosphofructo-kinase deficiency it may not be possible for the subject's muscle to utilize glucose at all, yet moderate exercise is perfectly possible. Thus carbohydrate is not essential for muscular activity, unless it necessarily involves fast twitch fibres.

Most of the carbohydrate store of the body is muscle glycogen, which is much greater in total than liver glycogen: it is doubtful whether there is much gluconeogenesis (except from lactate) in exercise. Nevertheless, most of the *anaerobic* breakdown of glycogen to lactate, in fast fibres, will only occur in strenuous exercise.

Lactate is continuously being produced by erythrocytes in blood. The rate is very slow, 1–2 mmole/litre packed cells/hr, but it is sufficient to account for the removal of a significant fraction of the glucose secreted into blood by the liver in fasting states. It is not, however, lost to carbohydrate metabolism (see p. 238).

Lactataemia can occur in several disorders of glucose metabolism, in some glycogen storage diseases, and occasionally in fructose overfeeding. Like ketonaemia (p. 107) it is potentially serious because of its effect on the acid–base status of the blood. As a pathway of energy metabolism it is of little importance. Raised blood lactate in the absence of exertion may however be diagnostic as, for example, in ischaemia (over-production), or in failure of liver function caused by disease or drugs (under-utilization).

Several iso-enzymes of lactic dehydrogenase exist (p. 42) and are characteristic for various tissues. The 'heart' enzyme LDH(H_4) is strongly inhibited by pyruvate, and it has been suggested that this form favours the oxidation of lactate. However, as already pointed out in connection with aldolase iso-enzymes (p. 72), catalysts cannot alter the position of equilibrium. An enzyme to which an inhibitor (in this case, pyruvate) is bound, is unavailable for catalysis in *either* direction.

The pentose phosphate pathway

This is the second of the possible fates for glucose-6-P shown in Fig. 6.15. It is an alternative pathway of glucose catabolism which only indirectly involves the enzymes of the Embden–Meyerhof sequence (except aldolase). The overall importance of this pathway in mammals is probably small, but in certain tissues, notably liver and adipose tissue, a considerable fraction (perhaps as much as 50%) of the glucose consumed may be metabolized in this way.

The pathway seems to serve two functions. One is the synthesis of ribose-5-phosphate, which is the starting point (via phosphoribose pyrophosphate, chapter 5) for the synthesis of all the purine nucleotides, and is also essential for the synthesis of the pyrimidine nucleotides. Some of these nucleotides are important in intermediary metabolism in their own right, notably ATP and also the dinucleotides, such as NAD, NADP, FAD and Coenzyme A, that contain an adenylyl residue. In addition the nucleotides are required for the synthesis

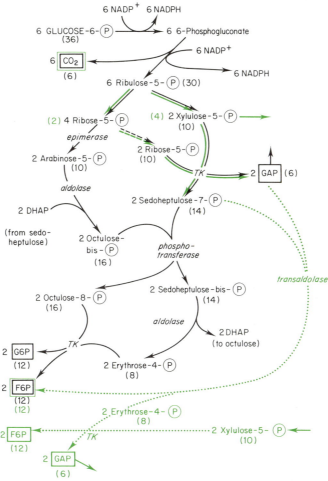

Fig. 6.19 The pentose phosphate pathway of glucose catabolism. The reactants and transformations delineated in black are those which are specific to the 'L' pathway. Those delineated in green are specific to the 'F' pathway. Arrows in both colours indicate that the steps so indicated are common to both pathways. End products are enclosed in boxes, and the numbers in parentheses under many of the intermediates serve to show how the sum of carbon atoms is partitioned between the various branches of each pathway, assuming that, as is conventional in discussing this pathway, it starts with 6 molecules of glucose or G-6-P (36 C atoms). The pathway has been slightly fragmented in the interests of overall clarity.

of RNA and, after the reduction of the sugar moiety to the deoxy form, for the synthesis of DNA. It is possible therefore that all nucleated cells contain at least a trace of the enzyme complement of the pentose phosphate pathway.

The other function is to serve as a source of NADPH. This is certainly so in erythrocytes, where NADPH is used to reduce methaemoglobin (chapter 14), and it is probably also important in liver and adipose tissue, where NADPH is required for fatty acid synthesis (chapter 7). In addition, mixed function oxidases (mono-oxygenases, see chapter 9) are frequently specific for NADPH; these enzymes are important in liver and in some other tissues (e.g. adrenal cortex). Because there are other modes of formation of NADPH, for example by export of isocitrate from mitochondria (chapter 9), it has always been difficult to establish a quantitative balance between the flux through the pentose phosphate pathway and the flux of reducing equivalents through $NADP^+/NADPH$, but there is a general consensus that provision of reducing power is an important function of the pentose phosphate pathway in several tissues. Fig. 6.19 shows that, in contrast to the glycolytic pathway, no direct formation of ATP is associated with the oxidation of glucose-6-phosphate or subsequent metabolites, and this pathway is therefore, although catabolic, not directly energy-producing.

Recent research has shown that there are two forms of the pathway: one (the 'F' pathway) that occurs only in fat cells, and the other (the 'L' pathway) which is found in liver and other tissues, but not in fat cells. Both variants catalyse the same overall reaction, which may be summarized

$$6\,G\text{-}6\text{-}\textcircled{P} + 12\,NADP^+ \rightarrow 12\,NADPH$$
$$+ 4\ \text{hexose phosphate} + 2\,GAP + 6\,CO_2$$

The difference between the two variants has been established by differences in enzyme patterns, by the occurrence of specific sugar phosphates, notably arabinose-5-$\textcircled{P}$ and the eight-carbon sugar octulose-8-$\textcircled{P}$ in 'L' type cells, and through different labelling patterns in the intermediates and products when 14-C-glucose is presented as primary substrate. Both pathways are summarized in Fig. 6.19.

The early stages are common to both variants. There are two oxidation steps:

$$G\text{-}6\text{-}\textcircled{P} + NADP^+ \rightarrow 6\text{-phosphogluconolactone}$$
$$+ NADPH + H^+ \quad (1)$$

The lactone ring is split to give free 6-phosphogluconic acid (6-$\textcircled{P}$G) by a *lactonase*, and the acid is then oxidatively decarboxylated:

$$6\text{-}\textcircled{P}G + NADP^+ \rightarrow \text{ribulose-5-}\textcircled{P}$$
$$+ NADPH + H^+ + CO_2 \quad (2)$$

The equilibrium for both these reactions, but particularly for reaction (2), lies to the right, so that the decarboxylation is effectively irreversible. Control of the pathway must therefore lie with reactions (1) or (2); probably it lies with the demand for NADPH. In contrast, the subsequent reactions are freely reversible, so that in principle the pentose phosphates may be synthesized from the trioses DHAP and GAP.

Two types of reaction in Fig. 6.19 will be unfamiliar. Although the splitting of a single sugar phosphate into two smaller sugars, which occurs twice in the 'L' pathway, is essentially the same as the splitting of fructose bisphosphate by aldolase in the Embden–Meyerhof pathway (Fig. 6.16), the transfer of a fragment

$$\begin{array}{c} OH \\ | \\ -C-CO-CH_2OH \\ | \\ H \end{array}$$

from sedoheptulose-7-$\textcircled{P}$ to GAP, to form fructose-6-$\textcircled{P}$ with erythrose-4-$\textcircled{P}$ left behind (green arrows at right of Fig. 6.19) has not been encountered before. It is nevertheless also the reverse of an aldol condensation (cf. Fig. 6.5). It is catalysed by a specific *transaldolase*.

The enzyme *transketolase* (TK in Fig. 6.19) also transfers a fragment from one sugar residue to another, i.e. from xylulose-5-$\textcircled{P}$ to ribose-5-$\textcircled{P}$ to give sedoheptulose-7-$\textcircled{P}$, and also from octulose-8-$\textcircled{P}$ to erythrose-4-$\textcircled{P}$ to give fructose-6-$\textcircled{P}$ ('L' pathway). The fragment transferred is $-CO-CH_2OH$, which is formally similar to the $-CO-CH_3$ that is transferred to lipoamide after the decarboxylation of pyruvate (Fig. 6.17). Thiamine pyrophosphate (TPP) is indeed the prosthetic group of transketolase, as it is of pyruvate dehydrogenase, and the mechanism of the reaction (condensation of the fragment to be transferred with the carbanion on the thiazole ring of TPP, Fig. 6.17) is very similar in both cases.

Gluconeogenesis (glucose synthesis)

The reactions by which hexoses are built up from smaller molecules are mainly reversals of those in the Embden–Meyerhof sequence. As already pointed out, three of the reactions in this sequence are practically irreversible, and other enzyme-catalysed reactions exist to bypass them. These are:

1. The formation of phospho-enol pyruvate from oxaloacetate

$$HOOC \cdot CH_2 \cdot CO \cdot COOH + GTP \rightleftharpoons$$
$$CH_2 = C \cdot COOH + CO_2 + GDP$$
$$| \\ O\textcircled{P}$$

where GTP stands for *guanosine* triphosphate. The reaction is catalysed by phosphoenolpyruvate carboxykinase (PEPCK).

2. The conversion of fructose bisphosphate and glucose-6-phosphate into fructose-6-phosphate and glucose respectively is carried out by two phosphatases, e.g.:

Fructose 1,6-bis ℗ → Fructose 6- ℗ + ℗

These enzyme reactions are again irreversible. The phosphatases, unlike those of blood and bone, are specific, and only kidney besides liver contains a significant amount of the specific *glucose-6-phosphatase*. Liver can form G-6-P from many compounds by the pathways outlined above, and can thus secrete free glucose into the blood when necessary, but glucose is not secreted by any other organ (except for an unimportant amount from the kidney). Hexose phosphates formed from small molecules in tissues other than liver can be converted only into glycogen, as described below.

The three 'by-pass' enzymes permit the possibility of a bidirectional flux (shown in Fig. 6.20). This would be normal in an 'equilibrium' enzyme step, e.g. enolase, but in the three instances shown in the figure, there would be an effective hydrolysis of ATP (or GTP) with

no chemical gain. Such a process is called *futile cycling*, and it is minimized in various ways.

There is no known suppression of futile cycling for reaction (1), but the opposing rates of reaction (2) are controlled by several mechanisms. The substrates and products are not identical, as they appear to be in Fig. 6.20. Fructose biphosphatase requires the α-anomer of FBP (cf. p. 61), and produces α-F6P, while phosphofructokinase requires β-F6P, and produces β-FBP. The two anomers interconvert spontaneously but only slowly. In addition, AMP activates PFK, but inhibits FBPase; finally, both enzymes are reversibly phosphorylated by protein kinase (see chapter 13), with opposite effects on their activity.

Suppression of futile cycling in reaction (3) is much simpler. As already mentioned, liver pyruvate kinase is subject to control by reversible phosphorylation. The effect is not to alter V_{max}, but to reduce the sigmoidicity of the V vs $[S]$ curve (chapter 13), so that the enzyme has much less affinity for its substrate PEP.

Substrates for gluconeogenesis include lactate and alanine (mainly from muscle), glycerol (mainly from adipose tissue) and glucogenic amino acids or their carbon skeletons (see chapter 8). These precursors are at various levels of oxidation. Glycerol is more reduced than glucose, while pyruvate and substances convertible to it in cytosol (which includes alanine and other amino acids) require cytosolic reducing equivalents at the level of GAPDH. These have to be provided by export from mitochondria (see discussion in chapter 9). Oxidation of fatty acids ultimately provides the reducing equivalents for this. In addition pyruvate may be generated in the cytosol, and then has to be transported into the mitochondrial matrix before it can be carboxylated. In the liver of man, unlike the liver of the rat, phosphoenol pyruvate carboxykinase (see p. 160 and also reaction (2), p. 78) is mitochondrial, so that PEP has to be transported out into the cytosol. There are many possibilities of regulation of the transport mechanisms. The hormone glucagon is very important in many of these regulation points in liver, chiefly by stimulating the synthesis of cAMP, and thus activating the hormone-sensitive protein kinase.

Whatever the substrate, ATP is necessary for the reversal of glycolysis, and this must come from mitochondria.

The kidney contains the requisite enzymes for gluconeogenesis, but it is difficult to detect net production of glucose from kidneys in situ. Although white muscle contains a little FBPase, no muscles contain glucose-6-phosphatase, and therefore cannot add glucose to the bloodstream. Thus the liver is the only effective source of blood glucose, and hypoglycaemia rapidly follows functional hepatectomy.

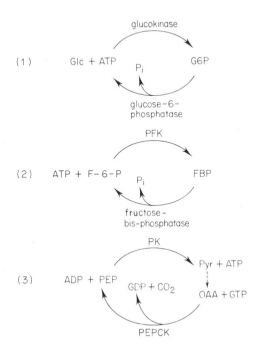

Fig. 6.20 Futile cycling in gluconeogenesis. These are the reactions that are 'by-passed' in the reversal of glycolysis. It can also be appreciated that if both catabolic and anabolic enzymes at each step are active, the pairs of substrates may just recycle, with no net flux through the pathway, but with a continuous loss of ATP.

Glycogen

This is a branched chain polysaccharide whose overall structure is shown on p. 69. It occurs in most mammalian cells, but particularly in liver and muscle. It functions as a store of glucose which is especially convenient from the osmotic point of view. Quantitatively, fat is more important than glycogen for the storage of those carbon atoms of ingested glucose which are not oxidized immediately after absorption. Nevertheless, glycogen is an effective store of immediately available energy. In muscles the amounts are rarely more than 2% and often under 1% of the wet weight. It is not very readily broken down unless the muscle is worked very hard, as for instance in long-distance running. Liver glycogen is much more labile. Its concentration can vary from 5%, or on special diets even 10%, of the total liver weight immediately after a meal, down to 0.1% after 24 hours fasting.

In digestion, both starch and glycogen are broken down by hydrolytic amylases some of which attack internal glycosidic bonds. In cells, on the other hand, glycogen is broken down by a *phosphorylase* which catalyses the reaction:

$$\text{Glycogen} + n\,\text{H}_3\text{PO}_4 \rightleftharpoons n\,\text{Glucose-1-}\textcircled{P}$$

Only terminal residues attacked by α-1:4 bonds are attacked, and these only if they are not too near a branch point (see below). The isomerization of the product, glucose-1-phosphate, to glucose-6-phosphate is catalysed by a *phosphoglucomutase*, and by this means glycogen is brought into the main stream of glucose metabolism

$$\text{Glycogen} + \text{H}_3\text{PO}_4$$
$$\updownarrow$$
$$\text{Glucose-1-}\textcircled{P}$$
$$\updownarrow$$
$$\text{Glucose-6-}\textcircled{P}$$
$$\downarrow$$

Further
metabolism

The reaction catalysed by phosphorylase is reversible, but the cellular concentrations of the reactants glycogen, P_i and G-1-$\textcircled{P}$ always favour breakdown. Net synthesis of glycogen cannot be forced to occur in vivo in such a freely reversible system. A distinct system of enzymes exists, coupled (ultimately) to the hydrolysis of ATP, and this favours the basic reaction in the synthesis of glycogen, i.e. the extension of the α-1:4-linked chain. A very similar system is found in plants for the synthesis of starch.

The glycogen-synthesizing enzymes catalyse the following reactions:

$$\text{Glucose-1-}\textcircled{P} + \text{UTP} \rightarrow \text{UDPG} + \textcircled{P}\text{-}\textcircled{P} \qquad (1)$$

UDPG is the coenzyme *uridine diphosphate-glucose* (formula on p. 83).

$$\underset{\text{glycogen}}{\text{UDPG} + [n\,\text{Glucose}]} \rightarrow \underset{\text{glycogen}}{\text{UDP} + [(n+1)\,\text{Glucose}]} \quad (2)$$

$$\text{UDP} + \text{ATP} \rightleftharpoons \text{UTP} + \text{ADP} \qquad (3)$$

Reactions (1) and (2) are reversible, but the system as a whole is strongly exergonic because pyrophosphate, a product of reaction (1), is promptly hydrolysed to inorganic phosphate by a *pyrophosphatase*. The enzyme catalysing reaction (2) is called *glycogen synthetase*.

Both phosphorylase and the UDPG-transferring enzyme can synthesize glycogen only by adding glucose residues to pre-existing glycogen molecules, or primers, and both enzymes also catalyse only the synthesis of α-1,4 bonds between the glucose residues. Glycogen proper, however, has a large number of *branch points*, at which a chain of glucose residues is attached by the reducing end to C-6 of another residue (see p. 69). This α-1,6 link must be synthesized by a special *branching enzyme* which transfers a section of an already formed chain from a C-4 of one glucose residue to a C-6 on another (the α-1,6 link is shown by $\downarrow$):

The optimum length of chain for transferring is 7 glucose units long.

When breaking down glycogen, phosphorylase can only split α-1,4 links, furthermore it is unable to act between branch points so its action ends with the production of a *limit dextrin* in which short chains (usually 4 units long) of α-1,4-linked residues are left attached to the branch points. Debranching is brought about by two reactions; firstly the transference of a piece of α-1,4 chain from one side chain to another chain, leaving a single glucose residue linked α-1,6; secondly hydrolysis of the α-1,6-linked 'stump' to give a molecule of free glucose (see Fig. 6.21.) Phosphorylase can now attack the remaining α-1,4-linked chain. The two reactions are both catalysed by the same protein (*debranching enzyme*).

Lysosomes contain an *acid glucosidase* (with optimum activity at pH 4) which hydrolyses only glycogen which has found its way into the lysosomes, possibly by proximity to the endoplasmic reticulum when the lysosomal vesicle was being formed. It will hydrolyse both terminal α-1,4 and α-1,6 bonds, and is thus capable of hydrolysing glycogen completely to free glucose. It is not important in normal glycogen degradation. The

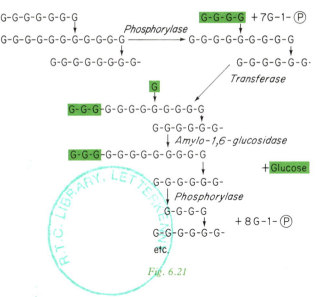

Fig. 6.21

enzyme was discovered as a result of studies on glycogen storage diseases, and is described in more detail in the next section.

Glycogen metabolism

The details of the cascade of phosphorylation/dephosphorylation reactions that control the rates of glycogen breakdown and synthesis are discussed in chapter 13. Some general comments are given below.

1. The control processes are much better worked out for phosphorylase than for glycogen synthetase. It is generally accepted (a) that G-6-P is not a physiologically important activator of the synthetase (D-form); and (b) that the sequential action of protein phosphatase, causing inactivation of phosphorylase *a* and subsequently activation of glycogen synthetase, that follows a rise in intracellular glucose concentration, is an important control mechanism in liver. It cannot be a control mechanism in muscle, where the intracellular glucose concentration is always very low. The fact that no less than four protein kinases are known to phosphorylate the I form of the synthetase suggests that the metabolic trigger of primary importance for the latter enzyme has still not been discovered.

2. Changes in $[Ca^{2+}]$, independent of hormone binding to receptors, are probably more important for glycogen metabolism in muscle than in liver. This is as may be expected, in view of the importance of Ca^{2+} in muscle contraction.

3. Insulin stimulates glycogen synthesis markedly both in liver and muscle but the mechanisms are still unknown.

Glycogen storage diseases

These are inborn errors of metabolism in which one of the enzymes of glycogen metabolism is missing or inactive.

Table 6.3 shows some of the most common glycogenoses.

Diseases of glycogen metabolism are not common; there are a number of sub-types not shown in the table which are very rare indeed. Nevertheless, investigation of these diseases has advanced and modified our ideas about the importance of glucose metabolism and its non-hormonal controls very considerably. Inability to metabolize glycogen is not fatal in itself, although there may be physiological handicaps, particularly those due to organ enlargement. The most harmful of these diseases is type II glycogenosis; in the most severe form death usually occurs before 1 year. For some reason the heart is most susceptible to acid glycosidase deficiency, and cardiac lysosomes become strikingly enlarged. On the other hand, in type V glycogenosis (McArdle's disease) it is established that muscle glycogen cannot break down during contraction and in adults there is

Table 6.3   Glycogenoses

Cori type	Organs affected	Hypoglycaemia	Glycogen structure	Glycogen content %	Enzyme defect
I	Liver, kidney	+	Normal	5–15 (L)	G-6-P-ase
II	Generalized	–	Normal	5–15 (L) 5–15 (M)	Lysosomal acid glucosidase
III	Liver, muscle	+	Abnormal, excessive branching	10–20 (L) 2–6 (M)	Debranching enzyme
IV	Generalized	–	Abnormal, very little branching	3 (L)	Branching enzyme
V	Muscle	–	Normal	2–5 (M)	Phosphorylase
VI	Liver	+	Normal	5–20 (L)	Phosphorylase 50% of normal
VII	Muscle	–	Normal		Phosphofructokinase
VIII	Liver, muscle	+ +	Normal	0.5 (L)	Glycogen synthetase
IX	Liver		Normal		Phosphorylase kinase reduced
X	Liver, muscle		Normal		cAMP-sensitive protein kinase

(L) = liver; (M) = muscle.

pain and weakness in the muscles to an extent depending on the severity of the exercise. However, up to the age of puberty this distress is absent. In fact some children later found to have type V glycogenosis have been satisfactory members of football teams.

Especially in type I disease, the blood lactate and pyruvate levels may be very high, particularly when adrenaline secretion is likely to have occurred. In the hypoglycaemic states, which may occur with some forms of the disease, ketosis is common.

Type VII glycogenosis is not strictly a defect of glycogen metabolism, since all forms of glucose breakdown, at least in muscle, are prevented. The implications of this are discussed at the end of this chapter. The clinical picture is very similar to that of type V glycogenosis: marked weakness and stiffness after vigorous or prolonged exertion, but no other symptoms.

Uridine diphosphate glucose

UDPG is an important substrate. It can be transformed into UDP-Galactose (p. 83), and can also be dehydrogenated by $2NAD^+$ to give UDP-Glucuronic acid, important in mucopolysaccharide synthesis and in detoxication (chapter 9). Thus we have:

G-6-(P) $\rightleftharpoons$ G-1-(P) $\rightarrow$ UDPG $\rightarrow$ Glycogen

UDP-Glucuronate   UDP-Galactose

Glucuronides   Chondroitin   Lactose

The metabolism of other sugars

Pentoses

Ribose derivatives, i.e. the nucleotides and nucleic acids, are present in all cells. They do not usually originate from ingested ribose, which is a variable constituent of foodstuffs and is rather slowly phosphorylated in vivo, but from glucose by way of the pentose phosphate pathway, or by synthesis from smaller molecules. The deoxyribose of DNA is formed by the reduction of ribose already in nucleosides or nucleotides (p. 56).

Free ribose, and also xylose and arabinose, are found in small quantities in blood from time to time. They come from the intestine during the digestion of plant polysaccharides, particularly those of fruit (cf. p. 65). They are rather slowly metabolized, since pentosuria is quite common in people who eat a great deal of fruit. *Essential pentosuria* is a harmless inborn error of metabolism in which about 1 g of xylulose is found each day in the urine. Its stereochemical configuration

indicates that it cannot have originated in the pentose phosphate pathway; it is probably excreted as the result of an enzymic defect in the metabolism of glucuronic acid.

Fructose

Considerable quantities of this sugar are ingested each day, chiefly combined in sucrose. Fructose has frequently been used as a substitute for glucose in conditions such as diabetes in which glucose utilization is impaired, because it is known that fructose metabolism is not subject to the same controls as that of glucose. This may be unwise, because the lack of control can be potentially dangerous.

Hexokinase is capable of phosphorylating fructose, but glucose is such a powerful competitive substrate that little if any F-6-P is ever formed, except in adipocytes. Moreover, glucokinase, which is the most important initiator of carbohydrate metabolism in liver, does not attack fructose. The sugar is metabolized by the following sequence of reactions:

$$\text{Fructose} + \text{ATP} \rightarrow \text{F-1-} \textcircled{P} + \text{ADP} \tag{1}$$

$$\text{F-1-} \textcircled{P} \rightleftharpoons \text{Glyceraldehyde} + \text{DHAP} \tag{2}$$

Glyceraldehyde can be either oxidized to glyceric acid or reduced to glycerol, and in either case the product can be phosphorylated, and so introduced into the main stream of carbohydrate metabolism. The fact that fructose phosphorylation is uncontrolled means that unpredictable quantities of intermediates can enter the Embden–Meyerhof pathway, leading, for example to lactataemia (see p. 76). If the activity of fructose-1-phosphate aldolase, which catalyses reaction (2), is low, it can happen that sufficient F-1-P accumulates in liver to remove most of the intracellular inorganic phosphate from the cytosol. This raises the phosphate potential (p. 186) and hinders oxidative phosphorylation. This is the biochemical lesion in *hereditary fructose intolerance*, where, in addition, the activity of the normal aldolase is reduced by 80%. The functions of kidney and of the small intestine are also severely affected, which suggests that this pathway of fructose metabolism is widely distributed in the body.

Fructosuria is a condition in which a small part of any fructose eaten is found in the urine. In fetal blood, and in seminal fluid, fructose is normally found as well as glucose.

Galactose

This sugar is found in milk-containing diets. For the inability to hydrolyse lactose in the gut, see p. 70. Galactose is rather slowly tranformed into glucose in the liver. It is made from glucose in large quantities in actively secreting mammary glands, and the blood and urine of pregnant and lactating women often contain both galactose and lactose.

Galactosuria or galactosaemia is an inborn error of metabolism which can have fatal results in infants if they are not quickly put on a lactose-free diet. For reasons largely unknown, galactose in excessive quantities is toxic. This is a problem that does not arise in older children or adults but can be very serious in infants. The symptoms are failure to gain weight, lethargy, vomiting, liver enlargement, and often jaundice, and at a later stage, cataract and mental defect. Some of these symptoms have been duplicated in adult rats by feeding diets containing 35% lactose.

The transformation of galactose into glucose has the following steps:

$$\text{Galactose} + \text{ATP} \rightarrow \text{Galactose-1-}\textcircled{P} + \text{ADP} \qquad (1)$$

$$\text{Uridine di-}\textcircled{P}\text{-glucose} + \text{Galactose-1-}\textcircled{P} \rightleftharpoons$$
$$\text{Uridine di-}\textcircled{P}\text{-galactose} + \text{Glucose-1-}\textcircled{P} \qquad (2)$$

$$\text{Uridine di-}\textcircled{P}\text{-galactose} \rightleftharpoons \text{Uridine di-}\textcircled{P}\text{-glucose} \qquad (3)$$

Reaction (3) is the isomerization (epimerization) of the —OH on C-4 of the galactose molecule. The uridine diphosphate glucose so formed can then react with more galactose-1-phosphate.

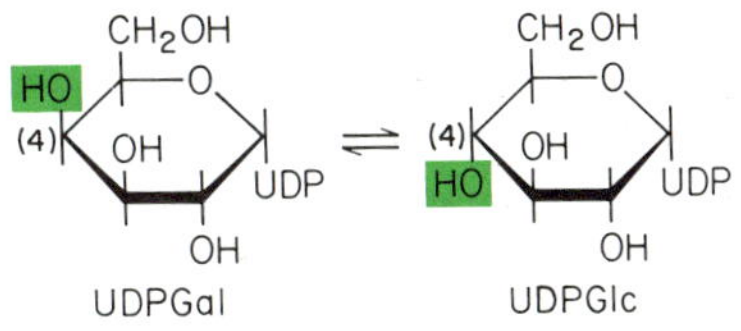

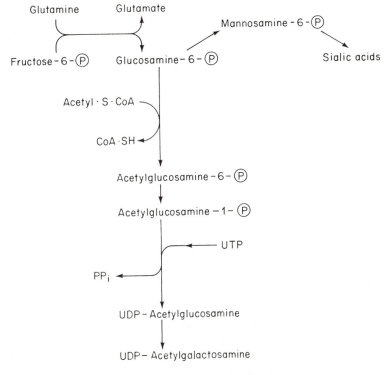

Fig. 6.22.

It has been established that in galactosurics the enzyme for reaction (2) is missing, so that there arises an accumulation of galactose-1-phosphate, and eventually of galactose, in the cells, with toxic consequences. The galactosuria can be diagnosed soon after birth and the infant placed on a lactose-free diet.

Lactose is synthesized in lactating breast tissue by the following steps:

$$\text{G-1-}\textcircled{P} + \text{UTP} \rightarrow \text{UDPG} + \textcircled{P}\text{-}\textcircled{P}$$

$$\text{UDPG} \rightarrow \text{UDPGal}$$

$$\text{UDPGal} + \text{G-1-}\textcircled{P} \rightarrow \text{Lactose-1-}\textcircled{P} + \text{UDP}$$
$$\qquad (\textit{or glucose}) \qquad (\textit{or lactose})$$

Amino sugars

The two most important members of this class are 2-amino-2-deoxyglucose and 2-amino-2-deoxygalac-

Fig. 6.23 The synthesis of amino sugars.

tose which are found (usually with the amino group acetylated) in the mucopolysaccharides. They are synthesized by the pathways shown in Fig. 6.23. For the structures of the major heteropolysaccharides, see pp. 65–68. The regular alternation of amino sugar and glucuronic acid residues in many of these polysaccharides requires sophisticated enzymic mechanisms, which will not be described here.

Glucosamine   Galactosamine

A General Consideration of Carbohydrate Metabolism

This is the first confrontation, in this book, with the dynamic aspect of biochemistry, and the first lesson to learn is the immense flexibility of the metabolic responses to the body's demands, whether they are constant or varying. The quantity of carbohydrate metabolized in a day may be 90% of the total energy needs, as in some of the diets that produce kwashiorkor in children (chapter 16). Here, apart from the essential fatty acids and amino acids, carbohydrate is the ultimate source of all the body's organic carbons, since in principle all the non-essential amino acids could be manufactured from carbohydrate, together with a single source of amino groups—glutamate, say. This extreme is more familiar in agriculture, in the use of cheap carbohydrate for fattening farm animals.

At the other extreme is the inborn absence of phosphofructokinase from liver and muscle (Type VII glycogenosis, p. 82). In this defect, the Embden–Meyerhof pathway is completely out of action in muscle, and since this tissue has very little of the pentose phosphate pathway enzymes, pyruvate formation must be nil at all times. It is true that in liver, some carbohydrate could be degraded by the hexose shunt, and gluconeogenesis should be unaffected. Yet the only physical effects of the absence of the enzyme are a certain muscular stiffness and weakness after exercise. There is not even hyperglycaemia after meals.

Mention of weakness after exercise does suggest, however, that carbohydrate is needed by muscles as a fuel for exercise, and this is generally true although Eskimos, living on a diet containing practically no carbohydrate, are capable of hard physical effort. Such a diet is only acceptable to others after adaptation to it.

In man, although not in all animals, the blood sugar level must always be maintained at a constant level, whether or not carbohydrate is immediately available from food. This is primarily because, in normal circumstances, the brain uses only carbohydrate as a fuel. Two separate processes in liver maintain the blood sugar level—*glycogenolysis* and *gluconeogenesis*.

Glycogenolysis, the breakdown of liver glycogen to free glucose, is the first process to come into play. It is under hormonal control by adrenaline and glucagon. The contribution which it can make to the body's glucose needs must depend on the previous dietary history, but it is usual to find that glycogen has almost completely disappeared from the liver after a 24-hour fast. Muscle glycogen does not usually disappear, however, even after a long fast and it can make no direct contribution to blood glucose.

Gluconeogenesis becomes vitally important when liver glycogen is exhausted. The extent of gluconeogenesis in a prolonged fast can be estimated. Urea-N excretion is maintained at about 6 g per day, even in the absence of all food. Since proteins, on the average, contain about 16% N, this is equivalent to $6 \times 100/16$, or 37 g protein or about 40 g amino acids. It is usual to regard about half the amino acids of an average tissue protein as glucogenic. Thus of the 40 g amino acids formed from endogenous protein per day, about 15–20 g glucose can, as a maximum, be produced. The glycerol moiety, but not the fatty acids, of triglycerides can also be a source of glucose carbons. In prolonged fasting, of the order of 25 g glucose can be formed per day from glycerol, giving a total of about 40 g, rather less than 10% of the total energy requirements. In a less prolonged fast, about 55 g glucose might be formed by gluconeogenesis. This accords fairly precisely with the estimated energy requirements of the brain, leaving nothing over for the rest of the body. Red blood cells use about 35 g/day of glucose (cf. p. 76), and this is sometimes subtracted from the 55 g, suggesting that production of new glucose in fasting falls seriously short of the energy needs even of the brain. Since we are dealing here, however, with a *carbon* balance, and not an *energy* balance, we must remember that the glucose catabolized by red cells is not lost to the carbohydrate pool of the body, because the liver can re-convert the lactate to glucose as soon as it is formed, at the expense of energy from fat oxidation. It does appear, however, that 55 g/day is an upper limit to the rate of gluconeogenesis. One must reckon with the possibility that the brain can use another fuel during fasting. This point is discussed at the end of chapter 7.

There is very little gluconeogenesis in the fed state, and gluconeogenesis from protein can be prevented to a very considerable extent by the consumption of a small amount of carbohydrate, even if it is quite inadequate as an energy source. This is known as the *protein-sparing* action of carbohydrate.

Fat is a much more important storage form of carbohydrate carbons than glycogen; indeed obesity—the result of excessive carbohydrate storage as fat—is a major clinical problem today. The teleological answer to this paradox may be the superior efficency of fat as an energy store when it is compared with carbohydrate on a weight basis. In a mobile organism, such as man, this is important; for example, the fetus in utero has a large store of glycogen but very little fat; the latter begins to be synthesized only after birth.

The control of carbohydrate metabolism

It is very noticeable that chronic accumulation of intermediates of glucose catabolism is very rare. Almost the only examples are the raised blood pyruvate of vitamin B_1 deficiency (when pyruvate decarboxylation is deficient), and the raised blood pyruvate and lactate of type I glycogen storage disease, when liver glycogen can be broken down to hexose phosphates which cannot be hydrolysed to free glucose, but are catabolized to pyruvate in excess.

The occurrence of a high blood and tissue lactate in ischaemia is an exception to this generalization above. The rapid production of lactic acid in these circumstances is an example of a phenomenon of general occurrence in biology called the *Pasteur effect*, which makes possible the rapid catabolism of hexoses in anaerobic conditions by arranging that the hydrogen atoms removed to NAD^+ in the oxidation of glyceraldehyde phosphate are transferred to a suitable acceptor. In vertebrates the acceptor is pyruvate, in some insects dihydroxyacetone phosphate, and in yeasts acetaldehyde.

In most cells that have a cytochrome system, pyruvate and lactate do not accumulate in aerobic conditions, although some tumour cells have a high rate of aerobic glycolysis. This is good evidence of a very effective control mechanism. In the short term, the coupled state of activity of phosphofructokinase and pyruvate kinase, together with the supply of cofactors to triose phosphate dehydrogenase, appears chiefly to determine the rate of flux through the Embden–Meyerhof pathway, but this in itself might only cause hexose monophosphates to accumulate. The relative activation and inactivation of phosphorylase and glycogen synthetase helps to regulate the disposal of these units, but this is of minor importance in brain and kidney, which normally contain little glycogen. In liver, glucose-6-phosphate may be hydrolysed to free glucose, thus setting up a 'futile cycle'. It seems that in muscle, at least, the rate of flux through the pathway must determine the rate of glucose uptake into the tissue. It is not at present known how this control is exerted; a direct inhibition of hexokinase by its product glucose-6-phosphate seems unlikely since the latter is a rather weak inhibitor of all except brain hexokinase. It is also difficult to explain fully why glucose uptake from plasma by muscle is inhibited by ketone bodies and by free fatty acids. The suppression, by fat, of carbohydrate oxidation through the TCA cycle in the post-absorptive state, and the converse immediately after carbohydrate feeding, both suggest a control at the pyruvate dehydrogenase level which has not been completely worked out.

In the control of metabolism, there are short-term effects, in which allosteric effectors and covalent modifications to enzyme protein are important, and longer-term effects, in which synthesis of new enzyme protein is the chief factor. Since some enzymes have a very short half-life, the boundary between the two types of control (both of them influenced by hormones) is not a clear-cut one.

7

Lipid Structure and Metabolism

A. Structure

Lipid Structures

Unlike carbohydrates and proteins, lipids are not assembled from monomers of uniform basic structure; the term lipid includes any biological compound that is soluble in a 'lipid' (i.e., non-polar organic) solvent. A widely-used 'universal' lipid solvent is chloroform + methanol (2:1, v/v). Within this definition, based on a physical property, one may distinguish two major classes. The first consists of substances in which long-chain carboxylic acids ('fatty acids') are important, and includes the triacylglycerols (fats), as well as the phospho-, glyco- and sulpho-lipids. The second class is made up of substances synthesized from isoprene units, such as cholesterol and other sterols and steroids, dolichol (p. 180), ubiquinone (p. 36) and the fat-soluble vitamins A, D, E and K (chapter 17).

Naturally-occurring fatty acids mostly have even numbers of carbon atoms (see Table 7.1), and—except for those of bacterial origin—they may be highly unsaturated. Apart from the number and reactivity of the double bonds, which will be discussed later, two

physical properties dominate lipid structures: the bulk of the unit components, which is highly related to the *melting temperature* or range (see p. 176), and the hydrophobic/hydrophilic balance of the molecule. The latter can be drastically altered by chemical changes that are apparently small. For example ionization of a carboxylic acid completely alters its partition between water and an organic solvent. A triacylglycerol has no hydrophilic character, whereas a diacylglycerol, and even more so a monoacylglycerol, is a powerful emulsifying agent. Similar differences exist between normal and lyso-phospholipids.

Since these general qualities of lipid molecules are so important, considerable variation in detail, particularly of the fatty acid residues, may be neglected, so long as the importance of double bonds is borne in mind. It should be added that wherever binding to a receptor is important, as with steroid hormones, very small variations in structure can be overwhelmingly important in determining biological activity (see p. 92).

Table 7.1 Some common fatty acids

Common name	Formula	Short-hand symbol	Systematic name	Melting point (°C)
Acetic	$CH_3 \cdot COOH$			16.6
Propionic	$C_2H_5 \cdot COOH$			−22
Butyric	$C_3H_7 \cdot COOH$			−19
Caproic	$C_5H_{11} \cdot COOH$	6:0	Hexanoic	−2
Caprylic	$C_7H_{15} \cdot COOH$	8:0	Octanoic	16
Capric	$C_9H_{19} \cdot COOH$	10:0	Decanoic	31
Lauric	$C_{11}H_{23} \cdot COOH$	12:0	Dodecanoic	44
Myristic	$C_{13}H_{27} \cdot COOH$	14:0	Tetradecanoic	54
Palmitic	$C_{15}H_{31} \cdot COOH$	16:0	Hexadecanoic	63
Stearic	$C_{17}H_{35} \cdot COOH$	18:0	Octadecanoic	69
Oleic	$C_{17}H_{33} \cdot COOH$	18:1	Octadec-9-enoic	4
Linoleic	$C_{17}H_{31} \cdot COOH$	18:2	Octadeca-9,12-dienoic	−12
Linolenic	$C_{17}H_{29} \cdot COOH$	18:3	Octadeca-9,12,15-trienoic	−16 to −17
Arachidic	$C_{19}H_{39} \cdot COOH$	20:0	Eicosanoic	77
Arachidonic	$C_{19}H_{31} \cdot COOH$	20:4	Eicosa-5,8,11,14-tetraenoic	−49

Structures of fatty acids

Table 7.1 shows the structures and nomenclature of some of the more common fatty acids. Those occurring in higher animals tend to have chain lengths of 16, 18 or 20 carbon atoms, except in milk fat, where the saturated acids C_{10}–C_{14} make up 10% of the molar total. Oleic acid is the most abundant fatty acid in humans, both in depot fat and in milk (see Table 7.2).

Table 7.2 Component acids of fats (moles %)

	Maize oil	Ox	Human
Saturated:			
12:0 + 14:0	0	2.6	2.6
16:0 (palmitic)	13.0	33.4	24.7
18:0 (stearic)	4.0	21.4	7.7
Unsaturated:			
14:1 + 16:1	0	2.5	7.7
18:1 (oleic)	29.0	35.2	45.8
18:2 (linoleic)	54.0	3.5	10.0
18:3 (linolenic)	0	0.1	0.2
All others	0	1.3	1.3

Fig. 7.1 shows the overall shape of some unsaturated fatty acids (the most likely conformation of a *saturated* carbon chain is linear). It is evident that even one double bond gives the molecule a more irregular shape. This fact has an important bearing both on the structures of phospholipids (p. 90), and on the way in which different kinds of lipids, and proteins, are packed together in membranes (chapter 10).

The anions of the long-chain fatty acids are not completely insoluble—maximum solubility may be taken as 0.5 mM at 35°C—but they may be dispersed in much higher concentration than this in aqueous solution, because they aggregate to form *micelles*. These are small droplets, typically 5–10 nm in diameter, having the ionized carboxyl groups dispersed at the circumference, and the non-polar tails packed into the interior. A dispersion of soap in water, in the micellar form, may exceed 1 M in concentration. The micelles are prevented from coalescing into larger droplets by a shell of cations around each one. *Mixed micelles*, containing more powerful emulsifiers such as mono-acylglycerols and bile salts, are very important in the digestion and absorption of lipids (see p. 94).

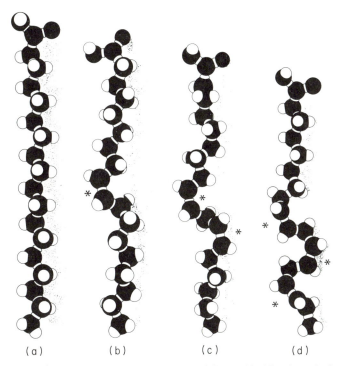

(a)	(b)	(c)	(d)

Fig. 7.1 The shapes of the long-chain saturated and unsaturated fatty acids. Note how the hydrocarbon tails become both shorter and less linear as the degree of unsaturation increases. This difference is an important factor in the 'fluidity' of membrane bilipid layers (see chapter 10). (a) Stearic; (b) oleic; (c) linoleic; (d) linolenic. The asterisks show the positions of the double bonds.

The chemistry of carboxylic acids

Saturated fatty acids

These are rather unreactive compounds. They are progressively less soluble in water as the chain length increases, and those with chain lengths of 10 or more are solids at room temperature (Table 7.1). In vivo, they can be oxidized (in the middle of the chain, giving oleic acid, p. 100), but this requires a complex mechanism for which O_2 is also necessary.

Reduction of the —COOH to —CH$_2$OH is possible in the laboratory, but requires a vigorous reagent. Similarly, in vivo reduction does not take place unless the carboxyl group has been activated in some way (cf. the formation of mevalonic acid, p. 113).

Unsaturated fatty acids

A double bond confers much greater reactivity on an aliphatic molecule, and the molecule grows more reactive the greater the number of double bonds. Greatest lability occurs when the bonds are *conjugated*, i.e. alternative double and single, as in vitamin A (p. 274); this is not the case with unsaturated fatty acids. Nevertheless, the poly-unsaturated acids such as arachidonic are much more reactive than oleic acid, as shown in their conversion to prostaglandins (p. 119). This reactivity includes shifting of double bonds and cyclization after attack by oxygen.

All fatty acids containing double bonds have four characteristic reactions:

1. *Reduction.* This is not important in vivo, but the hydrogenation of unsaturated plant oils (hardening) is very important commercially.

2. *Oxidation.* In biochemical terms this means direct attack by O_2. The initial product is often a peroxide:

$$-CH=CH \cdot CH_2 \cdot CH=CH-$$
$$\downarrow$$
$$-CH=CH \dot{-} CH \cdot CH=CH- + H \cdot$$
$$\downarrow O_2$$
$$-CH-CH=CH-CH=CH-$$
$$\overset{|}{O_2^{\cdot}}$$

The initial radical can then activate another acyl chain and so propagate the reaction. Note the shift of the double bond.

Subsequent products may be epoxides:

$$-CH=CH- + O \rightarrow -CH-CH-$$
$$\diagdown \diagup$$
$$O$$

which may be hydrolysed to diols with breaking of the chain. The short-chain fatty acids produced have a characteristic smell and in food fats the products are said to be rancid.

Attack by O_2 is also frequent in vivo. Since poly-unsaturated acids are, as we shall see, frequently components of membrane lipids, oxidation can lead to significant decreases in membrane function or rigidity. Mechanisms exist for keeping peroxide attack to a minimum (see particularly glutathione, p. 140). In foodstuffs, unsaturated compounds are often protected by the addition of anti-oxidants, of which the best known is tocopherol (chapter 17). It is not known whether tocopherol performs precisely this function in vivo.

3. *Addition.* Apart from an atom of oxygen, the only compound of importance to be added across a double bond in vivo is H_2O. This is *hydration*:

$$-CH=CH- + H_2O \longrightarrow -CH-CH_2-$$
$$\overset{|}{OH}$$

Cf. β-oxidation, p. 97.

4. *Isomerization.* By this term is meant either the migration of a double bond within a molecule (an example of which was given in para. (2) above), or interconversion between the *cis* and *trans* forms of a compound which contains a double bond. This latter use of the word is more frequent, and as *cis–trans* isomerism is an important topic in biochemistry, it is discussed separately below.

Cis–trans isomerism

An asymmetric arrangement of groups around a single tetrahedral C atom can lead to optical isomerism, discussed in chapter 6. If two C atoms share two bonds the arrangement of two or more groups on the remaining bonds can be asymmetric, as shown in Fig. 7.2.

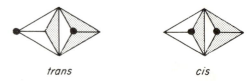

trans   *cis*

Fig. 7.2 The two possible arrangements of two identical substitutes attached one to each of two carbon atoms linked by a C=C bond.

As shown on a planar projection, the two compounds

$$\underset{\underset{cis}{X \quad X}}{-C=C-} \quad \text{and} \quad \underset{\underset{trans}{X}}{\overset{X}{-C=C-}}$$

are not spatially identical, although there is no polarization of light.

The *cis* and *trans* isomers have different physical properties and chemical reactivity. The simplest examples are maleic acid

$$
\begin{array}{c}
\text{H--C--COOH} \\
\parallel \\
\text{H--C--COOH}
\end{array}
\longrightarrow
\begin{array}{c}
\text{H--C--CO} \\
\parallel \qquad\qquad \diagdown \\
\text{H--C--CO} \diagup
\end{array}
\text{O} + \text{H}_2\text{O}
$$

which readily forms an anhydride (above), and fumaric acid,

$$
\begin{array}{c}
\text{H--C--COOH} \\
\parallel \\
\text{HOOC--C--H}
\end{array}
$$

which does not. Only fumaric acid occurs in biological systems.

In general, the unsaturated fatty acids occurring naturally have the *cis* configuration. This has particularly important consequences for membrane structure, because chains containing *cis* double bonds take up greater space than would the *trans* isomers (see Fig. 7.1). *Trans* and *cis* isomers are in equilibrium with one another, but except in conjugated poly-unsaturated systems (e.g. retinal, p. 275), interconversion is very slow at body temperature. Food processing can speed up isomerization, so that elaidic acid, the *trans* isomer of oleic acid, is found to some extent in partially hydrogenated fats.

A form of isomerism also called *cis* and *trans* can occur in fully saturated compounds, especially if there are fused rings. This is important in steroid structures and is dealt with in that section. *Cis* and *trans* isomers also occur in the inositols, and in the insecticide benzene hexachloride. Here also the different isomers have different biological activity.

Fatty acid esters

Both ester and amide bonds occur in lipids that contain fatty acids, but the most important group of such compounds are the triesters of glycerol, the triglycerides, or triacylglycerols (see Fig. 7.3). These are known as neutral fats when they are solid, and oils when liquid (*waxes* are defined as solid esters of fatty acids with alcohols other than glycerol). The most important determinant of physical properties is the degree of unsaturation of the acyl groups, because of the effect on the bulkiness, as shown in Fig. 7.1. For example, tristearin does not melt until 77°C, while triolein melts at −5°C. However, it is unusual for naturally-occurring triacylglycerols to contain three identical acyl residues. As a general rule, almost every molecule of a natural fat contains at least one unsaturated fatty acid residue.

$$
\begin{array}{c}
\text{CH}_2\text{O} \cdot \text{CO} \cdot \text{R}' \\
| \\
\text{R}'' \cdot \text{CO} \cdot \text{OCH} \\
| \\
\text{CH}_2\text{O} \cdot \text{CO} \cdot \text{R}'''
\end{array}
$$

(a)

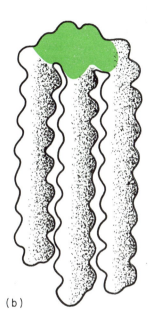

(b)

Fig. 7.3 The structure of a triacylglycerol. (a) Fischer projection formula. (b) Three-dimensional representation based on a model.

During hydrolysis, both in adipose tissue and in the intestine, diacyl- and monoacylglycerols are formed. Pancreatic lipase does not readily hydrolyse monoacylglycerols, which therefore accumulate in the intestinal fat micelles.

Derived lipids

Phospholipids

Stereospecific notation for glycerol derivatives. Glycerol itself is not optically active, but any derivative involving one of the —OH groups is bound to be so. To prevent ambiguity in nomenclature, a stereospecific notation (sn) is used. This specifies that if the central —OH points to the *left* in the Fischer projection, the top —OH is given the number 1-, and the bottom —OH the number 3- (see diagram)

(1)

$$CH_2OH$$
$$HO—C—H$$
(3) $$CH_2O\text{Ⓟ}$$

sn-3-glycerolphosphate

$$CH_2O\text{Ⓟ}$$
$$HO—C—H$$
$$CH_2OH$$

sn-1-glycerolphosphate

Naturally-occurring phospholipids are derivatives of sn-3-glycerolphosphate.

1. Phospholipids with a glycerol backbone

Quantitatively the most important of the phospholipids are those based on glycerol; they possess two acyl groups and a phosphodiester group on the third (sn-3) hydroxyl (Fig. 7.4)

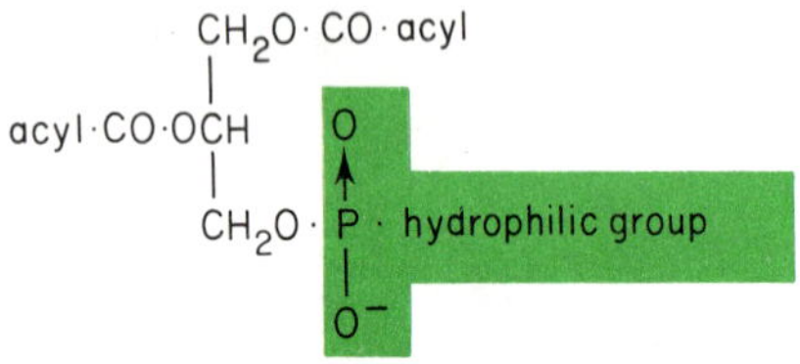

$$CH_2O \cdot CO \cdot acyl$$
$$acyl \cdot CO \cdot OCH \quad O$$
$$CH_2O \cdot P \cdot \text{hydrophilic group}$$
$$O^-$$

Fig. 7.4 Generalized diagram of phospholipid structure. The portion of the molecule shaded in green is hydrophilic, the rest hydrophobic.

The moiety esterified to the phosphoric acid is always hydrophilic, and is most commonly choline, giving *lecithin* or *phosphatidyl choline*. *Phosphatidyl ethanolamine* and *phosphatidyl serine* are also important. *Phosphatidyl inositol* contains the cyclic hexahydric alcohol inositol instead of a nitrogenous base. Two or more of the —OH groups of inositol may be phosphorylated (Fig. 7.5) giving an exceptionally high charge density to the hydrophilic moiety of the molecule.

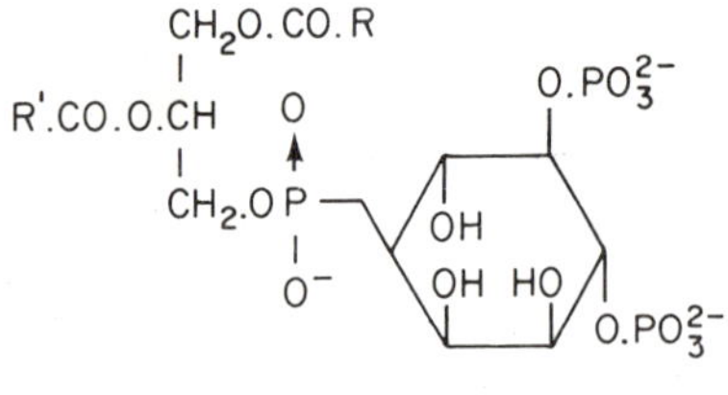

$$CH_2O \cdot CO \cdot R$$
$$R' \cdot CO \cdot O \cdot CH \quad O$$
$$CH_2 \cdot O \cdot P$$
$$O^-$$

$$O \cdot PO_3^{2-}$$
$$OH$$
$$OH \quad HO$$
$$O \cdot PO_3^{2-}$$

Phosphatidyl–(*myo*)inositol–4,5–diphosphate

Fig. 7.5 Structure of a phosphatidyl inositol. Note the high charge density on the inositol ring provided by the two ester groups. Various isomeric arrangements of the —OH groups on the inositol are possible; the one shown is the *myo*-isomer.

Plasmalogens (see Fig. 7.6) are similar in overall appearance to other phospholipids, but the hydrocarbon chain at the 1-position is linked to the glycerol backbone

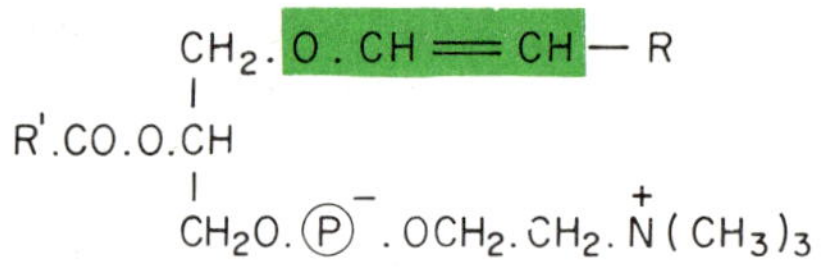

$$CH_2 \cdot \boxed{O \cdot CH = CH} — R$$
$$R' \cdot CO \cdot O \cdot CH$$
$$CH_2O \cdot \text{Ⓟ}^- \cdot OCH_2 \cdot CH_2 \cdot \overset{+}{N}(CH_3)_3$$

Fig. 7.6 Structure of a plasmalogen. Note that the grouping shaded in green is an unsaturated ether which cannot be hydrolysed by a phospholipase.

through an ether, rather than an ester, bond. There may in addition be a double bond in the α,β-position relative to the ether oxygen. Plasmalogens occur particularly in brain.

It is common for phospholipids to contain at least one poly-unsaturated fatty acid, preferentially at the 2-position. There is a good deal of exchange of acyl residues at this position, so that phospholipids are in a sense 'tailor-made' for a particular function. Another aspect of this adjustment is that the unsaturated residue may be both bulkier and shorter than a saturated one. Thus when it is incorporated into a membrane there may be a hole, into which a hydrophobic portion of a polypeptide chain may fit (cf. chapter 10).

An exception to this generalization about unsaturated fatty acyl residues is the so-called *lung surfactant*, which is secreted into the alveoli of the lungs. Its function is to lower the surface tension of the aqueous phase, and so to make it possible for the alveoli to remain filled with air. The surfactant is first secreted around the time of birth, and its absence can cause respiratory distress in the newborn. In composition it is almost entirely dipalmitoyl phosphatidyl choline, which has a higher melting point than the average phospholipid. The two saturated acyl residues enable the molecule to pack in a regular 'stiff' monolayer at the air–water interface.

A different aspect of the arrangement of saturated and unsaturated residues in phospholipids arises from the fact that phospholipases, unlike the neutral fat lipases (see p. 94), preferentially attack the 2-position. The removal of a bulky unsaturated fatty acyl residue from this position causes a marked diminution in the bulk of the molecule, in addition to increasing its hydrophilic component. Thus membranes which come to contain

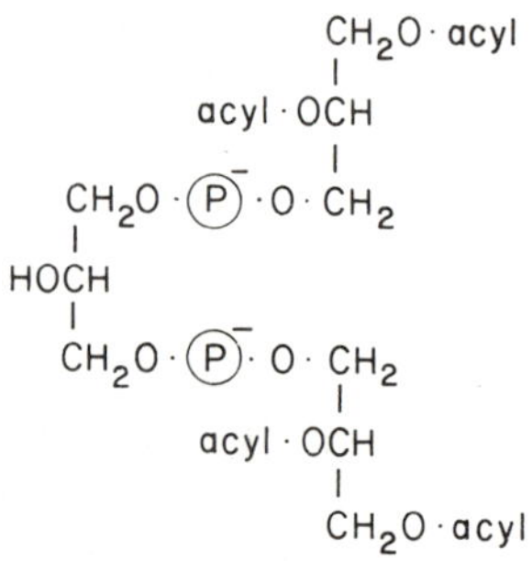

$$CH_2O \cdot acyl$$
$$acyl \cdot OCH$$
$$CH_2O \cdot \text{Ⓟ}^- \cdot O \cdot CH_2$$
$$HOCH$$
$$CH_2O \cdot \text{Ⓟ}^- \cdot O \cdot CH_2$$
$$acyl \cdot OCH$$
$$CH_2O \cdot acyl$$

Fig. 7.7 Structure of cardiolipin.

much partially-hydrolysed phospholipid often lyse spontaneously. Molecules which have a free —OH at position 2 are consequently called *lyso*-phospholipids (lysolecithin, etc.).

Cardiolipin is found especially in mitochondrial membranes. It has the structure shown in Fig. 7.7, and is metabolically rather stable.

2. Phospholipids based on sphingosine

Sphingosine (Fig. 7.8a), unlike glycerol, is lipophilic. Moreover, the acyl residue in sphingosine derivatives is attached by an amide, rather than an ester link. In dihydrosphingosine, which is of minor importance, the double bond has been reduced.

The only phosphate-containing derivatives of sphingosine are the *sphingomyelins*. They contain phosphorylcholine (Fig. 7.8b) and are found both in plasma membranes and in the complex membrane of myelinated nerves. Unlike other phospholipids, sphingomyelins contain mostly saturated fatty acids, some of them C_{20}–C_{24} in length. Sphingomyelin bilayers are thus more rigid than most others, and the myelin membrane contains relatively little protein.

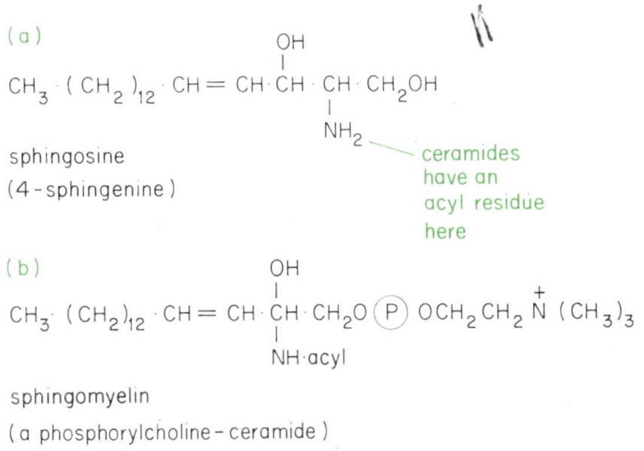

(a)

sphingosine
(4-sphingenine)

ceramides have an acyl residue here

(b)

sphingomyelin

(a phosphorylcholine – ceramide)

Fig. 7.8 Sphingolipid structures. (a) Sphingosine (*N*-acyl sphingosines are *ceramides*). (b) Sphingomyelin—a phospholipid.

Glycolipids

Phosphatidyl inositols are sometimes classed as glycolipids, although inositol is not a carbohydrate. True glycolipids are derivatives of sphingosine, or more precisely of *N*-acyl sphingosine (*ceramide*). The attachment of the sugar residue(s) is by a hemi-acetal bond to the terminal —OH of sphingosine.

Cerebrosides contain a straight-chain oligosaccharide, usually Gal-Gal-Gal-Glc-, attached to the terminal —OH of a ceramide. *Gangliosides*, of which some 15 variants are known, can have a branched chain (see Fig. 7.9a). NAN, or *N*-acetyl neuraminic acid (sialic acid), is a 9-carbon atom condensate of pyruvate and mannosamine. The ring form is shown in Fig. 7.9b.

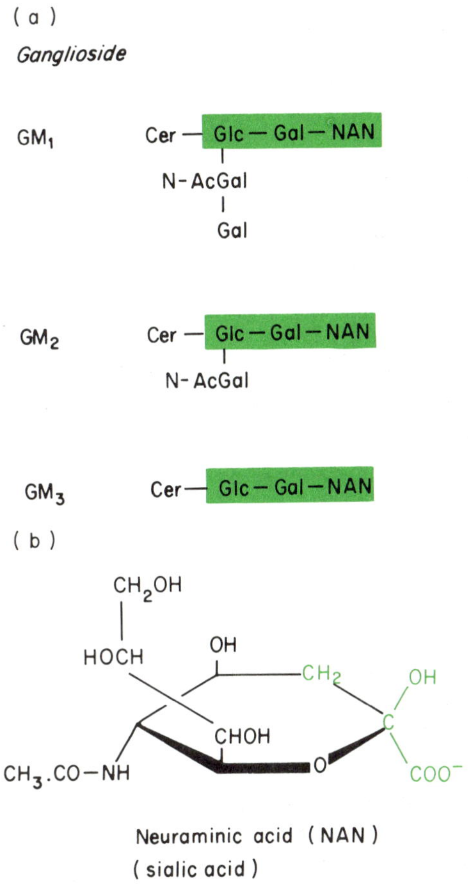

(a)

Ganglioside

(b)

Neuraminic acid (NAN)
(sialic acid)

Fig. 7.9 Glycolipid structures. (a) The relationship between the various classes of gangliosides. The portion of the molecule shaded in green—and the ceramide—is invariable. (b) The structure of *N*-acetyl-neuraminic acid (NAN). This compound is also called sialic acid.

It is an important constituent of glycoproteins as well as of glycolipids. It often seems to play a protective role, so that plasma glycoproteins which lose their sialic acid (through a *neuraminidase*) are very rapidly taken up by the liver and degraded.

Gangliosides are very extensively distributed on the external surface of plasma membranes (see chapter 10), but their physiological function is not yet precisely defined. It is known that cholera toxin must bind to a ganglioside of the GM_1 type before the toxin can interact with the adenyl cyclase receptor, and that tetanus toxin binds specifically to a different ganglioside. The cholera vibrio contains a neuraminidase that will remove all but the last sialic acid residue from GD- and GT-type gangliosides. However, no examples of binding of normal extracellular effectors to cell surface receptors, with gangliosides playing a specific role, are at present known. (For inherited diseases of ganglioside metabolism, see p. 111).

Sulphatides characteristically have the structure Cer-Gal-SO$_3^-$.

Steroids

These compounds, whose general shape and ring numbering system is shown in Fig. 7.10, have a completely reduced (non-aromatic) fused ring system (although one or more isolated double bonds may exist). The four rings are labelled A, B, C, and D as shown in Fig. 7.10(b). Because the rings are fully reduced, two configurations are possible at each of the linking carbon–carbon bonds (5, 10), (8, 9) and (13, 14). Normally the conformation is *trans* between rings B and C, and C and D, but may be *cis* or *trans* between A and B, as shown in Fig. 7.10(b). Any substituent that replaces an H atom elsewhere in the ring system can also have one of two configurations, namely axial or equatorial. In the conventional nomenclature, this is related to the configuration of the two methyl groups at C-18 and C-19. If the substituent is on the same side of the ring system as these two carbons (the 'front' of Fig. 7.10(a)), the orientation is β; if the substituent projects to the 'back' it is α. Such a difference can be very important; for example, oestradiol 17β (Fig. 7.11) has biological activity, whereas the 17α isomer is inactive.

It may be assumed that there is always an oxygen function at C-3. If this is —OH, and if the side chain, R, at C-17 has 8 or more carbon atoms, the compound is called a *sterol*, otherwise it is a *steroid*. Five groups of sterols and steroids are of importance. They are shown in Fig. 7.11.

A: *Cholesterol and its derivatives.* The latter include vitamin D and its relatives (see chapter 17). Cholesterol is the only compound in Fig. 7.11 that fits the description of a sterol. Note that the molecule contains one double bond ($\Delta6, 7$); this removes the possibility of *cis*–*trans* isomerism between rings A and B. The effect is to make the molecule both flatter and more rigid than it would otherwise be. When cholesterol is incorporated into membranes their fluidity is reduced (see chapter 10).

B: *Bile acids.* These are both metabolites of cholesterol that can be excreted (in the bile), and also important emulsifiers for lipid absorption in the small intestine. They have a side chain of 5 carbon atoms ending in a —COOH, which is usually conjugated with glycine or taurine, to form glycocholic or taurocholic acid, respectively. The most common unconjugated bile acids are *lithocholic*, with one —OH group at C-3; *deoxycholic* (—OH at C-3 and C-12); and *cholic acid* itself (3,7,12-trihydroxy-), but see also Table 7.6.

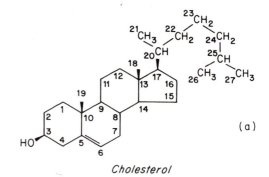

(a)

Cholesterol

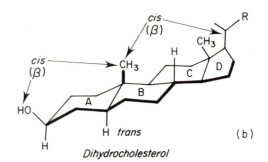

(b)

Dihydrocholesterol

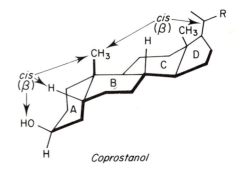

Coprostanol

Fig. 7.10 The structure of cholesterol. (a) Conventional planar representation. In this convention, groups which project above the plane of the paper (i.e. have a β-configuration) are joined to the nucleus by a heavy line. The reference methyl (C-19, joined to C-10) always projects upwards. Groups which project below the plane of the paper (α-configuration) are joined by a dotted line (as in Fig. 7.11). (b) Three-dimensional representation of steroid molecules, showing the puckered shape of the rings and the *cis*- or *trans*-junctions between them. In *dihydrocholesterol*, all the junctions between the rings are *trans*: note how the —H atom at C-5 projects 'downwards', while the —CH$_3$ at C-10 projects 'upwards'. Cholesterol itself is rather like this, but slightly altered by the presence of a double bond between C-5 and C-6. In *coprostanol* the junction between rings A and B is *cis* (the —H at C-5 projects on the same side of the molecule as the —CH$_3$ at C-10). The bile acids have this configuration.

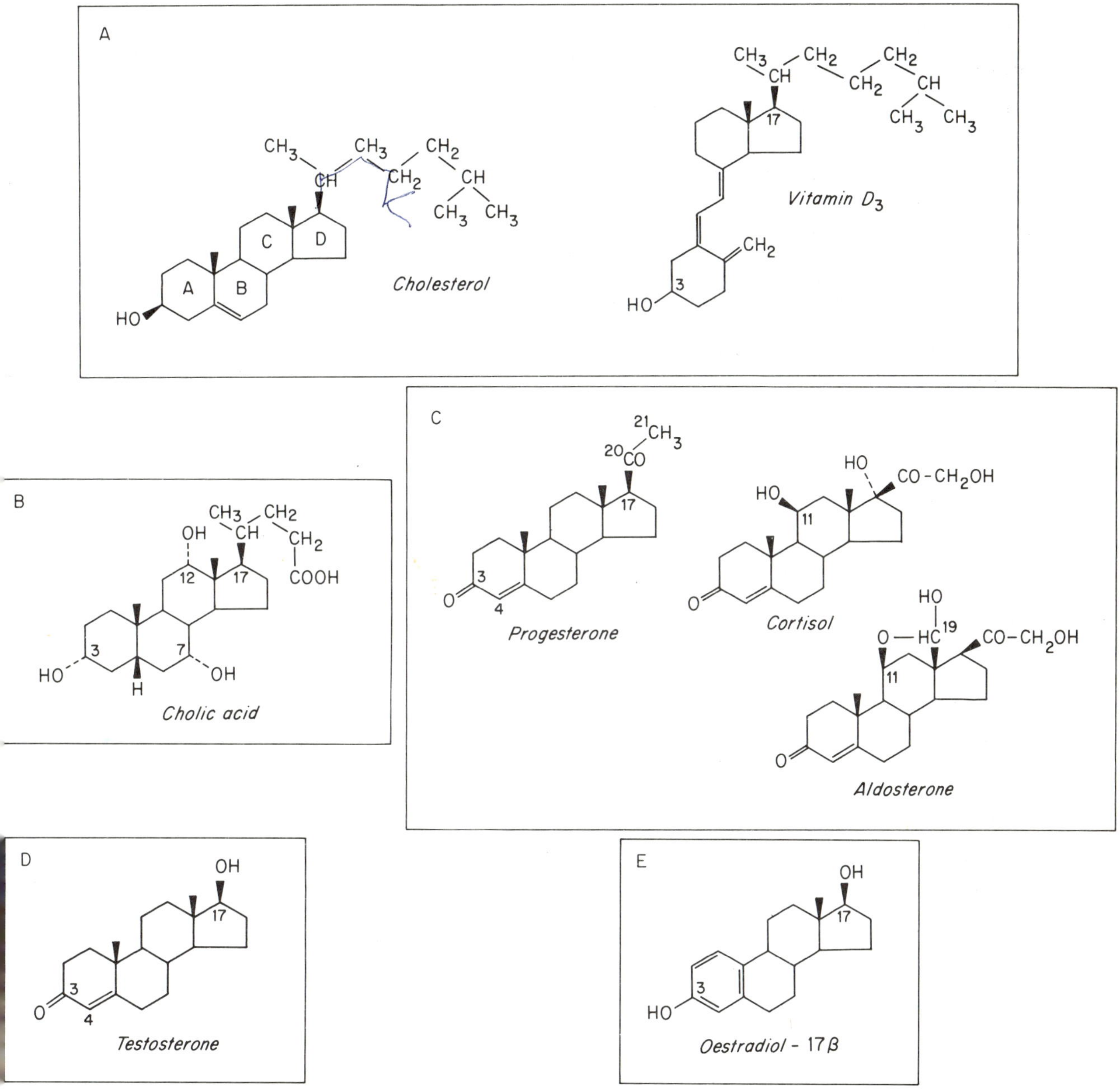

Fig. 7.11 Structures of some important derivatives of cholesterol. For details see text on pp. 92–94.

C: *Progesterone and the adrenal cortical steroids.* These are often called C_{21} steroids, because the side chain has been reduced to two carbon atoms in length. In *progesterone* the side chain has the structure —$CH \cdot CH_3$, and in the adrenal steroids, derived from cholesterol via progesterone, the side chain is —$CO \cdot CH_2OH$. The latter structure is quite readily removed from ring D in vitro by mild oxidizing agents (e.g. bismuthate), leaving an oxo function at C-17, so that the adrenal steroids are sometimes called the *17-oxogenic steroids.* The conjugation shown by the oxo group at C-3 and the double bond in ring A, i.e. the alternation of double and single bonds, is a characteristic feature of this group of steroids.

In *aldosterone*, the methyl group at C-19 has been oxidized to an aldehyde.

D: *Androgens*. The male sex hormone *testosterone* and related compounds have no side chain at all at C-17. Urinary metabolites of the androgenic hormones usually have an oxo group at C-17. This can also arise from oxidation in vivo of the side chain of the adrenal cortical steroids in the way described above. Thus the urinary *17-oxosteroids* are found in the urine of both males and females. About 50% of the total oxosteroid in adult male urine is of adrenal origin; in women and children almost all comes from the adrenal.

E: *Oestrogens*. The female sex hormones are characterized by the fact that ring A is aromatic. In order to introduce 3 double bonds into ring A it is necessary, as a preliminary, that the methyl group at C-19 should have been oxidized away completely. As the —OH group at C-3 is attached to an aromatic ring, it is phenolic, and the oestrogens, unlike all other steroids, can be extracted from solution in organic solvents by aqueous alkali.

Steroid esters

Only esters at the C-3 —OH group need concern us.

(a) *Cholesterol esters*. About two thirds of the cholesterol in plasma is esterified with long-chain fatty acyl residues, which are characteristically poly-unsaturated. High-density lipoproteins (HDL) are particularly rich in cholesterol and its esters (cf. Fig. 7.29).

Cholesterol esters are much more non-polar than free cholesterol, and are found in the neutral lipid core of lipoprotein particles, whereas the free sterol occurs in the outer shell. For a similar reason, the cholesterol that is found in many membranes is not esterified (for a discussion of the role of cholesterol in membrane structure, see chapter 10).

(b) *Steroid conjugates*. Many derivatives of steroid hormones are esterified in the liver and excreted in bile. Esterification with suitable small anions makes the steroids more water-soluble. Esterification with sulphate, $—OSO_3^-$, is quite common. The glucuronide conjugates of steroids are often put in the same category, but strictly speaking they are usually acetals rather than esters, i.e. the linkage is through C-1 of the sugar residue. True glucuronate esters, through the carboxyl at C-6, are possible however.

B. Metabolism

Digestion and Absorption

Triacylglycerols are the chief form of storage of energy-providing compounds in most terrestrial vertebrates. Many plant seeds and fruits also contain high concentrations of fat, e.g. avocado, olive, soya bean, most kinds of nuts; thus fat is a normal constituent of the diet except for those whose food is restricted to root vegetables or cereals. Ordinary diets contain 40% of fat or more.

Intestinal hydrolysis

Fat is not readily absorbed unless it is liquid at body temperature. Mechanical action then converts it into an emulsion, containing droplets of perhaps 500 nm in diameter. No hydrolysis takes place until the fat reaches the duodenum, where pancreatic lipase and bile salts are secreted from separate organs, but come together in a common duct. The role of the bile salts is first to stabilize the lipase, secondly to emulsify the droplets, and thirdly to help form mixed micelles (see p. 87) of 20 nm in diameter or less. A single droplet can form 10^6 micelles, so that their concentration in the lumen is high, and they diffuse to the brush border of the cells lining the lumen, where their contents are taken up.

The pancreatic lipase, which needs a small protein called *colipase* to keep it effectively attached to the substrate droplet, hydrolyses 1- and 3-acyl groups of triacylglycerols. Thus its products are free fatty acids and 2-acylglycerols, which are themselves powerful emulsifying agents. Micelles are consequently composed, in varying proportions, of free fatty acids, mono- and diacylglycerols, perhaps some unhydrolysed triacylglycerol, and bile salt anions, together with minor components of the diet such as phospholipids (unless hydrolysed) and fat-soluble vitamins.

Failure to secrete either the lipase or the bile salts (as for example in bile duct obstruction) leads to a failure in fat absorption and consequent *steatorrhoea*. The failure to absorb is only partial if bile salts are absent, since other emulsifiers, e.g. monoglycerides, are present. It is probable that bile salts are most important for the initial stages of lipase action.

Absorption of neutral fat

It seems likely that the components of the micelles are absorbed through the brush border separately; in particular, bile salt absorption is concentrated in the ileum, and the micelles in the lumen become richer in bile salts the further down the duodenum they travel. The various micellar components can in any case be treated separately from a metabolic point of view. In the mucosal cells the most important event is the resynthesis of triacylglycerols.

$$\text{Fatty acid} + \text{CoA} + \text{ATP} \rightarrow \text{Acyl-CoA} + \text{AMP} + \text{P-P}_i \tag{1}$$

$$\text{Acyl-CoA} + \text{MAG} \rightarrow \text{DAG (diacylglycerol)} + \text{CoA} \tag{2}$$

$$\text{DAG} + \text{Acyl-CoA} \rightarrow \text{TAG} + \text{CoA} \tag{3}$$

Diacylglycerols can also be used for the synthesis of phospholipids (see p. 108). Much phospholipid synthesis takes place in the mucosal cells of the small intestine, but this happens also during fasting (see VLDL, below). The intestine is therefore also an active site of fatty acid synthesis *de novo*. One consequence of this is that the TAG leaving the mucosa are not identical in composition to the food fats. Medium chain length fatty acids, e.g. those arising from milk products, pass without esterification straight into the portal blood stream.

The pancreatic juice contains several phospholipases and a phosphodiesterase, so that dietary phospholipids can in principle be hydrolysed completely to their component groups before absorption. It seems to be accepted, however, that a good deal at least of lyso-phospholipid (p. 91) is absorbed as such.

Intestinal mucosa secretes into the lymph two types of lipid particles, *chylomicrons* and *very low density lipoproteins* (VLDL). The latter are discussed more fully on p. 116. Chylomicrons, which enter plasma through the thoracic lymph duct, have a very short life in plasma, although they can be present in sufficient concentration to make the blood 'milky'. The chylomicrons, and for that matter the VLDL, are the substrates for a *lipoprotein lipase* or clearing factor which is attached to the endothelium lining the capillaries, particularly those of muscle and adipose tissue, by a heparin-like polysaccharide. Heparin will, in fact, displace the enzyme from its binding site, so it then becomes measurable in plasma, but it is not thought that this post-heparin lipase activity (PHLA) is other than a convenient artefact for measuring purposes.

The lipase needs a C-II apoprotein (see p. 116) to activate it. The chylomicrons in plasma, but not in lymph, do contain C-II. The enzyme preferentially hydrolyses the outer acyl residues; the 2-monoglyceride, and the FFA, are transferred into the surrounding cells, either to be used for immediate oxidation or for synthesis into storage TAG. In either case the monoglyceride must be hydrolysed by an intracellular

esterase. A good deal of the FFA is probably absorbed on to plasma albumin (see below) and removed from the site of hydrolysis. Liver also contains an extracellular lipoprotein lipase, but this does not readily hydrolyse the TAG of chylomicrons or VLDL, although it will break down the fat of lipid emulsions.

The VLDL secreted by intestinal mucosa is treated very much like that secreted by the liver, so the metabolism of both is dealt with under Plasma Lipoproteins (p. 115 ff.).

Absorption of steroids and fat-soluble vitamins

The body is capable of making more cholesterol than it needs. As with most lipid substances, the route for excretion of cholesterol from the body (see p. 114). difficult to maintain a balance in cholesterol if it were well absorbed from the gut, and it is in fact absorbed to only a limited extent. Only about 5% of plant sterols, which are chemically very similar to cholesterol, are absorbed. Bile salts, on the other hand, are readily reabsorbed in the jejunum, although not completely. The fraction not absorbed represents a large part of the excretion of cholesterol from the body (see p. 114). Secretion and re-absorption of bile salts is an example of *entero-hepatic circulation*, which happens also with many other lipid-soluble substances.

Vitamins A and D are absorbed with fat, as is vitamin K. Carotenes are also well absorbed by man. It is probable that the absorption of these three fat-soluble vitamins as well as cholesterol and carotene is completely dependent on the fine dispersion of fat into micelles in the intestine. The absorption of all these substances is markedly reduced when micelle formation is depressed. Thus patients with obstructive jaundice, in which the concentration of bile salts in the gut lumen is decreased, frequently suffer from deficiencies of vitamins A and D and particularly of vitamin K (the latter deficiency being more important as vitamin K is stored in the body to only a very limited extent).

In summary, neutral fat (TAG) taken in food is partially hydrolysed in the gut, resynthesized to TAG in mucosal cells, and partially hydrolysed again in the capillary bed. Monoglycerides and FFA are transported into various types of cell. The role of the liver in this is relatively small, although it can take up FFA from plasma and convert them into TAG for export (see p. 117). It does not, however, take up the acylglycerols, and would not be able to hydrolyse them, since hepatic cells do not possess a cytosolic TAG lipase. The liver does, however, play a very important part in lipid metabolism, partly through its involvement in cholesterol metabolism and partly through the synthesis of various classes of lipoprotein. Discussion of these topics will be postponed until we have considered the mobilization and oxidation of the major lipid substrate for energy production—free long-chain fatty acids.

The Metabolism of Fatty Acids

Fatty acid oxidation

Mobilization of fatty acids from adipose tissue

Adipose tissue cells contain separate lipases for triacylglycerols and diacylglycerols, and a very active enzyme for hydrolysing monoacylglycerols. Thus the end product of intracellular lipolytic action in adipocytes is free glycerol as well as fatty acids, in distinction to the extracellular events in absorption of fat. The glycerol is not re-utilized within the tissue and diffuses out. Figure 7.12 shows that the quantity of FFA entering the bloodstream and becoming bound to plasma albumin is the resultant of a dynamic balance between hydrolysis and re-esterification of DAG and MAG, and that this balance can be altered by several factors, but particularly by changes in activity of the hormone-sensitive lipase. This enzyme is regulated through the release of cAMP from adenyl cyclase, in conjunction with a membrane receptor site for the hormone. For details see chapter 10. Although adrenaline and noradrenaline are usually thought of as short-acting stimulants of lipolysis, there is evidence that steady release of catecholamine from sympathetic nerve endings in or near fat deposits helps to regulate the release of FFA in fasting. Other hormones only react with the adenyl cyclase receptors at concentrations too high to be physiological. The function of growth hormone is probably to stimulate synthesis of the lipase itself.

Free fatty acids in plasma

The concentration of long-chain non-esterified fatty acids (FFA or NEFA) in plasma is rather less than 1 μmole/ml—about 20–30 mg/100 ml. They come almost entirely from adipose tissue, except during the chylomicron phase of fat absorption, and are bound to albumin. The association constant for at least the first two molecules of fatty acid bound to each molecule of albumin is very large indeed, and there is no satisfactory explanation at present of how FFA dissociate from the protein as they enter the cells in which they are metabolized. Although the concentration of FFA is small, the turnover rate is high (half-life 2–3 minutes) and it is calculated that their rate of entry into and disappearance from plasma would be sufficient to account for all the fat oxidized by a fasting animal. They are the immediate substrate for fat oxidation in all organs (brain excepted).

Intracellular transport

In animal cells, the fatty acid oxidizing system is in the mitochondrial matrix. Moreover, the substrate for the oxidation cycle is not the free acid, but as shown below, its coenzyme A ester. There are two ways of introducing this into the matrix. The first, which is not thought to be normally of great importance, is diffusion of the free acid, followed by the activation reaction:

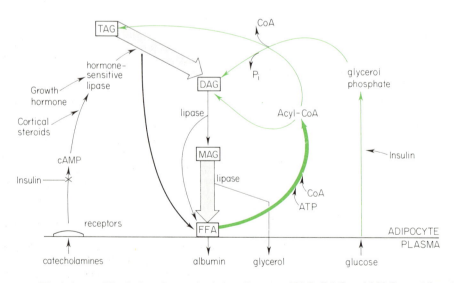

Fig. 7.12 The balance of lipolysis and resynthesis in adipocytes. TAG, DAG and MAG stand for tri-, di- and monoacylglycerol, respectively.

$$R \cdot COOH + HS \cdot CoA + GTP \rightarrow R \cdot CO \cdot S \cdot CoA \\ + GMP + P\text{-}P_i \quad (1)$$

GTP can be re-formed by means of an enzyme found only in mitochondria, which catalyses the reaction

$$GMP + ATP \rightleftharpoons GDP + ADP \quad (2)$$

The second mechanism involves synthesis of the acyl-CoA thioester in the cytosol, i.e. reaction (3) below. The acyl residue is then transferred across the mitochondrial membrane by the action of two *acyl-carnitine transferases*. Because this mechanism is very important in the control of fatty acid metabolism, it is described fully in a later section (p. 102).

The β-oxidation system

For the sake of completeness, the preliminary activation reaction in the cytosol is given here. For the structure of coenzyme A (CoA) see p. 36.

$$R \cdot COOH + HS \cdot CoA + ATP \rightleftharpoons \\ R \cdot CO \cdot S \cdot CoA + AMP + P\text{-}P_i \quad (3)$$

where $P\text{-}P_i$ represents pyrophosphate. Hydrolysis of pyrophosphate pulls the reaction to the right (cf. glycogen synthesis, p. 80).

The activated fatty acid, as a *thioester* of coenzyme A, is then subjected to the following series of reactions which result in the production of the CoA thioester of an acid with 2 carbon atoms less in the chain. This is known as *β-oxidation*.

$$R \cdot CH_2 \cdot CH_2 \cdot CH_2 \cdot CO \cdot S \cdot CoA \rightarrow \\ R \cdot CH_2 \cdot CH = CH \cdot CO \cdot S \cdot CoA + 2\,H \quad (4)$$

The hydrogen acceptor here is flavin-adenine dinucleotide (FAD), the prosthetic group of the enzyme. The double bond has the *trans* configuration.

$$R \cdot CH_2 \cdot CH = CH \cdot CO \cdot S \cdot CoA + H_2O \rightleftharpoons \\ R \cdot CH_2 \cdot \underset{\underset{OH}{|}}{CH} \cdot CH_2 \cdot CO \cdot S \cdot CoA \quad (5)$$

$$R \cdot CH_2 \cdot \underset{\underset{OH}{|}}{CH} \cdot CH_2 \cdot CO \cdot S \cdot CoA + NAD^+ \rightleftharpoons \\ R \cdot CH_2 \cdot CO \cdot CH_2 \cdot CO \cdot S \cdot CoA + NADH \quad (6)$$

This dehydrogenase is specific for the L-form of the hydroxy-acid.

$$R \cdot CH_2 \cdot CO \cdot CH_2 \cdot CO \cdot S \cdot CoA + HS \cdot CoA \rightleftharpoons \\ R \cdot CH_2 \cdot CO \cdot S \cdot CoA + CH_3 \cdot CO \cdot S \cdot CoA \quad (7)$$

This last reaction deserves particular attention for two reasons. One of the products is acetyl-CoA, which can be readily oxidized (see chapter 9), while the other product is also an acyl-CoA, which can undergo β-oxidation by reactions (4) to (7). This means that the process can be repeated again and again until the original long-chain acid has been completely split into acetyl groups. As almost all natural fatty acids have an even number of carbon atoms, the final product of reaction (7) will be two molecules of acetyl-CoA. When odd-numbered fatty acids are oxidized, the final products are acetyl-CoA and propionyl-CoA, which has to be oxidized by a special pathway (see p. 106). The normal fate of acetyl-CoA is condensation with oxaloacetic acid, with subsequent oxidation of the citric acid thus formed (chapter 9).

Fatty acid oxidation by this sequence of reactions is the most important mechanism for all saturated straight-chain acids of whatever chain length. Enzymes which catalyse ω-oxidation (i.e. the introduction of an —OH into the terminal methyl group) are known to exist, but their activities are of little importance in man. The β-oxidation enzyme complex is located in the mitochondria, and the reduced FAD of the *acyl-CoA dehydrogenase* (which catalyses (4)) is re-oxidized by an electron transferring factor (ETF) which transfers reducing equivalents to the ubiquinone region of the hydrogen transport system (chapter 9). The NADH from reaction (6) is, of course, also produced inside the mitochondria, and the β-oxidation system will only remain operative if the reduced FAD and NAD$^+$ are quickly re-oxidized. This means that the hydrogen transport system must be working: there is no equivalent, in fat metabolism, to anaerobic glycolysis of carbohydrates.

Each successive β-oxidation by this four-step cycle requires a fresh molecule of CoA·SH for the thiolytic splitting in step (7). The acetyl-CoA formed in this step must therefore be quickly metabolized, since the total concentration of coenzyme A molecules in any cell is rather small, and there can be no significant accumulation of any intermediate.

β-Oxidation is coupled with ATP regeneration; each turn of the cycle should result in the regeneration of 5 ATP molecules. The formation of 9 molecules of acetyl-CoA from a molecule of stearic acid can therefore produce 38 molecules of ATP from ADP and inorganic phosphate ($8 \times 5 - 2$ (for activation) = 38).

The oxidation of unsaturated fatty acids

As already mentioned (p. 87), oleic acid is the most commonly occurring long-chain fatty acid in the body. Although it turns over more slowly than the saturated fatty acids its catabolism is nevertheless important. β-Oxidation of $\Delta 9,10\text{-}C_{18}$ (oleic) leads in turn to

$\Delta 7,8\text{-}C_{16}$ (a)

$\Delta 5,6\text{-}C_{14}$ (b)

$\Delta 3,4\text{-}C_{12}$ (c)

Comparison with the unsaturated acyl residue formed in β-oxidation shows that the double bond in the latter is in the 2,3-position. Moreover, its configuration is *trans*, while $\Delta 3,4$-dodecenyl-CoA ((c) above) is *cis*.

An *isomerase*, present in mitochondria, has to catalyse the interconversion

$$\Delta 3,4 \; cis\text{-acyl} \rightleftharpoons \Delta 2,3 \; trans\text{-acyl}$$

The *trans* form is the more stable, but does not form in the absence of the catalyst (cf. p. 89).

A second difficulty arises in the oxidation of poly-unsaturated fatty acids, e.g. linoleic. The distal double bond arrives at the hydrating enzyme ((5) on p. 97) correctly in the $\Delta 2,3$-position, but with *cis* configuration. The hydration enzyme is not specific, but reacts on a *cis* double bond to give a D-hydroxyacyl-CoA, while the subsequent dehydrogenase ((6) on p. 97) is specific for the L-form. A second mitochondrial enzyme, a *racemase*, converts the D-hydroxy ester into the L-isomer, which can then be oxidized. The presence of these two enzymes enables all naturally-occurring unsaturated fatty acids to be completely oxidized.

A β-oxidation system not coupled to oxidative phosphorylation is found in the small organelles called *peroxisomes*, found in the cytosol notably of liver. The FAD-linked dehydrogenase transfers its 2 H directly to oxygen, forming H_2O_2. There is some dispute about the fraction of fatty acid that is completely oxidized in this way; the most recent estimates are lower ($< 10\%$ of the total) than earlier ones. Possibly the function of this system is to provide NADH in the cytosol for gluconeogenesis (cf. p. 162).

The synthesis of saturated fatty acids

The requirements for fat synthesis are acetyl-CoA units (from carbohydrate or the β-oxidation process), reducing equivalents (as reduced NADP produced in tissue oxidations, particularly in the hexose shunt, see chapter 6), energy (as ATP), and glycerol phosphate (from carbohydrate). It is evident that a plentiful supply of carbohydrate is a prerequisite for net synthesis of fat.

The β-oxidation process is perfectly reversible (except in peroxisomes), and an example appears on p. 99. However, the equilibria are not favourable to synthesis; in particular, K_{eq} for the equilibrium

$$\frac{[\text{AcAc-CoA}]\,[\text{CoA}]}{[\text{Ac-CoA}]^2}$$

(reaction (7), p. 97) is 10^{-5}. This unfavourable reaction

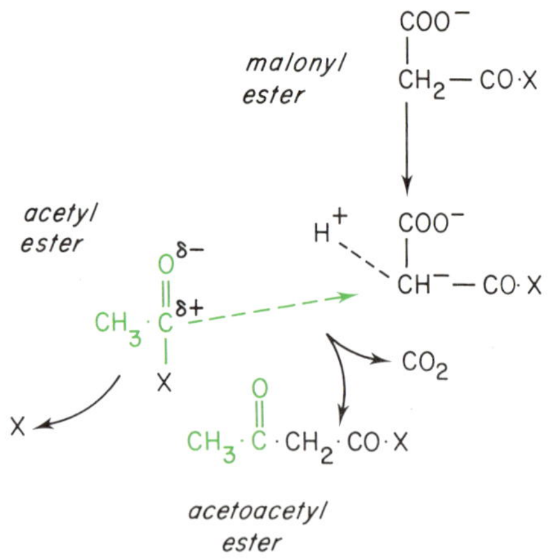

Fig. 7.13 The principle of condensation of a malonyl ester with an acetyl (or acyl) ester. The leaving group 'X' could be coenzyme A, but in the fatty acid synthase complex it is a protein-bound thiol group of the complex (see Fig. 7.15).

is avoided by condensing an acyl residue with a much more reactive group, the *malonyl* group (see Fig. 7.13).

The condensation is accompanied by splitting off CO_2, as shown above. The nature of the activating group X is deliberately unspecified, for the moment.

The equilibrium of this reaction is much more in favour of condensation, but a price has to be paid, in energetic terms. This is done by the expenditure of ATP during the synthesis of malonyl-CoA (Fig. 7.14).

The whole cycle of fatty acid synthesizing reactions—apart from the synthesis of malonyl-CoA—takes place in a *synthetase complex*. In bacteria the complex is formed by a fairly loose association of seven enzymes, together with a separable *acyl carrier protein* (ACP). This is a fairly small polypeptide, carrying phosphopantotheine, the 'specific' end of coenzyme A (formula on p. 36) which is bound to a seryl residue of the peptide.

The animal complex cannot be split so easily. Indeed it appears that there are only two proteins, each of which carries three or more enzymic sites. Examples of this are known in other areas of metabolism. In addition, a dissociable ACP cannot be extracted from the complex, but it is assumed that the acyl carrier is a covalently bound phosphopantotheine group. Certainly CoA is not directly involved. The abbreviation ACP is convenient to use.

The advantages of a complex of enzymes are that the concentration of reactive intermediates can be kept high within the complex, while at the same time reducing the possibility of unproductive side reactions with CoA esters. It is a peculiarity of fatty acid synthesis in animal cells that the synthetase is located in the cytosol, whereas the site of production of acetyl-CoA—the ultimate precursor—is in the mitochondria. This

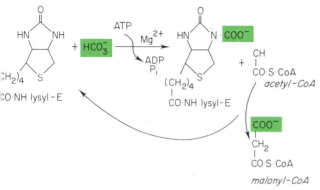

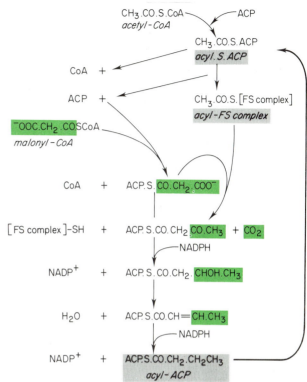

Fig. 7.14 The mechanism of acetyl-CoA carboxylase, showing the role of biotin.

Fig. 7.15 The mechanism of synthesis of saturated fatty acids. The path of the incoming malonyl residue is indicated by green shading, while the path of the lengthening acyl residue is shown by grey shading. When the latter reaches 16 C atoms (palmitoyl) it dissociates from the acyl carrier protein to give free palmitic acid.

requires a transport system, which is discussed in detail in a later section. In addition, the reducing steps require NADPH, part of which must also come (via a transport mechanism) from mitochondria (see also chapter 6, p. 78).

Acetyl-CoA carboxylase is essential to operation of the synthesis cycle. The enzyme is highly regulated, and this will be discussed later in this chapter. Here we concentrate on the mechanism (Fig. 7.14), because this is the first example of a class of reactions, the biotin-dependent carboxylases (or ligases) which occur in other areas of metabolism also. For biotin as a coenzyme, see also p. 38.

The overall mechanism of the synthetase cycle is shown in Fig. 7.15.

1. The *acyl transferase*, transferring acetyl residues to ACP (or possibly to a cysteinyl —SH), is not completely specific. Propionyl, and in some tissues even butyryl residues are accepted, to form the 'tail' carbons of the completed fatty acid. However, the overwhelming majority of 'tail' residues are acetyl, and this accounts for the fact that almost all the naturally-occurring fatty acids have an even number of C atoms (Table 7.1).

2. *Malonyl-transferase*, transferring a malonyl residue to a separate ACP, is completely specific however.

3. The characteristics of the condensing reaction have already been described (p. 98). The CO_2 which was incorporated into malonyl-CoA is inevitably lost, so that not even a radioactive label (from $^{14}CO_2$) would be found in the final fatty acid. Note that the already-formed butanoyl, hexanoyl, or other residue is transferred *to* the incoming malonyl group, whose carbons are thus always nearest the ACP.

4. Reduction, both of the 3-oxo- and of the enoyl intermediates, specifically requires NADPH, not NADH. This is discussed later (p. 102).

5. The cycle, or spiral, ceases when a fully-reduced intermediate is hydrolysed from the ACP to yield a free fatty acid (*not* a CoA ester). The hydrolysis is fairly

specific, and the chief product is palmitic acid. This happens to be the chief saturated fatty acid in human fats, but a great deal of stearic acid must nevertheless be made because stearic is the precursor of oleic acid. There are also C_{20} and longer acids present in the body.

All these are formed, chiefly from palmitic acid, and to a large extent on the endoplasmic reticulum. Palmityl-CoA is formed first, using the activation reactions shown on p. 97. One acetyl group is then added, using the malonyl-CoA sequence that has just been discussed, and the product then dissociates. The intermediates are not bound to ACP. This demonstrates the importance of a multi-enzyme complex for getting organized elongation from short-chain precursors to a long-chain product.

The condensing enzyme is not very specific, so that a variety of long-chain fatty acyl groups can be elongated by two carbons. Probably most of them, apart from palmitic, are unsaturated.

The outer mitochondrial membrane also contains an elongation system which works by reversal of β-

oxidation, and is completely reversible (for long-chain acyls). It is not at present thought to be very important.

The formation of unsaturated fatty acids

These are not synthesized *de novo*, but are formed by manipulation of long-chain saturated acyl-CoAs. The enzyme system, bound to the endoplasmic reticulum, that introduces double bonds, is often called a 'mixed function oxidase' (cf. p. 167), because the components of the system are very similar to those of true mono-oxygenases. A description of the electron-transfer sequence for this system is given in chapter 9, p. 169. However, there is no evidence that any intermediate in the reaction ever contains an atom of oxygen, and so the enzyme is more accurately called a *desaturase*. The complete mechanism is at present unknown.

While the desaturation of stearyl-CoA

$$\text{stearyl-CoA} + \text{NADH} + \text{H}^+ + \text{O}_2 \rightarrow$$
$$\text{Oleyl-CoA} + 2\text{H}_2\text{O} + \text{NAD}^+$$

occurs readily in mammalian liver, desaturation of oleyl-CoA occurs very slowly, if at all. Thus octadeca-dienoic acid (linoleic acid, see Table 7.1) is an essential component of the diet. It is estimated that an adult person requires some 10 g/day, but this is probably only a recommendation, rather than a true estimate. See also chapter 16, p. 264.

When one comes to consider the further desaturation of linoleic acid, two products are possible, as Fig. 7.16 makes clear. In one, the new double bond is on the 'tail' side of those already present, giving Δ-9,12,15-octatrienoic. This is the linolenic acid formed in plants, sometimes called α-linolenic acid. Animal desaturases cannot introduce a double bond beyond the 9,10-position, so that α-linolenic acid is entirely a plant product.

However, there is a second possibility—the double bond may be introduced on the carboxyl side of the two present in linoleic acid. This gives Δ-6,9,12-linolenic acid, sometimes called γ-linolenic acid. Animal tissues can carry out this reaction, probably rather slowly. γ-Linolenic acid is found only in traces in animal fats, but it is an intermediate in the formation of arachidonic acid (below).

Arachidonic acid is important in its own right as a

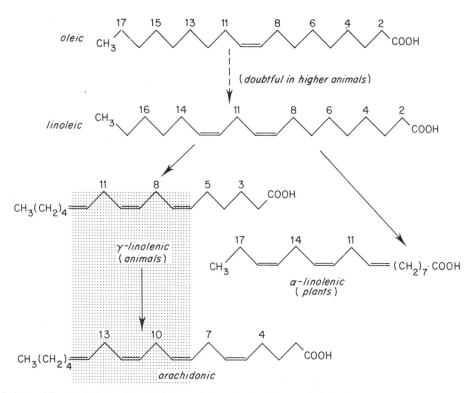

Fig. 7.16 Relationships between unsaturated fatty acids in animals. Note particularly that desaturase enzymes in animals do not operate beyond C-12, thus leaving a characteristic $\text{CH}_3(\text{CH}_2)_4$-saturated tail. Since the enzymes can introduce a double bond on the —COOH side of C-9, animal tissues can produce Δ-6,9,12 linolenic (γ-linolenic) acid from Δ-9,12 linoleic, but cannot produce the latter from Δ-9 (oleic). Δ-6,9,12 linolenic acid is the precursor of arachidonic acid in animal tissues. First the chain is lengthened to give Δ-8,11,14 eicosatrienoic acid, and then a fourth double bond is introduced between C-5 and C-6. The shaded area indicates the relationship between γ-linolenic and arachidonic acids.

component of liver and plasma phospholipids, and of cholesterol esters. It is more important as the precursor of the prostaglandins, thromboxanes and prostacyclins whose metabolism is discussed later in this chapter. Arachidonic is a C_{20} acid; the chain lengthening takes place before the final desaturation. The intermediate is Δ-8,11,14-eicosatrienoic acid, which is found in phospholipids. The desaturation of this compound takes place, as expected, on the —COOH side of the existing double bonds, to give Δ-5,8,11,14-eicosatetraenoic or arachidonic acid. Consideration will show that plant linolenic acid would not give this arrangement of double bonds, so it is the γ-linolenic acid formed in animals from dietary linoleic that is the precursor.

Many other tri-, tetra- and even penta-unsaturated fatty acids are found in traces, but hardly at all in depot fats.

Triacylglycerol synthesis

The immediate precursor for the glycerol backbone is sn-3-glycerol phosphate, never free glycerol. There are two possible sources: glycerol phosphate formed by the action of *glycerol kinase*

$$
\begin{array}{lll}
CH_2OH & & CH_2OH \\
| & & | \\
CHOH & +ATP \rightarrow HOCH & +ADP \\
| & & | \\
CH_2OH & & CH_2O\,\textcircled{P}
\end{array}
$$

This reaction is limited to liver and small intestine, because the enzyme is not found in adipocytes. Conversely, glycerol diffuses from adipose tissue during lipolysis, and is scavenged by the liver.

The second source is dihydroxyacetone phosphate, an intermediate of the Embden–Meyerhof pathway (p. 71).

$$
\begin{array}{lll}
CH_2OH & & CH_2OH \\
| & & | \\
C{=}O & +NADH+H^+ \rightarrow HOCH & +NAD^+ \\
| & & | \\
CH_2O\,\textcircled{P} & & CH_2O\,\textcircled{P}
\end{array}
$$

This enzyme, *glycerol phosphate dehydrogenase*, is active in most tissues and is probably the most important source of TAG and phospholipid glycerol.

The esterification mechanism shown in Fig. 7.17 leads via phosphatidic acid to diacylglycerols (DAG). Both compounds are precursors of phospholipids as well as triacylglycerols. In some tissues there is little phospholipid synthesis, so that the branch point shown in the figure has little significance. In liver, however, phospholipid turnover is rapid; it is not known how the relative fluxes are controlled.

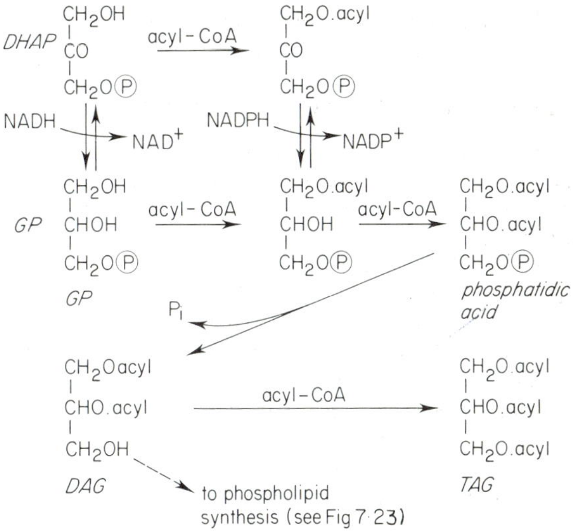

Fig. 7.17 Stages in the synthesis of triacylglycerols. Diacylglycerol (DAG) is a common intermediate in the synthesis of both TAG and phospholipids.

A relatively less important pathway starts from dihydroxyacetone phosphate, which is first esterified, then reduced:

$$
\begin{array}{lll}
CH_2OH & & CH_2O\cdot CO\cdot R \\
| & \xrightarrow{R\cdot CO\cdot S\cdot CoA} & | \\
CO & & CO & \xrightarrow{NADH} \\
| & & | \\
CH_2O\,\textcircled{P} & & CH_2O\,\textcircled{P}
\end{array}
$$

$$
\begin{array}{lll}
CH_2O\cdot CO\cdot R & & CH_2O\cdot CO\cdot R \\
| & \xrightarrow{R\cdot CO\cdot S\cdot CoA} & | \\
CHOH & & CHO\cdot CO\cdot R \\
| & & | \\
CH_2O\,\textcircled{P} & & CH_2O\,\textcircled{P}
\end{array}
$$

Although the acyl groups have been shown here as identical, triacylglycerols are mixed, and there is usually a tendency for unsaturated fatty acid residues to occupy the 2-position. The intermediates in both pathways appear to be bound to the endoplasmic reticulum, and are not found free in the cytoplasm.

In intestinal mucosa, both diacylglycerols and monoacylglycerols can be reesterified by acyl-CoA to neutral fat. 2-Monoacylglycerols are transformed to diacylglycerols faster than are the 1-monoacylglycerols.

There are three main tissues in the body in which triacylglycerol synthesis takes place: the small intestine, particularly the lower half (resynthesis of fat digested in the intestinal lumen), the liver (together with the synthesis and secretion of β-lipoproteins), and adipose tissue.

Transport processes in fatty acid metabolism

Because lipid metabolism involves both cytosol and mitochondria, transport across membranes becomes important (in contrast to carbohydrate metabolism, which is almost entirely cytosolic). This transport involves not only the acyl residues themselves, but also sources of reducing equivalents and the like. All are considered together in this section.

(a) *Transport in the degradation of fatty acids*

Liver contains some proteins that bind FFA in the cytosol; it is not known whether they are important in the translocation of fatty acids across the plasma membrane, nor whether they are present in all fat-oxidizing cells. In any case, the free fatty acid is soon converted to an acyl-CoA, if it is long-chain. Medium (C_{10}) and short-chain fatty acids are not esterified, and on the basis of experiments with isolated mitochondria, readily diffuse across the mitochondrial membrane, to be converted into CoA esters by reaction (1) (p. 97).

Long-chain acyl groups are translocated into mitochondria by a different mechanism, depicted in Fig. 7.18. This involves *carnitine*

$$(CH_3)_3N^+ \cdot CH_2 \cdot {}^{\star}CH \cdot CH_2COO^-$$
$$|$$
$$OH$$

The acyl group is transferred from CoA to the —OH on the C atom marked with an asterisk. This C atom is asymmetric, and only the L-isomer of carnitine is a substrate; acyl esters of D-carnitine competitively inhibit translocation. It is interesting that the equilibrium for the reaction

$$\text{Carnitine} + \text{acyl-CoA} \underset{}{\overset{\text{CAT}}{\rightleftharpoons}} \text{acyl-carnitine} + \text{CoA} \cdot \text{SH}$$

is approximately unity, which qualifies the ester link in acyl-carnitine as a 'high energy bond'.

There are two distinct locations for the carnitine acyl-transferases (CAT): in the cytosol (CAT 1), and on the inside of the inner mitochondrial membrane (CAT 2). In addition, there are at least three varieties of each enzyme, with specificities for long-chain acyl, medium-chain acyl, and acetyl, respectively. It is nevertheless not generally accepted at present that the latter two enzymes play any important role in metabolism.

A congenital deficiency of CAT 1 has been found which leads to difficulty in oxidizing fats (although medium-chain fatty acids can be oxidized). The serious consequences of this defect, e.g. muscular weakness on exercise, indicate how important is the carnitine–acylcarnitine antiport through the mitochondrial membrane in fatty acid degradation.

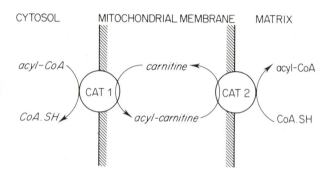

Fig. 7.18 The translocation of acyl residues through the mitochondrial membrane by means of carnitine acyl transferases 1 and 2.

(b) *Transport in fatty acid synthesis*

Here there are two problems: the supply of acetyl building blocks for the synthesis, and the supply of reducing equivalents.

Apart from the few acetyl-CoA molecules generated in the cytosol (e.g. in peroxisomes, p. 98), all the acetyl-CoA required for fat synthesis is formed within mitochondria. A little will be formed during degradation of ketogenic amino acids (chapter 9); the bulk comes from the oxidation of pyruvate (p. 73). Recycling of acetyl groups from β-oxidation would be energetically unrewarding.

A mechanism using CAT 2 and CAT 1 carnitine acetyl-transferases, in the reverse sense from that shown in Fig. 7.18, can be postulated but it is generally agreed at the present time that the mechanism involving *citrate lyase* is that which is most important.

$$
\begin{array}{ll}
\text{COOH} & \\
| & \\
\text{CH}_2 & \\
| & \text{ATP, CoA} \\
\text{HO—C—COOH} & \xrightarrow{} \\
| & \text{ADP} \\
\text{CH}_2 & +\text{P}_i \\
| & \\
\text{COOH} &
\end{array}
\qquad
\begin{array}{llll}
\text{COOH} & & \text{CH}_3 & \\
| & & | & \\
\text{CH}_2 & & \text{CO} & \\
| & + & | & \\
\text{CO} & & \text{S} & \\
| & & | & \\
\text{COOH} & & \text{CoA} &
\end{array}
$$

The integration of this reaction with the citric acid cycle is also discussed in chapter 9 (p. 162).

The antiport ion for citrate is malate, so that the oxaloacetate formed in the cleavage of citrate must be reduced to malate, by cytosolic NADH coming from glycolysis. Most of the acetyl-CoA is converted to malonyl-CoA, as described earlier (p. 98).

There is in addition a considerable demand for reducing equivalents. Each turn of the synthesis cycle requires 2 NADPH, which can be provided from various sources. It is perhaps as well to remind the reader that, although the $NADH/NAD^+$ and NADPH/

$NADP^+$ couples have almost the same E_0', they are not in equilibrium with one another, either in the cytosol or in mitochondria. In cytosol, in particular, $NADH/NAD^+ \ll 1$, while $NADPH/NADP^+ \sim 1$. The reducing pressure of this latter couple helps to drive synthesis in the direction of a long-chain acyl product.

1. 'Malic enzyme' catalyses the reaction

$$HOOC \cdot CH_2 \cdot CHOH \cdot COOH + NADP^+ \rightarrow$$
$$\text{\textit{malate}}$$
$$CH_3 \cdot CO \cdot COOH + CO_2 + NADPH + H^+$$
$$\text{\textit{pyruvate}}$$

and it is cytosolic, but the enzyme is found only in very low activity in mammalian liver. It seems unlikely that it could provide a large enough flux of NADPH.

2. The two oxidation steps of the pentose phosphate pathway (chapter 6, p. 77) provide 2 NADPH per mole of glucose metabolized. The overall reaction may be written

$$6 \text{ G-6-P} + 12 \text{ NADP}^+ \rightarrow 5 \text{ F-6-P} + 6 \text{ CO}_2$$
$$+ 12 \text{ NADPH} + 12 \text{ H}^+$$

It is difficult to estimate how fast the flux through this pathway is, but the most reliable results suggest that 30–50% of the total NADPH requirement, in liver, comes from this source. The proportion might be greater in adipose tissue. It is completely cytosolic.

3. The isocitrate/glutamate shuttle described in chapter 9 (p. 159) operates to provide cytosolic NADPH from mitochondrial isocitrate, using the cytosolic NADP-linked isocitrate dehydrogenase.

It seems likely that sources (2) and (3) are the most important, with some contribution from (1). The proportions will vary from tissue to tissue.

Overall regulation of fatty acid metabolism

A major determinant of fat metabolism is that, while glucose in excess of the body's energy needs can be converted to fat, fatty acids cannot in any circumstances be converted to glucose. Only the glycerol of TAG is available for gluconeogenesis. Inspection of the citric acid cycle (Fig. 9.1, p. 157) will show that, although 2 C atoms are incorporated into citrate with every molecule of acetyl-CoA that enters the cycle, an equivalent 2 C atoms are lost (as CO_2) by the time that succinate is formed. Thus, although radioactive tracer atoms can reach glucose from labelled fats, by way of succinate/malate/oxaloacetate, there is *no net gain* of carbohydrate from fatty acid.

A second point that needs perhaps to be stressed relates to the role of the liver. Much of what is said in this section will refer to fat metabolism in liver, but it must always be borne in mind that the liver is a major multi-functional organ in its own right. Table 7.4 (p. 107) shows that O_2 consumption by the liver accounts for about one-third of the total O_2 uptake of the body in the basal (resting) state. Liver has been described as an organ 'always on the verge of oxygen starvation', because the splanchnic circulation is not highly regulated. Liver produces most of the plasma proteins (albumin, β-lipoproteins, fibrinogen)—even in starvation; it synthesizes glucose from non-carbohydrate precursors; it is solely responsible for the detoxication of NH_3 (urea synthesis), as well as the modification and excretion of many other metabolites. In addition, mitosis in liver is rapid, and the turnover of cellular components (including the plasma membrane) is even more rapid. These are all energy-requiring processes. It may be taken for granted that the first priority of hepatocytes is to oxidize enough primary nutrient to satisfy their continuing demand for ATP. To suggest that mitochondrial oxidations may be slowed down, or speeded up, in order to regulate fatty acid metabolism is perhaps to put the cart before the horse.

Detailed discussion of endocrine function has been omitted from this book. Incidental reference to hormone action is made, but it should be borne in mind that many hormones have been implicated in the control of fat metabolism.

(a) *Adipocytes*

The activity of adipose tissue cells is quite high, per milligram of cellular protein. In addition to long-chain fatty acid synthesis *de novo* from glucose by way of acetyl-CoA, a process that is stimulated by insulin, they also carry out a futile cycle of partial hydrolysis and re-synthesis of TAG (Fig. 7.12). Some fatty acid residues are lost to the surrounding plasma, and, if the monoacylglycerol lipase comes into play, the glycerol liberated must be lost to the cells, because they do not contain glycerol kinase (p. 101). Replenishment of glycerol ester content then depends on a supply of glycerol phosphate, from carbohydrate catabolism (stimulated by insulin).

The hormone-sensitive TAG lipase, on the other hand, is subject to strong stimuli from a variety of primary causes, particularly exercise. These are mediated by hormone binding to receptor sites on the adipocyte membrane; in the longer term, growth hormone stimulates the synthesis of more lipase protein. The resultant of these stimuli is a tendency for plasma FFA concentration to swing violently over a far wider range than does the blood sugar—from 200 μM (after a glucose meal) to 2000 μM (prolonged exercise while fasting). The half-life of plasma FFA is also shorter than that of glucose—perhaps 2 to 3 min. These relatively violent fluctuations help to explain an apparent lack of regulatory control at sites of uptake.

(b) *Peripheral tissues—chiefly muscle*

One may presume that uptake of FFA by muscle from plasma is proportional to the plasma concentration, but as muscular activity is one of the main stimuli for release from adipose tissue, this helps to ensure a rough balance. It is at all events noteworthy that there is little accumulation either of FFA or of TAG in muscle, under normal circumstances. Perhaps if muscle contains a cytosolic FFA-binding protein, release from the inner side of the plasma membrane is hindered if there are no free binding sites on the protein. Free fatty acids, and also ketone bodies, inhibit glucose oxidation in muscle; conversely, older observations that glucose inhibits fat oxidation in muscle, independently of any effects on adipose tissue, are currently being re-investigated.

Kidney is an active tissue that can get all its energy from fat oxidation. Fatty acids are not taken up by brain, as already discussed. The mucosa of the small intestine is an active tissue which both synthesizes fatty acids and reconstitutes triacylglycerols. It is almost incapable of oxidizing fatty acids, its major fuel being glutamine (p. 133). However, the mucosa of fasting animals does abstract ketones from the blood supply, so that acetoacetate becomes an important fuel for this tissue in fasting.

(c) *Liver*

The situation in liver is different from the other tissues mentioned in that it is both an active site of fatty acid synthesis, and a tissue which actively takes up FFA from plasma during digestion/absorption. Regulation of the cytosolic FFA concentration consequently presents difficulties, particularly since uptake of FFA is proportional to their concentration in plasma. For a long time it appeared that the peculiarities of fatty acid metabolism in liver were in effect a passive response to 'flooding' with substrate, but more recently it has become apparent that there is an intracellular control, which centres around acetyl-CoA carboxylase (p. 105). The general picture will however be discussed first.

It is easiest to start with fasting metabolism (Fig. 7.19). Uptake of FFA from plasma is in excess of requirements for complete oxidation. There is some synthesis of DAG and TAG, because the export of VLDL (which has a short half-life) and HDL is not automatically diminished by fasting. This implies a continuing flux of glycerol phosphate. This is possible because gluconeogenesis, with its common set of triose intermediates, is almost certainly occurring at this time, but there is a possible limitation on the rate of DAG synthesis. Nevertheless, TAG accumulation in liver is common in fasting or uncontrolled diabetes.

Acetyl-CoA carboxylase is inhibited, but in any case there is little export of acetyl-CoA from the mitochon-

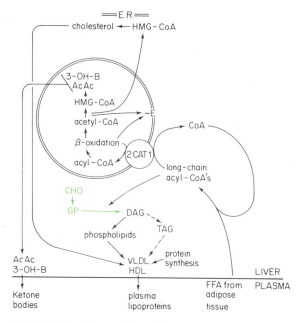

Fig. 7.19 Regulation of fatty acid metabolism in liver in the fasting state. Because there is little fatty acid synthesis in fasting animals, the cytosolic malonyl-CoA concentration is low, and CAT 1 is not inhibited. Thus all acyl-CoA arriving at the mitochondrial membrane is translocated into the matrix, and acetyl residues surplus to the energy requirement of the tissue are converted to ketone bodies which pass out into the plasma.

dria. The malonyl-CoA concentration in the cytosol is low (Fig. 7.19).

Carnitine acyl-transferase is working rapidly, and the carnitine/acyl-carnitine antiport loads the mitochondria with long-chain acyl-CoA. The acetyl-CoA residues that cannot be oxidized are converted to acetoacetyl-CoA, and thence to HMG-CoA, which is split to give acetoacetate. Note that cholesterol synthesis in liver is depressed in fasting.

It is not quite clear why citrate export from liver mitochondria is suppressed so strongly in fasting. It is often said that this is due to a lack of oxaloacetate for citrate synthesis, but a lowered OAA concentration in liver in fasting has never satisfactorily been demonstrated, and one would not expect there to be a significant reduction because of the fluxes through oxaloacetate for gluconeogenesis (p. 78) and urea synthesis (p. 129).

The citric acid cycle may be slowed down for a rather different reason. A significant proportion of the ATP produced in fatty acid oxidation comes from the β-oxidation stage. For palmitate oxidation the proportion is about 25%. Table 7.3 shows that doubling the rate of β-oxidation could give the same amount of ATP, while releasing two-thirds of the carbon of the palmitate

Table 7.3 Theoretical energy balances for fatty acid oxidation in liver

A 1 mole palmitate			B 2 moles palmitate		
β-oxidation to 8 moles acetyl-CoA	7 × 5 ATP	35	β-oxidation to 16 moles acetyl-CoA	14 × 5 ATP	70
Complete oxidation of 8 AcCoA in citric acid cycle	8 × 12 ATP	96	Oxidation of 5 AcCoA in citric acid cycle	5 × 12 ATP	60
Sum		131			130
			Surplus: 9 AcCoA, equivalent to 4.5 moles acetoacetate		

(Activation of palmitate neglected.)

as ketone bodies. It is unlikely that such an extreme state is reached in a 24-hour fast.

Figure 7.20 shows the situation in the post-absorptive (carbohydrate-rich) state. The most notable differences are that fatty acid uptake by the mitochondria is equal

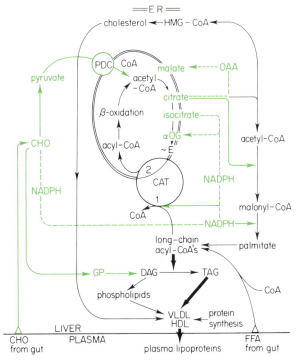

Fig. 7.20 Regulation of fatty acid metabolism in liver in the fed state. The situation is now more complex because (a) pyruvate is providing precursors for acetyl residues that are formed in the mitochondrial matrix (left-hand side of diagram), but are exported via citrate as precursors for fatty acid synthesis; (b) reducing equivalents for the synthesis come both from oxidation of glucose-6-phosphate and from the mitochondrial isocitrate shuttle (upper right centre); (c) the cytosolic malonyl-CoA concentration is high (as a consequence of (a)), so CAT 1 is inhibited, and long-chain acyl residues synthesized in the cytosol are steered towards ester (TAG) formation and export, and away from re-oxidation in the mitochondria.

to its oxidation rate, that acetyl units (from pyruvate oxidation) are being exported to the cytosol as citrate, and that acetyl-CoA carboxylase is fully active. The cells may still be taking up FFA from plasma, depending on the fat content of the preceding meal, but this is now being channelled towards TAG synthesis followed by export in VLDL. There is no limit to the rate of TAG (and phospholipid) formation, because glycerol phosphate is readily available from the Embden–Meyerhof pathway. In addition, glucose oxidation through the pentose phosphate pathway provides NADPH for fatty acid synthesis. The part played in this shift of emphasis by acetyl-CoA carboxylase is so important that it will be discussed separately.

Acetyl-CoA carboxylase and malonyl-CoA. It has been known for a long time that acetyl-CoA carboxylase has unusual physical properties. The most active form of the enzyme is a fibrous polymer, made up of an indefinite number of protomers (mol. wt 100000) which themselves have very little activity. Citrate, a precursor of the substrate acetyl-CoA, favours polymerization, while long-chain acyl-CoAs, which are ultimately products of the carboxylation reaction, favour dissociation of the polymer. Thus the balance between active and inactive forms varies with accumulation of precursors or product.

It has recently been realized that enzyme activity is also regulated by phosphorylation/dephosphorylation (see chapter 13). The hormone glucagon activates the kinase which inhibits acetyl-CoA carboxylase; glucagon causes a short-term stimulation of ketogenesis in liver. These complex interrelationships are shown in Fig. 7.21.

It has also recently been found that malonyl-CoA itself is a potent inhibitor of the cytosolic carnitine acyl-transferase (CAT 1). Thus as [malonyl-CoA] rises (conversion of CHO to fat), it inhibits FFA translocation into mitochondria and shifts the balance towards triacylglycerol synthesis. Conversely, when [malonyl-CoA] is low (CHO-free diet or fasting), the CAT 1–CAT 2 system becomes fully active, and the mitochondria may receive more FFA than they can oxidize completely, with consequent ketogenesis.

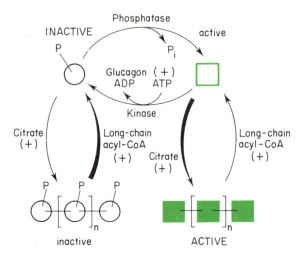

Fig. 7.21 Factors which regulate the activity of acetyl-CoA carboxylase.

These relationships are indicated in Fig. 7.20. The flux of glycerol phosphate may also aid or hinder TAG synthesis, as already described.

Propionate catabolism

If an odd-numbered fatty acid is oxidized, propionyl-CoA is formed. Propionate is also formed in the catabolism of a number of amino acids, namely valine, isoleucine, methionine and threonine (chapter 8). The formation of propionyl-CoA in the shortening of the side chain of cholesterol (p. 113) is of minor importance.

In herbivorous animals free propionic acid is a product of bacterial digestion of the carbohydrate in the fodder, and has to be converted to propionyl-CoA before catabolism can begin. Free propionate is a much rarer metabolite in omnivores (see, however, pp. 113, 138 and 144). The intermediate with which catabolism begins is always propionyl-CoA. This is carboxylated by a biotin-containing enzyme (cf. p. 99) to methyl-malonyl-CoA, which is isomerized to succinyl-CoA, an intermediate of the citric acid cycle. Via succinate, fumarate and malate, the propionyl residue is thus converted to oxaloacetate and can be used for gluconeogenesis. Propionate has long been known to be glucogenic.

This simple outline has a number of complicating features, some of which are apparent in Fig. 7.22:

1. Carboxylation of propionyl-CoA yields D-methylmalonyl-CoA, while the substrate for the isomerase is the L-isomer. A racemase catalyses the interconversion.

2. A specific form of the prosthetic group is needed for the mutase. This is adenosyl-cobalamin; no other reaction is known in animals for which this particular cofactor is needed. Several inborn defects of adenosylcobalamin synthesis are known (for further discussion of cobalamin, see chapter 17, p. 279).

3. In these and other inborn errors related to propionyl-CoA metabolism, methylmalonic acid and other substances are found in blood, and up to 6 g/day may be excreted in urine. There is often a severe acidosis, with blood pH 6.9–7.1, but the concentration of methylmalonate in blood (max. 2 mM) does not seem high enough to account for this degree of acidaemia. The reason is not yet known.

4. In many of the inborn errors, glycine also accumulates in blood and is found in urine. No explanation for this hyperglycinaemia is at present known.

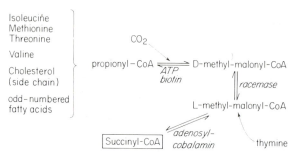

Fig. 7.22 The metabolism of propionate (propionyl-CoA).

Ketone bodies

Acetoacetic acid

Acetoacetic acid and its reduction product *β-hydroxybutyric acid* (3-hydroxybutyric acid) accumulate in the blood in certain states, particularly of carbohydrate deficiency such as starvation or diabetes. Accumulation may also occur spontaneously, especially in children. Free acetoacetate is formed by the splitting of HMG-CoA:

$$CH_2 \cdot CO \cdot S \cdot CoA \qquad\qquad CH_3 \cdot CO \cdot S \cdot CoA$$
$$CH_3 - \underset{\underset{OH}{|}}{\overset{|}{C}} - CH_2 \cdot COOH \rightarrow \qquad +$$
$$\qquad\qquad CH_3 \cdot CO \cdot CH_2 \cdot COOH$$

Hydroxymethyl-glutaryl CoA   *Acetoacetic acid*

For the synthesis of HMG-CoA, see cholesterol synthesis, p. 112.

A deacylase found only in liver can also produce free acetoacetate, by hydrolysis of acetoacetyl-CoA.

Acetoacetate is readily reduced to D(−)-3-hydroxybutyrate by a NADH-requiring dehydrogenase found in most tissues, so that both acids accumulate together.

$$CH_3 \cdot CO \cdot CH_2 \cdot COOH + NADH \rightleftharpoons$$
Acetoacetic
acid

$$CH_3 \cdot CHOH \cdot CH_2 \cdot COOH + NAD^+$$

3-hydroxybutyric
acid

The ratio of the two acids in plasma depends on the $NADH/NAD^+$ ratio *in mitochondria*.

Acetoacetate is a β-oxo acid and spontaneously undergoes decarboxylation:

$$CH_3 \cdot CO \cdot CH_2 \cdot COOH \rightarrow CH_3 \cdot CO \cdot CH_3 + CO_2$$
Acetoacetic acid   *Acetone*

Traces of acetone are found in the blood in ketotic states, hence the name *ketone bodies* for the three related compounds. As acetone is volatile it may be smelt on the breath.

Although ketone body production need not be a purely hepatic phenomenon, it is nevertheless certain that the ketone bodies found in blood in ketonaemia come from the liver. This was established many years ago, on the basis of the fall in ketone body concentration after experimental hepatectomy. The concentration of the two acids in blood at any time will be the resultant of the rates of production, excretion, and utilization. Acetoacetate is readily utilized by non-hepatic tissues, particularly muscles, in exercise as well as at rest although in vitro experiments suggest that ketone bodies, as sole fuel, allow only about 60% of the power development that glucose will give. It is extremely unlikely that ketone bodies are ever the sole fuel in vivo.

Most tissues contain a transferase, which reactivates acetoacetate by CoA transfer from succinyl-CoA, a normal intermediate of the citric acid cycle (chapter 9):

$$HOOC \cdot CH_2 \cdot CH_2 \cdot CO \cdot S \cdot CoA$$
Succinyl-CoA

$$+ CH_3 \cdot CO \cdot CH_2 \cdot COOH \rightarrow$$
Acetoacetic acid

$$HOOC \cdot CH_2 \cdot CH_2 \cdot COOH$$
Succinic acid

$$+ CH_3 \cdot CO \cdot CH_2 \cdot CO \cdot S \cdot CoA$$
Acetoacetyl-CoA

There is a rather low concentration of *acetoacetyl-CoA transferase* in brain; the utilization of ketone bodies by brain is negligible at low blood ketone levels, but it has been shown, by A–V difference studies, that a significant fraction of the energy needs of brain can be met from acetoacetate in moderate or severe ketonaemia.

The enzyme is absent from liver, so that any free acetoacetate which is formed in this tissue must necessarily diffuse out into the blood. 3-Hydroxybutyrate is not activated directly in this way and is rather slowly metabolized, since *3-OH-butyrate dehydrogenase* is, in many tissues, not very active.

There are always traces of ketone bodies in blood and also in urine, since there is no renal threshold for the two acids. A fast of less than 24 hours may produce a detectable ketonaemia, and after 3–4 days the blood ketone level may be around 2 mM, of which 3-OH-butyrate will make up 80%. After this the ketones only increase slowly, and an average excretion of 60 mmol 3-hydroxybutyrate per day at the end of a 31-day fast has been recorded. In low carbohydrate–high fat diets the ketosis is proportional to the fat loading. Ketosis does not develop on high protein–high fat diets because the dietary amino acids can be used for gluconeogenesis.

Table 7.4 gives a representative set of figures for the basal state, i.e. resting, and after a 12-hour fast. After a more prolonged fast, and with exercise, the proportion of ketone bodies oxidized rises significantly, perhaps to 25% of the total substrate oxidized. This would include a substantial fraction of the total substrate utilization in brain; as discussed in chapter 6 (p. 84), gluconeogenesis in fasting is not enough to account for the total O_2 consumption of brain tissue. Various hypotheses have been put forward to explain why brain does not oxidize long-chain fatty acids, but it seems likely that the permeability characteristics of the 'blood–brain barrier' are the most important factors.

The ketosis of starvation is not severe enough to cause any physiological disturbance, e.g. of acid–base status. In uncontrolled juvenile-type diabetes, however, the blood ketone body concentration may be 50–100 mM with 500 mmol/day appearing in the urine. The glycosuria in itself creates problems, and the reduction in plasma bicarbonate and the loss of cations in urine in such severe ketosis are very dangerous (see chapter 15).

Table 7.4

Substrate	Site of oxidation	Required O_2 consumption (ml/min)	Percentage of total
Glucose	Chiefly brain, but all tissues	45	18
Glycerol*	Chiefly liver	5	2
Amino acids	Chiefly liver	77	31
FFA	Liver		
FFA	Extrahepatic tissues	107	43
Ketone bodies	Extrahepatic tissues	15	6
	Total body	249	100

* If oxidized; probably used for gluconeogenesis. Independent measurements show that 15% of the CO_2 production of a fasting animal at rest comes from glucose.

Metabolism of Structural Lipids

Phospholipids

Synthesis

There are several pathways of phospholipid synthesis; the one which is thought to be most important in animal tissue is outlined in Fig. 7.23. The fundamental feature is the use of cytidine diphosphate (CDP) as a leaving group, from which choline phosphate or ethanolamine phosphate are transferred, the acceptor being a 1,2-diacylglycerol. The reactions can be summarized as

$$\text{Choline} + \text{ATP} \rightarrow \text{Choline-}\textcircled{P} + \text{ADP}$$

$$\text{Choline-}\textcircled{P} + \text{CTP} \rightleftharpoons \text{CDP-Choline} + \text{P-P}_i$$

$$\text{CDP-Choline} + \text{DAG} \rightarrow \text{Phosphatidyl-choline} + \text{CMP}$$

The reactions for ethanolamine are similar. In this series of reactions, CDP-choline plays a part formally similar to that of UDP-glucose in the synthesis of polysaccharides (chapter 6).

Phosphatidyl-serine is not, in animals, formed directly in this way but, as shown in Fig. 7.23, by exchange of serine for the ethanolamine of phosphatidyl-ethanolamine (PE). The free ethanolamine must be immediately phosphorylated, because very little of it is found in tissue. Phosphatidyl-serine can be decarboxylated to form PE, thus making a complete cycle, but this is probably a rather slow reaction. Successive methylation of PE to give phosphatidyl-choline (PC) is not as important in higher animals as incorporation of choline *de novo*. There is also a good deal of recycling of choline phosphate and ethanolamine phosphate between liver and plasma phospholipids (lipoprotein phospholipids).

Cardiolipin and phosphatidyl-inositol are formed in a different way, which is a variation of the same

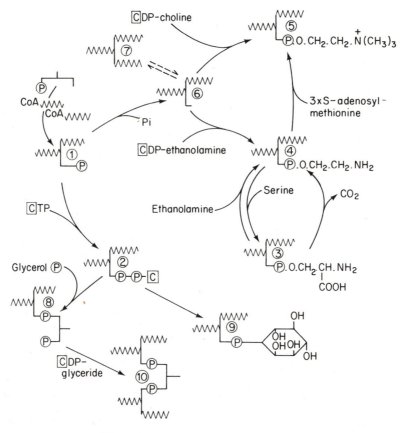

Fig. 7.23 The synthesis of phospholipids. ⓒ stands for cytosine. The numbered compounds are as follows: ① phosphatidic acid; ② cytidine diphosphate diacylglycerol; ③ phosphatidyl-serine; ④ phosphatidyl ethanolamine; ⑤ phosphatidyl choline (lecithin); ⑥ diacylglycerol; ⑦ triacylglycerol; ⑧ phosphatidyl glycerophosphate; ⑨ phosphatidyl inositol; ⑩ cardiolipin.

mechanism. CTP and phosphatidic acid (① in Fig. 7.23) react to form CDP-diacyl glycerol (② in Fig. 7.23) which is then transferred to glycerol phosphate or to inositol monophosphate, respectively (cf. p. 90). The inositol group often contains several phosphate residues. It is not clear whether they are attached before or after synthesis of the complete phospholipid.

Rearrangements

This term means rearrangement of acyl residues. Membrane phospholipids have a high proportion of highly unsaturated residues, particularly arachidonyl, at C-2 of the glycerol backbone, whereas the diacylglycerol precursor is likely to have at most an oleyl residue in this position. In order to accomplish this, hydrolysis to form a lysophospholipid, followed by re-acylation, e.g. from arachidonyl-CoA, must occur. Presumably this occurs while the membranes are being assembled within the cell, because the exterior of the cell membrane is not accessible to acyl-CoA molecules. The phospholipids of plasma lipoproteins do not regularly have such a highly unsaturated residue, which suggests that, in liver at least, the phospholipids are 'tailored' after their initial formation, to suit their eventual function. This extensive re-modelling, and concomitant rapid turnover, is particularly characteristic of liver, and to a lesser extent of intestinal mucosa, another site of active diacylglycerol metabolism. In other active organs such as kidney, phospholipid turnover is distinctly slower, and in brain many of the membrane lipids are very stable indeed.

For the transfer of acyl groups from phosphatidyl-choline to cholesterol, catalysed by LCAT, see p. 115.

Degradation

There are probably many phospholipases in tissues: most of them are bound to membranes, and so not easily purified. The best-known enzymes degrading glycero-phospholipids are shown in Fig. 7.24. *Sphingomyelinase* (see Table 7.5 and Fig. 7.25b) is a separate enzyme which hydrolyses sphingomyelin to sphingosine and phosphorylcholine.

Phospholipases A remove a fatty acyl residue from an intact phospholipid. PL-A₁ attacks only the *sn*-1 position. Two enzymes are known: one is lysosomal, and has a preference for phosphatidyl ethanolamine; the other is found on the plasma membrane, and hydrolyses MAG (p. 96) more readily than phospholipids. It seems to be identical with the plasma TAG lipase (not the lipoprotein lipase, p. 95). Neither of these enzymes attacks phosphatidyl serine or phosphatidyl inositol.

Phospholipase A₂ has been intensively studied, because one form is activated by a release of Ca²⁺ to set free arachidonic acid as a precursor of prostaglandins (see Fig. 7.34, p. 119). Two PL-A₂ enzymes have been

(a)

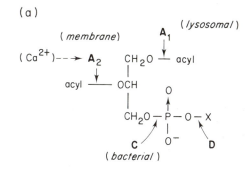

(b)

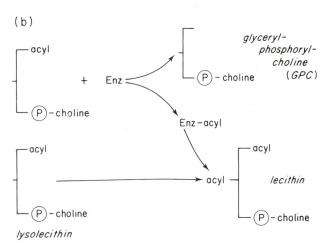

lysolecithin

Fig. 7.24 The specificity of the phospholipases. (a) General summary. (b) Mechanism of action of the acyl transferase, acting on 1-lysolecithin, which forms glycerol-phosphoryl-choline as one product, and so can be regarded as a phospholipase B.

isolated, one from red cell membranes, and one from liver mitochondrial membrane. They both have an absolute requirement for Ca²⁺, and a marked preference for phosphatidyl choline (PCh) as a substrate.

Either PL-A₁ or PL-A₂ will set free one acyl residue and leave a lysophospholipid which is not further attacked by either enzyme. The enzyme which removes the remaining acyl group used to be called a *phospholipase B*, but it has now been re-named *lysophospholipase-transacylase* (LPL-TA), which is clumsy but more accurate. As Fig. 7.24(b) shows, the enzyme removes an acyl group from the substrate, and transfers it either to a water molecule (hydrolysis), or to a second lysophospholipid molecule (transacylation). This enzyme is widely distributed in tissues, but it is not known whether it acts in plasma to remove the lyso-PC which is produced by the action of LCAT on high-density lipoproteins (p. 118). LPL-TA acts on lyso-PE and lyso-phosphatidyl glycerol (a product of cardiolipin metabolism) as well as lyso-PC.

The product of LPL-TA action is glycerol-phos-phorylcholine (GPCh)

$$CH_2OH$$
$$|$$
$$CHOH$$
$$|$$
$$CH_2O—\text{P}·O·CH_2·CH_2N^+(CH_3)_3$$

One must presume that this compound is further degraded by a phosphodiesterase. However, it has recently been found in high concentration as a component of normal muscle, so it appears that in this tissue, at least, the attack on GPCh is very slow.

Phospholipase C: There are many enzymes of this class, but they appear to be all bacterial. One of these enzymes, that from *Clostridium perfringens*, is lethal in animals; it is necrotizing and haemolytic, and is closely associated with the development of gas gangrene. None of the PL-C enzymes from other bacteria is toxic, and it is not known precisely why the *C. perfringens* enzyme has these virulent properties.

Phospholipase D: It used to be thought that enzymes of this class were confined to plants, but a PL-D acting on PC and PE has been isolated from brain. The products are phosphatidic acid and the free base.

Sphingolipids

Sphingosine is the long-chain alkyl base (p. 91) which is the common component of the sphingolipids, both sphingomyelin and the ceramides, the latter being found in gangliosides and other glycolipids (p. 91). In general, the turnover of sphingolipids is very slow, particularly in brain, but the catabolism of these lipids is nevertheless affected by a series of inborn errors, many of which are fatal in early life.

Synthesis

Sphingosine itself (4-sphingenine) is formed by condensation between an acyl-CoA and serine (see Fig. 7.25a).

The synthetase has pyridoxal phosphate as prosthetic group, and will accept acyl-CoAs of chain length from 14 to 18, so that the sphingenines are a class of compounds rather than a single entity. The precise mechanism by which the double bond is introduced is unknown.

The amide bond in the sphingolipids is formed when an acyl group is transferred from an acyl-CoA. The acyl residue is often 2-OH-lignoceric (C_{24}). The acyl amide so formed is a *ceramide* (p. 91; see also Fig. 7.25(b)).

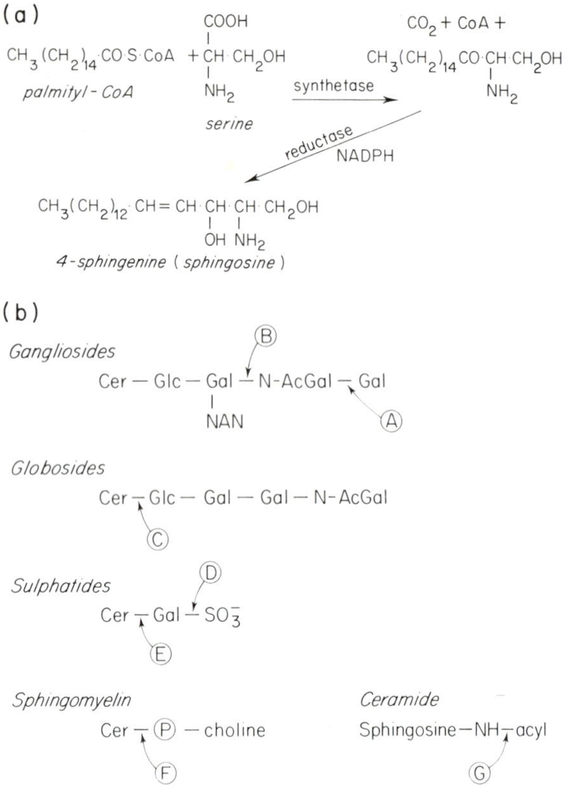

Fig. 7.25 (a) The synthesis of sphingosine. (b) The points of action of the missing hydrolytic enzymes in the heriditary lipidoses. For key to lettering, see Table 7.5.

The group occupying the primary —OH group is phosphoryl-choline in the sphingomyelins.

Ceramide-CH_2OH + CDP-choline →

ceramide-$CH_2O·\text{P}$ -choline + CMP

sphingomyelin

In the cerebrosides, and the more complicated gangliosides (p. 91), the transfer of the monosaccharide residues takes place from the UDP-sugar, e.g.

Ceramide-CH_2OH + UDP-Gal →
Ceramide-$CH_2O·$Gal + UDP

cerebroside

The sulphate residue in the sulphatides (p. 92) comes from PAPS (chapter 8, p. 140).

Degradation

By study of various inborn errors of metabolism in which sphingosine derivatives accumulate, particularly

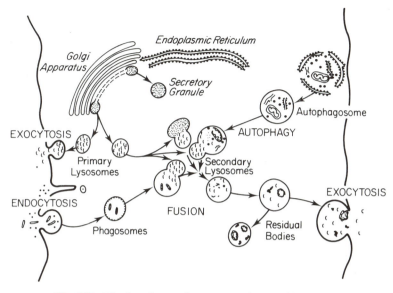

Fig. 7.26 The functions and turnover pathways of lysosomes.

in nervous tissue, it has become clear that the membrane sphingolipids are degraded by a series of hydrolases located in *lysosomes*. Indeed it is probable that the function of these small organelles, packed with hydrolytic enzymes and with an internal pH ~ 4.5, is the turnover of membranes. They are not the cellular 'suicide bags' they were once thought to be. Fig. 7.26 is a diagrammatic representation, based on electron-microscopic observation, of ways in which lysosomes can attack membranes. In chapter 10, the continuous renewal of plasma membranes, particularly in liver, is discussed. Other cell components do find their way into lysosomes; in chapter 6, the serious consequences of glycogen accumulation in lysosomes (Type II glycogenosis) were described. Although the full details of lysosome life history are not known, it seems to be the case that they cannot entirely disintegrate until their contents have been completely hydrolysed. In Type II glycogenosis, and in the lipidoses to be summarized below, new lysosomes continue to be formed, and the number in each cell increases until they interfere with cell function. In Type II glycogenosis it is muscle, particularly heart muscle, cells that are disrupted; in the lipidoses it is mainly cells of the central nervous system, retina as well as brain. Hunter's syndrome, a rapidly fatal accumulation of dermatan polymers (p. 111) arising from the absence of an enzyme hydrolysing iduronic acid residues, and cholesteryl ester storage disease (p. 115), are other lysosomal defects.

There are too many genetic defects of sphingolipid metabolism, each due to the absence of a single enzyme,

Table 7.5 Sphingolipidoses

Disease	Missing enzyme	Tissues most affected
Defects in ganglioside breakdown		
A. Generalized gangliosidosis	β-Galactosidase	Brain. Rapidly fatal
B. Tay-Sachs	Hexosaminidase A	Brain, retina. Rapidly fatal
C. Gaucher	β-Glucosidase	Liver and spleen; brain
Defects in sulphatide breakdown		
D. Metachromatic leucodystrophy	Sulphatase 'Arylsulphatase A'	Nervous system
E. Krabbe's leuco-dystrophy	Galactocerebrosidase	Brain. Rare. Rapidly fatal
Defect in sphingomyelin breakdown		
F. Niemann-Pick	Sphingomyelinase	Spleen and liver; brain. Rapidly fatal
Defect in ceramide breakdown		
G. Farber	Ceramidase	Joints

for them all to be catalogued here. Table 7.5 gives details of some of the best-known diseases, and Fig. 7.25b indicates where the missing enzymes normally act.

The sphingolipidoses are of interest in that they are one of the few groups of diseases arising from genetic defects in which enzyme replacement therapy looks promising (perhaps because the rate of accumulation of the intermediates is so slow). At the present time, success has been greatest with Gaucher's disease. When the defective enzyme, ceramyl-glucosidase, is injected into a patient's bloodstream, it rapidly disappears, mostly taken up by liver, and within 24 hours all its activity has been lost. Within that period, however, it has been able to hydrolyse as much ceramyl glucoside as has been accumulated over a 4-year period, so that widely-spaced injections may be effective throughout life. Because the enzyme is not completely absent from the affected patient, and because the replacement is purified from human tissues (placenta), immunological problems have been minimal in tests so far. This fortunate combination of circumstances may not apply to many of the other inborn errors caused by lack of a single enzyme.

Cholesterol

Cholesterol is synthesized in all nucleated cells. About one-third of the total daily synthesis—about 800 mg—is formed in liver, and much of it is exported in β-lipoproteins (see p. 116), and may be taken up by other cells. In addition, liver is the organ which excretes excess cholesterol, either directly or as bile acids, and therefore cholesterol metabolism in the body comes to be equated with hepatic cholesterol turnover, perhaps unrealistically.

Synthesis

The acetyl carbons of acetyl-CoA provide all the C atoms of cholesterol and other steroids, but the real starting point of cholesterol synthesis is 3-hydroxy-3-methylglutaryl-CoA (HMG-CoA). The synthesis of this compound, from acetyl-CoA and acetoacetyl-CoA, has already been mentioned in the section on ketone bodies (p. 106). There are in fact two HMG-CoA synthetases in liver cells, one in mitochondria and one on the endoplasmic reticulum. The former makes AcAc-CoA by reversal of β-oxidation, while the latter makes it by condensation of acetyl-CoA and malonyl-CoA, with loss of CO_2.

It is presumed that the endoplasmic reticulum system is the more important for cholesterol synthesis, partly because later stages are also concentrated on the endoplasmic reticulum, and partly because the rate of cholesterol synthesis is reduced in fasting, when the malonyl-CoA concentration in cytosol is low (p. 104). However, leucine is a good precursor both of ketone bodies and of cholesterol, and its catabolism, which gives HMG-CoA directly (chapter 8, p. 145), is confined to mitochondria.

The determining reaction in cholesterol synthesis is the reduction of HMG-CoA to mevalonic acid

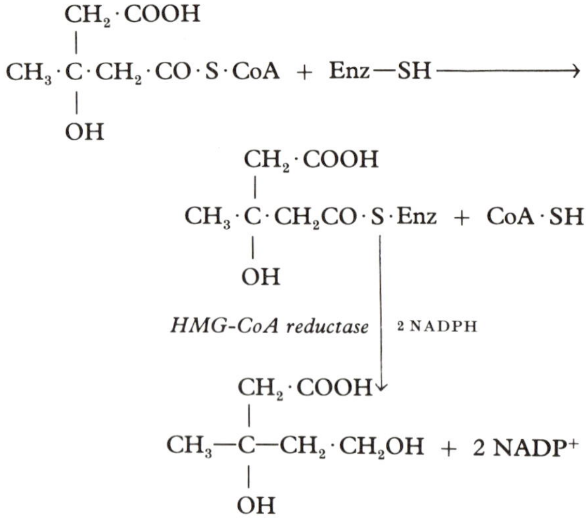

$$CH_3 \cdot C \cdot CH_2 \cdot CO \cdot S \cdot CoA + Enz\!-\!SH \longrightarrow$$

with the structure:

$$\begin{array}{c} CH_2 \cdot COOH \\ | \\ CH_3 \cdot C \cdot CH_2 \cdot CO \cdot S \cdot CoA \\ | \\ OH \end{array}$$

$$\begin{array}{c} CH_2 \cdot COOH \\ | \\ CH_3 \cdot C \cdot CH_2 CO \cdot S \cdot Enz + CoA \cdot SH \\ | \\ OH \end{array}$$

HMG-CoA reductase | 2 NADPH

$$\begin{array}{c} CH_2 \cdot COOH \\ | \\ CH_3\!-\!C\!-\!CH_2 \cdot CH_2OH + 2\ NADP^+ \\ | \\ OH \end{array}$$

This is an unusual 4-electron reduction.

The next stages in the synthesis are shown in Fig. 7.27. The primary alcohol group of mevalonate is successively reacted twice with ATP to give the pyrophosphate ester, and subsequently the phosphate ester of the tertiary alcohol is formed. When this compound is dephosphorylated, a double bond is formed, and at the same time the bond to the —COOH is weakened, so that CO_2 is lost. The resulting compound, *isopentenyl pyrophosphate*, is the building block of the isoprene and terpene compounds, which occur very widely in plants.

A non-reactive isopropyl group at one end of the molecule is assured by isomerization of isopentenyl pyrophosphate to *dimethylallyl pyrophosphate*, and chain lengthening then takes place by nucleophilic attack as shown in the figure, with the pyrophosphate acting as leaving group. This linear extension can go on indefinitely (as in the synthesis of natural rubber), but is not common in animal cells. Examples are the side chains of ubiquinone (p. 36)—10 units—and dolichol (p. 180)—16 units. For the synthesis of cholesterol the lengthening reactions stops after 3 units, the product being *farnesyl pyrophosphate*.

The final stages are shown in Fig. 7.28. Condensation of two farnesyl groups to give a C_{30} compound (squalene) is not a simple nucleophilic reaction, and several intermediate stages are involved. At least one intermediate, pre-squalene pyrophosphate, has been identified. The details are not shown in the figure.

In the cyclization of squalene, folding of the molecule is very important. No doubt binding at the active site of the enzyme ensures this; note that almost all the double bonds are *trans*. A mixed function oxidase

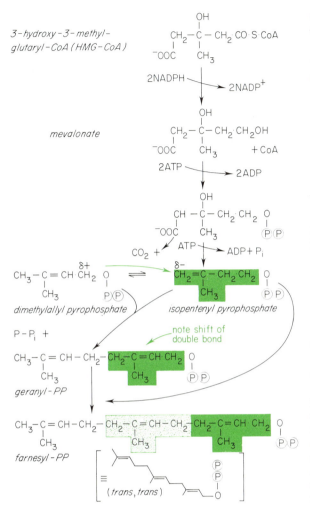

3-hydroxy-3-methyl-glutaryl-CoA (HMG-CoA)

mevalonate

dimethylallyl pyrophosphate

isopentenyl pyrophosphate

note shift of double bond

geranyl-PP

farnesyl-PP

(trans, trans)

Fig. 7.27 The first stages in the biosynthesis of cholesterol (from HMG-CoA to farnesyl-pyrophosphate).

(co-substrate unknown) converts squalene to squalene epoxide, and a concerted set of ring closures follows; at the same time there are 2 methyl shifts (rings C and D). The product, *lanosterol*, occurs in wool fat.

The final conversion to cholesterol involves oxidation away of 3 —CH_3 groups, and rearrangement of double bonds. The first step in the oxidation is attack by a mixed function oxidase to give —CH_2OH, with further oxidation to —COOH, which is then lost as CO_2.

Almost all the reactions in cholesterol synthesis take place on the endoplasmic reticulum. The products become successively less water-soluble, and a carrier protein is required to transport the intermediates from one enzyme site to another.

Cholesterol synthesis is limited to eukaryotes, and not all animals can carry it out, e.g. housefly larvae get their cholesterol from the meat on which they feed.

Degradation to bile acids

Catabolism, as opposed to conversion to steroid hormones, takes place in liver. The major change is the shortening of the side chain. This takes place first by oxidizing one of the terminal —CH_3 groups by a mixed function oxidase:

followed by β-oxidation of the side chain to give a 24-oxoacid:

which is then cleaved by a thiolase to give propionate (cf. p. 106) and a 24-ester coenzyme A. This is the active group for conjugation with glycine or taurine.

Before these oxidation and cleavage reactions take place, the nucleus is hydroxylated. The enzyme catalysing the first reaction, *cholesterol 7-α hydroxylase*, is subject to feedback inhibition (by chenodeoxycholate) and this seems to regulate the rate of cholesterol degradation in liver (see below). Loss of unchanged cholesterol into bile is not as great as the efflux of bile acids, but because the latter are efficiently reabsorbed, the absolute losses of cholesterol and bile acids are about equal (Fig. 7.29).

The complex reactions of bile acid metabolism, including bacterial attack on the primary bile acids (chenodeoxycholic and cholic) in the gut, and the enterohepatic circulation, are summarized in Table 7.6.

Regulation of cholesterol metabolism

Although the overall turnover of cholesterol in the body is no more than 5 g/day, regulation has been for long the subject of intensive research because of its connection with atherosclerosis, cerebral haemorrhage and coronary heart disease. The atheromatous plaques in aorta and elsewhere do contain a high concentration of cholesterol esters, but there is no simple relationship between plasma cholesterol and coronary artery disease. Moreover the relationships are quite different in many animals other than man, making experimental research difficult. Plasma cholesterol is discussed in more detail under 'Plasma Lipoproteins' (p. 115 ff).

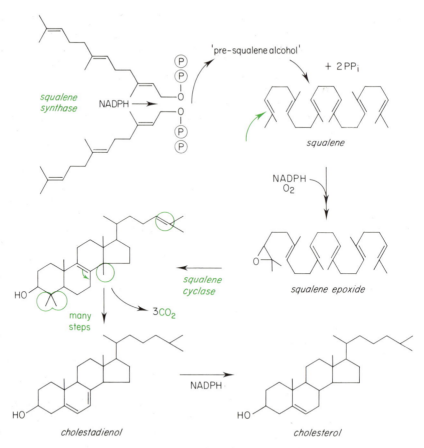

Fig. 7.28 The later stages in the biosynthesis of cholesterol (from farnesyl-pyrophosphate to cholesterol). *N.B.* Several stages, especially in the formation of lanosterol, and in the conversion of lanosterol to cholesterol, have been condensed or omitted.

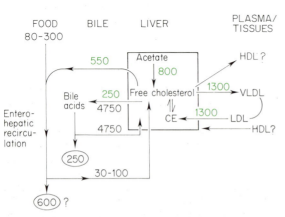

Fig. 7.29 The balance between synthesis and destruction of cholesterol in man. Note that the liver is the sole organ capable of eliminating cholesterol from the body once it has been absorbed or synthesized.

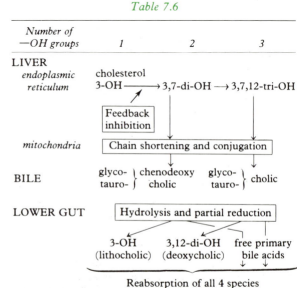

Table 7.6

Number of —OH groups	1	2	3
LIVER *endoplasmic reticulum*	cholesterol 3-OH ⟶	3,7-di-OH ⟶	3,7,12-tri-OH
	Feedback inhibition		
mitochondria	Chain shortening and conjugation		
BILE	glyco- ⎫ tauro- ⎭ chenodeoxy cholic		glyco- ⎫ tauro- ⎭ cholic
LOWER GUT	Hydrolysis and partial reduction		
	3-OH (lithocholic)	3,12-di-OH (deoxycholic)	free primary bile acids

Reabsorption of all 4 species

So far as cholesterol itself is concerned, metabolism in liver is of major importance, particularly since liver is the only site of cholesterol excretion. Two-thirds of cholesterol is synthesized outside the liver, but much of this is in the adrenal glands, and will not affect hepatic turnover. In many ways, regulation appears to have evolved primarily to control the body pool of bile acids, rather than cholesterol. HMG-CoA reductase is indeed strongly inhibited by cholesterol and by chenodeoxycholate, and synthesis of the enzyme itself is repressed by cholesterol, but it is possible that the non-hepatic and non-adrenal sites of cholesterol synthesis are not so strongly affected, as LDL binding sites (see p. 118) may be saturated before the tissues are loaded with cholesterol. Thus depression of cholesterol synthesis in the body as a whole may not be so complete as in the liver. In addition, there is a variable burden provided by dietary cholesterol; although only 30–40% of the sterol in the diet is absorbed, a diet rich in meat and milk products may bring 100 mg/day to the liver cholesterol pool.

It is at this point that the inhibition of cholesterol degradation by bile salts (chenodeoxycholate) becomes significant. It regulates bile salt production, but is not sensitive to the hepatic cholesterol burden *per se*. Another difficulty is that interfering with the enterohepatic bile salt circulation may increase the rate of cholesterol conversion to bile salts, but at the same time it partly removes the brake on cholesterol synthesis *de novo*. The response of the liver is often to export more cholesterol esters to peripheral tissues in VLDL, with potentially serious consequences in man.

Cholesterol esters

Long-chain acyl esters of cholesterol behave differently, both structurally and metabolically, from free cholesterol. They do not occur in membranes, and in plasma lipoproteins they segregate with the triacylglycerols in the lipid core, rather than with phospholipids in the outer shell, as free cholesterol does. Interconversion between the free sterol and its acyl esters is therefore important and is summarized in Fig. 7.30. There are two esterification mechanisms:

1. This operates in plasma, and transfers acyl groups from the 2-position of lecithin (*lecithin-cholesterol acyl transferase* or LCAT). The enzyme is activated by apoprotein A1 (normally found in high-density lipoproteins). Because plasma lecithin has highly-unsaturated acyl groups in position 2, the cholesterol ester so formed is also highly unsaturated (at least 2 double bonds, linoleyl). It is the degree of unsaturation of the acyl moiety of these esters that has been implicated in atherosclerosis or its prevention.

2. When the lipoproteins (whether HDL or LDL) containing the cholesterol esters described above are taken up by cells, the esters are hydrolysed by a *CE hydrolase* located in lysosomes. The free cholesterol so released eventually accumulates in the plasma membranes (perhaps in equilibrium with bound cholesterol in the cytosol). The membrane cholesterol—apart from its structural function—has two effects:

(a) it suppresses HMG-CoA reductase (p. 112);
(b) it activates *acyl-cholesterol acyl transferase* (ACAT) which catalyses the reaction

$$cholesterol + oleyl\text{-}CoA \rightarrow CE\ (oleyl) + CoA$$

Thus the intracellularly stored cholesterol ester (mostly in liver) is much less highly unsaturated than that in plasma.

The resultant of the effects (a) and (b) described above is to regulate the concentration of free cholesterol in the membrane, which is an important determinant of membrane fluidity (chapter 10).

In the congenital absence of the lysosomal cholesteryl ester hydrolase, CE accumulates, chiefly in reticuloendothelial cells, with serious but not fatal consequences. There is a similar accumulation in Tangier disease (p. 117), but this is unconnected and is possibly due to the uptake of chylomicron remnants by RE cells.

The role of plasma lipoproteins

The lipoproteins of plasma have two well-established functions: the transport of triacylglycerol and the transport of cholesterol and its esters (free fatty acids are transported by albumin). The composition of the major classes of lipoproteins is indicated in Fig. 7.31. The circles should be imagined as cross-sections through a sphere. This method of depiction shows that the protein shell has much the same thickness for each class, although the proportion of protein varies from 2% for chylomicrons to 40–50% for high-density lipoprotein. The shell is not made up of a single protein, but certain proteins predominate in each class, and they have specific functions. The nature of the lipoproteins is summarized (in both current nomenclatures) in Table 7.7.

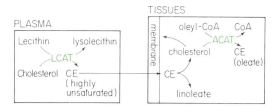

Fig. 7.30 Interconversions between cholesterol and cholesterol esters. *LCAT* is lecithin-cholesterol acyl transferase; *ACAT* is acyl-CoA-cholesterol acyl transferase.

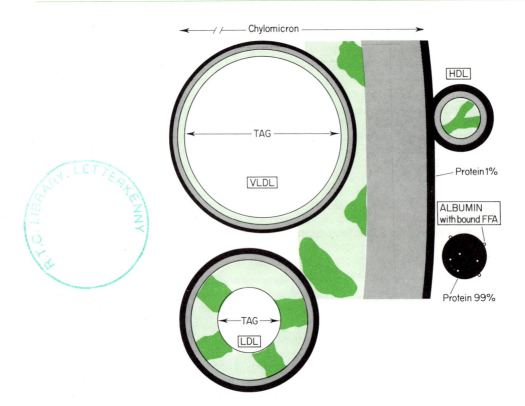

Fig. 7.31 The relative sizes of the lipid-carrying proteins in plasma. The circles in the diagram are sections through spheres of the appropriate size. The sphere corresponding to chylomicrons is so large, when drawn to the same scale, that only a small portion of it can be included in the diagram. This method of representing the lipoproteins emphasizes that, although the proportion of protein in the various particles varies from 1% (for chylomicrons) to 40% (for HDL), the thickness of the outer protein–phospholipid shell is about the same for each particle.
Key: black = protein; grey shading = phospholipids; solid green = free cholesterol; green shading = cholesteryl esters. The uncoloured area represents triacylglycerol.

Table 7.7(a)   Plasma lipoproteins: physical properties

Type	Density	Electro-phoresis	Equiv. mol. wt	Character-istic proteins	Turnover time
Chylo-microns	< 0.95	origin	3×10^{10}	B (+CII)	5 min
VLDL	0.95–1.006	pre-β	5×10^{6}	B, E (+CI–III)	2 hr
[IDL]	—	inter-mediate	—	B, E, CII	2 hr
LDL	1.01–1.06	β	2×10^{6}	B, E	4 days
HDL	1.06–1.21	α	2×10^{5}	AI–III	? 10 hr
also Lp(a)	1.05–1.12	pre-β	5×10^{6}	B, Lp(a)	?*
Lp X	1.01–1.06	β–γ	$\sim 10^{6}$?	—	?**

* A minor lipoprotein found in most normal plasmas.
** An abnormal constituent, very rich in free cholesterol, which accumulates in plasma in obstructive liver disease.

Table 7.7(b)   Plasma lipoproteins: chemical composition

	Protein	Phospho-lipid	Free cholesterol	Chol. esters	TAG
Chylo-microns	1	10	0–10		80–90
VLDL	10	15	—	15	60
LDL	20	20	10	40	0–10
HDL	45	30	5	20	0

Exogenous lipid transport

This is summarized in Fig. 7.32, and has already been discussed from the purely lipid point of view on p. 104 ff. Here we look at the role of the proteins. The shell protein in chylomicrons is apoprotein B. Without it, mucosal cells cannot secrete chylomicrons, and become loaded with fat; the liver also fails to secrete VLDL, as discussed below. This has beeen established by study of the genetic defect *abetalipoproteinaemia*, in which apo-B is completely missing from plasma. Sufferers frequently die from secondary causes before reaching adult life.

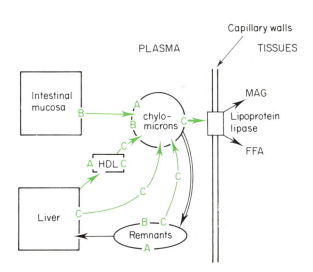

Fig. 7.32 Transport of exogenous lipid by lipoproteins. The green letters refer to the various species of apoprotein, whose functions are described in the text.

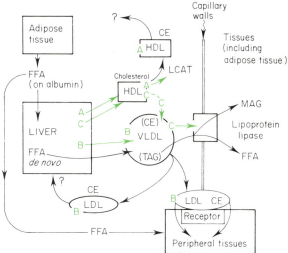

Fig. 7.33 The transport of endogenous lipid in plasma; the roles of lipoproteins and plasma albumin. *N.B.* The role of high-density lipoproteins is certainly understressed in this diagram.

During their passage through the thoracic duct and the bloodstream, chylomicrons pick up another protein, apo-C, either from high density lipoproteins (HDL) or from plasma itself. Apo-C is made only in liver, and the variant apo-CII is a powerful activator of the lipoprotein lipase, already mentioned on p. 95. The hydrolysis of chylomicron TAG is extremely fast ($t_{\frac{1}{2}} \sim 3$ min); the depleted particle, or remnant, which contains cholesterol, phospholipids and fat soluble vitamins, binds to receptors on the surface of hepatocytes. It is then internalized and degraded in lysosomes (Fig. 7.26).

Endogenous lipid transport (Fig. 7.33)

This may be looked at first with reference to transport of triacylglycerols. Endogenous TAG is made in liver, from fatty acids which have been synthesized *de novo* from carbohydrate, from FFA taken up from plasma, and also, during digestion, from long-chain acyls made by elongation of the medium-chain length fatty acids which are absorbed straight into plasma. This TAG is secreted by the liver in particles (very low density lipoproteins, VLDL) rather smaller than chylomicrons, and containing a good deal of phospholipid and cholesterol, most of it esterified (Table 7.7 and Fig. 7.31). About 40% of the protein is apo-B (without which VLDL cannot be secreted) and 60% is apo-C, but apo-B rises and apo-C falls as the TAG is removed.

The hydrolysis of this TAG by the vascular lipoprotein lipase is moderately rapid; the half-life of

VLDL is about 2 hours. So far as triacylglycerol transport is concerned, this completes the story, as the TAG is oxidized in peripheral tissues, or stored in adipose tissue.

However, the VLDL particles themselves do not disappear from the plasma. When much of the TAG has gone, but some still remains, they begin to interact with another lipoprotein in plasma, the high-density lipoprotein (HDL). The density of the VLDL has increased, and they are now frequently called intermediate-density lipoproteins (IDL). These also have a short half-life (2 hours), and are transformed into low-density lipoproteins (LDL). In man, but not in many experimental animals, LDL have a much longer half-life (4 days). Discussion of the transformation VLDL → IDL → LDL will be postponed until the high-density lipoproteins have been described.

HDL are much smaller than the other lipoproteins, and less is known about their function. About 90% of their protein consists of two forms of apoprotein A —apo-AI and apo-AII. Most about their function has been learned by study of two genetic defects: *Tangier disease*, in which apo-A is not made, and HDL are not found in plasma; and *familial LCAT deficiency* (see below). Patients with Tangier disease have a low plasma cholesterol, but deposits of cholesterol ester in all the reticulo-endothelial cells of the body; there is an increased risk of vascular disease. It is concluded that either HDL scavenges cholesterol from peripheral tissues, or that it helps in the destruction of IDL and

117

LDL. If the latter were true, then in the absence of HDL, LDL would persist in plasma, and ultimately be taken up by RE cells. If the former were true, HDL would be limited to picking up free cholesterol from plasma membranes, since the outer shell of such membranes contains only free cholesterol and not cholesterol esters (cf. p. 115). The latter, indeed, are stored in the interior of cells, away from direct reaction with HDL particles.

LCAT, or *lecithin-cholesterol acyl transferase*, catalyses the reaction

$$\text{Cholesterol} + \begin{array}{l} \text{CH}_2\text{OCOR} \\ | \\ \text{CH}_2\text{OCOR}' \\ | \\ \text{CH}_2\text{O}\textcircled{P}\text{—choline} \end{array} \longrightarrow$$

lecithin

$$\begin{array}{l}\text{Cholesterol} \\ \text{ester}\end{array} + \begin{array}{l} \text{CH}_2\text{OCOR} \\ | \\ \text{CHOH} \\ | \\ \text{CH}_2\text{O}\textcircled{P}\text{—choline} \end{array}$$

lysolecithin

It is secreted by liver, and is found in plasma, but its effective substrate is HDL. That is to say, cholesterol in the outer shell of HDL is readily esterified by lecithin on the same particle. Moreover, apo-AI is an activator of LCAT, while apo-AII is an inhibitor. In the congenital absence of LCAT, the HDL particles have an abnormal shape, and contain only free cholesterol, which is also found in excessive amounts in erythrocyte membranes. These findings suggest that when HDL is secreted from liver, it contains only free cholesterol, which is later esterified by LCAT. As suggested earlier, HDL may subsequently pick up from (some) cell membranes free cholesterol rather than cholesterol

ester. Experiments in vitro have indeed shown that HDL, when equilibrated with some cells (erythrocytes, ascites cells), takes up cholesterol. Moreover, HDL readily takes up both cholesterol and the other substrate, lecithin, from VLDL. After the reaction the products are translocated. The lysolecithin becomes bound to plasma albumin, while the cholesterol ester moves partly to the lipid core of HDL, and partly returns to the VLDL particles, where it also moves into the lipid core. Since the acyl group is transferred from the 2-position of the lecithin, it will be linoleate, or an even more unsaturated fatty acyl residue.

It cannot be said that this role as a passive vehicle for the LCAT reaction is the only, or even the most important, function of HDL. It is known, for instance, to act as a reservoir for apo-C (p. 117). It appears very important that HDL has a *negative* correlation with coronary heart disease; that is, the higher the plasma HDL concentration, the lower the risk of coronary disease. The reasons for this, and for the sex difference in HDL concentration, are still unknown.

Let us now return to the low-density lipoproteins. The overall cholesterol concentration in plasma is a reflection of the LDL concentration. In many mammals, the plasma LDL level is low and it is rapidly cleared from the blood. In rats, for example, it is taken up by the liver, whereas this is not important in man. It has been established that, in order for LDL to be cleared, it must first be bound to specific cell-surface receptors. These have been most intensively studied in fibroblasts, but are thought to be fairly widely distributed in tissues, although not demonstrated in liver or intestine (in man). The rapid turnover of LDL in other animals follows from the fact that the receptors are never saturated, but in man (except in the new-born) the receptor sites are saturated with LDL, which accumulates in plasma and

Table 7.8 Abnormal lipoprotein patterns in familial hyperlipoproteinaemia

Type	Plasma lipid peculiarity	Cholesterol and TAG		Ratio C/TAG	Associated factors
I (rare)	Marked increase and persistence of chylomicrons	C↑	TAG↑	< 0.2	Deficiency in lipoprotein lipase
II (most common)	LDL increased	C↑	TAG↑ or normal	> 1.5	Severe xanthomata. High percentage of ischaemic heart disease
III (uncommon)	'β-VLDL' present (LDL with very high percentage of TG)	C↑	TAG↑	~ 1	With xanthomata. Responds well to restrictive diet
IV (common)	VLDL increased. Few chylomicrons	C↑ or normal	TAG↑	variable	'Carbohydrate-induced' (excessive synthesis of TG from dietary CHO)
V (uncommon)	VLDL increased. Chylomicrons present	C↑	TAG↑	0.15–0.6	Genetically heterogeneous. 75% have ketotic diabetes

These types have been distinguished on the basis of the appearance and composition of plasma 12–16 hours after the last (evening) meal, and on investigation of familial incidence of similar symptoms.

is only slowly cleared. It then binds to less specific sites (e.g. to RE cells in Tangier disease). One class of such sites is where the wall of an artery has been damaged, and platelets have formed a plug. LDL adhesion, forming a plaque, is not the primary event in atherosclerosis, but it is very important in the occlusion of the blood vessel that precedes a coronary or cerebral event.

Various drugs are used in an attempt to treat raised plasma cholesterol, i.e. raised plasma LDL. Among them are *nicotinic acid*, which depresses VLDL secretion by liver, and also increases plasma HDL, but has unwelcome side-effects; *clofibrate*, which lowers plasma lipids by an unknown mechanism; and various substances which sequester bile salts in the intestine, and so increase the rate of conversion of cholesterol to bile acids. Unfortunately these latter drugs may also increase the rate of hepatic cholesterol synthesis.

The analytical patterns of cholesterol and triacyl-glycerol that are found in hyperlipoproteinaemias of familial origin are shown in Table 7.8. These data must be interpreted with some caution, because they give no clear indication of the defects which cause these abnormalities to become apparent.

Prostaglandins

These are derivatives of arachidonic acid. They have physiological and pharmacological effects too complex to discuss here, but one of the most consistent actions is to cause smooth muscle to contract; this is probably mediated by changes in intracellular cyclic AMP concentration. Recent research has implicated two new arachidonic acid derivatives in the mechanism of platelet aggregation and thrombus formation, as well as transient effects on artery contraction and blood pressure. The actions on platelets may also involve

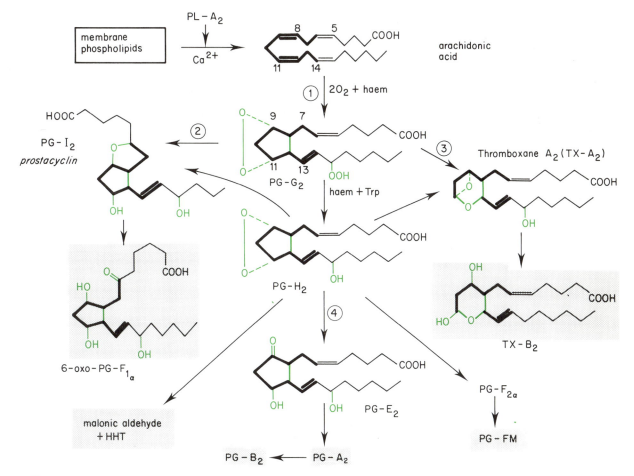

Fig. 7.34 The synthesis of prostaglandins and related substances. The grey shading indicates end products with little or no biological activity.

contractile processes, because these fragments of megakaryocytes contain a good deal of actomyosin, as well as granules and mitochondria.

There is good evidence to connect platelets with both early and late stages of atherosclerotic plaque formation. Platelets, as part of their normal function, adhere strongly to collagen exposed after internal damage to the endothelium of blood vessels. If the stimulus is severe enough, the platelets aggregate and release a smooth muscle cell growth factor. After infiltration of plasma lipoproteins into the damaged area, the proliferating cells accumulate cholesterol and then degenerate, to form a plaque of amorphous lipid, which may later calcify. There is then always a risk that platelets will again aggregate at the affected spot, to form a thrombus which occludes the artery.

Figure 7.34 shows the main pathways of prostaglandin formation. The newer compounds are (1) *thromboxane A_2*, which is formed in platelets from PGG_2. It has a half-life of about 3 min in blood, being transformed into the inactive thromboxane B_2. It is significant that one of the early events in platelet stimulation is the activation of a phospholipase A_2, which releases arachidonic acid from a membrane phospholipid. The physiological effects of thromboxane are shown in Fig. 7.35.

(2) *Prostacyclin* (PGI_2) is formed in aorta and probably other artery walls and has a very powerful antagonistic effect to thromboxane. As little as 1 ng/ml

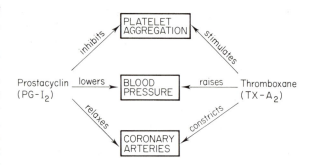

Fig. 7.35 The physiological antagonism between thromboxane and prostacyclin.

will prevent platelet aggregation. Since this phenomenon is completely separate from the fibrogen → fibrin response to vascular damage, partial inhibition of platelet aggregation might help to minimize the risk of thrombosis without causing excessive bleeding.

A number of anti-inflammatory drugs are known to interfere with prostaglandin formation. Aspirin, the most general of them, inhibits the cyclo-oxygenase irreversibly by acetylating it. A search is now being made for drugs which will selectively inhibit thromboxane synthesis; it is already known that nicotine selectively inhibits prostacyclin synthesis, thus enhancing platelet stickiness.

General Considerations

The metabolism of lipids has to be divided into two parts: the first, which concerns the long-chain fatty acids and the triacylglycerols, of which the net throughput may be ~ 100 g/day; and the second, concerned with structural lipids, for which the total turnover is of the order of 2 g/day (Fig. 7.29). Some structural compounds, such as the sphingolipids, may have a turnover much lower even than this, and yet run into an imbalance with serious, even fatal, consequences (Table 7.5).

The metabolism of triacylglycerols, apart from the details of the enzymic pathways, is fairly straightforward. It is, however, in higher animals, inextricably entangled with carbohydrate metabolism, mainly because dietary intake of carbohydrate which is surplus to immediate requirements is converted to fat and laid down in long-term storage, while fatty acids cannot be reconverted to carbohydrate. Moreover, the glycerol component of triacylglycerols must come from carbohydrate or a precursor, no matter what other circumstances may be. In addition, both carbohydrate and fat

may be the major energy nutrients for most cells (except brain), and each exerts a controlling influence on the extent to which the other is selected for complete oxidation. An estimate of the proportions of the major nutrients metabolized in a fasting animal at rest is given in Table 7.4.

Two relatively minor topics are the extent to which ketone bodies can replace long-chain fatty acids as fuel for extra-hepatic tissues, and the importance of poly-unsaturated acids in dietary fat. The first was not regarded as a minor topic 25 years ago, because of the danger to acid–base balance in the severe ketosis of uncontrolled diabetes, but it is probable that diabetes is nowadays better managed, and interest has switched elsewhere. Only in children can a ketotic episode still present problems (that is to say, leaving aside disorders of amino acid metabolism and the like). With regard to poly-unsaturated fatty acids, it is noteworthy that human depot fat contains a higher percentage of linoleic acid than the major dietary fats (in the U.K. at all events). This suggests that linoleic acid turns over rather slowly, and does not readily come into short supply. Only if linolenic acid is considered to be essential might dietary poly-unsaturates come into question. On this, consult Fig. 7.16.

The 'structural' lipids, of which cholesterol is the most important example, have metabolic pathways that are complex in themselves, and doubly so because they interact with the metabolism of triacylglycerols, particularly at the level of transport. Indeed, if one studies Figs. 29–33, one can see how important a role vascular transport plays in the metabolism of lipid substances, in contrast to the relatively minor role it plays in glucose or amino acid metabolism. Much of the evidence that implicates lipoproteins in the aetiology of atherosclerosis and related diseases is epidemiological, so that a purely biochemical explanation cannot yet be given. At the time of writing, the inter-relationship between cholesterol and triacylglycerol is most clear-cut with VLDL. Nevertheless, we still do not know the reason for the (epidemiologically-derived) inverse relation between HDL levels and coronary artery disease, so that we are far from being able to give a complete picture of plasma lipoprotein interactions.

The prostaglandin/thromboxane/prostacyclin group of compounds are not structural lipids, but in effect local hormones. Many of them have very short half-lives. They are included here because their relationship to one another becomes clearer if their common derivation from arachidonic acid (a component of a structural lipid) is stressed. In addition, they play an important part in the regulation of platelet activity, a topic discussed elsewhere in this volume (chapter 14).

8

Amino Acid Metabolism

The Dynamics of Amino Acid Metabolism

It is well known that many carbohydrate and lipid species are turned over rapidly within the body, but it is difficult to grasp that an equally rapid turnover may be true also of amino acid metabolism. Several enzymes are known to have turnover times of less than an hour. Although these enzymes make up only a fraction of the total protein of any cell, the total protein turnover of an adult male is estimated to be about 400 g/day, of which 50 g is accounted for by synthesis of digestive enzymes, and another 15 g by haemoglobin synthesis. By comparison, the amino acid concentration in plasma is very small, about 3.2 mM altogether, of which one-quarter is due to the amide-N of glutamine. Calculations show that the turnover of 400 g/day of protein is equivalent to the uptake, and release back into the plasma, of 4.6 moles of α-amino-N, so that the average lifetime of an amino acid molecule in plasma is about 5 min. This agrees with the observation that as much as 5–10 g of a single amino acid, injected into the bloodstream, will disappear from the plasma within 5 min.

This average turnover time for plasma amino acids, although illuminating, may be misleading, because some of the amino acids that are metabolically most active in cells are present in plasma in the lowest concentration. Aspartic acid, for example, is present at only 0.002–0.004 mM. Nevertheless these calculations show that plasma amino acids are turned over with the same kind of rapidity as plasma glucose or free fatty acids, even if one neglects the additional intake from the digestion of dietary protein.

In view of this, the overall plasma amino acid concentration remains remarkably stable; although the level may rise 30–100% for a few hours after a meal containing protein, there is no fall below normal on fasting, even if this is prolonged. In reality, there are several tissue amino acid pools, not all readily communicating with one another. In addition, the ability of tissues to take up particular amino acids varies widely. There are at least six carrier systems for various groups of amino acids, some of which are Na$^+$-dependent, while others are not. In addition, absorption of amino acids in kidney and intestine, and possibly in other cells, may involve a γ-glutamyl transferase system, which is, in effect, energy-linked (see Fig. 8.2, p. 124). A very simplified picture would show that there is a group of amino acids, such as aspartate, glutamate and glutamine, and also taurine, which are concentrated 10–50 times in cells with respect to plasma. There is a much larger group whose concentrations in cell water and in plasma are roughly the same. Finally, there is a group, which includes the branched-chain amino acids, whose concentration in cells is distinctly less than it is in plasma. The extreme members of this group are methionine and tryptophan, whose intracellular concentrations are vanishingly small.

It is thus very difficult to know how the plasma amino acid levels are maintained. It is known that overall protein synthesis is much faster if an energy-providing nutrient is fed at the same time as protein. This is one facet of the protein-sparing action of carbohydrate or fat, discussed more fully in chapter 16. Moreover, at least in liver, protein synthesis slows down relatively rapidly after a meal, and appears to depend significantly on the circulating tryptophan level, although tryptophan never disappears from the blood. It must be said that we know very little about this aspect of the control of protein synthesis. Since an amino acid residue that has been incorporated into protein is not immediately available for catabolism, it follows that—for a given protein intake in the diet—conditions that maximize new protein synthesis will minimize urea synthesis, i.e. they will favour retention of N. This is in agreement with the observations of nutritionists.

The foregoing analysis has been based on the overall flux of N. Amino acid metabolism again differs from that of fat or carbohydrate, in that the fluxes of the carbon skeletons can often be considered separately from those of the amino groups. Here also there are marked differences in metabolism. The early isotopic tracer work with ^{15}N showed that while glutamate, and also aspartate, very readily lose their α-amino groups, followed presumably by rapid removal and re-synthesis of the carbon skeletons, other amino acids, notably lysine, do not transfer their amino groups in this way, and the whole structure is irreversibly catabolized as a unit. There are too many differences between the amino acids for any useful generalization to be made on this topic.

Digestion of Proteins

The proteolytic enzymes of the digestive tract are of two kinds; the *endopeptidases*, which attack polypeptide chains at peptide bonds away from the ends, and the *exopeptidases*, which attack the terminal peptide bonds. Proteins are much more readily attacked if they have first been denatured by cooking or the action of gastric HCl.

Endopeptidases

Pepsin is secreted in the gastric juice as an inactive precursor, *pepsinogen*. Pepsinogen (mol. wt 42500) is converted into a pepsin–inhibitor complex and various small peptides spontaneously at pHs below 6.0. The reaction is very slow at pH 6.0 but almost instantaneous at pH 2.0. The reaction is autocatalysed by pepsin. At pHs below 5.4 the inhibitor (mol. wt 3100) dissociates from pepsin (mol. wt 34500), but the complex will re-form at pHs above 5.4. Both pepsin and the inhibitor are hydrolysed to peptides by pepsin itself.

Pepsin is a very acidic protein with an isoelectric point less than pH 1 and an optimum pH of 1.5–2.5 depending on the substrate. It is stable in acid solution but is rapidly inactivated in neutral or alkaline solutions; it has no prosthetic group. When acting on synthetic substrates it attacks peptide bonds between an acidic amino acid and an aromatic one ((Asp or Glu)-(Tyr or Phe)). From its action on insulin, however, it is known to attack, among others, links between two aromatic amino acids (Phe-Phe or Phe-Tyr) and links adjacent to leucine (Leu-Val and Tyr-Leu). Pepsin, like rennin, will coagulate milk by converting the phospho-protein caseinogen to casein which forms an insoluble complex with calcium.

The *gastric HCl* is secreted by the oxyntic cells of the gastric mucosa; at the same time the blood coming from the mucosa is made more alkaline. The cells contain carbonic anhydrase whose presence appears necessary for HCl secretion; about 80% of the H^+ secretion into gastric juice may fail to appear in the presence of the carbonic anhydrase inhibitor acetazolimide (diamox). The juice then contains much more Na^+ ion than usual. It is probable that the hydrogen ion, secreted together with chloride, is derived from water and the remaining hydroxide ion reacts with CO_2 under the influence of carbonic anhydrase to form bicarbonate (see Fig. 8.1). The H^+ almost certainly traverses the membrane by

Fig. 8.1.

means of a proton pump; whether there is a specific Cl^- pump as well is not known (cf. chapter 15, p. 255).

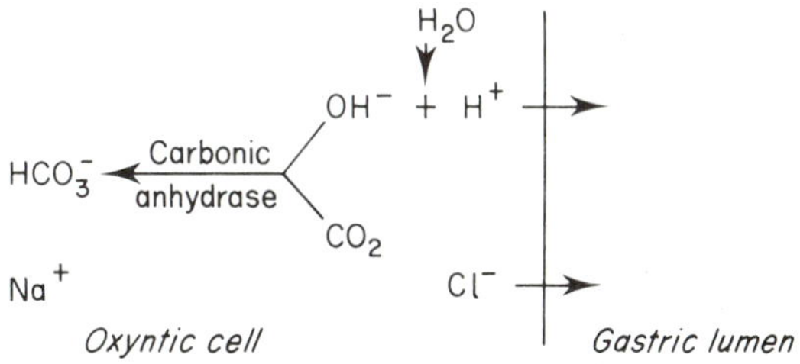

It is very doubtful whether non-ruminant gastric juice contains *rennin*, an endopeptidase with a particularly powerful milk-clotting action. It has been reported that human milk is not clotted by calf rennin.

Trypsin is secreted by the pancreas as an inactive precursor, *trypsinogen*, which is activated by the enzyme *enterokinase*, secreted by the intestinal mucosa, and then autocatalytically by trypsin itself. A hexapeptide, $Val \cdot (Asp)_4 \cdot Lys$, is split off the N-terminal end of trypsinogen during activation by trypsin, leaving an N-terminal isoleucine.

Trypsin has no prosthetic group; it is relatively stable to heat in acid solution but less so in alkaline solution. The optimum pH is in the range 7–9 and it has a low Michaelis constant, indicating that the substrate is firmly bound to the enzyme. It catalyses the hydrolysis of peptides, amides, and esters where a diaminomonocarboxylic amino acid (Lys or Arg) provides the carboxyl group. When acting on a natural substrate it also splits other bonds, e.g. Arg-Gly, Lys-Ala, Phe-Tyr, Lys-Tyr, Arg-Arg, Arg-Ala, and Tyr-Leu.

The *chymotrypsin* group of enzymes is all derived from a common precursor *chymotrypsinogen* secreted by the pancreas. The activation is initially brought about by trypsin to give an active chymotrypsin which may be converted to other chymotrypsins by autolysis; there is no change in the molecular weight in the first step, but there are differences in the subsequent products. The optimum pH is 7–8, the Michaelis constant is high, and the enzymes have no prosthetic group. These enzymes attack peptides or esters of a number of amino acids, but particularly non-polar ones (Leu-, Tyr-, Phe-, Met-, Trp-).

Elastase (pancreatopeptidase E), from the pancreas,

hydrolyses peptide bonds adjacent to small neutral amino acid residues such as Ala, Gly and Ser.

These endopeptidases (pepsin, trypsin, the chymotrypsins, and elastase) bring about the hydrolysis of large protein molecules to smaller peptide fragments. Their further hydrolysis then depends on the action of a number of exopeptidases and dipeptidases either secreted by the pancreas or to be found within the cells lining the intestinal mucosa.

Exopeptidases

A number of these enzymes, in contrast to the endopeptidases, require a metal ion as activator.

The two *carboxypeptidases* are secreted as precursor *procarboxypeptidases* which are activated by trypsin. Carboxypeptidase A contains firmly bound Zn^{2+} and hydrolyses off the carboxy-terminal amino acid unless this is lysine or arginine. Carboxypeptidase B hydrolyses peptides with carboxy-terminal lysine or arginine. Neither will attack dipeptides.

Leucineaminopeptidase is the best characterized of the intracellular exopeptidases that complete the digestion of protein in the gut. It brings about the hydrolysis of amino-terminal residues from peptides, but not from dipeptides. In spite of the name, it is rather unspecific. There is also a *prolidase* which catalyses the hydrolysis of proline peptides which are mainly derived from the breakdown of collagen.

Absorption

From experiments involving the feeding of isotopically labelled proteins, it is known that absorption is very rapid in man. The extent of hydrolysis of the food protein is low in the stomach, 10–15%, but quickly reaches values of 50–60% in the duodenum. In the duodenal contents the enzymes trypsin and chymotrypsin are present at concentrations of 200–800 μg per ml of fluid, within a short time of stimulation. These concentrations of enzymes are capable of rapidly hydrolysing food proteins to small peptides. Absorption of fed protein fragments takes place in the duodenum and the jejunum, most of it absorbed as di- and oligopeptides. Although it is generally believed that all ingested protein is completely split to amino acids (either in the lumen or in the mucosa) before passing into the body fluids, it must be pointed out that this has never been rigorously demonstrated by quantitative experiments.

In *coeliac disease* a peptide of molecular weight about 1500 formed by enzymic hydrolysis from the gliadin fraction of cereal gluten has a toxic effect on the mucosal cells of the jejunum. Although it is far from established that this is the primary cause of the disease, it indicates that quite large peptides can penetrate the brush border.

In the new-born infant considerable amounts of colostrum proteins can be absorbed, particularly in the first 48 hours of life. This process is, however, very selective, greatly favouring globulins at the expense of albumins. It is believed that absorption of whole proteins ceases after the first 2 weeks of life.

Transport into cells

The absorption of L-amino acids and some L-peptides is an active process and requires the metabolism of the mucosa to be intact. This absorption is interfered with by the presence of the D-isomers. Two mechanisms are known to exist.

1. The major mechanism is by carriers that are Na^+-dependent, so the amino acids are not taken up unless Na^+ is also present in the lumen. The Na^+ then has to be pumped out of the mucosal cell by a Na^+/K^+ pump (see also chapter 15). A similar mechanism accounts for the transport of some, but not all, amino acids into tissue cells (cf. p. 122).

2. *The glutamyl-transfer mechanism.* This is widespread, i.e. not confined to mucosal cells, but it is not easy to estimate what proportion of amino acids absorbed from the gut, or in general transferred across cell membranes, are moved by this mechanism. Indirectly it is highly energy-dependent, since glutathione—the glutamyl donor—has to be completely resynthesized after each transfer (for details see p. 140), while the disposal of the by-product pyroglutamate also requires ATP (p. 133).

The details of the mechanism are shown in Fig. 8.2. The glutamyl group, which in glutathione is attached through the γ-COOH group, rather than the α-carboxyl as is usual, is transferred to the amino acid to be absorbed. This presupposes that the glutathione, and

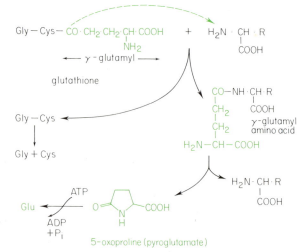

Fig. 8.2 The γ-glutamyl transfer mechanism for the translocation of amino acids across plasma membranes.

the transferase, are at least exposed to the outer surface of the cell. It is presumed that the γ-glutamyl amino acid, and the residue of the glutathione (cysteinylglycine), diffuse back into the cytosol, where the latter dipeptide is hydrolysed, since it is not an intermediate in glutathione synthesis (cf. p. 140). The γ-glutamyl amino acid is also hydrolysed, but the glutamate is cyclized to pyroglutamate (5-oxoproline) during the course of this hydrolysis.

Amino Acid Catabolism: General

It has been mentioned in the preceding section that metabolism of the amino group of amino acids can often be looked at separately from that of the carbon skeleton. This chapter is therefore divided into two sections: a consideration of pathways common to all amino acids, and of the ways in which the nitrogen is excreted, and then some discussion of the individual pathways and some of the more important syntheses of non-protein nitrogenous compounds from amino acids.

Some eight of the 20 amino acids commonly found in proteins are *essential* (see chapter 16); that is to say, their carbon skeletons cannot be synthesized in the animal body. A discussion of the way in which they are synthesized by plants or micro-organisms will not be given here.

The general catabolic pathways involving amino acids are indicated in Fig. 8.3.

The first stage in the catabolism of almost all amino acids is the formation of the corresponding oxo-acid

$$R—CH—NH_2 \atop \underset{COOH}{|} + \tfrac{1}{2}O_2 \rightleftharpoons R—CO \atop \underset{COOH}{|} + NH_3 \qquad (1)$$

The last stage is the formation of urea, $NH_2 \cdot CO \cdot NH_2$, and of CO_2 and water.

An L-*amino-acid oxidase*, directly catalysing reaction (1) above, occurs in animal tissues. The enzyme, however, is present in very low concentration even in tissues, such as liver or kidney, in which amino acid catabolism is vigorous. Most of the amino groups must

thus be removed by another route, discussed in the next paragraph. Curiously, D-*amino-acid oxidase* is present in high concentration in kidney in particular. D-Amino acids occur infrequently in some peptides of bacterial origin but are not quantitatively important. In addition, some amino acids, e.g. methionine, racemize quite readily. It is probable that the D-amino acid oxidase is a detoxifying enzyme.

Transamination

If to a liver or kidney homogenate is added an amino acid and α-oxoglutaric acid, the reaction mixture will soon contain, besides the original components, the oxo-acid corresponding to the amino acid and also glutamic acid; the enzyme carrying out this reaction is a *transaminase*:

$$\underset{A}{\underset{COOH}{\overset{R}{|}} \atop CH \cdot NH_2} + \underset{B}{\underset{COOH}{\overset{R'}{|}} \atop CO} \rightleftharpoons \underset{C}{\underset{COOH}{\overset{R}{|}} \atop CO} + \underset{D}{\underset{COOH}{\overset{R'}{|}} \atop CH \cdot NH_2} \qquad (2)$$

In the example mentioned, B was α-oxoglutaric acid. The reaction is reversible, so that no matter from which direction it is started, the reaction mixture will finally contain all four components.

This *transamination* system is now accepted as the most important way of deaminating amino acids.

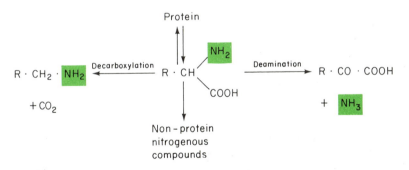

Fig. 8.3.

Confusion may arise unless certain points are understood:

1. *Specificity*. The amino *donor* A may be almost any one of the normally occurring amino acids but the *acceptor* B can only be α-oxoglutaric acid. Confusion may occur if the reaction is thought of as running from *right to left* in equation (2); now the transaminases are specific for a single amino donor (D)—glutamic acid—while the acceptor (C) may be any one of a number of oxo-acids.

In several plants and micro-organisms, transaminases have been found that are specific for acids other than oxoglutaric as amino acceptor (B), namely pyruvic acid or oxaloacetic acid. These acceptors have never been found to be important in mammalian metabolism, and may be neglected. Of course there are, in animals, two specific transaminases for which the acceptor (C) is pyruvic acid or oxaloacetic acid, respectively. This underlines the importance of specifying the direction of the reaction. In the alternative nomenclature for these enzymes—*aminotransferases*—this ambiguity is removed, because the two enzymes mentioned above are alanine aminotransferase and aspartate aminotransferase. Unfortunately they both have the same acronym (AAT), so that the older nomenclature with the acronyms GPT and GOT is often retained.

2. *Reversibility*. Although this reaction produces an oxo-acid (C) from an amino acid (A), it is reversible. (A) will not continue to be deaminated unless the reaction products (C) and (D) are continuously removed. There must, therefore, be *another* mechanism for deaminating glutamic acid, i.e. one which does not depend on a transamination.

Transamination is also important as a means of changing the proportions of amino acids available for synthesis. It is clear that the proportions of the constituent amino acids in the proteins of food are not necessarily related to the proportions, or absolute amounts, of each of the amino acids required by the body for synthetic purposes. If two transamination reactions are coupled together as shown in Fig. 8.4, the overall effect has been to synthesize some of the amino acid with side chain B at the expense of the amino group of the amino acid with side chain A. This process must

obviously depend on the availability of the oxo-acid $B \cdot CO \cdot COOH$, and cannot therefore be used to any significant extent for the formation of the essential amino acids. Glycine, too, although non-essential is not synthesized in this way. Nevertheless, the isotopic evidence suggests that amino transfer of this kind plays a part in the body's amino acid economy.

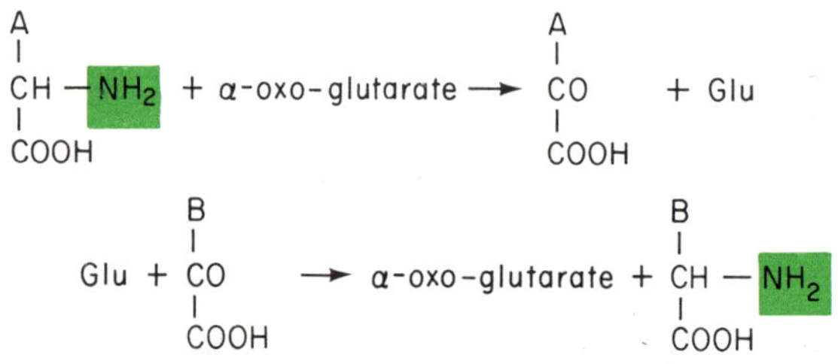

Transaminases require, as a prosthetic group, pyridoxal phosphate (vitamin B_6). The mechanism of this and similar reactions is described below.

The mechanism of transamination

The deamination of lysine, via saccharopine, to α-aminoadipic semialdehyde (Fig. 8.28, p. 146), illustrates the general principle of transamination. The movement of an amino group from a donor to an acceptor molecule proceeds by formation of a *Schiff's base*

$$R \cdot CH_2 \cdot NH_2 + O = C \cdot R' \rightleftharpoons R \cdot CH_2 \cdot N = C \cdot R' + H_2O$$

which is an unstable structure, easily reconverted to an amine and a carbonyl by the addition of water. The mechanism consists in moving the double bond from one side of the central N atom to the other, before the link is broken. Study of the mechanism of lysine deamination (in which the Schiff's base is reduced in an intermediate stage) shows that the nature of the enzyme to which the central intermediate saccharopine is bound, determines which of the two C—N bonds is attacked. This dependence on the nature of the catalytic site is also true for the more common transamination mechanism described below.

Nature has developed a cofactor (pyridoxal phosphate) with which the amino donor can form a Schiff's base that can remain at the active site of the enzyme, and subsequently react with the amino acceptor. This removes both the requirement for two enzymes and the release of a free intermediate. Moreover, pyridoxal phosphate has an aromatic ring that can act as an electron sink. Given a suitable enzyme, this gives the capacity to labilize *any* atom or group on the carbon atom proximal to the Schiff's base, thus widening the range of the enzyme + cofactor very considerably. We find, in fact, that pyridoxal (or pyridoxamine) phosphate is the prosthetic group of over 100 enzymes, catalysing the following main classes of reaction:

(a) transamination;
(b) decarboxylation of amino acids;
(c) loss of side chain, as in transhydroxymethylation, or threonine cleavage;
(d) racemization.

In addition, pyridoxal phosphate is found at the active site of some enzymes, such as glycogen phosphorylase, where its function is unknown.

A
|
CH — NH₂ + α-oxo-glutarate ⟶ CO + Glu
| |
COOH   COOH

          B   B
          | |
Glu + CO ⟶ α-oxo-glutarate + CH — NH₂
          | |
          COOH   COOH

Fig. 8.4.

The mechanism of transamination is summarized diagrammatically below, and then the other possibilities inherent in the labilization of bonds at the α-carbon are briefly outlined. The diagrams are simplified, since it is known that in some enzymes at least, when substrates are absent, the pyridoxal phosphate is bound, again by a Schiff's base, to the terminal —NH₂ of a lysyl residue near the active site. This explains why the cofactor does not readily diffuse away from the enzyme, and has more of the properties of a prosthetic group than of a coenzyme. This extra step in the mechanism is omitted here.

Nomenclature. The approved name for enzymes that catalyse the reactions outlined below is 'aminotransfer-ases'. There is, however, no equivalent noun that describes the *process*, which is consequently still known as 'transamination', and many people still refer to the enzymes as 'transaminases', rather than aminotransfer-ases. In this book, the older nomenclature is preferred, without the approved form being rigorously excluded.

The steps $(1) \rightarrow (4)$ in Fig. 8.5 are stages in the half-reaction which leads to the release of the oxoacid R·CO·COOH corresponding to the amino acid substrate R·CH(NH₂)·COOH. Note that the pyridoxal phosphate is left as an amine, pyridoxamine phosphate. It does not diffuse away from the enzyme, probably because of electrostatic binding between the phosphate group and a positively-charged group at the active site.

After the departure of the oxoacid R·CO·COOH, a second oxoacid (not a random visitor, but the second substrate) attaches to the active site, and the reactions are repeated in the reverse order, $(4) \rightarrow (1)$, ending up with the second product R'·CH(NH₂)·COOH.

This mechanism, in which the first product leaves before the second substrate binds, with the enzyme (i.e. its prosthetic group) temporarily in an altered form, is called a 'ping-pong' mechanism. It is characteristic of aminotransferase reactions. Note that in the intermediate stages, between steps (2) and (3), a base is needed at the active site to act as proton acceptor and donor. The geometry of the site determines the specificity of the enzyme for the amino donor and acceptor.

If we look again at the aldimine form of the cofactor, supposing an amino acid to be the substrate, we can appreciate that it is not obligatory that the C—H bond on the α-C should be the one to be labilized. If the C—C bond between the α-C and the carboxyl group is labilized instead, the products will be CO₂ together with an amine. If the bond between α- and β-carbons breaks, the products will be glycine and the appropriate side chain. This may be released as a free entity, as with threonine, which is split to glycine and acetaldehyde (see p. 145). More frequently, tetrahydrofolic acid acts as an acceptor (see p. 138). Finally, if the C—H bond is broken and then re-formed, the asymmetry of the α-carbon may be changed, so that the other enantiomer will be formed (see Fig. 8.6). If this goes on for long

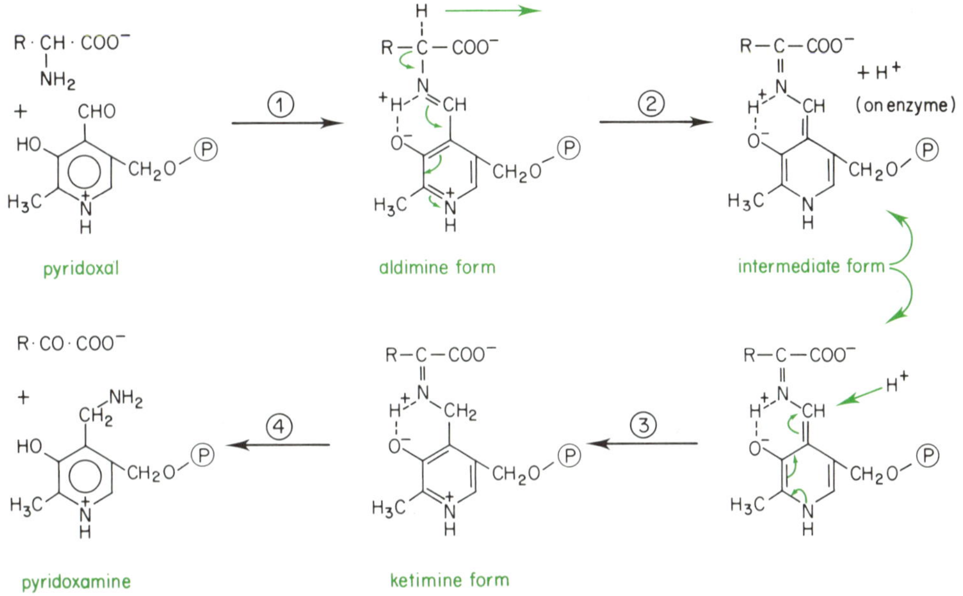

Fig. 8.5 The mechanism of transamination. The cofactor to which the —NH₂ group is first transferred is *pyridoxal phosphate*. When the amino transferase is not catalysing a transamination, the cofactor is bound through a Schiff's base to a lysyl group near the active centre. *Pyridoxamine phosphate* is the stable form of the cofactor formed halfway through the complete transamination cycle.

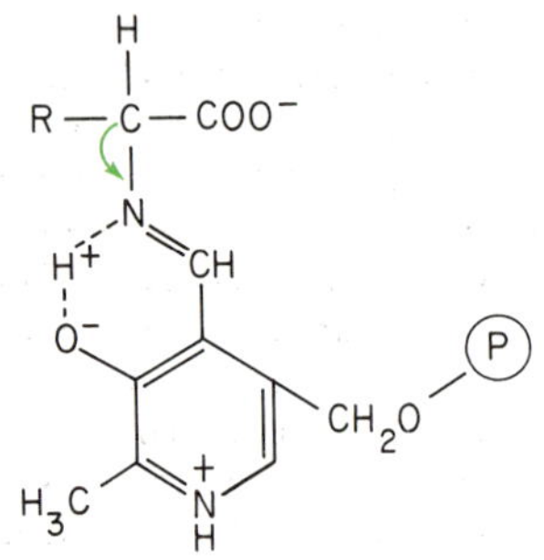

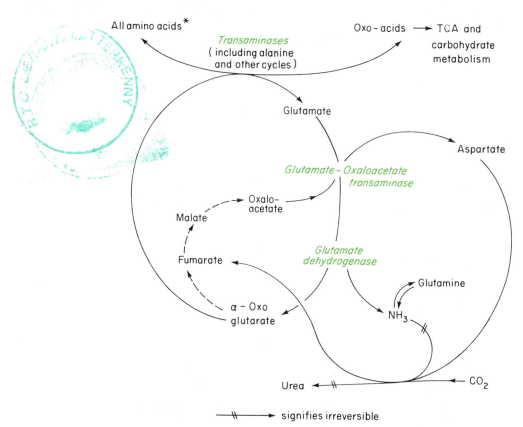

Fig. 8.6.

enough, the system will contain an equilibrium mixture of both isomers; this is *racemization*. Note that in the types of reaction discussed in this paragraph, there is no second substrate.

Decarboxylations apart, reactions catalysed by pyridoxal phosphate are reversible, so that glycine together with an acceptor molecule may be used in a ligase reaction. Examples are the formation of serine from glycine (Fig. 8.23), and the condensation of glycine with

succinyl-CoA to give δ-aminolaevulinic acid (p. 44). The ambiguity arising from the reversibility of aminotransferase reactions is discussed on p. 126.

There are many drugs that inhibit pyridoxal-phosphate-containing enzymes. *Isonicotinic acid hydrazide* is very effective in the treatment of tuberculosis. Other drugs are analogues of amino acid substrates, e.g. *cycloserine*. Many such compounds inhibit the enzymes in human tissues also, and are therefore too toxic for antibacterial or antitumour use.

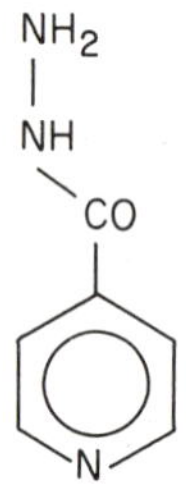

Isonicotinic acid hydrazide

⟶╫⟶ signifies irreversible

Fig. 8.7 The production of ammonia from amino acids and its excretion as urea (overall view).
* With some exceptions (see pp. 133–149).

The major products of transamination: glutamate and aspartate

Glutamate is an inevitable product of transamination, so its further metabolism is of primary importance. In addition, the aminotransferase found in highest activity in most tissues is aspartate aminotransferase (GOT), suggesting that transamination between glutamate and oxaloacetate is of particular importance. This is indeed the case.

Glutamate dehydrogenase. This enzyme is not a flavoprotein like the other amino acid oxidases, but requires either NAD or NADP as coenzyme. It is present in high concentration in several tissues; no comparable dehydrogenase for any other amino acid is known. It is, like most NAD (or NADP)-linked dehydrogenases, reversible. The reaction which it catalyses is:

$$HOOC \cdot (CH_2)_2 \cdot CH(NH_2) \cdot COOH + NAD(P)^+ + H_2O$$
$$\rightleftharpoons HOOC \cdot (CH_2)_2 \cdot CO \cdot COOH + NH_3 + NAD(P)H$$

The equilibrium lies fairly strongly in the direction of glutamate synthesis, but is pulled in the direction of deamination by the continuous removal of NH_3.

The enzyme is therefore of central importance in any scheme for the deamination of all amino acids (Fig. 8.7).

Fig. 8.8 shows that the amino group of aspartate is used directly in providing one of the $—NH_2$ groups of urea. Thus, so far as the synthesis of urea—the major end product of amino acid catabolism—is concerned, glutamate and aspartate are of equal importance. It is significant that in isotopic tracer experiments, ^{15}N is always found in highest concentration in glutamate and aspartate. Aspartate amino transferase is found in both mitochondria and the cytoplasm. One explanation for this (at least for liver) will be found on p. 159; the dual location of this enzyme is also made use of in the shuttling of reducing equivalents between cytosol and mitochondria (see p. 162).

The transport of ammonia equivalents

In liver, ammonia (more precisely NH_4^+) is set free as such by the glutamate dehydrogenase reaction described above. It is doubtful whether a great deal of free NH_4^+ is ever produced in routine amino acid metabolism in other tissues. In anaerobic muscle, prolonged contraction will cause AMP to be produced, which is readily

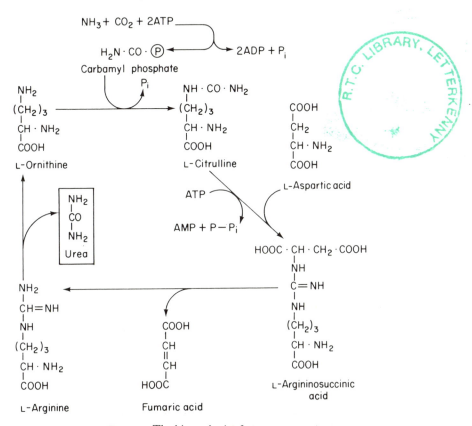

Fig. 8.8 The biosynthesis of urea.

deaminated to IMP (p. 57), but this is not a continuous mechanism. Most tissues have a glutaminase, but except in two tissues (kidney and small intestine) glutamine metabolism seems to be rather sluggish. Nevertheless, there is significant amino acid catabolism in tissues other than liver; leucine uptake by muscle is mentioned below (p. 143). It follows that pathways for transport of —NH_2 groups in blood, other than as NH_4^+, must exist (ammonium ion is toxic to brain, as discussed below). Two such pathways are described in this section.

The alanine cycle

Alanine is present in plasma at a concentration (0.3–0.4 mM) second only to that of glutamine, and the concentration rises and falls with the amino acid status

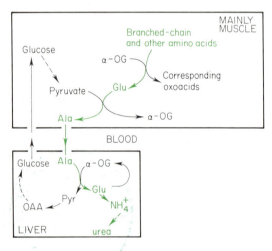

Fig. 8.9 The alanine cycle, which transfers amino groups from peripheral tissues to liver.

of the subject. Measurements of arteriovenous amino acid differences in the forearm of human volunteers show that there is a positive A-V difference only for alanine, i.e. this amino acid alone is released into the blood. In vitro experiments show that there is a net formation of alanine in muscle tissue at a rate proportional to the intracellular concentration of pyruvate. Conversely, in vitro liver preparations take up alanine, with a concomitant increase in urea output. On the basis of these findings a glucose–pyruvate–alanine cycle has been postulated to serve as the chief vehicle for —NH_2 transport from muscle to liver, as outlined in Fig. 8.9. Note that, since pyruvate production by muscle is much reduced in fasting, this cycle must be much less active in such conditions. It is not clear whether amino acid catabolism is equally reduced in muscle; see a further discussion under leucine (p. 145).

A similar alanine transport operates from small

intestine to liver, but as this is interlocked with glutamine transport, it is discussed below.

Since muscle constitutes such a large proportion of total body weight, this cycle can account for a considerable proportion of total non-hepatic amino acid catabolism. Nevertheless, it must be borne in mind that rather little is known about —NH_2 export in other active tissues, especially brain, where maintenance of a low free NH_4^+ ion concentration is especially important.

Glutamine cycles

The compound methionine sulphoximine

$$CH_3 \cdot \underset{\underset{O}{\downarrow}}{\overset{\overset{NH}{||}}{S}} \cdot CH_2 \cdot CH_2 \cdot \underset{NH_2}{\overset{|}{CH}} \cdot COOH$$

is a powerful inhibitor of *glutamine synthetase*:

$$Glu + NH_4^+ + ATP \rightarrow Glu—NH_2 + ADP + P_i + H^+$$

In many species intravenous injection of the inhibitor causes a fall in plasma glutamine and a rise in plasma NH_4^+ concentrations, followed by rapid development of coma. This and other evidence has led to the supposition that glutamine is an essential carrier of —NH_2 groups from peripheral tissues to liver for detoxication. However, in muscle glutamine synthetase is present in only trace amounts, so it is difficult to see how glutamine could be as important as alanine in peripheral → hepatic transfer. It now seems more likely that the glutamine synthesis that is inhibited by methionine sulphoximine is in liver itself, and that glutamine acts in hepatic cells as a buffer to keep ammonia concentrations low in the stage between deamination of glutamate and synthesis of citrulline.

In addition to this, there are two transport functions of glutamine that are important; to call them cycles is not strictly true. The first is the export of glutamine from liver to kidney, where the glutamine is deaminated to provide free NH_4^+ ions for buffering urine (chapter 15). The fact that the glutamine can be shown to come from liver adds point to the suggestion that glutamine is not, to any extent, synthesized in extra-hepatic tissues and moved to the liver. This transport is shown in Fig. 8.10(a); it will be seen that although the resulting glutamate could in principle be returned to the liver, in all probability much of it is oxidized in kidney mitochondria, incidentally providing a second NH_4^+ ion for secretion into urine.

The second transport mechanism is basically concerned with the provision, not of —NH_2, but of carbon skeletons. This is the provision of glutamine as a major energy nutrient for small intestinal mucosa (see chapter 9, p. 158). Glutamine appears to be used because it readily penetrates membranes, unlike

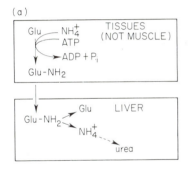

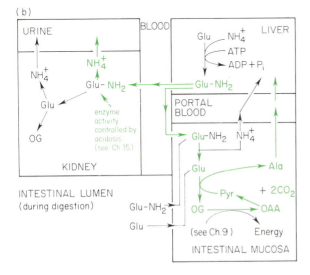

Fig. 8.10 Glutamine cycles. (a) Transport of amino groups from some peripheral tissues to liver. (b) Transport from liver to other tissues. The left-hand box indicates how glutamine is used as a carrier of —NH$_2$ groups to kidney, as a source of NH$_3$ (see chapter 15). The lower right-hand box shows how glutamine is used as a carrier of carbon skeletons to be oxidized as a source of energy. This is particularly important for the cells of intestinal mucosa, but other cell types can also use glutamine carbons for oxidation.

glutamate. Much of it no doubt comes from the intestinal lumen during absorption of protein-containing meals, but in any case, the utilization of glutamine means that two —NH$_2$ groups have to be disposed of per carbon chain partially oxidized. One of them can be accounted for by transamination between glutamate and pyruvate, as shown in Fig. 8.9. The other may be excreted in the blood, since it will normally be taken up quickly from (portal) vein blood by the liver, but there is some suggestion that citrulline may be used to provide part of the transport.

In this specialized cycle, it is the —NH$_2$ groups which cycle between liver and intestine. Of the five carbon atoms leaving the liver as glutamine, three will return as pyruvate, but two will be oxidized to CO$_2$.

Glutamine is also taken up from the lumen of the intestine, and oxidized by the mucosa, when available.

The overall balance of amino group catabolism in ureotelic animals is shown in Fig. 8.7.

Ammonium ion is toxic to brain, because when its concentration rises, the glutamate dehydrogenase reaction (p. 129) is forced backwards, and the concentration of α-oxoglutarate in the tissue drops, until the catalytic function of the TCA cycle is impaired. Oxygen consumption by the brain can be shown to fall as the blood ammonia concentration rises, and this is the dangerous consequence of impaired urea formation in failure of liver function (hepatic coma). Chronic ammonaemia leads to irreversible brain damage, especially in children. Curiously, the concentration of NH$_4^+$ in other organs is quite high (about 0.5 mM), much higher than in plasma, and they do not seem to suffer from TCA cycle inhibition. The intense ammonia production from AMP in anaerobic muscle has already been mentioned.

The synthesis of urea

The pathway of urea formation in mammals is shown in Fig. 8.8. It may be noted that the carbon atom and one of the nitrogen atoms of urea arise from CO$_2$ and ammonia while the second nitrogen atom comes via aspartic acid. The enzymes concerned in urea synthesis are all found in the liver, although they are not confined to this organ. Experimental hepatectomy has, however, shown that the liver is the site of production of blood urea. Urea synthesis is an energy-requiring process and 3 molecules of ATP are used in the formation of each molecule of urea.

Detailed study of the urea cycle in liver shows that the steps leading to the synthesis of citrulline from ornithine are located in liver mitochondria, while the other reactions are cytosolic. At first sight this seems unnecessarily wasteful, since it means that the enzymes converting fumarate to oxaloacetate have to be duplicated in the cytosol. In addition citrulline must be transported out of mitochondria, to be balanced by an energy-dependent inward transport of ornithine.

Further consideration suggests that this arrangement arises from the necessity of keeping the fumarate concentrations low within the urea cycle. The reaction which splits argininosuccinic acid has a K_{eq} of 10^{-2} M, which means that it can readily be inhibited by its products, fumarate and arginine. Within mitochondria, fumarate (and malate) concentrations are high (ca. 1 mM) because they are essentially linked to the citric acid cycle, and to the mitochondrial NADH/NAD$^+$ ratio. In the cytosol, fumarate is not continuously produced from succinate. Moreover, the malate concentration is much lower because the cytosolic NADH/NAD$^+$ ratio is $\sim 10^{-3}$, instead of ~ 1 as it is in

131

mitochondria (cf. chapter 9). Thus the fumarate concentration will also be lower, and much less likely to interfere with the splitting of argininosuccinate.

The synthesis of argininosuccinate is quite reversible, and is probably only pulled in the direction of synthesis by the hydrolysis of pyrophosphate (cf. p. 80).

By far the greatest proportion of nitrogen excreted in man is in urea. The actual amount of urea varies with the protein intake but is between 10 and 25 g/day. It is a very soluble substance, distributed throughout the body water, both intracellular and extracellular. It is completely harmless to tissues, even at higher than normal concentrations, and is so far as we can tell metabolically inert. *Uraemia* is usually a symptom of renal dysfunction; it may possibly pose problems because of the retention of water which ensues.

Many plants and micro-organisms, however, contain an enzyme, *urease*, which will hydrolyse urea to $2NH_3$ and CO_2. The nitrogen thus becomes available for the growth of these organisms. Urea may be used as a source of protein nitrogen in ruminants, and a small amount may be tolerated by man, the flora in the alimentary tract hydrolysing it to CO_2 and NH_3, which is absorbed.

Decarboxylation

Removal of the —COOH group as CO_2, leaving a primary amine, is not in man a major pathway of amino acid metabolism. Several of the resultant amines or derivatives of them are, however, physiologically important. Examples are *histamine* from *histidine*, *adrenaline* from *tyrosine* by way of *tyramine*, *serotonin* or *5-hydroxytryptamine* (5-HT) from *5-hydroxytryptophan*, *ethanolamine* from *serine*. *Putrescine* and *cadaverine*, which give the characteristic smell and taste to tainted meat, are the amines corresponding to *ornithine* and *lysine* respectively. The most active of all these enzymes is probably brain *glutamate decarboxylase* (p. 133). The amino acid decarboxylases have pyridoxal phosphate as a prosthetic group. *Amine oxidases* initiate the further oxidation of these compounds.

The fate of the oxo-acid residues

As already described, an oxo-acid is the first product of amino acid catabolism in nearly all cases. Each oxo-acid or 'carbon skeleton' follows a unique metabolic pathway to a compound which can be completely oxidized by way of the tricarboxylic acid cycle. The detailed pathways for most of the amino acids will be given later in this chapter; in this section two general categories will be described.

It was observed many years ago that when single amino acids were fed to diabetic or phlorhizinized animals, some caused the excretion of extra glucose in the urine, others the excretion of extra ketone bodies. According to the results of this type of experiment, the amino acids then known were classified as *glucogenic* or *ketogenic*. It is now realized that the glucose excreted after feeding a glucogenic amino acid does not necessarily come directly from the carbon residue itself. The residue can be directly incorporated into the TCA cycle and so 'spare' glucose, i.e. leave more glucose in the blood to be (in diabetes) excreted. In starvation, on the other hand, the glucose found in the blood must have been formed by *gluconeogenesis*; almost all of it comes directly from the carbon skeletons of glucogenic amino acids.

The ketogenic amino acids give rise to acetoacetic acid which cannot *directly* be oxidized in the tricarboxylic acid cycle. It is, of course, oxidized completely, in normal conditions, unless carbohydrate is absent, just as in the oxidation of fats.

The usual classification is:

Glucogenic: Gly, Ala, Ser, Thr, Cys, Met, Val, Glu, Asp, Arg, His, Pro, Hypro.

Ketogenic: Leu.

Glucogenic and ketogenic: Ile, Phe, Tyr.

Unclassified: Lys, Trp.

The interrelationships between the pathways of amino acid metabolism and the terminal common pathway are shown in Fig. 8.11.

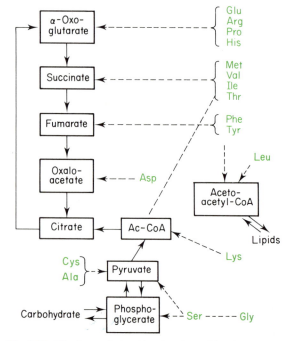

Fig. 8.11 The interrelationship of amino acids with the tricarboxylic acid cycle.

Metabolism of Individual Amino Acids

Alanine

Alanine is glucogenic and non-essential. The alanine-oxoglutarate transaminase (GPT), and the alanine cycle, have already been mentioned on p. 130.

Aspartic acid and asparagine

Aspartate is very readily metabolized by isolated tissues; although its concentration in plasma is very low, it is much more concentrated in cells. Transport across the mitochondrial membrane is energy-linked, and it is possible that this is also true for the plasma membrane. It first undergoes transamination with α-oxoglutarate to oxaloacetate and glutamate. The former is at the beginning of the pathway of glucogenesis (chapter 6), so that aspartate is strongly glucogenic. Besides this, the nitrogen atom of aspartate is used specifically in the synthesis of purines (Fig. 5.3.) and pyrimidines (Fig. 5.2), and of arginine from citrulline (Fig. 8.8). Thus half the nitrogen in all urea excreted has passed through the amino group of aspartate.

Asparagine, the amide of aspartic acid, occurs widely in plants but its concentration in animal tissues is fairly low. More than one ATP-dependent asparagine synthetase appears to exist, but the total rate of synthesis may be very low. Asparagine supplementation increases the growth rate of young rats which puts it in the semi-essential category, like arginine and histidine. Nothing is known about human demands.

Some tumours appear to lack asparagine synthetases, and to depend entirely on the host. These tumours regress rapidly when the host is injected with bacterial asparaginase, but so far complete regression has not been achieved, partly because the rate of asparagine synthesis in the host responds, even if slowly, to the increased rate of loss.

Amino acids related to glutamate

Glutamate and glutamine

Most of the free amino acid pool in the body, both in tissues and in plasma, consists of glutamine and glutamic acid. The importance of the latter as a metabolic intermediate in amino acid metabolism cannot be over-emphasized.

The formation of glutamic acid by transamination and its reversible conversion to α-oxoglutaric acid and ammonia have already been discussed.

Glutamate decarboxylase, which is very active in brain, converts glutamate to γ-aminobutyric acid (GABA), which is important in the function of the central nervous system. GABA is transaminated to succinic acid semi-aldehyde, which is oxidized to succinate.

Although glutamate is readily metabolized, excess in the diet can cause unpleasant neurological disturbances in sensitive persons. It is widely used to enhance flavour, particularly in Chinese cookery. About 100 000 tons are made industrially each year for use as a food additive.

Glutamine is synthesized in most tissues by the following reaction

$$HOOC \cdot CH(NH_2) \cdot CH_2 \cdot CH_2 \cdot COOH + ATP + NH_3$$

$$\rightarrow HOOC \cdot CH(NH_2) \cdot CH_2 \cdot CH_2 \cdot CONH_2 + ADP + P_i.$$

but except in liver the rate is usually slow. As discussed earlier in this chapter, the probability is that most glutamine is transported *from* liver *to* other organs, to be used as a source of either NH_4^+ or oxoglutarate. Unlike glutamate, glutamine readily crosses both plasma and mitochondrial membranes. Glutamine is a major fuel for the mucosal cells of the small intestine, see chapter 9.

A phosphate-activated glutaminase, particularly concentrated in kidney, hydrolyses glutamine to glutamate and ammonium ion. It is part of the renal system for the control of acid–base balance (see Fig. 15.5). Besides this, the amide-N of glutamine is used in the synthesis of purines (see Fig. 5.1) and histidine.

Pyroglutamic acid

This substance, also called 2-oxoproline, is not a normal constituent of protein. Some polypeptide chains (notably the heavy chains of immunoglobulins) have it as their N-terminal residue in place of Met; it is thought to arise by cyclization of glutamine after the chain has been synthesized.

In addition, it is formed in the γ-glutamyl transpeptidation mechanism for amino acid transport, which has been discussed earlier in this chapter (p. 124). Twenty-five to fifty grams of the acid may be excreted per day in a rare inborn error of metabolism, but it seems likely that this is due to over-production, and the normal turnover in the body is much less than this.

Pyroglutamate is the internal amide (lactam) of glutamate. The five-membered ring is very stable, and opening it requires energy from ATP:

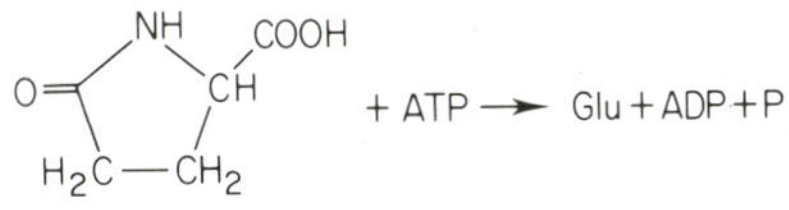

133

Proline and hydroxyproline

Proline is a non-essential amino acid, required in quantity for the synthesis of collagen. It is synthesized from glutamic acid semialdehyde

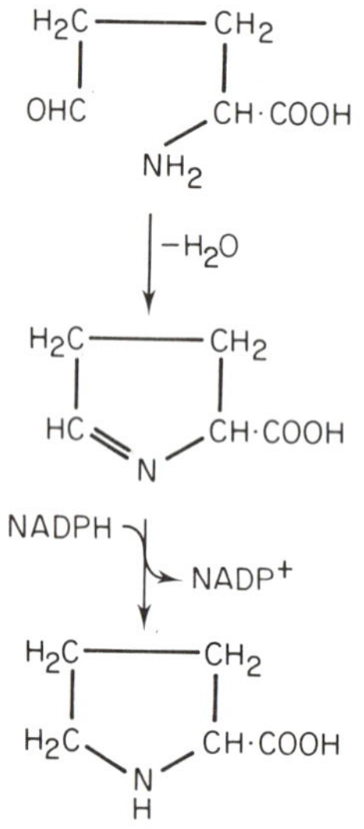

Fig. 8.12 The biosynthesis of proline.

For the formation of this precursor, see under *ornithine*.

Proline oxidation involves first a mitochondrial flavoprotein-catalysed dehydrogenation in the ring. The product is equivalent to glutamate semialdehyde, but a second oxidation step takes place without the ring being opened. Hydroxyproline can be oxidized by the same system, but the product is 4-OH-glutamate, which is split to glyoxylate and pyruvate.

Hydroxyproline is found only in collagen: the intake from food is probably low. It is formed by oxidation of prolyl residues in procollagen. The enzyme is a mixed function oxidase in which α-oxoglutarate acts as oxidizable co-substrate (see chapter 9, p. 169).

A fraction of degraded collagen is excreted as small Pro-Hypro-peptides (mol. wt $\leqslant 1000$). The endogenous release of hydroxyproline in an adult is about 2 g per day; about one-sixth of this is excreted in peptide form.

Histidine

This is another amino acid part of whose skeleton is converted to glutamic acid, but details of its metabolism are more complex than for ornithine and proline. It is not essential for man, but is so for many animals and for young children. It is synthesized from 5-phosphoribosyl-pyrophosphate and ATP (see Fig. 8.14). The oxidation of histidinol to histidine (bottom line of Fig. 8.14) is one of the few examples of a dehydrogenase which uses 2 NAD$^+$ (i.e. catalyses a 4ϵ-transfer). The oxidation of UDPG to UDP-glucuronic acid (p. 82) is another.

Histidine is catabolized by opening the ring to produce formimino-glutamic acid. The formimino

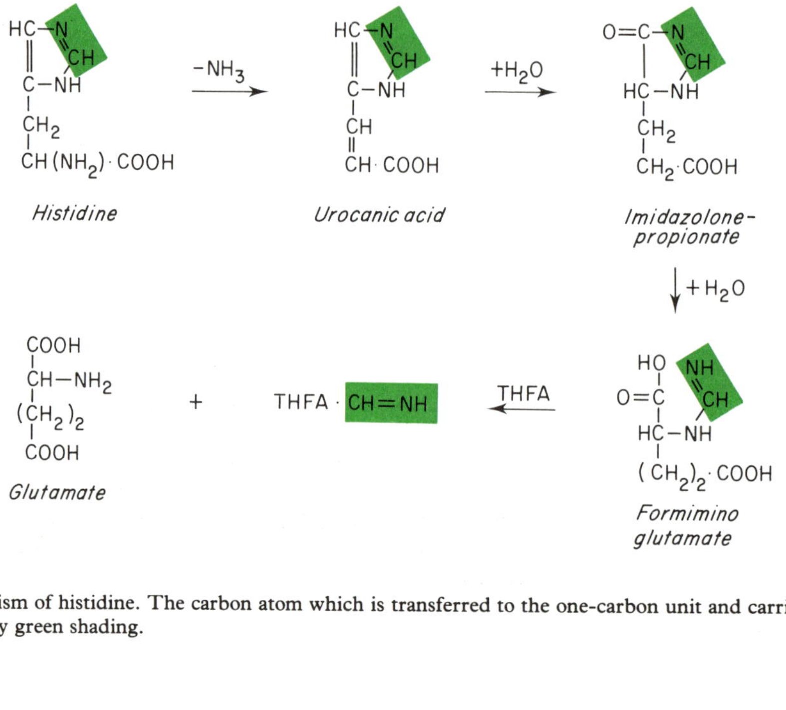

Fig. 8.13 Catabolism of histidine. The carbon atom which is transferred to the one-carbon unit and carried by THFA (p. 142) is shown by green shading.

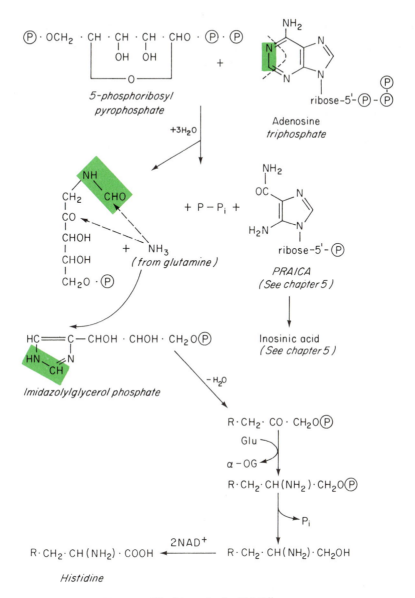

Fig. 8.14 The biosynthesis of histidine.

group, —HC=NH, is transferred to THFA and converted into formyl-THFA and NH_3. See Fig. 8.13.

An important derivative of histidine is the local hormone *histamine*, $R \cdot CH_2 \cdot CH_2 \cdot NH_2$, formed by decarboxylation of the amino acid. This is oxidized in tissues by *histaminase* to $R \cdot CH_2 \cdot CHO$. It is further oxidized in liver to $R \cdot CH_2 \cdot COOH$, imidazolylacetic acid, and excreted as the ribotide of the latter. In subjects suffering from folic acid deficiency, the methylated derivatives of histamine and imidazolylacetic acid are excreted in considerable quantities, suggesting that the pathway of histidine catabolism via formimino-glutamic acid is inhibited.

A number of histidine derivatives of unknown function are found in the body. They include ergothioneine, an —SH derivative, found in red blood cells, and the β-alanyl dipeptides carnosine and anserine, found in muscle.

In addition, a specific histidine residue in myosin is methylated (from SAM) on the N-3 of the ring, after translation. When the myosin is broken down, this 3-Me-His is not further metabolized, and is excreted. It can be used for estimating muscle protein turnover, as explained on p. 150.

Arginine and ornithine

The synthesis of arginine from ornithine has been shown in Fig. 8.8. In adult man arginine is non-essential, but there is a demand for supplementation in the diet of the young. It appears, then, that ornithine can be made slowly by humans; other animals, e.g. the rat, are completely dependent on dietary arginine for their ornithine requirements.

Ornithine is known to be converted to glutamic acid by the pathway shown in Fig. 8.15 and it is formed, slowly, essentially by reversal of the same pathway.

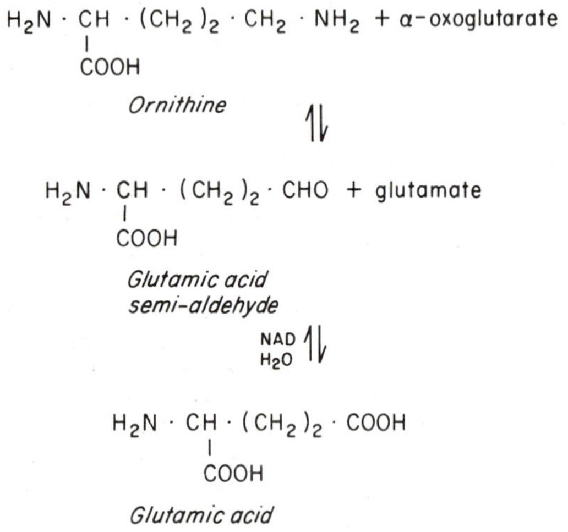

$$H_2N \cdot CH \cdot (CH_2)_2 \cdot CH_2 \cdot NH_2 + \alpha\text{-oxoglutarate}$$
$$|$$
$$COOH$$

Ornithine

$$H_2N \cdot CH \cdot (CH_2)_2 \cdot CHO + glutamate$$
$$|$$
$$COOH$$

Glutamic acid
semi-aldehyde

NAD
H_2O

$$H_2N \cdot CH \cdot (CH_2)_2 \cdot COOH$$
$$|$$
$$COOH$$

Glutamic acid

Fig. 8.15 The interconversion of ornithine and glutamate.

Polyamines

Ornithine decarboxylase converts ornithine into *putrescine*. This enzyme has the shortest half-life (11 min) of any enzyme whose turnover has been measured. A rise in its activity precedes new RNA synthesis in many situations in which cell division or protein synthesis is stimulated. There is thus a great deal of interest in the possibility that the product of this reaction is in some way essential, or regulates, transcription or translation (see chapter 13, p. 234). It is not, however, present in equivalent amounts to RNA and DNA.

Unlike cadaverine, putrescine is formed in all cells during life. It is the precursor of *spermidine* and of *spermine* (the latter present only in eukaryotes). All cells in the body appear to be able to make at least spermidine. The mechanism involves the decarboxylation of a methionine residue (Fig. 8.16) (cf. p. 138).

Copper-containing *diamine oxidases* oxidize putrescine, spermidine and spermine, so that there is a turnover of these compounds. It is presumed that the 15 % of methionine which cannot be used for cysteine synthesis in minimal diets has to be used for polyamine synthesis.

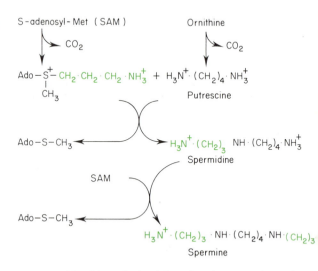

Fig. 8.16 The biosynthesis of the polyamines spermine and spermidine. The parts of the polyamine molecules that originated from methionine are shown in green.

Amino acids involved in one-carbon unit metabolism

Glycine

As might be expected from the simplest amino acid, glycine takes part in a number of synthetic pathways, as well as being catabolized for energy. Isotope labelling studies show that glycine is strongly glucogenic, but it does not readily produce hyperglycaemia in experimental animals. This suggests that the incorporation of glycine carbons into glucose precursors is indirect.

Glycine metabolism is summarized in Fig. 8.17.

1. *Conjugation*: The bile salt glycocholic acid, and also hippuric acid (benzoyl-glycine) and other detoxication products, are peptides formed between a —COOH group and the —NH₂ group of glycine.

2. *Creatine*: The transfer of the guanidino group from arginine to glycine forms guanidino-acetic acid (glycocyamine)

$$NH_2-C-NH-CH_2-COOH$$
$$\|$$
$$NH$$

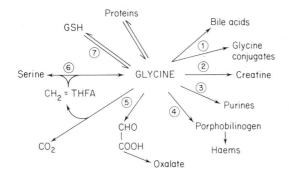

Fig. 8.17 Overall view of the metabolism of glycine. The numbers refer to the paragraphs in the text on pp. 136–137.

Creatine is formed from this compound by methylation. The donor is *S*-adenosyl-methionine (p. 138). The methylation takes place in liver, while guanidino-acetic acid is formed in kidney. Creatine is taken up from plasma into muscles by an unknown mechanism:

$$
\begin{array}{ccc}
& CH_3 & Adenosine \\
& | & + \\
& {}^+S{-}Adenosyl & SH \\
NH_2 & | & NH_2 & | \\
| & CH_2 & | & CH_2 \\
C{=}NH & | & C{=}NH & | \\
| + & CH_2 & \rightarrow & | + & CH_2 \\
NH & | & N \cdot CH_3 & | \\
| & CHNH_2 & | & CH \cdot NH_2 \\
CH_2 & | & CH_2 & | \\
| & COOH & | & COOH \\
COOH & & COOH & \\
\end{array}
$$

Glycocyamine   *S-Adenosyl methionine*   *Creatine*   *Homocysteine*

The reaction is not reversible, so creatine is not a source of methyl groups.

Creatine is found in striated muscle in a concentration of about 1 g/100 g. About half of this creatine is combined with phosphate as creatine phosphate or phosphocreatine,

$$
\begin{array}{ll}
CH_2 \cdot N(CH_3) \cdot C \cdot NH \cdot \textcircled{P} & CH_2 \cdot N(CH_3) \cdot C \cdot NH \cdot PO_3^{2-} \\
| \quad\quad\quad || \quad \text{or} & | \quad\quad\quad\quad || \\
COOH \quad\quad NH & COO^- \quad\quad\quad NH_2^+ \\
& \quad\quad\quad \text{at neutral pH}
\end{array}
$$

which has been mentioned in chapter 6 as a phosphate energy store in this tissue. Smaller amounts of creatine and phosphocreatine are found in smooth muscle, and traces in other tissues.

Phosphocreatine slowly but spontaneously cyclizes to form *creatinine*

$$
\begin{array}{l}
CH_2 \cdot N(CH_3) \cdot C{=}NH \\
| \quad\quad\quad\quad\quad | \\
CO{-}\!\!-\!\!-\!\!-\!\!-\!\!-\!\!-NH
\end{array}
$$

with simultaneous formation of inorganic phosphate. This internal amide is quite stable and diffuses from the muscle into the blood, from which it is excreted into the urine. The amounts found in urine per day depend on the muscle mass, and are reasonably constant from day to day for any one individual, so that it is common to express values for urinary constituents as 'x g/g creatinine' rather than 'x g/24 hours'.

An adult man excretes about 1–1.5 g creatinine per day, but only traces of creatine. Women may excrete more of the latter, since it is eliminated during the breakdown of the smooth muscle cells of the endometrium during menstruation. Creatine is also found in urine in other conditions of muscular wasting, such as muscular dystrophy and thyrotoxicosis. It is not uncommon to find creatine in the urine of children.

Creatine must therefore be continuously synthesized to replace the lost creatinine; it appears to be formed in liver and kidney and not in muscle itself.

3. *Purine* synthesis: The N—C—C skeleton of one molecule of glycine is incorporated into each molecule of the purines adenine and guanine (p. 54). There is no incorporation of glycine into the pyrimidines.

4. *Porphyrin* synthesis: Glycine is incorporated into the precursor of protoporphyrin IX in such a way that its amino group provides all four N atoms (see p. 44).

5. Glycine transamination seems to be important only in plants. In animals it is decarboxylated and transformed into a formimino unit (p. 142), and it is also oxidized by a flavoprotein glycine oxidase to *glyoxylic acid*, $OHC \cdot COOH$. This is further oxidized to CO_2 and formic acid, $H \cdot COOH$. The latter is used for various synthetic purposes (see the one-carbon unit pool, p. 141). Formate formation appears to be blocked in *oxaluria*, when the glyoxylate is oxidized to oxalate.

6. Glycine can be formed from, and converted to, *serine* (see below).

7. Glycine is a component of glutathione (p. 140). Degradation of GSH releases glycine back into the free amino acid pool.

Serine

Serine is non-essential, being synthesized from 3-phosphoglycerate by oxidation to phosphohydroxy-pyruvate, followed by transamination to phosphoserine, and hydrolysis of the phosphate ester bond.

Transamination yields hydroxypyruvic acid, $CH_2OH \cdot CO \cdot COOH$, which can be converted to phosphoglyceric acid, and so to carbohydrate. Serine is also metabolized by a dehydratase which converts it to pyruvic acid:

$$CH_2 \cdot C(NH_2) \cdot COOH \xrightarrow{-H_2O} CH_2{=}C(NH_2) \cdot COOH$$
$$\underset{\overline{OH\ \ H}}{}$$
$$\xrightarrow{+H_2O} CH_3 \cdot CO \cdot COOH + NH_3$$

This enzyme also attacks threonine (see p. 145).

The most important fate of serine is conversion to glycine. The reversible reaction is catalysed by *serine hydroxymethyl transferase*. The enzyme requires THFA (p. 141) as acceptor of the hydroxymethyl group, and pyridoxal phosphate (p. 127) as prosthetic group. See one-carbon unit metabolism (pp. 141–143).

Serine is exchanged for ethanolamine in the formation of phosphatidyl serine (p. 108).

Methionine

Methylation

The methylation of guanidinoacetic acid to form creatine has been described. The activation of methionine, i.e. the formation of *S*-adenosyl methionine, a sulphonium ion which acts as a methylating agent, takes place by means of the following reaction:

Met + ATP →

$$\underset{\underset{Ribose\text{-}Adenine}{|}}{CH_3{-}S^+{-}CH_2 \cdot CH_2 \cdot CH(NH_2) \cdot COOH} + P_i + P\text{-}P_i$$
$$SAM$$

Quantitatively the most important of the other methylations is the synthesis of *choline* from ethanolamine (in phosphatidyl-ethanolamine). This requires three successive methyl transfers from 3 molecules of *S*-adenosyl methionine. About 0.5–1 g of choline is usually present in the diet, and another 0.5–1 g is synthesized in the body each day during the synthesis of lecithin (chapter 7).

Choline is degraded by oxidation to *betaine*

$$(CH_3)_3\overset{+}{N} \cdot CH_2 \cdot CH_2OH \xrightarrow{Ox.} (CH_3)_3\overset{+}{N} \cdot CH_2 \cdot COOH$$

One of the methyl groups may be transferred to homocysteine yielding methionine. The other two are probably oxidized, forming hydroxymethyl-THFA. *N*-Methylglycine or *sarcosine* is found in muscle; complete demethylation of betaine gives glycine.

Several other methylations may be mentioned; among them are the formation of adrenaline from noradrenaline, several detoxication reactions, and the formation of some 'unusual' methylated bases (but *not* thymine) in RNA (cf. p. 212). Transmethylation is also discussed on p. 141.

It seems probable that most of the dietary methionine is used in this way. The homocysteine which is the product of the reaction can be re-methylated from methyl-THFA (p. 141). Probably this happens several times before the carbon skeleton of methionine is converted to cysteine by the reactions shown in Fig. 8.18. These reactions are not reversible.

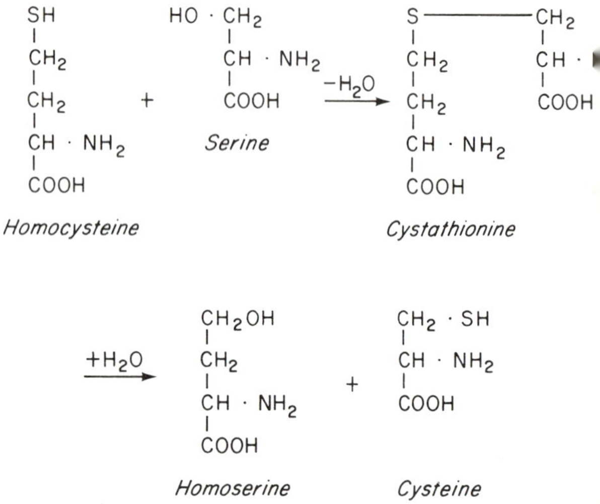

Fig. 8.18.

Because the reaction is not reversible, methionine is an essential amino acid, but about 85% of cysteine requirements can be spared if a minimal amount of methionine is present in the diet. The residual demand for cysteine suggests that some methionine is metabolized by pathways that do not lead to the formation of homocysteine. Some of this may be by transamination, followed by irreversible reactions, but some methionine is undoubtedly used for the synthesis of spermidine (see p. 136).

Homoserine is converted to α-oxobutyric acid,

$$CH_3 \cdot CH_2 \cdot CO \cdot COOH$$

which is oxidatively decarboxylated to propionic acid (chapter 7).

In addition to the normal requirement for methionine in protein synthesis, the initiation codon for *every* polypeptide requires t-RNA-Met (chapter 11). Thus dietary methionine deficiency may have more severe consequences for the overall rate of protein synthesis than a deficiency of any other amino acid. It seems probable, however, that the initial Met in polypeptide chains is usually scavenged by post-translational modifications, and returned to the methionine pool.

As a result of these demands on the available methionine, it is likely that, although it can undergo transamination, this is not an important pathway on diets containing an average amount of protein.

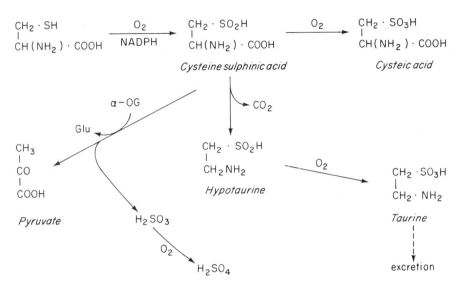

Fig. 8.19 The catabolism of cysteine. Taurine is the major end product (other than the sulphate esters, see Fig. 8.20) in man.

Cysteine

If enough methionine is present in the diet, cysteine is not an essential amino acid.

It is converted to pyruvic acid in liver by a 'desulphydrase' enzyme (cf. serine catabolism).

$$HS \cdot CH_2 \cdot CH(NH_2) \cdot COOH \xrightarrow{+H_2O}$$
$$CH_3 \cdot CO \cdot COOH + H_2S + NH_3$$

However, probably more important is the oxidation of the —SH of cysteine in situ, which can lead to the formation of *taurine*,

$$NH_2 \cdot CH_2 \cdot CH_2 \cdot SO_3H$$

This compound is conjugated with cholic acid to form the bile acid taurocholic acid (p. 114) and it is also found free in muscle, particularly heart. It is, in fact, present free in cells in higher concentration than any other amino acid, but no irreplaceable function has been found for it in man. In some animals, e.g. the cat, it is essential and without it the retina degenerates and the animal goes blind. This suggests some kind of transmitter function. The major route of taurine excretion is as taurocholic acid. The proportion of taurine to glycine used for conjugation probably regulates the body's taurine pool: it is not metabolized in any other way.

Cysteine catabolism proceeds by several pathways (see Fig. 8.19). The first step is oxidation by *cysteine dioxygenase* to cysteine sulphinic acid, a complex reaction for which NADPH is required as well as O_2. There are considerable species variations in the metabolism of cysteine sulphinate; in man, oxidation to cysteic acid seems to be unimportant, so that taurine is formed mainly by oxidation of hypotaurine. Transamination, leading to the formation of pyruvate, is the third pathway. The presumed intermediate, β-sulphinyl pyruvate, has never been isolated, and probably liberates sulphite while still bound to the transaminase. It is interesting that the latter enzyme may be identical with glutamate-oxaloacetate transaminase, while cysteine sulphinate decarboxylase is very similar to, if not identical with, glutamate decarboxylase.

Cystine is probably the prevailing form of the cysteine/cystine pair in extracellular fluids. It is excreted in a congenital abnormality of renal function. As it is very insoluble, it may form harmful kidney stones.

Sulphate metabolism

The oxidation of sulphite to sulphate may be represented by the following equation

$$SO_3^{2-} + OH^- \rightarrow SO_4^{2-} + H^+ + 2\epsilon$$

In vivo, this is accomplished by a mitochondrial enzyme, containing both molybdenum (which is reduced from Mo^{6+} to Mo^{4+}) and a b_5-type haem group, which transmits electrons to cytochrome c. A few children have been studied who have had congenital sulphite oxidase deficiency. They died in infancy with severe neurological symptoms.

About 20–40 mmoles of sulphates are excreted in the urine each day, 90% as inorganic sulphate. There is nevertheless an exchange between inorganic and organic sulphate within the body, chiefly due to the synthesis and breakdown of sulphated glycosaminoglycans (see chapter 6). A series of sulphatases (A, B and C) have been characterized; deficiency of sulphatase A leads to death at about 4 years of age, with accumulation of cerebroside sulphates in brain. This time scale indicates the relative slowness of sulphate ester turnover in vivo. Sulphatase C may be an oestrogen sulphate-hydrolysing enzyme.

The formation of these esters is energy-linked, requiring the synthesis of a mixed sulphate–phosphate anhydride PAPS (see Fig. 8.20). It is not certain that

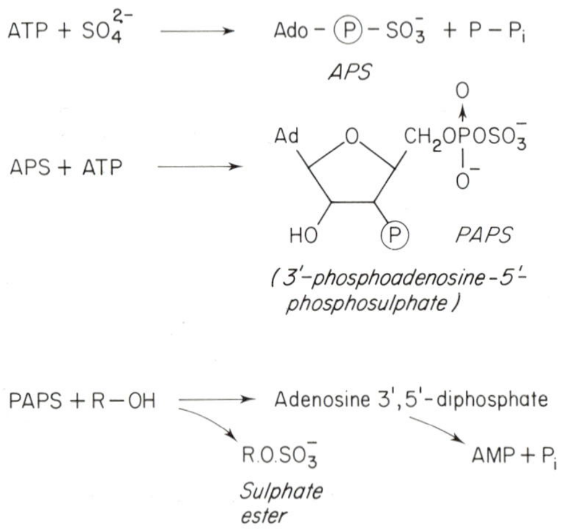

Fig. 8.20 The biosynthesis of the sulphate transferring coenzyme *phosphoadenosine phosphosulphate* (PAPS).

dietary sulphate can be used as efficiently in this activation process as that produced from cysteine oxidation.

Glutathione

This is a tripeptide containing cysteine, which is present in fairly high concentration in most animal cells (1–2 mM; up to 8 mM in liver). For its structure, including the unusual γ-glutamyl bond, see Fig. 8.21. It is quantitatively the most important intracellular —SH compound, and probably buffers the intracellular redox potential of compounds that react dynamically with the —SH pool; this can include some —SH groups on proteins, but not coenzyme A or lipoamide. The buffering is possible because glutathione, like the cysteine–cystine couple, can exist both as a monomer and as a dimer:

$$2 \text{ GSH} \rightarrow \text{GSSG} + 2\text{H}$$

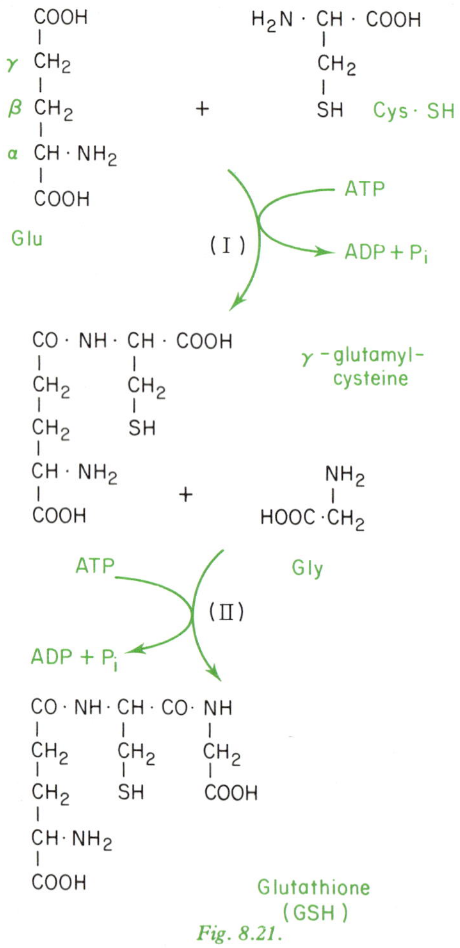

Fig. 8.21.

A *glutathione reductase* catalyses this reversible oxido-reduction. In most cells, NADPH is the hydrogen donor.

In addition to this general function, glutathione has two specialized functions: *glutathione peroxidase* protects cells from damage by H_2O_2 and organic peroxides (see chapter 14 and p. 167), while *γ-glutamyl transpeptidase* is concerned with the transport of amino acids across cell membranes (p. 124). This function is different in kind from those discussed above, since it does not depend on sulphhydryl activity, and both GSH and GSSG are substrates. Moreover, the transport must be followed by complete resynthesis of the glutathione molecule.

Two enzymes in sequence provide for the synthesis of GSH, as shown in Fig. 8.21. Enzyme (I) is inhibited by glutathione, giving a feedback control of the overall rate of production. The early ^{15}N labelling experiments indicated that glutathione turns over rapidly, at least in liver; the half-life of both glutamate and glycine moieties in rat liver was found to be about 5 hr. Part of the whole-body turnover may be due to breakdown of glutathione implicit in the γ-glutamyl transferase reaction (p. 124).

One-carbon unit metabolism

By this is meant the transfer of radicals containing one carbon atom (in various states of oxidation) from one compound to another. This is not quantitatively important by comparison with the overall flow of materials through the body, but it is qualitatively very important in the synthesis of some vital compounds and in the detoxication of others.

Methanol, CH_3OH, formaldehyde, $H \cdot CHO$, and formic acid, $H \cdot COOH$, are one-carbon compounds, and might be expected to participate in this metabolism, but they do not occur in the body; indeed, the first two are toxic. The fixation of CO_2, as in the carboxylation of pyruvate (chapter 6), or in the formation of malonyl-CoA (chapter 7) might strictly be called a part of one-carbon unit metabolism but is not usually so regarded. The name is reserved for the transfer of methyl ($-CH_3$), hydroxymethyl ($-CH_2OH$), or formyl ($-CHO$) radicals by means of carriers.

There are two carriers: *S*-adenosyl methionine (below) and tetrahydrofolic acid. Although there is a connexion between the two, they are essentially separate carriers concerned with the synthesis of different types of compound.

S-Adenosyl methionine

In the synthesis of compounds containing methyl groups, methionine itself is the most important donor. There is evidence that homocysteine can be remethylated. Isotopic evidence shows that hydroxymethyl-THFA can be reduced to 5-methyl-THFA. The transfer to homocysteine is catalysed by an enzyme for which methyl-cobalamin (see p. 278) is a cofactor. There probably has also to be a valency change in the cobalt atom, from Co^{3+} to Co^+. In addition to this, one of the methyl groups of betaine ($HOOC \cdot CH_2N^+(CH_3)_3$, the carboxylic acid equivalent to choline) can be used for methionine synthesis (for the complete oxidation of choline, see p. 138). Thus a single carbon skeleton of methionine can be used for the transfer of more than one methyl group. The number of transfers is nevertheless not very large, because of the competing claims of cysteine synthesis (p. 138). Fig. 8.22 summarizes these transfers.

The most important substances synthesized by transmethylation from *S*-adenosyl methionine are creatine, choline, adrenaline, and the methylated bases in RNA, especially tRNA and ribosomal RNA. Both adrenaline and noradrenaline are rendered inactive by methylation of one of their phenolic $-OH$ groups,

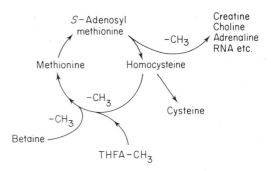

Fig. 8.22 Transmethylation reactions involving methionine.

and there are other examples of such detoxifying methylations.

Tetrahydrofolic acid

The second aspect of one-carbon unit metabolism is more complicated. The coenzyme, formed by reducing the vitamin folic acid (chapter 17) at C-5, -6, -7 and -8 on the pteridine ring, carries the one-carbon unit

Tetrahydrofolic acid (THFA or FH₄)

attached either to N-5 or N-10, or to both (in the latter case as a one-carbon bridge). It is helpful to consider the relationships between the formyl derivatives first. A reaction which is of little importance in higher animals leads to the formation of N^{10}-formyl-THFA:

$$H \cdot COOH + ATP +$$

$$ADP + P_i$$

THFA

N^{10}- formyl – THFA

This is interconvertible with a cyclic form, 5,10-methenyl-THFA:

5, 10 - methenyl - THFA

A third form of the carrier, N^5-formyl-THFA, can only be converted to the 5,10-methenyl form by the expenditure of ATP, since the free energy of hydrolysis of this form is considerably lower than that of the N^{10}-form. A fourth form, 5-formimino-THFA, which is formed during the catabolism of histidine (p. 134), is converted to 5,10-methenyl-THFA by losing NH_3:

N^5- *formimino - THFA*

$$\xrightarrow{H_2O} NH_3 + 5,10 \text{ - methenyl - THFA}$$

A closer look shows that interconversion between all these forms involves gain or loss of water (i.e. it does not involve oxidation or reduction), and consequently all of them, except perhaps formimino-THFA, can conveniently be referred to as 'formyl THFA', or THFA-CHO. Donation to formyl acceptors, as in purine synthesis (p. 55) can occur either from the 5,10-methenyl- or the 10-formyl-form.

A similar compound exists at the formaldehyde state of oxidation and can be referred to as hydroxymethyl-THFA, or THFA-CH_2OH, although in practice it exists as the methylene bridge compound shown here:

5, 10 - methylene - THFA

Finally, there is a form at an even lower stage of

oxidation, corresponding to methanol. This is 5-methyl-THFA, and its only known function is to donate a methyl group to homocysteine:

5-methyl-THFA

This brief summary shows that tetrahydrofolate can carry one-carbon units at three levels of oxidation, —CHO, —CH_2OH and —CH_3. These three forms are interconverted through oxidation or reduction, as shown in Fig. 8.23. This is important, because the demand for one-carbon units at the formyl level of oxidation, in particular, is greater than the supply from metabolites at that level of oxidation. The interconversions shown in Fig. 8.23 therefore provide flexibility for the system. Several minor pathways have been omitted from the figure, which does, however indicate that serine and glycine are probably the main sources of 1-C units in human metabolism.

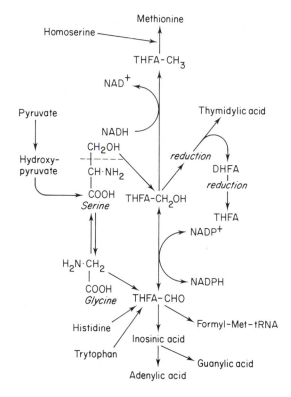

Fig. 8.23 Interrelationships between the various oxidation states of the one-carbon unit carried by tetrahydrofolic acid.

A unique reaction involving THFA is the synthesis of thymine (more precisely of thymidylate). This can be summarized as

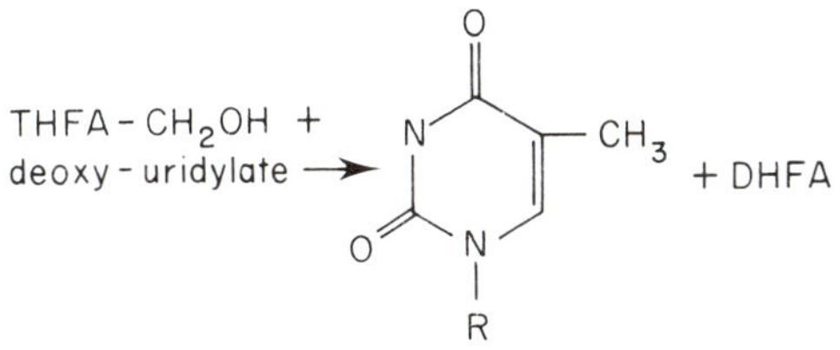

thymidylate

Clearly a reduction must have occurred, since the product has a methyl (—CH$_3$) group. The reductant is THFA itself, which is oxidized to dihydrofolate (DHFA) in the transfer. Before the folate nucleus can re-enter the 1-C unit pool, DHFA must be reduced to THFA by a reductase, using NADPH. This fact has been used in a variety of ways as a focus of drug action.

Inspection of Fig. 8.23 shows the importance of 1-C units in the synthesis of nucleotides. Since nucleotides are essential for synthesis of both DNA and RNA (chapter 5), derangements of 1-C unit metabolism affect both protein synthesis and cell reduplication. Moreover, the formylation of methionyl-tRNA is necessary for the initiation of peptide chain synthesis (chapter 12) in bacteria. Antagonists of folic acid, such as *aminopterin*, which has an amino group instead of the hydroxyl at position 4 of folic acid, and *methotrexate* (amethopterin), which is the N^{10}-methyl derivative of aminopterin, markedly inhibit cell division. Both these compounds bind very tightly, though reversibly, to folic acid reductase, thereby inhibiting this enzyme. Even high concentrations of folic acid do not reverse the inhibition.

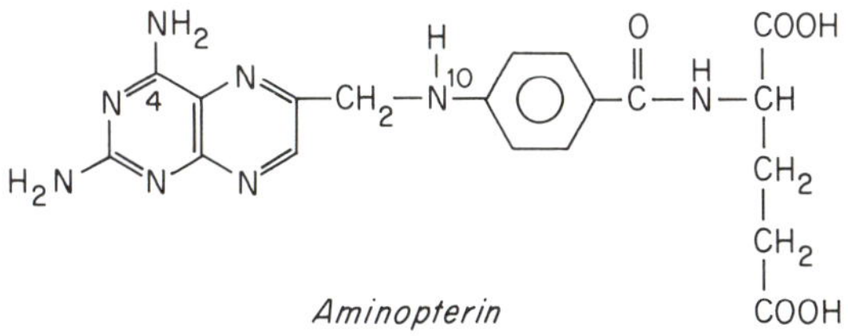

Aminopterin

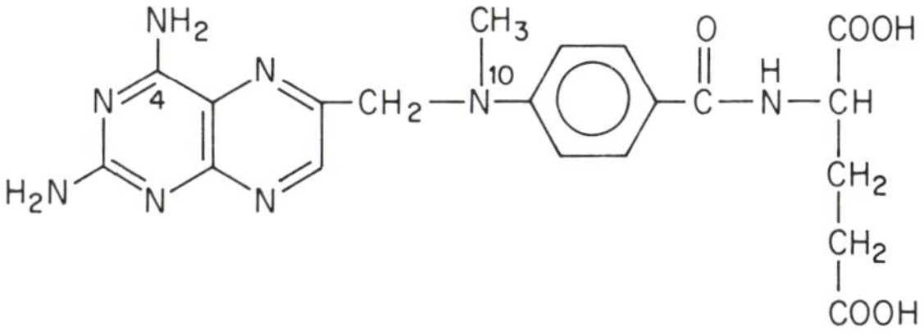

Methotrexate (amethopterin)

These drugs have been used as anti-tumour agents, although in the long run they prevent the patient from getting his normal requirements for THFA, and lead to, for example, macrocytic anaemia.

On the other hand, *trimethoprim* binds 10^4 times more strongly to the dihydrofolic acid reductase of bacteria than it does to the mammalian enzyme. This is an example of differential sensitivity, because drugs of this kind inhibit the growth of bacteria but not of animal cells, and can therefore be used for the control of bacterial infections in man.

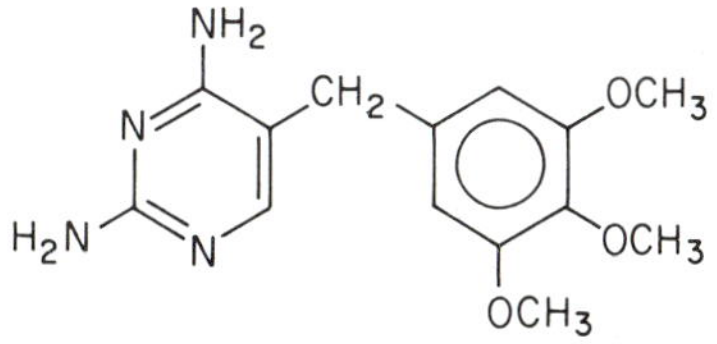

Trimethoprim

A third type of antagonism to folic acid is that produced by *sulphonamides* which competitively inhibit the incorporation of *p*-aminobenzoic acid into the folic acid molecule by bacteria normally able to synthesize the complete compound.

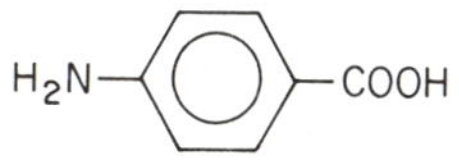

p-Aminobenzoic acid

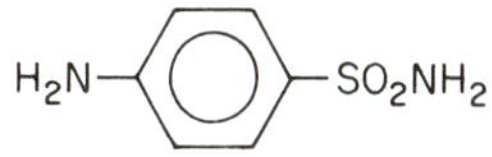

Sulphanilamide

The sulphonamide drugs are therefore also bacteriostatic agents.

The branched-chain amino acids

The three branched-chain amino acids valine, leucine, and isoleucine are all essential. Their catabolic pathways are in some ways very similar. All three are taken up predominantly by muscle after a protein meal, and are transaminated there. There is still controversy whether subsequent catabolism also takes place in muscle, or whether the oxo-acids are lost to plasma, and taken up by the liver.

It was originally thought that a single transaminase and a single oxoacid dehydrogenase processed all three amino acids, but evidence from inborn errors of metabolism shows that there is certainly a separate pair of enzymes for valine. Possibly the dehydrogenases for leucine and isoleucine are also separate. Nevertheless, the enzymes are physically and in mechanism very closely related, so that, as shown in Fig. 8.24, the early

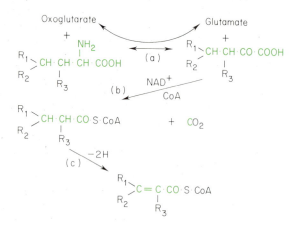

Fig. 8.24 The common preliminary stages in the catabolism of the branched chain amino acids. For valine, $R_1 = R_2 = H, R_3 = Me$; for leucine, $R_1 = R_2 = Me, R_3 = H$; for isoleucine, $R_1 = H, R_2 = R_3 = Me$.

stages of metabolism of all three amino acids may be treated as a single unit, i.e. (a) transamination followed by (b) oxidative decarboxylation and then (c) a dehydrogenation (cf. β-oxidation of fatty acids, p. 97). The enzymes catalysing reaction (b), the 'branched-chain oxoacid dehydrogenases', are mitochondrial, and have a structure and mechanism very similar to pyruvate and oxoglutarate dehydrogenases (see chapter 6, p. 73).

Valine

For this amino acid, the product of reaction (c) in Fig. 8.24 is *methacrylyl-CoA*. Hydration gives 3-

hydroxy-*iso*butyryl-CoA (I, Fig. 8.25), which loses its CoA to form free 3-OH-*iso*butyrate (II).

Subsequent stages in valine degradation are not so straightforward as might be expected. Partial oxidation of the —CH_2OH group gives methylmalonate semi-aldehyde (III), but when this is further oxidized, it decarboxylates, yielding propionate (IV). The latter is converted to succinate by the pathway detailed in chapter 7 (p. 106), but it is curious that the propionate has to be re-carboxylated in the course of the conversion.

Isoleucine

This amino acid has two asymmetric carbon atoms (marked by asterisks in the formula), and only one of the four possible isomers will support growth in experimental animals.

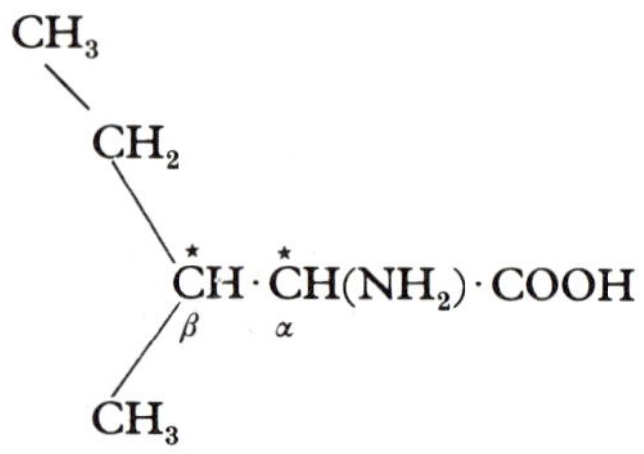

$$CH_3$$
$$\diagdown$$
$$CH_2$$
$$\diagdown$$
$$\overset{*}{CH} \cdot \overset{*}{CH}(NH_2) \cdot COOH$$
$$\diagup_{\beta} \quad _{\alpha}$$
$$CH_3$$

The molecule with the 'unnatural' configuration at the β-carbon atom is referred to as *allo*-isoleucine. In the Prelog convention (p. 60), the 'natural' isomer is (2S,3R)-isoleucine.

The degradation of isoleucine is more straightforward than that of valine. The oxidation step (c) in Fig. 8.24 is followed by hydration and completion of one β-oxidation cycle to give *methyl-acetoacetyl-CoA* (I, Fig. 8.26). This is split by a thiolase to give acetyl-CoA

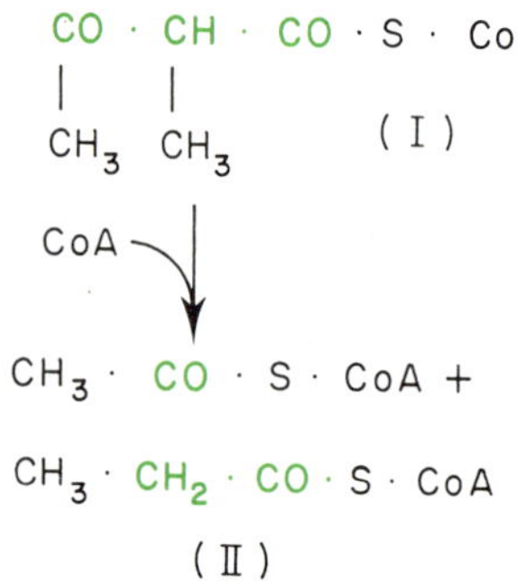

$$CH_2 \cdot CH \cdot CO \cdot S \cdot CoA$$
$$| \quad \quad |$$
$$OH \quad CH_3 \quad (I)$$

↓ CoA

$$CH_2 \cdot CH \cdot COOH$$
$$| \quad \quad |$$
$$OH \quad CH_3 \quad (II)$$

↓ NAD^+

$$CH \cdot CH \cdot COOH$$
$$\| \quad \quad |$$
$$O \quad CH_3 \quad (III)$$

→ $C \cdot CH_2 \cdot CH_3 + CO_2$ (IV)

Fig. 8.25 The catabolism of valine. The 'core' of three carbon atoms which is common to all three branched-chain amino acids (see Fig. 8.24) is green.

$$CO \cdot CH \cdot CO \cdot S \cdot CoA$$
$$| \quad \quad |$$
$$CH_3 \quad CH_3 \quad (I)$$

CoA ↓

$$CH_3 \cdot CO \cdot S \cdot CoA +$$
$$CH_3 \cdot CH_2 \cdot CO \cdot S \cdot CoA$$
$$(II)$$

Fig. 8.26 The catabolism of isoleucine. The 'core' of three carbon atoms which is common to all three branched chain amino acids (see Fig. 8.24) is green.

and propionyl-CoA (II). The latter is converted to succinate as described in chapter 7.

As the direct product of isoleucine degradation is acetyl-CoA rather than free acetoacetate, the amino acid is not strongly ketogenic, and it is also glucogenic in virtue of the succinate formed from one-half of the original carbon skeleton.

Leucine

For leucine the product of reaction (c) in Fig. 8.24 is *methyl-crotonyl-CoA* (I, Fig. 8.27). This has two methyl groups on the γ-carbon atom, and direct hydration would give a 3-hydroxy acid that could not be further oxidized (because the alcohol group would be a tertiary one). Instead, methyl-crotonyl-CoA is carboxylated to give (II). As usual for such reactions, the enzyme contains biotin as a prosthetic group, and ATP is required (cf. chapter 7, p. 99). Compound (II) is now hydrated to give hydroxymethylglutaryl-CoA (HMG-CoA, III). This compound is on the direct pathway of acetoacetate formation in liver (see chapter 7). Leucine is strongly ketogenic, so it must be presumed that at least the later stages of leucine catabolism occur predominantly within liver cells. In other tissues HMG-CoA can only be converted to cholesterol (p. 113).

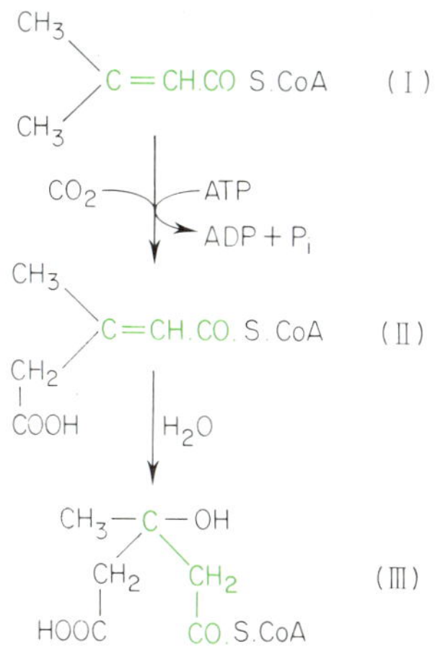

In an inborn error of metabolism called 'maple syrup urine disease' the oxidase catalysing step (b) for leucine and isoleucine is absent. The corresponding α-oxo-acids accumulate, and as they inhibit the α-oxo-isovaleric acid oxidase (p. 144), this oxo-acid also accumulates and the defect comes to involve all three branched-chain amino acids. The α-oxo-acids also inhibit α-oxoglutaric acid oxidase, and thus impair the working of the TCA cycle. There is rapid degeneration of the CNS and the disease is usually fatal within a few months of birth, since feeding an artificial diet low in branched-chain amino acids is not satisfactory for growth. The three branched-chain amino acids and the oxo-acids accumulate in blood and urine, and the latter are partially reduced to the corresponding hydroxy-acids, which have a characteristic odour of maple syrup.

Apart from its function as an essential amino acid, leucine has been specifically implicated in the regulation of metabolism, e.g. as an inhibitor of urea formation in liver, and as a regulator of the rate of protein synthesis in muscle. It is also one of the most powerful stimulators of insulin secretion from the islets of Langerhans.

Threonine

Like isoleucine, threonine has two chiral centres. The form found in proteins has the configuration $2S,3R$.

The metabolism of this essential amino acid is very unusual in that the accepted pathways are not specific for threonine. It does not undergo transamination; *threonine dehydrogenase* produces 2-amino-3-oxobutyrate, which spontaneously loses CO_2 to form aminoacetone. However, this pathway, which is specific, is not accepted as the major one for threonine in animals.

Serine dehydratase (see p. 138) reacts also with threonine, to give 2-oxobutyrate as a product. The *pyruvate dehydrogenase complex* (p. 73) will slowly oxidize this to propionyl-CoA. The conversion of the latter to succinate, a glucogenic intermediate, is dealt with on p. 106. There also exists a *threonine dehydratase*, which could initiate the same pathway.

A third possible route for threonine is via *serine hydroxymethyl transferase* (p. 138). This would form glycine and acetaldehyde (not a THFA derivative, surprisingly enough). This reaction is reversible, so that it ought to be capable of synthesizing threonine, at least in small amounts, and it is not known how significant this pathway is in normal threonine metabolism.

Fig. 8.27 The catabolism of leucine. The 'core' of three carbon atoms which is common to all three branched chain amino acids (see Fig. 8.24) is coloured green.

Lysine

The major pathway for lysine was established only after the isolation of a urinary metabolite *saccharopine* (I), which accumulates in a rare inborn error of lysine metabolism. As Fig. 8.28 shows, saccharopine is formed by the condensation of the terminal $-NH_2$ group of lysine with the $>C=O$ of oxoglutarate to form a Schiff's base, which is then reduced by NADPH. A separate enzyme, requiring NAD^+, reoxidizes saccharopine, and the Schiff's base is broken in the opposite sense to give the aldehyde of α-amino adipic acid (II), together with glutamate. The free Schiff's base is not released from either enzyme. The effect is a transamination, but this

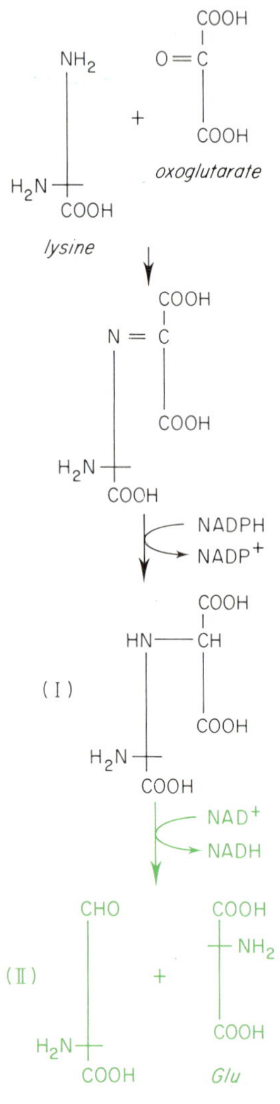

Fig. 8.28 The catabolism of lysine (lysine to 2-amino adipate semialdehyde).

146

description of the mechanism explains why lysine is not a substrate for any of the usual pyridoxal phosphate transaminases. The involvement of NADPH and NAD^+ emphasizes the oxidation–reduction character of transamination.

Aminoadipate semialdehyde (II) is oxidized to aminoadipate (III) itself (see Fig. 8.29, opposite) by an NAD^+-requiring enzyme, and this product is transaminated to α-oxoadipate (IV) by a 'normal' transaminase. The subsequent complete oxidation of oxoadipate contains a number of standard reactions. First comes oxidative decarboxylation by an enzyme complex containing TPP, lipoate and FAD. This is followed by β-oxidation to give glutaconyl-CoA (V), whose terminal $-COOH$ is sufficiently labilized by the double bond for it to be lost, producing crotonyl-CoA (VI). This is a normal intermediate of β-oxidation, and is readily converted to acetoacetyl-CoA, and thus to acetyl-CoA.

Since four of the six carbon atoms of lysine are converted to acetoacetyl-CoA, it might be expected to be strongly ketogenic, but it is not. No explanation has ever been offered for this.

Post mortem, lysine is decarboxylated to form *cadaverine*, which has an unpleasant smell. This reaction does not seem to be important in normal metabolism.

Tryptophan

The concentration of tryptophan in tissues is very low; in liver the supply of tryptophan may be the limiting factor in the rate of protein synthesis.

Tryptophan does not undergo transamination unless the tissue concentration of oxo-acids is abnormally high, as in phenylketonuria or 'maple syrup urine' disease. The resulting oxo-acid is converted into *indolylacetic acid*, which is excreted in larger than normal amounts.

A major pathway of tryptophan metabolism begins by opening the 5-membered ring to form formylkynurenine:

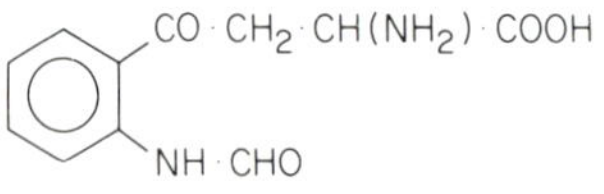

This involves the addition of 2 atoms of oxygen from a molecule of O_2 across the double bond in the pyrrole ring. The oxygenase responsible, *tryptophan pyrrolase*, is one of the few mammalian enzymes that is inducible.

Formylkynurenine loses a formyl group to the one-carbon unit pool, leaving *kynurenine*. This is hydroxylated to form 3-hydroxykynurenine (Fig. 8.30).

The conversion of oxoadipic acid into $2CO_2$ and crotonyl-CoA has been summarized under lysine catabolism (Fig. 8.29, formula IV onwards).

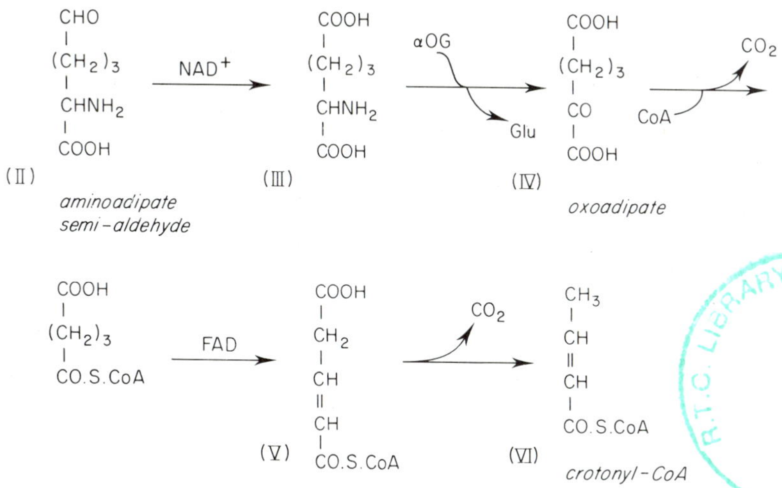

Fig. 8.29 The catabolism of lysine (continued): 2-amino adipate semialdehyde to crotonyl-CoA.

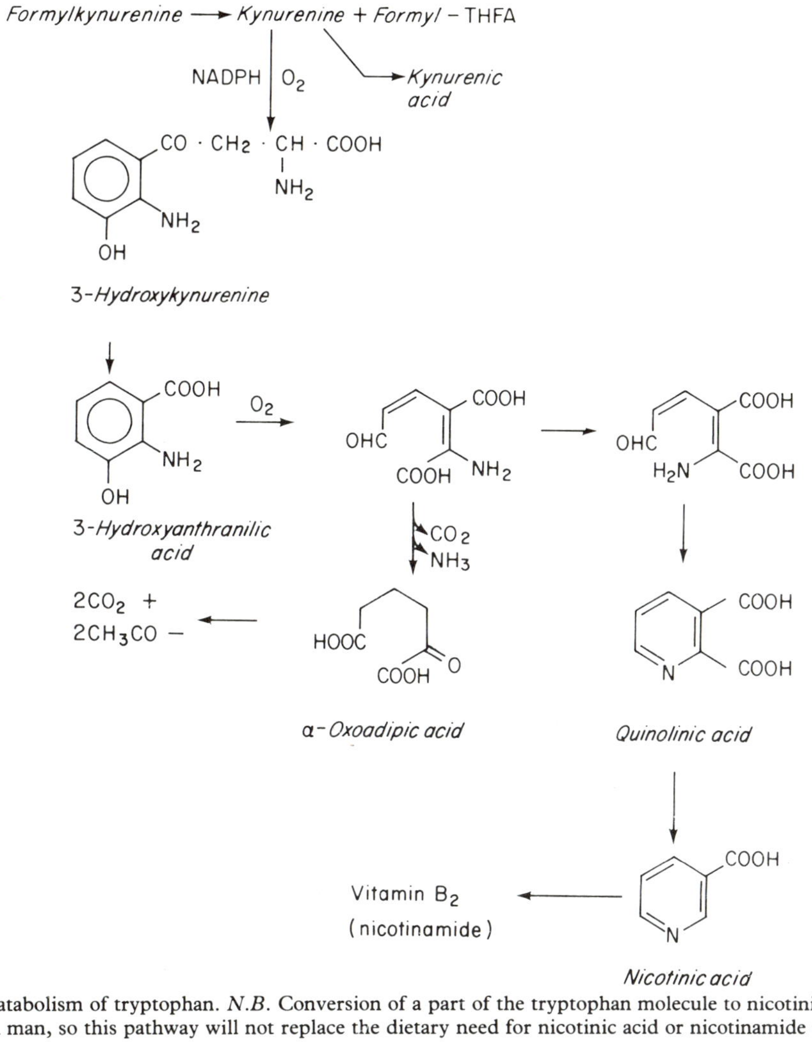

Fig. 8.30 The catabolism of tryptophan. *N.B.* Conversion of a part of the tryptophan molecule to nicotinic acid is not a major pathway in man, so this pathway will not replace the dietary need for nicotinic acid or nicotinamide (see chapter 17).

The series of reactions from 3-hydroxyanthranilic acid to nicotinic acid then leads, through nicotinic acid ribotide, to nicotinamide ribotide, the immediate precursor of NAD. This indirectly provides enough nicotinamide, one of the B vitamins (see chapter 17), in some species such as the rat, for the animal to be completely independent of dietary sources (other than tryptophan). Man does depend on his diet for the vitamin, but a fraction of his requirement can come from tryptophan. A deficiency of this amino acid can exacerbate a deficiency of nicotinamide; hence the peculiar prevalence of pellagra in maize-eating areas (the chief protein of maize is particularly deficient in tryptophan).

The conversion of 3-hydroxykynurenine into 3-hydroxyanthranilic acid requires another of the B vitamins, pyridoxine (B_6). Thus pyridoxine deficiency will accentuate the symptoms of pellagra (see chapter 17). In B_6 deficiency, 3-hydroxykynurenine and its precursors accumulate in the tissues or are excreted in greater quantity, and it is not uncommon to test for a pyridoxine deficiency by a tryptophan-load test. The excretion rate of an intermediate, usually xanthurenic acid, is measured after a standard dose of tryptophan.

5 - Hydroxytryptamine

Tryptophan can be decarboxylated to form *tryptamine*, or first hydroxylated and then decarboxylated to form 5-hydroxytryptamine (5-HT, or *serotonin*), a local hormone. This substance, or its derivative 5-hydroxy-indolyl acetic acid, is excreted in quite large amounts by subjects suffering from *carcinoid*, a tumour of the argentaffin cells of the intestinal mucosa. Estimation of these substances in urine is of great aid in the diagnosis of the disease.

Phenylalanine and tyrosine

The main catabolic pathway of these essential amino acids is delineated in Fig. 8.31. The figure makes it clear why these two acids are both glucogenic *and* ketogenic. A number of the reactions involved are unusual. The hydroxylation of phenylalanine is carried out by a mixed function oxidase for which the co-substrate is tetrahydro-biopterin (see chapter 9). The oxidation of *p*-OH-phenylpyruvate is carried out by a dioxygenase that both oxidatively decarboxylates the side chain, and introduces an —OH group into the ring, while the side chain is labilized and moved round the ring. Other examples of

this 'NIH shift' are known. Finally, the aromatic ring of homogentisate is opened by another dioxygenase that requires Fe^{2+}. Tyrosine is the starting point for the synthesis of several compounds of biological importance. Peptide-linked tyrosine is iodinated in the synthesis of *thyroxine* and other thyroid hormones. The two most important are thyroxine (T4) and 3,3',5-triiodothyronine (T3). Note that 'reverse T3' (3,3',5'-TIT) is inactive.

Thyroxine (T4)

T3

The copper-containing enzyme *tyrosinase*, which occurs in melanocytes, and also in plants, is a mixed function oxidase that catalyses the oxidation of tyrosine by oxygen to 3,4-dihydroxyphenylalanine ('dopa') and to the corresponding quinone.

Dopa　　　　　*Dopa quinone*

The co-substrate (H donor) for the reaction is DOPA itself. Thus we have:

$$Tyr + Dopa + O_2 \rightarrow Dopa + Dopa\text{-}quinone + H_2O$$

The latter is the starting point for the synthesis, in the adrenal medulla, of *noradrenaline* and *adrenaline*.

Noradrenaline

Adrenaline

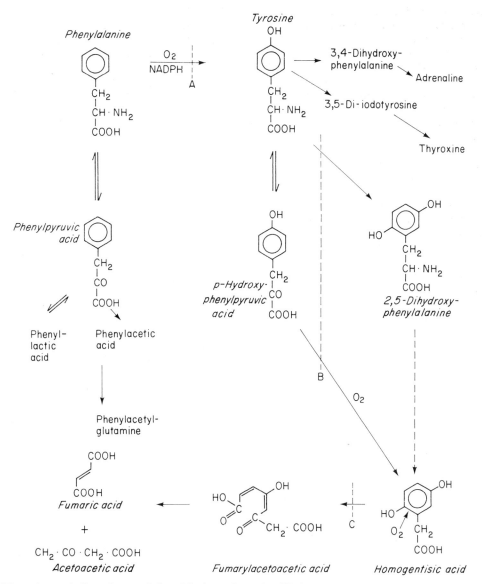

Fig. 8.31   The main metabolic pathways of phenylalanine and tyrosine. Blocks occur at: A in phenylketonuria, B in tyrosinosis, and C in alkaptonuria.

Dopa is also converted into an indole derivative, which condenses to form the high molecular weight pigment, *melanin*. In melanocytes, melanin is firmly bound to protein. The hereditary defect *albinism* appears to be due to a lack of tyrosinase in the melanocytes.

Phenylketonuria, alkaptonuria, and *tyrosinosis* are closely related inborn errors of metabolism (see Fig. 8.31) in which the normal metabolism of aromatic amino acids is blocked at different points.

Phenylketonuria is characterized by mental deficiency and the excretion in the urine of phenylpyruvic acid in excessive amounts (0.5–1.0 g per day). In addition to phenylpyruvic acid there are abnormally large amounts of phenylalanine, phenyllactic acid, and phenylacetyl-glutamine in the urine. The body fluids contain a high concentration of phenylalanine, but not of the other substances listed above, which appear to be rapidly cleared from the blood by the kidneys. The basic defect is the inability of the body to hydroxylate phenylalanine (Fig. 8.31): it has been demonstrated directly that part of the liver enzyme-complex which normally catalyses this reaction is absent.

Alkaptonuria is characterized by the continuous daily

excretion of several grams of homogentisic acid in the urine. The presence of this substance in urine causes it to darken on standing in contact with air, due to the oxidation of homogentisic acid to a pigment. The level of homogentisic acid in the blood is low, because of the low renal threshold. There are no serious effects on the sufferer. No other constituents are found in blood or urine in abnormal concentrations. Homogentisic acid is on the normal pathway of metabolism of tyrosine, the further catabolism of which is completely blocked. Such a block affects 80–90 % of the metabolism of phenylalanine and tyrosine, since the other pathways to dihydroxyphenylalanine, adrenaline, thyroxine, etc., are probably quantitatively limited.

Tyrosinosis is a very rare disease in which there is a continuous excretion in the urine of tyrosine and *p*-hydroxyphenylpyruvic acid.

General Considerations

There is a difference between carbohydrate and lipid metabolism, on the one hand, and amino acid metabolism, on the other, in that there is no storage polymer for amino acids. Reflection will show that the pattern of amino acids absorbed from the gut will vary according to the dietary protein, while every polypeptide is synthesized according to an inflexible sequence carried in the genome. Thus there is no possibility of synthesizing a protein of varying amino acid composition which would store varying proportions of amino acids. In this sense amino acid metabolism is immediate; if free amino acids are given by mouth, the major fraction is incorporated into protein, or catabolized, within 4 hours. The free amino acid content of cells rises significantly after a protein meal, but the total storage of monomers is not large. One must remember, however, that the liberation of amino acids from protein in the gut is much slower than the release of glucose from polysaccharides, so that after a protein meal amino acids may be taken up into the bloodstream for many hours.

Although there is no specific storage protein, many enzymes are present in cells in amounts far beyond catalytic requirements. Analysis of genetic defects shows that cells may often lose 95 % of a particular enzyme, without serious harm to the metabolic pathway. As indicated in the earlier part of this chapter, many enzymes have short half-lives, and in consequence, liver may lose 30 % of its protein in a 2-day fast. This burst of catabolism is supplemented by a much longer lasting breakdown of more stable proteins.

This very large reservoir of protein explains the discrepancy between the immediate utilization of dietary amino acids, and the fact that protein deficiency diseases may take months to develop, and that adults may live in a negative nitrogen balance for years. It is also significant that there are so many inborn errors of amino acid metabolism. Some, indeed, are lethal within a few months of birth, but others are compatible with an indefinite life-span, suggesting that the pathways in which they occur are of secondary importance. There are, by contrast, no inborn errors of tricarboxylic acid cycle metabolism.

The detailed descriptions of the metabolic pathways of the individual amino acids show that there are often several pathways, some of which are not very extensively utilized. For example, transamination, although theoretically possible for methionine, is unlikely to be very important in most circumstances. This multiplicity of options, and the differences in structure of the carbon skeletons, makes it difficult to summarize amino acid metabolism beyond the general outline of Fig. 8.7. It also makes detail very hard to remember.

It is striking how much of the metabolism of amino acids has been gleaned from the study of human genetic abnormalities. Much still remains to be established. It is to be hoped that the information summarized here may be of value in relating new and unexplained observations made in the future to the framework of present knowledge.

Measurement of protein turnover

One of the first uses made of stable isotope (^{15}N) methodology was to estimate amino acid turnover, but a technique suitable for clinical use is still not universally agreed upon. Part of the trouble is that 'body protein turnover' is a gross over-simplification; the digestive enzymes, of which about 50 g are synthesized per day, have a life of only a few hours, while at the other end of the scale, collagen, one of the most abundant proteins in the body, has a half-life of 3 months. The contractile protein of muscle has a half-life of about 1 month, and many of the measurements that are made are really distorted estimates of muscle protein turnover.

In view of the importance of muscle growth in childhood, or the wasting of muscle in confinement to bed, this limitation of the measurements is not entirely unsatisfactory. One method of estimating body protein turnover frankly limits itself to muscle turnover. It depends on the fact that one histidyl residue in each light

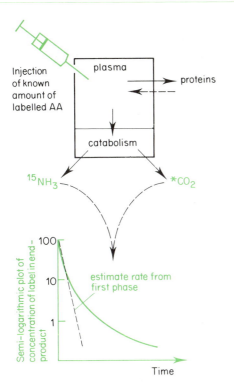

Fig. 8.32 A technique for estimating protein turnover in man. The amino acid which is injected into the bloodstream can be labelled either with ¹⁴C (which is radioactive), or with ¹⁵N (which is a stable isotope). The graph shows one method of interpreting the results that may be obtained.

myosin chain is always methylated after translation, to give 3-methyl histidine. This modified histidine is not metabolized at all, but is excreted in the urine when the myosin molecule breaks down. In a steady state, simple measurement of 3-Me-histidine output should give a measure of the rate of myosin synthesis. However, even this method has drawbacks. It cannot be used in a non-steady state, and the diet has to be free from meat. In addition, there is a lurking suspicion that there may be another protein in another tissue that contains 3-Me-histidine.

The more general methods use a labelled amino acid. Fig. 8.32 shows that, in a short-term experiment, measurement of the rate of excretion of the amino acid gives an estimate of the fraction that is not being excreted, i.e. is being immobilized in protein. In longer term experiments there are complications, because the labelled amino acid may be released again from rapidly turning-over proteins, and so distort the calculations.

Two kinds of label have been used: ¹⁴C, normally in the —COOH group of the amino acid, and ¹⁵N in the amino group. The ¹⁴C label is very easy to measure in the CO_2 of expired air, but giving relatively large doses of radioactive amino acid to patients is not attractive. The alternative is to give a ¹³C-labelled amino acid as ¹³C is not radioactive, but this needs special measuring equipment, as does the ¹⁵N method, for which urinary NH_3 (rather than urea) is the most suitable end product to assay. Stable isotope measurements are also time-consuming. The result is that the number of reliable estimates of protein turnover in human subjects, particularly in childhood or in recovery from surgical procedures, is sadly still very small. Some values are quoted in chapter 16.

9

Common Terminal Pathways of Metabolism

In the preliminary stages of metabolism, carbohydrate, fat, and the carbon skeletons of amino acids are partially oxidized and produce, apart from some carbon dioxide and water, one of six substances: (a) *acetate* as acetyl-coenzyme A (acetyl-CoA) from two-thirds of the carbon of carbohydrate and glycerol, all the carbon of fatty acids, and a part of the carbon of amino acids; (b) *α-oxoglutarate* from glutamate, histidine, arginine, citrulline, ornithine, and proline; (c) *oxaloacetate* from aspartate; (d) *fumarate* from part of the benzene rings of phenylalanine and tyrosine; (e) *succinate* from methionine, valine, and isoleucine; and (f) *acetoacetate* from parts of leucine, phenylalanine and tyrosine.

These substances are metabolically closely related in that they all take part in a common metabolic pathway—the *tricarboxylic acid cycle* or *citric acid cycle* (see Fig. 9.1); acetoacetate enters the cycle through acetyl-CoA. In the cycle both carbon atoms of the acetyl residue are oxidized to carbon dioxide at the expense of the oxygens of two water molecules, the hydrogen atoms

being transferred by dehydrogenases to the hydrogen transporting and oxidizing mechanism described later in this chapter. The cycle also has what has been called an *anaplerotic* function, i.e. part of it can be used for the synthesis of metabolites needed for other functions in the cell. Although this role is more important in bacteria than in animals, the continuous provision in liver cells of oxaloacetate for the formation of the aspartate necessary for the urea cycle (p. 131) is an example of anaplerosis.

The major function of the cycle in animals is to be the most important series of reactions in cellular economy in which energy from oxidative processes is released and trapped. One example of this importance is the brain damage that follows inhibition of α-oxoglutarate oxidation by ammonia (p. 129) or branched-chain oxo-acids (p. 143). Before discussing the reactions of the tricarboxylic acid cycle in detail, the theoretical background of biochemical energetics will be discussed.

Biochemical Energetics

Living organisms require a constant supply of energy and materials to maintain their structure and to perform their numerous functions; these requirements are supplied by the food taken in. The processes of digestion and metabolism are directed towards supplying energy and compounds necessary for growth and replacement of the tissues. Energy is not only required for mechanical work, maintenance of body temperature, and osmotic work, but also to drive numerous synthetic reactions.

Free energy

The free energy content of a substance, G, is a measure of the potential energy of that substance, but it is a quantity that cannot be measured directly. If, however,

substance A reacts chemically and is converted into substance B:

$$A \to B \tag{1}$$

the free energy content of B (G_B) differs from that of substance A (G_A) and a *change of free energy* (ΔG) has occurred as follows:

$$\Delta G = G_B - G_A$$

When ΔG is negative, consequently G_B is less than G_A; when A reacts to produce B the free energy content of the system (A plus B) decreases. This means that an equal amount of energy has been given up by the system (as heat, work, or change in entropy). When ΔG is positive, G_B is greater than G_A; thus when A reacts to give B the free energy content of the system increases, i.e. energy has been taken up. For such a reaction to proceed energy must be supplied to drive it.

When the free energy change in a reaction is negative (ΔG negative) the reaction is called *exergonic*, and when the free energy change is positive (ΔG positive) the reaction is called *endergonic*. It is often said that reactions with a negative ΔG can occur *spontaneously*, but the sign and magnitude of ΔG give no information on the rate of a reaction. Thus for the complete oxidation of glucose according to the equation

$$C_6H_{12}O_6 + 6O_2 \rightarrow 6CO_2 + 6H_2O$$

$\Delta G = -2870$ kJ per mole of glucose; in spite of this large negative ΔG, glucose is quite stable in air (which contains 20% oxygen), the rate of spontaneous reaction being negligibly small. The importance of biological systems is that they contain catalysts (enzymes), which reduce the activation energy needed to go from an initial to a final state.

Classical thermodynamics presupposes that the processes with which it deals are reversible, so that reaction (1) ought to be re-written

$$A \rightleftharpoons B \qquad (1a)$$

If $A \rightarrow B$ with a decrease in free energy, and $B \rightarrow A$ only with an increase, one can see intuitively that there is probably some point in a reversible system at which the two energy changes balance, and no net energy change in the system is occurring. This is in fact the *equilibrium* point of the system and for any reaction the conditions can be quantitated, in terms of free energy, by the treatment below. First, however, we should say something about reversible reactions in biochemistry. Enzymes are catalysts, and the latter by definition do not alter the equilibrium point of a reversible reaction, so that all enzyme-catalysed reactions are assumed to be reversible. There is some doubt whether this is really true, particularly for oxidases. In addition, some reactions are so far from equilibrium in the conditions pertaining in vivo, that the rate of the 'back' reaction must be very small. The use of the phrase 'effectively irreversible' means that such reactions are nevertheless taken to be in principle reversible, so that thermodynamic calculations may be applied to them.

For reaction (1a) the free energy change is given by:

$$\Delta G = \Delta G_0 + RT \log_e \frac{[B]}{[A]} \qquad (2)$$

where ΔG_0 is the *standard free energy change*, R is the gas constant, and T is temperature in kelvins. Thus when $[B] = [A]$ (implicitly both are 1 M), $\Delta G = \Delta G_0$, but this is not a convenient way of finding the standard free energy of a reaction.

The equilibrium constant, K, for reaction (1a) is given by:

$$K = \frac{[B]_{eq}}{[A]_{eq}} \qquad (3)$$

At equilibrium there is no change in free energy ($\Delta G = 0$); therefore (2) becomes:

$$0 = \Delta G_0 + RT \log_e \frac{[B]_{eq}}{[A]_{eq}}$$

which, using (3) and changing to logarithms to the base 10, becomes:

$$0 = \Delta G_0 + 2.303 RT \log K$$

or

$$\Delta G_0 = -2.303 RT \log K \qquad (4)$$

Equation (4) shows that if it is possible to determine the equilibrium constant for any reaction, the standard free energy change can also be calculated. Conversely, if ΔG_0 is known, the actual free energy change ΔG for any ratio of $[B]/[A]$ can be calculated.

The relationship between K and ΔG_0 for a number of values is shown in Table 9.1.

Table 9.1 Relationship between the equilibrium constant (K) and the standard free energy change (ΔG_0) of a reaction at 25°C

K	$\log K$	$\Delta G_0 = -2.3RT \log K$ (joules)
1000	3	-17121
10	1	-5707
1.0	0	0
0.1	-1	5707
0.001	-3	17121

For reactions with more than one reactant and product, the full form of (2) is

$$\Delta G = \Delta G_0 + RT \log_e \frac{[C]^c[D]^d}{[A]^a[B]^b} \qquad (5)$$

where a, b, c and d are the numbers of molecules of A, B, C and D taking part in the balanced reaction.

If the numbers of reactants and products are not equal, e.g. in the reaction $A \rightleftharpoons B + C$, their ratio is not a pure number, and the value of the logarithmic term in (5) depends very much on the absolute concentrations, i.e. whether thay are molar, millimolar or micromolar. This, in turn, means that the actual value of ΔG can be very different from ΔG_0.

Coupled reactions

Two or more reactions may be coupled in such a way that an exergonic reaction may be used to drive an endergonic one. Many instances of this sort of coupling are found in metabolism when the energy derived from an exergonic reaction (usually an oxidation) is used to drive an endergonic synthetic reaction. This sort of coupling requires both reactions to contain a common reactant. Thus if we have two reactions:

$$A \rightleftharpoons B, \quad K = 0.01, \quad \Delta G_0 = 11\,414 \text{ J/mol} \quad (6)$$

and

$$B \rightleftharpoons C, \quad K = 1000, \quad \Delta G_0 = -17\,121 \text{ J/mol} \quad (7)$$

reaction (6) is endergonic and [B]/[A] tends to 1/100; reaction (7) is exergonic and [C]/[B] tends to 1000/1. Since both reactions can be coupled through the common reactant B, (6) and (7) can be added together to give:

$$A \rightleftharpoons C, \quad K = \frac{[C]}{[A]} = 10, \quad \Delta G_0 = -5707 \text{ J/mol} \quad (8)$$

Thus, although the reaction of $A \rightarrow B$ is endergonic and therefore unfavourable, the sequence $A \rightarrow B \rightarrow C$ proceeds to the right because the reaction $B \rightarrow C$ is exergonic and favourable. Reaction (6) has been driven to the right by the favourable free energy change of reaction (7). It is common in biochemical systems for the exergonic (driving) reaction to be the hydrolysis of ATP (p. 156). In many instances, the mechanistic coupling of the two reactions may not be obvious, e.g. in glutamine synthesis (p. 130); nevertheless it exists, and the coupling is stoichiometric.

Heat of reaction

The change in the heat content of the reactants, ΔH, is related to the free energy change as follows:

$$\Delta G = \Delta H - T\Delta S$$

or

$$\Delta H = \Delta G + T\Delta S \quad (9)$$

where ΔS is the change in the *entropy* of the reactants. The entropy of a system is a measure of the degree of order of the system and a measure of the freedom of the components of the system to exist in all possible resonance forms, tautomers, etc.

When ΔH for a reaction is negative the heat content of the reactants diminishes as the reaction proceeds and heat is given out. Conversely, when ΔH is positive the heat content increases and heat is taken up. The magnitude and sign of ΔH for a given reaction may be measured in a calorimeter but, because of relation (9), the change in the heat content must not be equated with the free energy change.

Entropy

For any given reaction, the entropy of the system (e.g. a cuvette in which an enzyme reaction is being studied) may increase, decrease or remain constant, but the entropy of the system *together with its surroundings* will increase. An increase in the entropy of system plus surroundings is a more fundamental criterion than a negative ΔG, for the possible occurrence of a spontaneous reaction. The entropy of the surroundings may increase by a transfer of heat energy, but this does not conflict with what has been said above about heat of reaction, because ΔH applies only to the *system*.

Entropy is a difficult concept, and it is not always easy to label the system and its surroundings with confidence. The idea is however extremely useful in abstract terms, because it enables us to cope, at least partially, with non-equilibrium thermodynamics.

Non-equilibrium states

Any living organism, no matter how simple, is not able to attain equilibrium. As a matter of observation, it is continuously taking in substrates from, and returning waste products to, its environment. Thus we cannot strictly apply equilibrium thermodynamics to the chemical behaviour of such an organism.

If we analyse the organism plus its environment in terms of entropy, we see that there is a continual flux of entropy to the environment. Mathematical treatment shows that, in a steady state, the gain of entropy by the surroundings is at a minimum. Thus a biological organism is able to maintain itself in a defined state of order, at the expense of increasing the disorder of the environment.

In more familiar descriptive terms, we may say that in an organism, even at rest, proteins of rapid turnover may be degraded and have to be re-synthesized; membrane components may become oxidized and need replacement; Na^+ ions will leak through the membrane and have to be pumped out. All this requires a certain expenditure of energy (usually expressed in terms of ATP), for which a certain minimum amount of nutrient must be oxidized, per unit time. These processes are accompanied by a liberation of heat energy that in physiological terms may be identified with the Basal Metabolic Rate (chapter 16), but in physico-chemical terms is the minimal entropy flux.

It may be thought that, since biological systems in vivo are not in equilibrium, the thermodynamic expressions that were derived in equations (1) to (9) cannot be applied to them. The great merit of a thermodynamic treatment is that it can show which reactions are *likely* to occur 'spontaneously', and which probably need to be coupled to an exergonic reaction. Moreover, if the entropy changes are small by comparison with the free energy changes, which often seems to be the case, the error introduced by neglecting non-equilibrium conditions may be small.

Oxidation–reduction

It is useful here to give a short account of oxidation–reduction systems. *Oxidation* implies that the atom oxidized loses electrons and, conversely, *reduction* implies that the atom reduced gains electrons. Thus in the system:

$$Zn + Cu^{2+} \rightarrow Zn^{2+} + Cu$$

the metallic zinc atom is the reducing agent which is oxidized to a zinc ion, while the cupric ion is the oxidizing agent which is reduced to metallic copper.

Another, more complex, example is the oxidation of hydroquinone by Fe^{3+} (ferric) ions in which the hydroquinone loses not only 2 electrons but also 2 protons (H^+), thus:

Hydroquinone   *Quinone*

In a third example, two atoms of hydrogen may be transferred, as in:

$$CH_2(NH_2) \cdot COOH + O_2 \rightarrow CH(:NH) \cdot COOH + H_2O_2$$

Any reaction of this kind can be represented as if the transfer of $2H^+$ and two electrons had taken place.

A fourth type of oxidation is that in which the compound which is oxidized gains one or more oxygen atoms directly from molecular oxygen. Some important reactions of this type are considered later in this chapter.

Oxidation–reduction potential

The tendency of any particular atom, ion, or molecule to lose an electron, thereby being oxidized, is measured by its oxidation–reduction potential, E_0', which is the potential of a platinum electrode placed in a solution of equimolar concentrations of both the reduced and the oxidized forms of the substance and measured against the standard hydrogen electrode. The more strongly reducing the substance the more negative is its E_0'. Thus substance A with a more positive E_0' is able to oxidize substance B with a more negative E_0'. It is often, however, found in practice that some intermediate link or carrier is required to bring about the reaction.

The difference in oxidation–reduction potential, $\Delta E_0'$, between an oxidizing and a reducing agent is related to the free energy change which occurs when the oxidizing agent is reduced by the reducing agent, by the expression

$$\Delta G_0' = -nF\Delta E_0' \qquad (10)$$

where n is the number of electrons transferred and F is Faraday's constant (96 230 J/V equivalent). For instance, the oxidation of NADH by molecular oxygen (through the intermediate steps to be described) can be represented by the reaction

$$H^+ + NADH + \tfrac{1}{2}O_2 \rightarrow NAD^+ + H_2O \qquad (11)$$

In this reaction $n = 2$ and $\Delta E_0'$ is given by the difference between E_0' for NADH/NAD$^+$ (-0.320 V) and for water/oxygen (0.816 V) both at pH 7.0; $\Delta E_0' = 0.816 - (-0.320) = 1.136$ V. Therefore for reaction (11) numerical values can be inserted into equation (10) giving:

$$\Delta G_0' = (-2)(96 230)(1.136) = -218.6 \text{ kJ/mol}$$

Since ΔG for this reaction is large and negative the oxidation of NADH is strongly exergonic. This favourable free energy change does not, however, mean that NADH is rapidly oxidized by molecular oxygen in the absence of a catalyst.

When the ratio of reduced and oxidized members of a redox pair is not unity, an equation, analogous to equation (2), is used to calculate the actual redox potential.

$$E' = E_0' + \frac{2.3RT}{nF} \log_{10} \frac{(Ox)}{(Red)} \qquad (12)$$

where 'Ox' stands for oxidized form or electron acceptor, and 'Red' for reduced form, or electron donor. At 25°C, the value of the term $2.3RT/nF$ is about 59 mV for a 1-electron couple, and ~ 30 mV for a 2-electron pair. In any real electron transfer, other than to an electrode, there has to be a second redox pair involved, and equation (2) can be expanded, on the lines of equation (5), to accommodate this, but in order to avoid confusion over signs it is often better to set up a second

Table 9.2 Standard redox potentials

	*E_0' (volts)
$H_2 \rightarrow 2H^+ + 2\epsilon$	-0.42
Thioredoxin (1ϵ transfer)	-0.40
Isocitrate/Oxoglutarate + CO_2	-0.38
NADH $\rightarrow$ NAD$^+$ + H$^+$ + 2ϵ	-0.32
Lactate/Pyruvate	-0.19
Malate/Oxaloacetate	-0.17
GSH/GSSG + 2H$^+$ + 2ϵ	-0.10
Succinate/Fumarate	0.00
Ubiquinone (UH$_2 \rightarrow$ U + 2H$^+$ + 2ϵ)	$+0.10$
†Cytochrome c (Fe$^{2+} \rightarrow$ Fe^{3+} + ϵ)	$+0.26$
$O^{2-} \rightarrow \tfrac{1}{2}O_2 + 2\epsilon$	$+0.81$

* The use of the prime (E') indicates that the potential is calculated for some pH other than standard (pH = 0). The difference is -0.42 V for hydrogen at pH 7.

† For redox potentials of other cytochromes, see Table 9.5, p. 164.

equation of the form of (12), for the second redox pair, and then to equate the two right-hand sides, since E' will be the same for both systems if a catalyst (enzyme) is mediating between them.

Table 9.2 gives a list of some standard potentials of biological redox systems.

High-energy compounds

A large number of the coupled reactions discussed on p. 153 involve ATP or another nucleoside triphosphate. That is, an endergonic reaction is forced to go by being kinetically coupled with the hydrolysis of ATP, either to ADP and P_i, or to AMP and pyrophosphate, even if the product does not contain phosphate (p. 154). The standard free energy of hydrolysis varies with the pH and with [Mg], but for pH 7 the following values are representative

$$MgATP^{2-} + H_2O \rightarrow Mg^{2+} \quad + ADP^{3-} + HPO_4^{2-} + H^+ \quad -30.5 \text{ kJ/mol}$$
$$MgATP^{2-} + H_2O \rightarrow AMP^{2-} \quad + P\text{-}P_i^{3-} + 3H^+ \quad -32.2 \text{ kJ/mol}$$
$$P\text{-}P_i^{3-} + H_2O \rightarrow 2 \text{ } HPO_4^{2-} + H^+ \quad -30 \text{ kJ/mol}$$

Thus hydrolysis of ATP to AMP and pyrophosphate, as in activation of fatty acids, yields substantially the same amount of free energy as the more common hydrolysis to ADP and P_i. However, pyrophosphate is usually removed by hydrolysis to inorganic phosphate (note that the free energy of this reaction, by itself, is never coupled; it is an example of continuous disturbance of equilibrium).

If the concentration of ATP is not 1 M, but 1 mM (as it is intracellularly), the free energy of hydrolysis is increased considerably, to about 45–50 kJ, depending on conditions. The free energies of hydrolysis of other nucleoside triphosphates are very similar to that of ATP.

In the metabolic pathways so far outlined, ATP has always been formed by transfer of a phosphoryl group from a suitable donor; in the processes discussed later in this chapter, ATP is synthesized in reactions coupled to the transport of reducing equivalents from oxidizable substrates to oxygen, coupled reactions in which the synthesis of ATP is the endergonic process. Thus ATP can take part in two-fold coupled reactions, effectively transferring chemical energy between oxidative and synthetic reactions. This phenomenon is so widespread, and so important to the economy of the cell, that it is convenient to speak of the nucleoside triphosphates as 'high-energy compounds', or as 'the common energy currency of the cell'. The former term has been extended to the acyl thioesters of coenzyme A and acyl carrier protein, and to a few phosphate-containing compounds (e.g. creatine phosphate) for which the equilibrium constant approaches 1 for phosphoryl exchange with ADP.

In many cases the large free energy of hydrolysis may be explained on structural grounds (pp. 35 and 36), but several instances may be quoted of apparently simple compounds which could be classed as 'high-energy' on equilibrium grounds. Acylcarnitine, aminoacyl-tRNA and N^{10}-formyl-FH_4 all fall into this category.

It should also be remembered that endergonic reactions need not always be driven by ATP. The synthesis of long-chain fatty acids, for example, is largely driven by the energy inherent in the redox potential of the $NADPH/NADP^+$ couple.

Reactions of the Tricarboxylic Acid Cycle

Citrate synthase (condensing enzyme)

In this reaction the acetyl residue of the energy-rich thioester, acetyl-CoA, is condensed with oxaloacetate to form citrate, free $CoA \cdot SH$ being liberated. This is an exergonic reaction, ΔG being about $-33\,600$ J/mol, and the equilibrium lies very much in the direction of citrate synthesis (see p. 158). The coupling of the hydrolysis of acetyl-CoA with citrate synthesis is very necessary because the condensation of free acetate with oxaloacetate to form citrate (as catalysed, for example, by a bacterial enzyme) is not nearly so favourable to synthesis. The K_m for both substrates of the enzyme is very low, and in all normal circumstances the enzyme is saturated with acetyl-CoA.

Aconitase

This enzyme catalyses the attainment of an equilibrium between cis-aconitate

$$HOOC \cdot CH = C \cdot CH_2 \cdot COOH$$
$$|$$
$$COOH$$

citrate and isocitrate. As has been discussed on p. 41, citrate does not behave as a symmetrical molecule on this enzyme, and in the isocitrate formed by the aconitase reaction the hydroxyl group is on the methylene carbon atom which was derived from oxaloacetate.

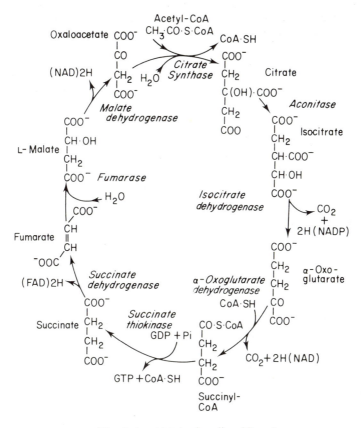

Fig. 9.1 The citric acid (tricarboxylic acid) cycle.

Isocitrate dehydrogenase

This enzyme catalyses the oxidative decarboxylation of isocitrate to α-oxoglutarate, CO_2 being liberated and 2H being transferred to a nicotinamide nucleotide, which may be either $NADP^+$ or NAD^+. The reaction catalysed by the $NADP^+$-requiring enzyme is reversible, that catalysed by the NAD^+-requiring enzyme almost irreversible (because of a very high K_m for CO_2).

The NAD^+ enzyme is entirely mitochondrial, while the $NADP^+$ enzyme is found both in mitochondria and cytoplasm. In most mammalian tissues the major fraction of the total isocitrate dehydrogenase in mitochondria is NADP-linked; in heart the $NADP^+$-linked enzyme constitutes 85% of the total. This is something of a puzzle. For allosteric effectors of the NAD^+-linked enzyme, see the end of this section.

Although this reaction can be considered to occur in two stages, free oxalosuccinate

$$HOOC \cdot CO \cdot CH \cdot CH_2 \cdot COOH$$
$$|$$
$$COOH$$

is not an intermediate.

α-Oxoglutarate dehydrogenase

This enzyme complex catalyses the oxidative decarboxylation of α-oxoglutarate with the formation of succinyl-CoA. The mechanism is analogous to the oxidative decarboxylation of pyruvate to give acetyl-CoA (see p. 74). NAD, thiamine pyrophosphate (TPP), lipoic acid, and CoA-SH are the necessary coenzymes. In this reaction an energy-rich thioester is formed and the reaction is not readily reversible. Indeed, it is this reaction, more than any other, that gives the cycle its unidirectional character.

Succinate thiokinase

In this coupled reaction the energy-rich thioester succinyl-CoA, formed in the oxidative decarboxylation of α-oxoglutarate, is hydrolysed, and at the same time the energy-rich compound guanosine triphosphate (GTP) is formed by the phosphorylation of GDP:

$$\text{succinyl-CoA} + \text{GDP} + P_i \rightleftharpoons \text{succinate} + \text{GTP} + \text{CoA} \cdot \text{SH}$$

The reaction is readily reversible, K being approximately 4.

It is not clear how the GTP is utilized. Although there is an enzyme in mitochondria which catalyses the reaction:

$$\text{GTP} + \text{AMP} \rightleftharpoons \text{GDP} + \text{ADP}$$

there is no reaction known to exist in mitochondria which would produce AMP at the requisite rate. *Nucleoside diphosphate kinase*, which catalyses the general reaction

$$\text{NDP} + \text{ATP} \rightleftharpoons \text{NTP} + \text{ADP}$$

appears to be largely cytoplasmic.

Succinate dehydrogenase

The next step in the cycle is the dehydrogenation of succinate to form the unsaturated compound fumarate. The oxidizing agent is FAD which is, in this enzyme, covalently bound to the dehydrogenase, which also contains three iron–sulphur centres (see p. 165). The reducing equivalents are probably transferred to ubiquinone. Succinate dehydrogenase is very strongly inhibited by the next lower homologue to succinate, i.e. malonate

Malonic acid

Fumarase (fumarate hydratase)

Water is next added across the double bond of fumarate to form the hydroxy compound L-malate. The free energy change in this step is small; K is between 4 and 5.

Malate dehydrogenase

In this oxidative step two hydrogen atoms are removed from L-malate to form oxaloacetate; NAD^+ is the hydrogen acceptor. At pH 7.0 the equilibrium is very much in favour of malate; K is about 10^{-5}. The reaction is, however, driven towards oxaloacetate because the latter reacts with acetyl-CoA in the exergonic citrate synthesis.

Partial reactions of the cycle

The citric acid cycle is a set of eight enzymes, all of which are at least partly located in mitochondria. It is not essential that they should act only as a single multi-enzyme complex, since they are surrounded by other enzymes with common substrates, and the flux of substrates into, and out of, mitochondria may at times dictate the use of some of the cycle enzymes, and not others. Some important instances of incomplete use of cycle enzymes are discussed below.

1. Glutamate oxidation

This pathway seems to provide a temporary increase in cycle flux very quickly. Glutamate is first transaminated with oxaloacetate:

$$\text{Glu} + \text{OAA} \rightarrow \alpha\text{-OG} + \text{Asp}$$

Oxoglutarate is then oxidized as far as oxaloacetate, so that the complete pathway is

$$\text{Glu} \rightarrow \text{Asp} + CO_2 + 3 \times 2H$$

Since glutamate does not readily traverse the cell membrane, this system is limited to the cell's own store of glutamate.

A more complex version is used by the cells of the intestinal mucosa, which have very little pyruvate dehydrogenase. They pick up glutamine readily from blood or from the intestinal lumen, which is first deaminated

$$\text{Glu-NH}_2 \rightarrow \text{Glu} + NH_3$$

$$\text{Glu} + \text{Pyr} \rightarrow \alpha\text{-OG} + \text{Ala}$$

and the α-OG is then oxidized.

$$\alpha\text{-OG} \rightarrow \text{OAA} + CO_2 + 3 \times 2H$$

The Ala diffuses into the bloodstream. Reticulocytes, and some tumour cells, are thought to use the same pathway.

Acetoacetate oxidation

Ketone bodies are secreted by liver, as described in chapter 7. Acetoacetate can be re-activated and oxidized in muscle cells, in intestinal mucosal cells, and after prolonged fasting, in brain. In each case the mechanism is the same:

$$\text{AcAc} + \text{succinyl-CoA} \rightarrow \text{AcAc-CoA} + \text{succinate}$$

$$\text{AcAc-CoA} + \text{CoA} \rightarrow 2 \text{ Ac} \cdot \text{CoA}$$

Urea synthesis

In liver of ureotelic animals, fumarate which arises from the splitting of argininosuccinate is converted to oxaloacetate, as described in detail in chapter 8. This takes place outside mitochondria, emphasizing that the citric acid cycle is not a purely mitochondrial multi-enzyme process.

Transport of reducing equivalents

This is discussed in more detail in the next section, but a cycle of reactions fairly well established for the outward transport of NADPH from mitochondria may be mentioned here. The central role is played by an NADP⁺-linked isocitrate dehydrogenase, located in the cytosol. As the diagram shows, the export of isocitrate from mitochondria is balanced by the intake of malate, so that the system acts as an NADH/NADP⁺ transhydrogenase (see Fig. 9.2). For the *malic enzyme*, see below.

Acetyl transfer

This is discussed more fully under fatty acid synthesis (p. 102). The export of citrate from mitochondria and its splitting into acetyl-CoA and oxaloacetate is balanced by the reduction of the latter to malate, which is exchanged against the citrate. The two important points here are that citrate cannot traverse the mitochondrial membrane alone, without a counter-ion to maintain electrical neutrality; and secondly, animal mitochondria are relatively impermeable to oxaloacetate, but are readily permeated by malate.

It is worth noting that citrate does not readily escape from the mitochondria of many tissues, notably muscle.

Oxidation of amino acids : gluconeogenesis

Fig. 8.11 in chapter 8 summarizes the relationships of the carbon skeletons of many amino acids with the citric acid cycle, and may seem to imply that the skeletons are at once completely oxidized. It must be stressed that in the first instance, compounds that are converted to any intermediate of the cycle can only be oxidized to oxaloacetate. For complete oxidation, the latter has to be converted to acetyl-CoA. Two mechanisms of achieving this are known. One works through 'malic enzyme' (NADP-linked malate dehydrogenase)

$$\text{malate} (\equiv \text{OAA}) + \text{NADP}^+ \rightarrow \text{pyruvate} + \text{CO}_2 + \text{NADPH} \quad (1)$$

but this is not thought to be of major importance, since the enzyme is only present in low concentration in liver and similar tissues.

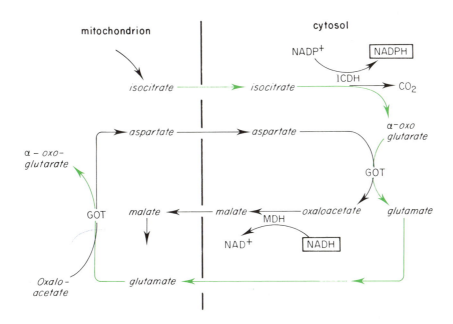

Fig. 9.2 The export of reducing equivalents from mitochondria. This shuttle, involving isocitrate and glutamate, is especially important in liver, because NADPH is required in the cytosol not only for fatty acid synthesis, but also for mixed function oxidation (cf. Fig. 9.8). Note that NADH disappears from the cytosol when oxaloacetate is reduced to malate, so that the shuttle effectively operates as a cytosolic transhydrogenase.

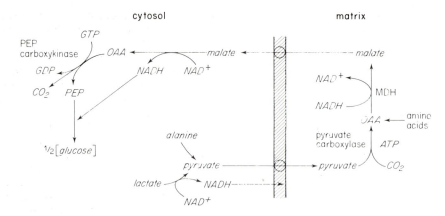

Fig. 9.3 Interrelationships between mitochondrial matrix and cytosol in gluconeogenesis. In this scheme, malate performs two functions: (a) to transfer carbon atoms from pyruvate, carboxylated within the matrix, back into the cytosol; (b) to provide reducing equivalents for the reduction of 1,3-diphosphoglycerate to glyceraldehyde phosphate. This second function may not be necessary, e.g. if lactate is the gluconeogenic substrate, but a balance of reducing equivalents has nevertheless to be attained.
The dashed arrow showing NADH entering the mitochondria is schematic only; for a complete mechanism see Fig. 9.5(b).

The second mechanism is identical with the conversion of amino acid skeletons to glucose precursors, namely

$$OAA + GTP \rightleftharpoons PEP + GDP + CO_2 \qquad (2)$$

$$PEP + ADP \rightarrow pyruvate + ATP \qquad (3)$$

As in the previous mechanism, the pyruvate can be converted to acetyl-CoA by oxidation, losing another CO_2 as it does so. Alternatively, the phosphoenolpyruvate can be converted to glucose or other hexose residues as described in chapter 6.

Reaction (2) is reversible, and may be used for replenishing the mitochondrial concentration of citric acid cycle intermediates. This is more easily accomplished if the enzyme catalysing reaction (2), *PEP carboxykinase*, is mitochondrial, as it is in some species. The more general mechanism for achieving the same end is by *pyruvate carboxylase*

$$pyruvate + CO_2 + ATP \rightarrow OAA + ADP + P_i$$

These relationships are summarized in Fig. 9.3.

Mitochondrial metabolite transport

Any description of the operation of the citric acid cycle, and in particular of partial reactions of the cycle, draws attention to the importance of the transport of metabolites into and out of mitochondria. All types of mitochondria in mammalian cells take up substances such as pyruvate and long-chain acyl-CoA, the substrates of the oxidative reactions which take place in the matrix. In addition, the mitochondria of tissues engaged in fatty acid synthesis, gluconeogenesis or urea production must translocate compounds such as citrate,

malate, ornithine and citrulline. Oxidative phosphorylation itself is an intramitochondrial process, and requires the uptake of phosphate and ADP, and the transport of ATP into the cytoplasm. The inner membranes of mitochondria therefore contain a variety of permeases, many of which are antiporters (p. 183). The best known are shown in Fig. 9.4.

The best-characterized of these permeases is adenine nucleotide translocase, a hydrophobic protein of molecular weight about 32 000. In mitochondria from tissues such as heart muscle, with a high capacity for oxidative phosphorylation, this one polypeptide may constitute as much as 15% of the inner membrane protein. It has been isolated, and reconstituted into vesicles in an active form. It is an antiporter which exchanges ADP for ATP, and therefore catalyses ADP uptake and ATP output by mitochondria which are phosphorylating. This exchange is electrogenic, each cycle involving the loss of one negative charge from the matrix: it is therefore an active process, which consumes some of the energy of electron transport, by dissipating the mitochondrial membrane potential (p. 185). This energy is not wasted, however, since the action of the adenine nucleotide translocase and the phosphate porter (see below) decrease the ratio $ATP/ADP \cdot P_i$ within the mitochondria, and thereby reduce the energy required for ATP synthesis.

Inhibitors of adenine nucleotide translocase, such as atractyloside and bongkrekic acid, block oxidative phosphorylation in mitochondria. Only ADP is phosphorylated by mitochondria, because of the specificity of the translocase for adenine nucleotides—other nucleoside diphosphates are not taken up. Phosphorylation of AMP is carried out by *adenylate kinase*

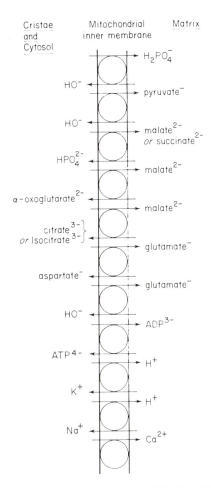

Fig. 9.4 Mitochondrial translocases. Only the well-established systems have been included in this list. All but the last are *antiporter* translocases, i.e. one ion is exchanged for another in a stoichiometric ratio, but the ionic charge on each species is not necessarily the same, e.g. citrate³⁻/malate²⁻ or ATP⁴⁻/ADP³⁻. In such cases the antiport is *electrogenic*.

Fig. 9.4 labels (cristae and cytosol side / matrix side): $H_2PO_4^-$; HO^- / pyruvate⁻; HO^- / malate²⁻ or succinate²⁻; HPO_4^{2-} / malate²⁻; α-oxoglutarate²⁻ / malate²⁻; citrate³⁻ or isocitrate³⁻ / glutamate⁻; aspartate⁻ / glutamate⁻; HO^- / ADP^{3-}; ATP^{4-} / H^+; K^+ / H^+; Na^+ / Ca^{2+}

(myokinase), an enzyme located between the inner and outer mitochondrial membranes:

$$AMP + ATP \rightleftharpoons 2ADP$$

Other nucleoside diphosphates are phosphorylated by nucleoside diphosphate kinase, a cytoplasmic enzyme which catalyses phosphate exchange between ATP and a variety of nucleoside diphosphates (see p. 158). The properties of the translocases for adenine nucleotides and for several other metabolites are summarized in Fig. 9.4. These processes are functionally and energetically linked: the phosphate permease exchanges phosphate for hydroxyl ions, a process which produces active uptake of phosphate, by dissipating the mito-

chondrial transmembrane proton gradient. Succinate and malate are actively accumulated, by electroneutral exchange of these ions for phosphate. In this way transport of most carboxylic acids draws on the transmembrane proton gradient: a significant fraction of the protons translocated by the respiratory chain may be back-translocated, directly or indirectly, by these metabolite permeases.

Most mitochondria also contain permeases for calcium ions; there is a ubiquitous electrogenic uniporter (Fig. 9.4) and also, in some tissues, antiporters which catalyse exchange of Ca^{2+} for H^+ or Na^+. The function of these is not fully understood, but it is thought that mitochondria play an important part in controlling the cytosolic concentration of Ca^{2+}.

Transport and anaplerosis. The export of acetyl-CoA, and of NADPH equivalents, for the synthesis of long-chain fatty acids, has been briefly mentioned in the previous section (p. 159), and is also discussed in chapter 7 (p. 102). The transfer of acyl residues into mitochondria by the carnitine exchange system is also described in that chapter (p. 102). The role of the citric acid cycle in gluconeogenesis has been mentioned in the previous section (p. 159), and the export of glucogenic precursors (including NADH equivalents) is discussed in chapter 6 (p. 79).

The transport of reducing equivalents *into* mitochondria is described in the section on p. 162.

Control of the tricarboxylic acid cycle

Although many inhibitory effects are demonstrable on particular enzymes in vitro, they may not be important in vivo. For example, succinate dehydrogenase is strongly inhibited by oxaloacetate, but the concentration of free oxaloacetate within mitochondria is too low for this to be significant; citrate synthetase is inhibited by ATP and activated by ADP, but the magnesium complexes of these nucleotides, which are the predominant forms within the cell, are much less effective.

As an example of integrated control of a pathway, the enzymes of the mitochondrial tricarboxylic acid cycle are considered. They are regulated by covalent modification, end product inhibition and by allosteric effects, but these controls may vary from one tissue to another. The principal controlling factor is the ratio ATP/ADP, decrease of which (by ATP utilization) affects some enzymes directly and also, more importantly, stimulates mitochondrial electron transport, resulting in an increase in the ratio $NAD^+/NADH$ (p. 162), which controls the dehydrogenases of the TCA cycle.

Important control enzymes and controlling factors are, in heart muscle:

1. *Adenine nucleotide translocase*
   Competition between ATP and ADP for entry into mitochondria (p. 160).

2. *Pyruvate dehydrogenase*
   (a) Covalent modification: Phosphorylation (inactivation) is increased by $NADH/NAD^+$, $AcCoA/CoA$; ATP/ADP. Dephosphorylation (activation) is increased by Ca^{2+}, $NADH/NAD^+$.
   (b) Product inhibition by AcCoA, NADH.

3. *Citrate synthetase*
   (a) Substrate availability (AcCoA, oxaloacetate).
   (b) Product inhibition (citrate).
   (c) Allosteric activation (ADP, AMP).

4. *Isocitrate dehydrogenase (NAD-linked)*
   (a) Substrate availability ($NAD^+/NADH$).
   (*b) Allosteric inhibition by ATP, activation by ADP.
   (*c) Activation by Ca^{2+}.

5. *2-Oxoglutarate dehydrogenase*
   (a) Substrate availability ($NAD^+/NADH$).
   (b) Product inhibition (succinyl-CoA).

The effectors in the list above are those that have been found to alter the activity of the isolated and purified enzymes. They are all genuine experimental observations, but it is obvious that it would be very difficult to single out one of the effectors as being of over-riding importance, or to say that changes in activity of one of the enzymes are more important than any of the others, giving that enzyme the status of a 'pacemaker'.

In the end, the most important factors controlling activity of the cycle are those located at the mitochondrial membrane. The first of these is the flux through the adenine nucleotide translocase. It is this that integrates the energy demand of the complete cell with the flux through the electron transport chain, and thus with the $NADH/NAD^+$ ratio, and so with the activity of individual enzymes of the cycle. The other controlling factors are those of the acyl-CoA transferase (CAT, chapter, 7, p. 102), and of pyruvate translocase and the pyruvate dehydrogenase complex. Between them (except in cells that oxidize much amino acid, see p. 125), these latter factors determine the proportions of the primary nutrients, fat and carbohydrate, that supply the energy demands of the cells and tissues.

When it comes to partial reactions of the cycle, control of individual enzymes may be more important. For example, relative inhibition of NAD^+-linked isocitrate dehydrogenase could promote build-up of citrate and isocitrate within the mitochondria, and promote transfer into the cytosol for fatty acid synthesis (p. 98) or NADPH export (see below).

The Transport and Oxidation of Reducing Equivalents

The NADH dehydrogenase component of the respiratory chain does not oxidize NADPH, yet some mitochondrial enzymes produce NADPH, e.g. glutamate dehydrogenase and the NADP-linked isocitrate dehydrogenase. It is not known precisely how re-formation of $NADP^+$ occurs. Theoretically the exchange

$$NADPH + NAD^+ \rightleftharpoons NADP^+ + NADH$$

should have a K_{eq} of ~ 1, but no mammalian enzyme is known to catalyse this simple reaction. An *energy-linked transhydrogenase* is known which uses the proton gradient of the mitochondrial membrane to force the reaction above from right to left. It appears at present that the same enzyme may catalyse hydrogen transfer from NADPH to NAD^+ without an energy-linked component.

Transport of reducing equivalents into mitochondria

In liver, NADH produced in the cytosol may be used directly or indirectly for other purposes in that part of the cell. Examples are the oxidation of lactate to pyruvate, concurrently with the use of NADH in gluconeogenesis from pyruvate (chapter 6); and the effective transhydrogenation from NADH to $NADP^+$, outlined on p. 159. In most other tissues, cytosolic NADH is oxidized in mitochondria. Since vertebrate mitochondria are impermeable to NADH, the transfer of reducing equivalents has to be made indirectly.

One mechanism, which is known to predominate in insect muscle, is a cycle involving dehydroxyacetone phosphate and glycerol phosphate (GP) (see Fig. 9.5a). The mitochondrial membrane enzyme is a flavoprotein (see below), which probably transfers its reducing equivalent to ubiquinone. A similar enzyme exists in mammalian mitochondria, but is not active enough to account for the reoxidation of a large proportion of the NADH produced by the Embden–Meyerhof pathway. Its activity (at least in liver) is increased, however, by thyroid treatment, which is known to increase O_2 consumption by tissues.

The generally accepted pathway at the present time is a shuttle involving malate dehydrogenase and glutamate-oxoloacetate transaminase, both of which are located in cytosol and in mitochondria. Fig. 9.5(b) shows the symmetry of the exchange.

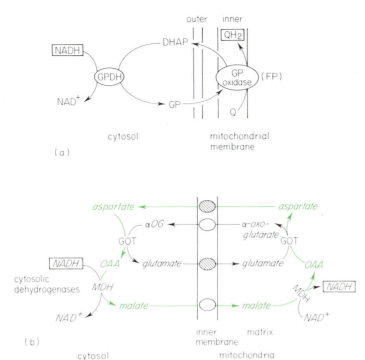

Fig. 9.5 Two possible mechanisms for transporting reducing equivalents from cytosolic NADH into mitochondria. (a) The glycerophosphate shuttle. Present evidence suggests that this is not the most important mechanism in most mammalian tissues. (b) The malate–aspartate shuttle. Although this shuttle could in principle operate in both directions, it probably works only in the direction shown because the antiport of aspartate is unidirectional (Fig. 9.3).

A problem about this mechanism is that the $NAD^+/NADH$ ratio is much larger (by a factor of 10^3) in cytosol than in mitochondria. This is equivalent to a potential difference of 0.09 V, with the cytosol more positive; reducing equivalents will not spontaneously flow 'uphill'. Mechanisms have been suggested for bringing energy into the exchange, but the net loss is still about 12 kJ, a significant fraction of the energy needed to synthesize 1 ATP. For the glycerol phosphate shuttle, the path from flavoprotein to O_2 would in any case only provide 2 ATP (see p. 188).

Components of the Electron Transport Chain

Nicotinamide-adenine dinucleotides

The structures and function of the coenzymes NAD and NADP have been described in chapter 3. The general reaction may be represented by

$$AH_2 + NAD(P)^+ \xrightarrow{\text{Dehydrogenase}} A + NAD(P)H + H^+$$

Table 9.3 shows some dehydrogenases known to be linked to NAD and NADP respectively. For E_0' values, see Table 9.2.

Table 9.3 NAD(P)-linked dehydrogenases

NAD-linked	NADP-linked
Isocitrate	Isocitrate
Triosephosphate	Glucose-6-phosphate
Malate	6-Phosphogluconate
Glutamate	Glutamate
β-Hydroxybutyrate	Malic enzyme
β-Hydroxyacyl-CoA	
Lactate	
Pyruvate (to acetyl-CoA)	
α-Oxoglutarate	
Ethanol	

Flavoprotein dehydrogenases (Table 9.4)

Several important dehydrogenation reactions of intermediary metabolism are carried out by flavoprotein enzymes. These have been discussed in chapter 7. Examples are *acyl-CoA dehydrogenase*, which catalyses the α,β-unsaturation of acyl-CoA (the first step in the breakdown of fatty acids), and *succinate dehydrogenase*, which catalyses the removal of two hydrogens from succinate to yield fumarate. The E_0' of this couple is less negative than that of the others in the citric acid cycle, and it is thermodynamically unlikely that it would be linked to a nicotinamide-adenine dinucleotide.

The ultimate electron acceptor for these and similar dehydrogenases is discussed below.

Other flavoproteins, not connected with the synthesis of ATP (cf. p. 184), are 'aerobic dehydrogenases', that is to say they can transfer 2H directly from a substrate to molecular oxygen, which is reduced to hydrogen peroxide, H_2O_2. H_2O_2 is toxic, and it is presumed that these enzymes usually work in conjunction with catalase, a haem-containing enzyme widely distributed in tissues, which catalyses the reaction

$$2H_2O_2 \rightarrow 2H_2O + O_2$$

However, in congenital acatalasia lesions are confined to

Table 9.4   Flavoprotein enzymes

Substrate	Enzyme	Other groups
(a) *FMN-containing enzymes*		
NADPH	'old yellow enzyme' (NADPH oxidase of red cells)	
NADPH	cytochrome P_{450} reductase	
NADH	NADH dehydrogenase	Fe/S
(b) *FAD-containing enzymes*		
NADH	cytochrome b_5 reductase	
lipoic acid (enzyme bound)	e.g. in pyruvate and oxoglutarate dehydrogenases	
succinate	*succinate dehydrogenase	Fe/S
acyl-CoA	acyl-CoA dehydrogenase	
L-amino acids	L-amino acid oxidase	
glycine	glycine oxidase	
glucose	glucose oxidase	
xanthine	xanthine oxidase	Mo, Fe
aldehydes	aldehyde oxidase	Mo, haem

N.B. The redox potentials of iron–sulphur proteins are very variable, ranging from -0.41 to $+0.34$ V. In succinate dehydrogenase there are three IS centres with E_0' values of -0.40, 0.0 and $+0.06$ V. The redox potentials of flavoproteins depend very much on the nature of the protein, and range from -0.3 to $+0.3$ V; E_0' for riboflavin itself is -0.21 V.

* Flavin covalently bound to the enzyme protein. E_0' for yellow enzyme is -0.122 V at pH 7.

the oral and nasal mucosa, and may be quite absent in some individuals.

The cytochromes

All aerobic cells possess members of a group of respiratory pigments called cytochromes, consisting of a protein with various iron-containing haem-like prosthetic groups. The structure of haem has been discussed in chapter 4. Except for the haem in cytochromes a and a_3, which has a bulky and lipophilic side chain attached to one of the pyrrole rings, and an aldehyde group, the haem groups have the same

Table 9.5   The cytochromes

		E_0' (pH 7.0)
*Cytochrome b	Firmly bound, not autoxidizable when bound	$+0.05$ V
Cytochrome b_5	Not autoxidizable, microsomal	$+0.02$ V
Cytochrome P_{450}	Autoxidizable, forms a compound with CO. Microsomal	$+0.24$ V
*Cytochrome c_1	Not autoxidizable, bound	$+0.22$ V
*Cytochrome c	Not autoxidizable, soluble	$+0.25$ V
*Cytochrome a	Not autoxidizable, firmly bound	$+0.29$ V
*Cytochrome a_3 ($a + a_3$ are in same protein)	Firmly bound. Autoxidizable, forms compounds with CO, HCN, and H_2S. Contains Cu	

* Component of mitochondrial respiratory chain.

structure as the haem of haemoglobin; the haem of cytochrome c appears different because it is attached covalently to its protein through covalent links with two cysteinyl residues. There are at least four cytochromes concerned in the 'mainstream' transfer of electrons from oxidizable substrate to oxygen. Many others are known but only cytochrome b_5, found in endoplasmic reticulum, need be mentioned here. The properties of some cytochromes are shown in Table 9.5. The nature of the protein moiety is important in two ways. First, unlike haemoglobin, the cytochromes act as electron carriers, i.e. the iron in the haem ring is oxidizable to Fe^{3+} and reducible to Fe^{2+}. Secondly, many of the proteins, other than cytochrome c, have large lipophilic domains so that the whole molecule fits readily into a lipid membrane structure; it is not easy to extract such intrinsic membrane proteins into aqueous solution, except with detergents. Some cytochromes, such as b_5 and P_{450}, are globular proteins which are anchored to membranes by hydrophobic 'tails'; such proteins can be solubilized with detergents or proteases (p. 184).

Iron–sulphur proteins

If mitochondria are deproteinized with acid, a smell of H_2S is observed. The inorganic sulphide responsible for this is normally held in coordinate linkage with iron in

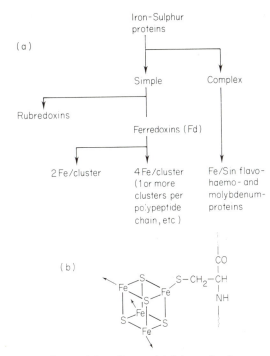

Fig. 9.6 Iron–sulphur clusters. (a) Scheme for the nomenclature of iron–sulphur proteins. (b) Three-dimensional view of a 4-Fe cluster. Each of the Fe atoms is coordinately bound to the —S⁻ of one cysteine residue of the peptide chain (shown explicitly on the right, and indicated elsewhere by arrows), and is also coordinated to three inorganic sulphide (S^{2-}) anions.

'iron–sulphur clusters', which are of importance in plants and bacteria, as well as in animals.

The best-characterized iron–sulphur clusters occur in proteins of small molecular weight (*ca.* 12 000) readily obtained from bacteria and plant tissues (ferredoxins). In these, the iron atoms are held in a rigid framework of S atoms provided partly by S^{2-} ions, and partly by cysteinyl residues coming from the polypeptide chain itself. Fig. 9.6(b) shows the basic structure of a $Fe_4S_4Cys_4$ complex characteristic of a bacterial ferredoxin.

Some of the iron–sulphur complexes of animal cells are of this kind: for example, a ferredoxin occurs in the system that provides electrons for the fatty acid desaturase complex (p. 100), and a special *adrenodoxin* is concerned in the hydroxylation of steroid residues in adrenal mitochondria. However, the main electron transport chain of mitochondria contains several iron–sulphur clusters that are incorporated into larger polypeptide chains. For example, of the three Fe–S clusters of succinate dehydrogenase, two are in a 70 000 mol. wt subunit, and one in a 27 000 mol. wt subunit. The nomenclature of iron–sulphur proteins is summarized in Fig. 9.6(a).

In terms of their redox function, there are two

important points. One is that in spite of the number of iron atoms present, each *cluster* only accepts or donates *one* electron, i.e. the iron–sulphur clusters behave more like cytochromes than flavoproteins (the electron is probably de-localized). The second point is that, although in general the redox potential of iron–sulphur clusters is much lower than that of Fe^{3+}/Fe^{2+} (+0.77 V), and in the ferredoxins can be as low as −0.42 V, the value depends on the nature of the polypeptide environment, rather than on the architecture of the Fe–S centre. Thus the redox potentials of particular centres in different environments may be as much as 0.7 V different from one another. This dependence on the nature of the protein is also true of flavin and haem prosthetic groups, but here the differences are even more striking.

Lipid hydrogen acceptors

There has been speculation about the possible role, as a redox intermediate, of vitamin K, which is a quinone cf. p. 282). At the present time, however, the only lipid-soluble compound known to be of importance in hydrogen transport is *ubiquinone*, which links complexes I and II (Fig. 9.7). The chemistry of ubiquinone has been described in chapter 3, p. 36.

The organization of the electron transport chain

The order of arrangement of components of the electron transport chain was deduced partly by the use of inhibitors. Thus rotenone and amytal inhibit the oxidation of NAD^+-linked substrates, while not affecting the oxidation of succinate. On the other hand, the antibiotic antimycin A inhibits both succinate and NADH oxidation, while permitting oxidation (in vitro) of added reduced cytochrome c. The order was partly elucidated by observing spectroscopically the extent to which components are 'dammed up' in the reduced form when oxidative phosphorylation of ADP is restricted.

These deductions have been substantiated and refined by the isolation of 'respiratory complexes', which can be obtained by treating the mitochondrial inner membrane with mild detergents. They are not multi-enzyme complexes in the sense of, for example, the pyruvate dehydrogenase complex, in that the respiratory complexes are not self-assembling (at least outside the membrane), nor will they reassemble into a membrane when placed together into a test tube. Nevertheless, they contain many polypeptide chains (see p. 186), together with many protein-bound redox groups (riboflavin derivatives, iron–sulphur centres, haems). The soluble cofactors such as NAD and cytochrome c are lost in the isolation, and ubiquinone can be extracted from them with non-aqueous solvents.

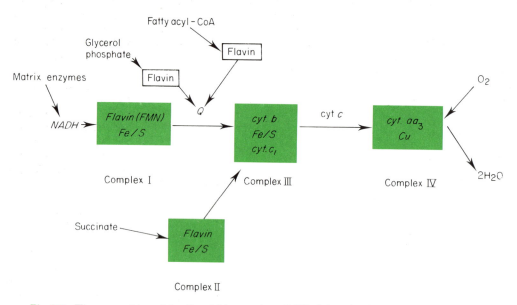

Fig. 9.7 The composition of the dissociable complexes I–IV of the mitochondrial respiratory chain.

Each of the respiratory complexes carries out a fairly defined and limited function in vitro, and the functions correspond to the earlier postulates of the electron carrier sequence, as shown in Fig. 9.7.

The three complexes I, III and IV appear to be the sites of linkage of electron transport with mitochondrial phosphorylation, as described in chapter 10.

As each complex contains several electron-accepting sites, the notion of a single electron, or a pair of electrons, passing sequentially from one carrier to the next, although useful in understanding the stoichiometry, is rather naive in practice. The system must be thought of in terms of a flow of electrons from a defined number of substrate molecules to a defined number of O_2 molecules, through the complexes. The associated protons do not necessarily follow the electrons in absolute physical proximity, but the overall H^+ gradient must obviously correspond to the electron gradient. This gradient is not identical to that established by the mitochondrial proton pump, since the latter can be abolished by uncouplers, while electron transport is unaffected or even increased ('uncoupled' respiration, see chapter 10).

The K_m of the system for oxygen is extremely low, about 0.5 μM. The whole pathway is blocked by cyanide, azide, carbon monoxide, or hydrogen sulphide at the oxidase step. Only cytochrome a_3 reacts with these reagents.

Oxidations not linked to the Respiratory Chain

About 15% of total O_2 uptake, in most tissues, is cyanide-insensitive, and thus presumed to be independent of the respiratory chain. A good deal of this oxidation occurs in so-called peroxisomes, small organelles about the size of mitochondria, which contain a variety of enzymes, often flavoproteins, that reduce oxygen directly (see also p. 164). Substrates include xanthine, uric acid, formate, ethanol and even long-chain fatty acyl CoA derivatives. Catalase will catalyse the oxidation of a variety of alcohols, using H_2O_2 produced by some of these aerobic dehydrogenases.

Some oxygen is used by *dioxygenases*, e.g. those oxidizing cysteine (p. 139) or tryptophan (p. 146). The mixed function oxidases of the endoplasmic reticulum (see p. 167 and chapter 10, p. 184) also account for some non-mitochondrial oxygen consumption.

Oxygen toxicity

The 'oxygen burst' in macrophages

In some specialized cells there is a large O_2 utilization not coupled to ATP production. This has been particularly studied in neutrophils, which, when stimulated either by the ingestion of bacteria or by external stimuli such as membrane binding of immuno-globulins, respond by increasing their O_2 consumption 50-fold over a short period. This increase is accompanied by a marked increase in glucose oxidation by the pentose phosphate pathway, producing large amounts of NADPH. The sequence of events can be summarized as:

$$H^+ + NADPH + 2O_2 \rightarrow 2O_2^- + NADP^+ + 2H^+ \quad \text{(a)}$$

(membrane-bound flavoprotein oxidase)

$$2O_2^- + 2H^+ \rightarrow H_2O_2 + O_2 \quad \text{(b)}$$

(superoxide dismutase)

$$H_2O_2 + Cl^- \rightarrow OCl^- + H_2O \quad \text{(c)}$$

(*myeloperoxidase*, in neutrophil granules)

possibly also

$$O_2^- + HOCl \rightarrow OH^\cdot + O_2 + Cl^- \quad \text{(d)}$$

Both hypochlorite and the hydroxyl radical $OH^\cdot$ have been shown to be formed by macrophages, and are powerful microbicides. In chronic granulatomous disease, the neutrophils are unable to express a respiratory burst, and the patient suffers from recurrent and intractable bacterial infections which are often fatal in early life.

Superoxide dismutase

The ion O_2^-, produced in reaction (a) above, is called the *superoxide* ion. It has a much higher redox potential than O_2 (1.2 V as against 0.8 V), and was for some time thought to be a dangerous oxidizing agent in cells. More recent research has shown that it is kinetically not very active, and functions best as an electron donor rather than as an acceptor as in reaction (d) above. Superoxide is also produced as a by-product by several oxygen-reducing enzymes, notably xanthine oxidase (chapter 5), and in cytochrome P_{450} oxidations (see Fig. 9.8).

The idea that O_2^- is a dangerous species was encouraged by the discovery of a family of enzymes catalysing reaction (b) above, called *superoxide dismutases*. They are all metallo-enzymes; in animal cells the metal is either Cu or Mn, while in bacteria it is often Fe. It appeared that the enzyme is universally distributed in aerobic cells, while being completely absent from obligate anaerobic bacteria. Further research has shown, however, that some aerobic cells, e.g. adipocytes, lack superoxide dismutase, while a few strict anaerobes possess the enzyme.

Whether or not superoxide dismutase is the most important protective agent, there is no doubt that dissolved oxygen at high concentration is toxic. Rats caused to breathe 2 atm pure O_2 undergo convulsions after 5 hours. X-irradiation of tumours is ineffective in the absence of oxygen, and attempts have been made to use 'hyperbaric oxygen' in conjunction with X-ray treatment, in order to increase the efficiency of tissue destruction.

X-irradiation is known to produce free radicals (basically, molecules with an unpaired electron) in cells. The $OH^\cdot$ radical (equation (d), above) is such a species. On the other hand, the appearance of convulsions as a symptom of O_2 toxicity suggests disruption of the lipid-rich membranes of neurones. In dietary fats, oxygen causes spoilage (rancidity) by forming peroxides of unsaturated fatty acid residues

$$-HC{=}CH- + O_2 \rightarrow \begin{array}{c} -HC-CH- \\ | \quad | \\ O-O \end{array}$$

These peroxides decompose, breaking the carbon–carbon bond, and giving rise to short-chain aldehydes. In man-made fats, such as margarine, it is common to add tocopherols (relatives of vitamin E, see p. 282) purely as anti-oxidants.

Glutathione peroxidase

Tocopherols prevent peroxidation of lipids by being themselves preferentially oxidized. Glutathione per-oxidase can remove the peroxide by providing reducing equivalents from glutathione (p. 140):

$$2GSH + \begin{array}{c} | \\ O-C \\ | \\ O-C \\ | \end{array} \rightarrow GSSG + \begin{array}{c} | \\ C \\ || \\ C \\ | \end{array} + H_2O_2 \quad \text{(I)}$$

The enzyme is fairly non-specific, and is the only one in animals known to contain selenium. In animal experiments, the requirements for selenium and tocopherols are inversely related, in respect of protection of (membrane) lipids from peroxidation.

The GSSG produced in reaction (I) can be reduced by *glutathione reductase*, which usually requires NADPH.

$$GSSG + NADPH + H^+ \rightarrow 2\,GSH + NAD^+ \quad \text{(II)}$$

while H_2O_2 is removed by catalase

$$H_2O_2 \rightarrow H_2O + \tfrac{1}{2}O_2 \quad \text{(III)}$$

In cells capable of producing NADPH, the set of reactions (I)–(III) therefore provide an important protection against some of the deleterious consequences of oxygen toxicity.

Mixed function oxidases (mono-oxygenases)

Oxidation by direct addition of an atom or molecule of oxygen to an organic compound is not in general coupled to ATP synthesis (see, however, sulphite oxidation, p. 139). Much of it is purely degradative, or leads to the detoxication of foreign compounds (or 'xenobiotics'), but there are several important classes of compound which can only be formed by mixed function oxidation. These include

Cholesterol derivatives:
  Bile acids
  Steroid hormones (including cleavage of side chain)
Phenylalanine:
  Hydroxylation of aromatic ring
Catecholamines:
  Hydroxylation of tyrosine ring
  Hydroxylation of side chain
Collagen residues:
  Hydroxylation of lysyl residues
  Hydroxylation of prolyl residues

The overall mechanism may be described as

$$RH + AH_2 + O_2 \rightarrow ROH + A + H_2O$$

where AH_2 is a reductant.

The enzymes catalysing these reactions are called 'mixed function oxidases' because it has been established that the O_2 which appears in the product comes from atmospheric oxygen (and not from water). The enzymes are also called 'mono-oxygenases', because only one atom from a molecule of O_2 appears in the product, the other being reduced to H_2O.

There is one important set of reactions which from its mechanism clearly belongs to the same type, although no oxygen appears in the product. This is the desaturation of fatty acid chains (described in chapter 7), thus:

Fatty acid derivatives:
  Production of oleic acid and its homologues
  Desaturation of linoleic and other unsaturated acids

There are several known reductants and probably several mechanisms of oxygen activation. Fig. 9.8 gives what is currently thought to be the mechanism of the

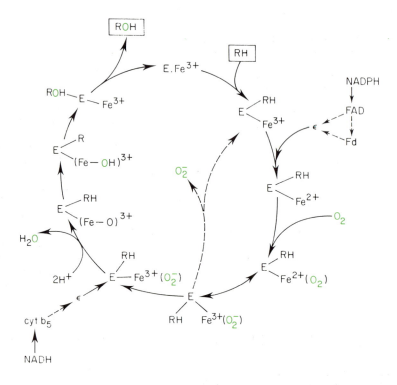

Fig. 9.8 The mechanism of operation of cytochrome P_{450}. During one cycle the substrate RH is bound by the enzyme (symbolized by $E \cdot Fe^{3+}$) and converted to ROH. The sources of the two electrons taken up by P_{450}, from NADPH and NADH, respectively, are summarized in the diagram. This is thought to be correct for the P_{450} of liver microsomes. The superoxide ion O_2^- shown in the centre of the diagram is a by-product that arises from the occasional breakdown of one intermediate form of the enzyme–substrate–oxygen complex.

most intensively studied system, centred on the haem-containing enzyme *cytochrome* P_{450}.

A strikingly similar mechanism, involving the sequential transfer of electrons one at a time to an active site containing Cu^+, which stabilizes the activated oxygen molecule O_2^- or $O_2^{\cdot}$, operates in the enzyme *tyrosinase*.

Table 9.6 summarizes what is known of the reductants and products for the most important of the reactions leading to natural products.

Table 9.6

Enzyme	Reductant	Product
Phenylalanine hydroxylase	Reduced biopterin	Tyrosine
Tyrosine hydroxylase	Reduced biopterin	DOPA
Tyrosinase	DOPA	DOPA [+DOPA quinone]
Prolyl hydroxylase	α-Oxoglutarate	4-OH proline (in collagen)
Lysyl hydroxylase	α-Oxoglutarate	5-OH lysine (in collagen)
Dopamine-β-mono-oxygenase	Ascorbate (and others?)	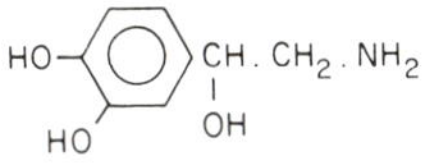
Cytochrome P_{450} series	Adrenodoxin (adrenal mitochondria) Cytochrome b_5 Ferredoxins Flavoproteins (endoplasmic reticulum)	Pregnenolone and adrenal steroid hormones Unsaturated fatty acids Steroids Drug derivatives Arene oxides

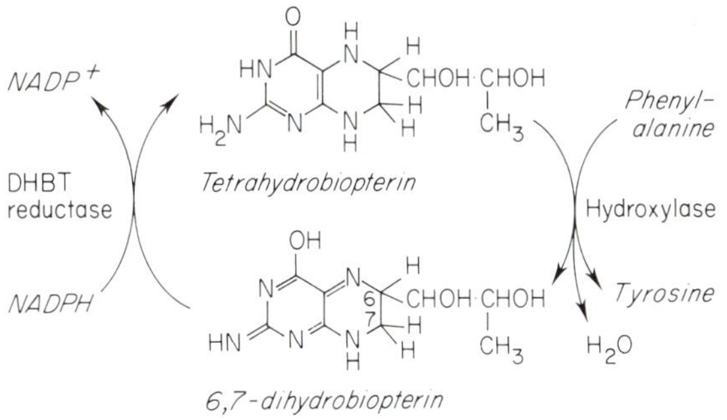

Fig. 9.9 The role of tetrahydrobiopterin in the hydroxylation of phenylalanine. Biopterin is a derivative of the pterin ring of folic acid (p. 277). In phenylalanine hydroxylase, it has been established that there are two enzymes, as shown in the diagram. One uses tetrahydrobiopterin (H_4-BT) as hydrogen donor for the hydroxylase itself, which contains two Fe^{3+} atoms. The other enzyme reduces 6,7-dihydrobiopterin ('quinonoid H_2-BT') back to H_4-BT with the aid of NADH. The hydroxylation of tyrosine to DOPA, and of tryptophan to 5-OH-tryptophan, proceed by a similar mechanism.

Cytochrome P_{450}

Three features of the cytochrome P_{450} system (named from the absorption maximum of its CO complex) are of interest:

1. Apart from the special steroid hydroxylases in adrenal mitochondria, the components are typically bound to the endoplasmic reticulum of many cells, notably liver. The substrates are normally lipophilic. There are probably at least 10 proteins that bind substrates rather unspecifically, although there is some demarcation, e.g. fatty acid desaturases do not hydroxylate drugs.

2. The fundamental electron donor is NADPH (sometimes NADH), as it is also when tetrahydrobiopterin is the proximate reductant (phenylalanine hydroxylation, see Fig. 9.9). In the P_{450} system, short (non-phosphorylating) electron transport chains, also bound to the ER, transfer reducing equivalents to the ultimate electron donor

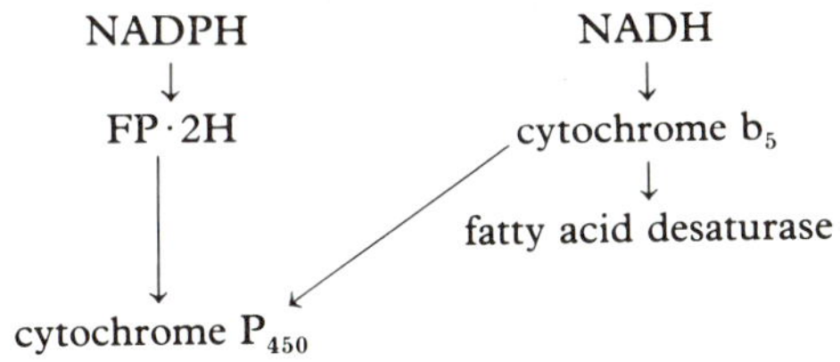

3. Some at least of the substrate-binding proteins and their associated cytochrome P_{450} molecules are inducible.

About 85% of the haem turnover in liver is for P_{450}, which suggests that these enzyme complexes, like endoplasmic reticulum itself, have a rapid turnover. Two important inducers are known:

(a) *Barbiturates*. Phenobarbitone (I) is hydroxylated to the inactive derivative (II). Feeding phenobarbitone over a period of a few days will increase the capacity of liver to carry out this hydroxylation several fold. The rates of many other drug hydroxylations are increased, but not those belonging to the group (b).

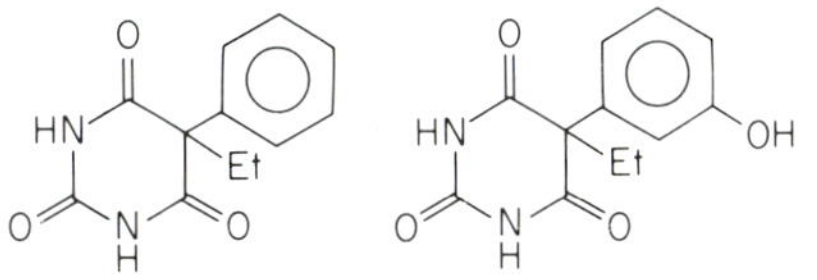

(b) *Poly-aromatic hydrocarbons* (*PAH*). Many compounds of this type taken into the body, not only by mouth (food additives, decomposition products of

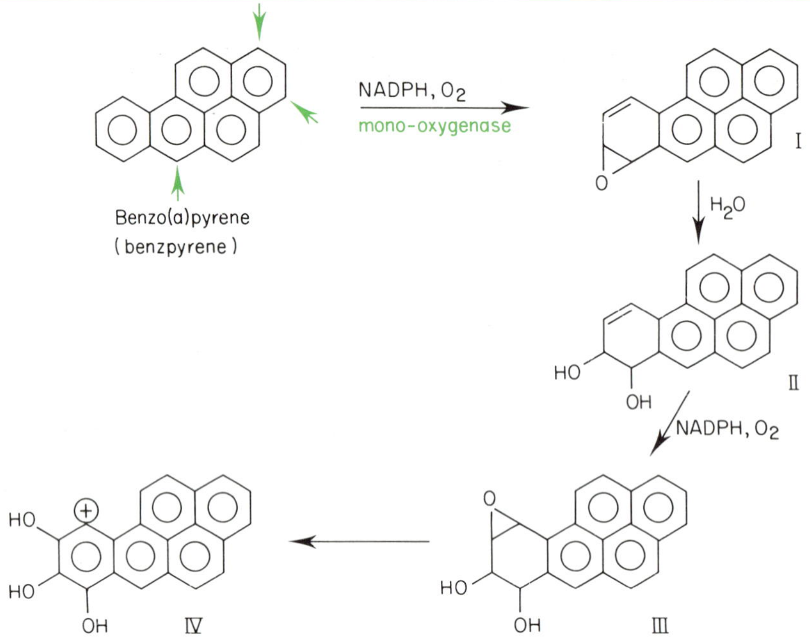

Fig. 9.10 Hydroxylation of benzpyrene. If the molecule is hydroxylated at any of the points shown by the arrows (upper left) an inactive derivative results. However, mono-oxygenase attack to give the epoxide (I), followed by hydrolysis to give a diol (II) and further attack to give a diol epoxide (III), leads to the formation of a carbonium ion (IV) that can form a covalent bond with a residue of DNA. It is thought to be the 'proximate carcinogen' derived from this polycyclic aromatic hydrocarbon.

fats), but also by inhalation (by-products of fuel combustion, particularly of coal). There is space here to deal with only one such compound, namely benzo(a)-pyrene (see Fig. 9.10).

The compound (I), formed by adding an atom of oxygen across a double bond, is an *epoxide*: in the case of PAH it is called an *arene oxide*. An *arene oxide hydrolase* converts this into the diol (II). It is thought that the compound (IV), related to the diol epoxide (III), is the 'ultimate carcinogen', i.e. that it binds covalently to a DNA base (usually guanine) to cause a misreading of the genetic code. It must be stressed that not all such mis-readings produce a transformation into a tumour cell; moreover, not all chemical carcinogens work through mechanisms described here. Moreover, PAH hydroxylation appears to be rather a hit-and-miss affair. Hydroxylations at other positions (indicated by the green arrows in Fig. 9.10) actually detoxify this and similar molecules, and after conjugation (e.g. with sulphate (p. 140)) they are excreted. In spite of such detoxifying reactions, it is worrying that so many lipophilic xenobiotics can induce one or another of the hydroxylases of endoplasmic reticulum. It is very difficult to predict how the metabolism of some quite different compound (e.g. a drug) will be affected at any moment. Some of the inducers are unexpected, e.g. insecticides.

Xenobiotic metabolism can take many forms other than straight hydroxylation. Examples are N-hydroxylation, de-alkylation, de-sulphuration, etc.

The Ames test. Because of the importance of mixed function oxidases in transforming foreign chemicals, an empirical test has been developed to screen new chemicals for potential carcinogenicity. Briefly, the chemical is incubated with a suspension of liver endoplasmic reticulum (microsomes) and a source of NADPH. The supernatant is then incorporated into a culture plate which is seeded with a suitable micro-organism, and the number of mutant colonies is later counted. Not all mutagens are carcinogens, but there is a high positive correlation between the two.

Dioxygenases

Not so much is known about the mechanism of these enzymes, which add a complete molecule of oxygen to a substrate, and therefore do not need a co-substrate to provide reducing equivalents.

One important group of dioxygenases are those which form the prostaglandins from arachidonic acid (p. 100). Two molecules of O_2 are successively added, and it is known that the cyclo-oxygenases have a haem-type prosthetic group.

Other dioxygenases occur in the metabolic pathways leading to the complete oxidation of phenylalanine and tyrosine (p. 148), and of tryptophan (p. 146). Here the addition of two oxygen atoms, one on either side of a double bond, leads to the opening of an aromatic ring.

Tryptophan pyrrolase is interesting in that the true substrate is superoxide ion O_2^-, rather than neutral oxygen, O_2. Another dioxygenase converts cysteine to cysteine sulphinic acid (p. 139).

Oxidative Phosphorylation

Current concepts of the mechanism of oxidative phosphorylation are described in chapter 10. Here it is only necessary to discuss the energetics.

During the oxidation of a mole of reduced NAD by $\frac{1}{2}O_2$, a considerable amount of energy is liberated (217.5 kJ). There are, however, several stages in the oxidation process (Fig. 9.7), and the decrease in free energy associated with a single step may be quite small. As pointed out on p. 156, the hydrolysis of ATP has one reactant and two products, so that the standard free energy change does not apply. Reversal of the hydrolysis, at the usual millimolar concentration of the reactants in vivo, requires 54–58 kJ, instead of the standard 31 kJ. It should be remembered that, by the laws of thermodynamics, this calculation is valid whatever the mechanism of phosphatic bond synthesis,

and however that mechanism is coupled to electron transport. From these energy considerations, it appears that there are rather few reactions in the respiratory chain which could effectively be coupled with phosphorylation. The experimental evidence shows that the number of molecules of ATP synthesized per atom of oxygen reduced ($\equiv 2H \equiv 2\epsilon$ transferred along the chain), is usually 3 for NAD-linked mitochondrial substrates, but only 2 for succinate and other FP-linked oxidations. As a result of such considerations, the three sites of phosphorylation from NADH to O_2 have been assigned to the approximate sites corresponding to Complexes I, III and IV in Fig. 9.7. Other kinds of experimental evidence confirm this (see chapter 10 for a fuller discussion).

General Considerations

Apart from the reactions in which oxygen is directly incorporated into organic compounds, the pathways considered in this chapter are almost all concerned with energy production and trapping in the chemically useful form of ATP. Anaplerotic uses of the TCA cycle, examples of which have been described on pp. 159–161, are of relatively little importance in animals other than for gluconeogenesis.

Many facts show that oxidative phosphorylation is of major importance in the energy metabolism of most tissues, e.g. the sensitivity of the brain to anoxia, the rapidly toxic effects of cyanide, and of atractyloside which prevents the egress of ATP from mitochondria. There are indeed cell types which obtain their energy from glycolysis, notably erythrocytes, and the cells of the cornea and the retina, but elsewhere fatty acid oxidation, even to ketone bodies, must always be kept going by respiratory chain activity.

It may be calculated that the conversion of one mole of glucose to two moles of pyruvate gives rise to 2 moles of ATP by substrate-level phosphorylation. At the same time 2 moles of NADH are formed which can yield 6 moles of ATP by oxidative phosphorylation at maximum, but see, however, p. 188. Conversion of the pyruvate to acetyl-CoA yields another 2 moles of NADH and thus 6 moles of ATP. The oxidation of each

acetyl group in the TCA cycle gives rise to 12 moles of ATP, from $3 \times NADH$, $1 \times FPH_2$, and one substrate level phosphorylation. Thus the total ATP yield from the oxidation of 1 mole of glucose is $(2 \times 12) + (2 \times 6) + (2 \times 6) + 2 = 38$ moles of ATP, of which only 1/19 is provided by the Embden–Meyerhof pathway. The results of a similar calculation for fatty acid oxidation have been given in chapter 7, and show for palmitic acid oxidation a total of 131 moles of ATP formed (less 2 for acyl-CoA formation), of which 96 are synthesized in the oxidation of the acetyl-residues in the TCA cycle. These calculations are somewhat over-precise, particularly in view of the uncertainty surrounding transfer of reducing equivalents from cytosol to mitochondria, but they give a useful indication of the scale of energy yield from mitochondrial oxidations. For amino acids the returns generally are lower and more variable, since not all the reactions are coupled to the respiratory chain.

This considerable synthesis of ATP may be directly coupled to synthetic reactions. For example, it may be calculated that 40 % of all ATP synthesized in growing *E. coli* is used for protein synthesis, particularly for the highly endergonic synthesis of short-lived messenger RNA. In animal cells, however, much of the ATP is coupled to more complex energy-requiring processes, such as muscular contraction, the sodium–potassium pump and tubular reabsorption processes in the kidney.

Most of the processes described in this chapter take place within or on the membrane of the mitochondria and biochemists become accustomed to taking into account the extremely restrictive and selective nature of mitochondrial permeability. The outer membrane is moderately permeable, but the inner membrane, although permeable to O_2, H_2O and CO_2, is impermeable to cations, and to most anions other than monocarboxylates. Inorganic phosphate, di- and tricarboxylate ions and amino acids are transported by complex permeases on which one ion has to be exchanged for another of the same charge, cf. Fig. 9.4. ATP has to be exchanged for ADP, and this translocase is absolutely specific for the adenine nucleotides. The impermeability of the membrane to CoA and acyl-CoA has already been mentioned (chapter 7).

The 'efficiency' of oxidative phosphorylation does not seem to change in physiological circumstances, except through deliberate use of uncouplers, such as dinitrophenol. It is certain, however, that the O_2 uptake of most tissues is increased, after a lag period, in thyrotoxicosis. So far as mitochondria are concerned, this is brought about partly by an increase in concentration of some enzymes, such as cytochrome oxidase, and partly by an increase in the capacity of the ATP/ADP translocase. This increased capacity of mitochondria to synthesize ATP, per unit of tissue, has to be balanced by extra ATP utilization. This seems to occur by various kinds of 'futile cycling', which it is outside the scope of this book to consider.

In thyroid deficiency (myxoedema), on the other hand, O_2 consumption may be reduced until it is not enough to provide for adequate heat output, and the patient may complain of feeling cold.

10

Membrane Structure and Function

Membrane Function

The simplest types of living cells, called prokaryotes, are surrounded by a membrane (the plasma membrane) but contain little or no intracellular membrane. Bacteria and blue-green algae are of this type. In addition to the plasma membrane, which is in contact with the cytoplasm, some bacteria possess a second, outer membrane, of different chemical composition. Such bacteria are termed gram-negative, gram-positive organisms having no outer membrane. The cells of animals, plants, fungi and protozoa are termed eukaryotic: they are invariably larger than prokaryotic cells (1 μm or more), and possess, as well as a plasma membrane, membranous intracellular organelles such as a nucleus, mitochondria, Golgi apparatus, endoplasmic reticulum and lysosomes.

The plasma membrane

The primary function of the plasma membrane is to provide a permeability barrier, which protects the contents of the cell from the surroundings. Most membranes are impermeable not only to molecules of high molecular weight, but also to many ions and low molecular weight metabolites, except those which are very lipophilic. The plasma membrane thus allows the maintenance of relatively high cytoplasmic concentrations of metabolic intermediates, without loss to the surroundings, and the creation of an intracellular space. Water, however, readily permeates the plasma membrane, which therefore offers no protection against osmotic effects—erythrocytes are stable in iso-osmotic solution, such as 0.15 M NaCl, but if suspended in a hypo-osmotic medium they swell and burst, as water enters. The cells of plants and micro-organisms are protected by cell walls, readily permeable to substances of low molecular weight, which surround the plasma membrane and provide mechanical protection against hypo-osmotic surroundings.

Since all cells need to communicate with their environment, to take up nutrients, the plasma membrane is made selectively permeable by the presence of permeases, proteins which specifically translocate particular molecules or classes of compounds through the membrane; the plasma membranes of most cells contain many different permeases. The metabolism of many cells can be altered in response to extracellular signals, in the form of molecules which bind to specific receptors on the cell surface. In multicellular organisms these signals (hormones) are produced in specialized organs and conducted to their target tissue through the blood. Bacteria can move towards high concentrations of attractant substances (chemotaxis), and the cells of slime moulds communicate with each other. These processes involve receptors, transmembrane proteins which bind the signal on the external face of the plasma membrane, and initiate a metabolic response in the cytoplasm (although the signal molecule itself does not usually cross the plasma membrane). Other plasma membrane components are concerned with cell recognition, e.g. the cell surface antigens of erythrocytes which are glycoproteins and glycolipids, inserted into the plasma membrane, with the carbohydrate portion exposed at the extracytoplasmic face.

Organelles

The intracellular membranes of eukaryotic cells form a number of closed compartments, called organelles. Location of particular metabolic pathways within these compartments increases the efficiency of metabolism by reducing the distances which substrates have to diffuse between enzymes, by creating favourable environments in which the conditions (of pH, metabolite concentration and so on) may be different from those in the cytoplasm, and by offering the possibility of control through modulation of the activity of the permeases through which metabolites enter. Intracellular membranes may also provide attachment sites for enzymes, especially those catalysing sequences of reactions; for example, in the hydroxylation systems of the endoplasmic reticulum (p. 167), electron transfer from one component to another is accelerated by the close proximity of the electron carriers. The intracellular cytoskeleton, concerned with cell shape and mobility, and with the movement of organelles within the cell, is also membrane-bound: microfibrils, containing the proteins actin and myosin and therefore involved in ATP-dependent contractile processes, are attached to many membranes, possibly through 'anchor' proteins such as α-actinin. Microtubules are composed principally of the protein tubulin, but may also contain an ATPase, although the nature of the membrane attachment sites is not known.

The Chemical Composition of Membranes

Membranes are composed of lipid, protein and carbohydrate. The relative content of these components varies widely from one type of membrane to another (Table 10.1), but typically 40% of the dry weight is lipid, about 60% protein, and 1–10% carbohydrate. In addition, membranes are extensively hydrated, up to 20% by weight being water. The variation in composition reflects the function of the membrane: since many membrane proteins are enzymes or permeases, membranes with a high protein content, such as mitochondrial inner membranes, are metabolically active, whereas the function of membranes which are very lipid-rich (such as myelin) is mainly insulation. Each type of membrane has a unique complement of proteins and glycoproteins, and a characteristic lipid and glycolipid composition; a particular lipid may be found in many different types of membrane (e.g. phosphatidyl choline is very widely distributed) or may be characteristic of a single type (e.g. cardiolipin is found almost exclusively in the inner membranes of mitochondria). All membrane carbohydrate is covalently attached to protein or lipid.

Lipids

Very many different lipids are found in membranes: they can be broadly classified according to structure, as phospholipids (derivatives of either glycerol or sphingosine), glycolipids (in eukaryotes, mainly glycosylated derivatives of sphingosine) or sterols (cholesterol and its derivatives).

Phospholipids

The structures of glycerophospholipids are given on p. 90. Phosphatidyl choline (lecithin) and phosphatidyl ethanolamine (cephalin) are the most abundant lipids in most eukaryotic membranes. Phosphatidyl serine and phosphatidyl inositol are often found as minor membrane components; inositol can carry additional phosphate groups, on the 4- or 4- and 5-positions. Phosphatidyl glycerol is abundant in the membranes of many bacteria, and is a minor component of mitochondrial membranes; cardiolipin is also found in the membranes of mitochondria and bacteria.

Table 10.1 Composition of different membranes. (a) Content of lipid, protein and carbohydrate, as percentage of dry weight; (b) content of lipids, as percentage of total lipid

(a)	Lipid	Protein	Carbohydrate
Plasma membrane (mammals)	43	49	8
Plasma membrane (bacteria)	30	70	—
Myelin	75	22	3
Mitochondrial outer membrane	48	52	(trace)
Mitochondrial inner membrane	24	76	(trace)
Nuclear membrane	35	59	3
Endoplasmic reticulum	44	54	2

(b)	Cholesterol	PC	PE	PS	PI	SM	PG	CL	GL
Plasma membrane (mammals)	20	19	12	7	3	12	0	0	10
Plasma membrane (bacteria)	0	0	75	0	0	0	25	0	0
Myelin	28	11	17	6	1	7	0	0	29
Mitochondrial outer membrane	8	45	20	2	7	4	2	4	0
Mitochondrial inner membrane	0	35	25	0	3	3	2	18	0
Nuclear membrane	3	45	20	3	7	2	0	0	0
Endoplasmic reticulum	5	48	19	4	8	5	0	0	0

PC, PE, PS, PI = phosphatidyl choline, -ethanolamine, -serine, -inositol. SM = sphingomyelin, PG = phosphatidyl glycerol, CL = cardiolipin, GL = glycolipid.

Plasmenyl choline and plasmenyl ethanolamine, in which there is a 1-alkenyl group at C-1 of glycerol, occur in heart muscle, myelin and some bacteria.

The acyl substituents of phospholipids vary considerably between membranes, and even within membranes, so individual types of phospholipid show heterogeneity. As discussed in chapter 7, these groups usually contain 14–24 carbon atoms (16 and 18 being the most common); in monounsaturated chains the double bond is usually between C-9 and C-10, and in the *cis*-configuration. Additional double bonds may occur nearer to the terminal methyl; these are never conjugated, but isolated by a methylene group.

Lysophosphatidyl-choline and -ethanolamine, which lack a 2-acyl group, are found in small quantities in many membranes, although, as discussed later, they have a detergent-like action which destabilizes membranes.

Sphingolipids

The most common phosphosphingolipid in mammalian cells is ceramide 1-phosphocholine (sphingomyelin), which is a major component of plasma membranes. Most glycolipids in mammalian membranes are also sphingosine derivatives. Galactosylceramide is a major component of myelin, and glucosylceramide occurs in plasma membranes. More complex glycosphingolipids, containing 2, 3 or more carbohydrate residues, are important constituents of plasma membranes; some of the blood-group substances of erythrocyte membranes are complex glycosphingolipids, others proteins.

Sterols

The major sterol in mammalian membranes is cholesterol (p. 93): it is particularly plentiful in plasma membranes, and is found in small amounts in many intracellular membranes, including the outer membrane of mitochondria, but it is absent from the inner mitochondrial membrane. 7-Dehydrocholesterol is found in small amounts in the intracellular membranes of some animal cells, and different sterols are found in plants and bacteria.

Proteins

Although most lipids occur in many different membranes, each protein is characteristic of its source; it is rare for any protein to be found in more than one subcellular location. Treatment of membranes with detergent and electrophoretic separation of the solubilized proteins reveals the number of protein components in the membrane. There appear to be seven major protein components in human erythrocyte membranes, about twenty in many other plasma membranes,

and at least forty in mitochondrial inner membranes. These are minimum estimates, as some proteins are not separated under the conditions of electrophoresis, and minor components may be undetected. Specific stains can be used to reveal which of the components are glycoproteins.

Intrinsic and extrinsic proteins

Some membrane-associated proteins are easily solubilized by washing of the membrane with solutions of high ionic strength, by changes of pH, or with chelating agents, such as EDTA, to bind divalent cations; these procedures do not extensively disrupt the membrane or solubilize lipid. Such proteins are known as *extrinsic proteins*, and they adhere to the membrane principally by electrostatic interactions: after removal from the membrane they behave like soluble proteins. Cytochrome c, spectrin in erythrocytes and the basic protein of myelin are examples. *Intrinsic proteins* are inserted into or through the membrane, being held there by hydrophobic interactions with lipids. They can be solubilized only by use of detergents (p. 178) or chaotropic agents, such as urea or large inorganic ions (I$^-$, SCN$^-$), and they aggregate and become insoluble if these agents are removed. The electron transport complexes of the inner mitochondrial membrane and glycophorin in erythrocytes are examples of intrinsic

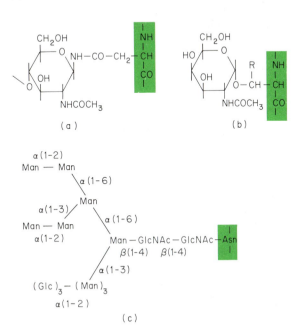

Fig. 10.1 Linkage of carbohydrate to glycoproteins: (a) *N*-acetylglucosamine linked to asparagine; (b) *N*-acetylgalactosamine linked to serine (R=H) or threonine (R=CH$_3$); (c) structure of the 'core' oligosaccharide of vesicular stomatitis virus G-protein.

membrane proteins. Other membrane proteins have largely hydrophilic surfaces, but are anchored to the membrane by a hydrophobic 'tail' inserted into the lipid layer. These are called amphiphilic proteins: an example is cytochrome b_5 of endoplasmic reticulum (p. 184).

Glycoproteins

Many membrane proteins are glycosylated, having one or more covalently attached polysaccharide chains. These chains may contain the monosaccharides D-galactose, D-mannose, L-fucose, N-acetyl glucosamine, N-acetyl galactosamine and N-acetyl neuraminic acid; they are attached to the protein either by an N-glycosidic linkage from N-acetyl glucosamine to asparagine or by an O-glycosidic linkage from N-acetyl galactosamine to serine or threonine (see Fig. 10.1). The amino acid sequences around the carbohydrate attachment sites of different proteins are often similar, presumably because they have to be recognized by glycosylating enzymes. The carbohydrate chains of many glycoproteins show structural variation from one molecule to another, a phenomenon known as microheterogeneity. Most membrane glycoproteins are intrinsic, with the glycosylated regions exposed on one side of the membrane (the extracytoplasmic face) only. Carbohydrate chains are found on the outer surface of plasma membranes and the luminal faces of endoplasmic reticulum and secretory vesicles.

Membrane Structure

The lipid bilayer

In the electron microscope all biological membranes have a similar 'tramline' appearance, showing two electron-dense bands about 4 nm apart, separated by a lighter region. This suggests that all membranes have the same basic structure, and that it is composed of two layers (the *bilayer*), although since the chemistry of the fixing and staining procedures used to prepare samples for electron microscopy is not fully understood, it is not immediately obvious what these layers are. Comparison of the measured surface area of membranes with the area calculated to be occupied by the phospholipids extracted from them suggests that the lipids themselves form a double layer. The most stable structure is formed when the hydrophilic head-groups of the lipids are external, and the hydrophobic acyl chains internal. The occurrence of this structure in biological membranes is supported by numerous physical and chemical studies, on membranes derived from many sources, and on synthetic bilayers, which can be produced from lipids such as phosphatidyl choline, when they are dispersed in water by shaking or ultrasonication. Such synthetic bilayers form closed vesicles, known as *liposomes*, which have the same electron miscroscopic appearance as biological membranes, and show similar permeability properties.

The major driving force for the formation of phospholipid bilayers is entropic: a hydrophobic group in water causes extensive ordering of surrounding water molecules, so there is a big gain in entropy if the hydrophobic alkyl chains of phospholipids cluster together in the hydrophobic interior of the membrane. All of the glycerophospholipids and phosphosphingo-

lipids mentioned on p. 90 can fit into a bilayer, as all have polar head-groups and long apolar tails. There is great variation in the size and charge of the head-groups: all phospholipids contain a negatively charged phosphate (pK_a about 2), the remainder of the head being uncharged (glycerol or inositol), positively charged (choline and ethanolamine), zwitterionic (serine) or negatively charged (phosphoinositol). The sphingosine-based glycolipids have large hydrophilic carbohydrate head-groups, which may be neutral, or carry positively-charged amino or negatively-charged sulphate groups.

There is also variation in the length and degree of unsaturation of hydrophobic tails. Double bonds (invariably in the *cis* conformation) introduce bends of about 30° into the linear hydrocarbon chain: this interferes with the close packing of the chains and increases the fluidity of the membranes. Lysophospholipids, in which the 2-hydroxyl of glycerol is not substituted, are wedge-shaped molecules which destabilize the bilayer. Phospholipase A_2, found in the venoms of some snakes, causes disruption of membranes by converting glycerophospholipids to lysophospholipids, and can be used to solubilize proteins. A number of anti-inflammatory steroid drugs act as inhibitors of phospholipase A_2, so production of lysolecithin may be important in the inflammatory response.

Cholesterol is an approximately planar molecule, with a protruding, polar 3-hydroxyl at one end. It can pack into both sides of phospholipid bilayers, with its 3-hydroxyl near the surface. Its presence affects the packing of the hydrophobic tails of phospholipids, and at physiological temperatures it tends to decrease the fluidity of the bilayer.

Association of proteins with the lipid bilayer

Although the structures of phospholipids and sterols suggest the way in which they are organized in membranes, it is less obvious how membrane proteins associate with lipid components. They were originally thought to interact electrostatically with the charged lipid head-groups on the surface of the bilayer, and although this is true of some extrinsic proteins, it does not explain why intrinsic proteins are so tightly associated with membranes. Most evidence supports the fluid mosaic model of membrane structure (Fig. 10.2)

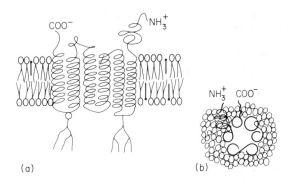

Fig. 10.3 Possible structure for a membrane channel formed by laterally-bundled α-helices in a transmembrane protein. (a) Two-dimensional side view; (b) view down the channel from the cytoplasmic side.

chrome b_5, are largely external to the membrane, but have a hydrophobic region inserted into the bilayer. The carbohydrate moieties of glycoproteins are, of course, wholly external, being too hydrophilic to interact with membrane lipids.

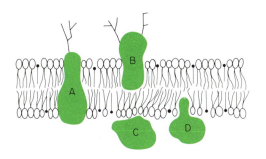

Fig. 10.2 The lipid bilayer. A and B are intrinsic proteins, and are shown bearing oligosaccharide chains. C is an extrinsic protein, and D is amphiphilic.

in which intrinsic membrane proteins are visualized as being inserted into the lipid bilayer, on one side or the other, or extending right through the bilayer and protruding on both sides. This can be seen directly by freeze-etch electron microscopy in which a frozen membrane sample is fractured with a microtome, cleaving the bilayer between the two lipid layers. Metal shadowing of the surface reveals intrinsic proteins as particles protruding from the smooth lipid surface. The number of visible particles correlates with the known protein content of the membrane under study, no particles at all being visible in liposomes, and very few in myelin. In some cases the particles can be identified with particular proteins, by their number or size.

Transmembrane proteins are thought to be folded into structures which expose hydrophobic amino acid side chains on the surfaces in contact with the bilayer; for example, a polypeptide chain could traverse the membrane several times, the bilayer segments of the protein being in the α-helical conformation, and the α-helices stacked together (Fig. 10.3). Such bundles of α-helices may be an important functional constituent of permeases (p. 180). Parts of the protein protruding from the membrane will have surface polar residues which interact with the solvent in the same way as those of soluble proteins. Amphiphilic proteins, such as cyto-

Membrane fluidity

A feature of the fluid mosaic model is the lateral mobility of proteins within the bilayer ($D \simeq 10^{-10}\ \mathrm{cm}^{-2}$), which has a viscosity about equal to that of olive oil. This can be demonstrated by inducing fusion of cultured mouse and human cells, and following the distribution of cell surface antigens (which are intrinsic membrane proteins) in the resultant hybrid cell. Immediately after fusion the antigens are segregated into different halves of the cell, but they rapidly become mixed over the entire cell surface. Mixing is independent of protein synthesis and of chemical energy (ATP), but is inhibited at low temperatures, and is therefore brought about by lateral diffusion of proteins. Lipids also have lateral mobility, which is much greater than that of proteins, because of their smaller size ($D \simeq 10^{-9}\ \mathrm{cm}^{-2}$). However it is not envisaged that all components are mobile: intrinsic proteins are surrounded by an annulus of lipid molecules which may be immobilized by their interaction with protein, and there is considerable evidence for lateral inhomogeneity of some membranes, with particular proteins or lipids concentrated in certain areas. Such regions may be maintained by interaction between membrane proteins, or by their binding to components in the cytoplasm, such as microfilaments and microtubules. These have been implicated in the movement of IgG molecules in lymphocyte membranes; on exposure of the cell to anti-IgG serum, these proteins cluster to form 'caps' at one end of the cell, in a process which requires metabolic energy.

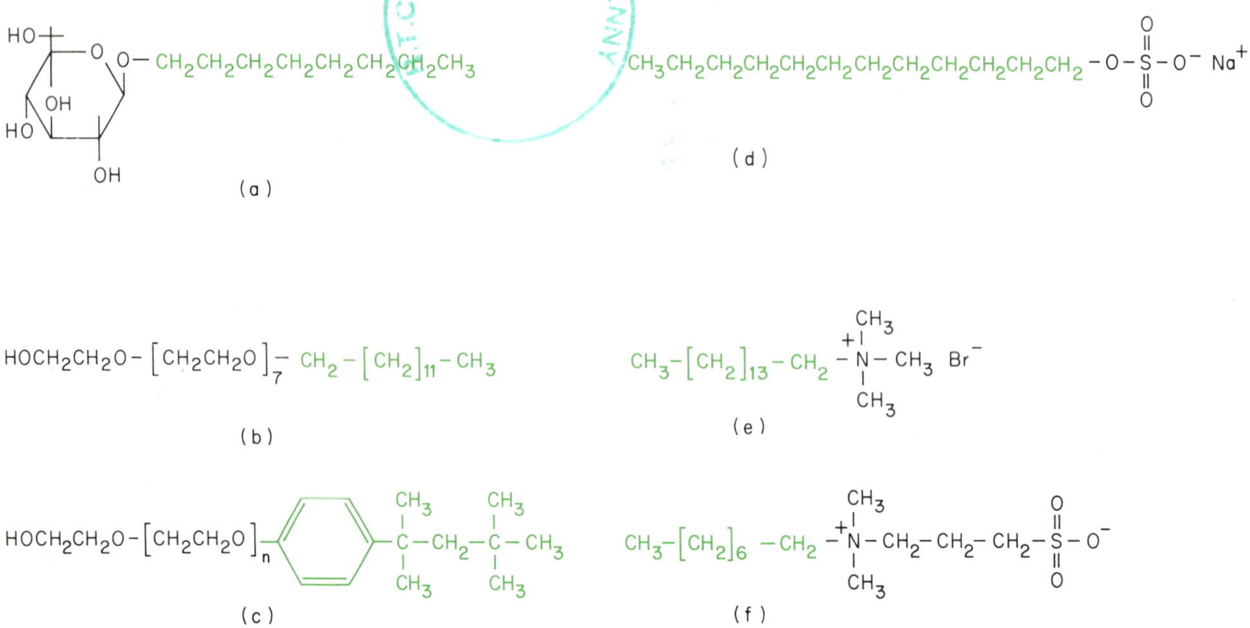

Fig. 10.4 Structures of some detergents: hydrophobic parts are shown in green. (a)–(c) nonionic detergents: β-octyl glucoside, octaethylene-glycol dodecyl ether, triton X-100; (d) the anionic detergent sodium dodecylsulphate (SDS); (e) the cationic detergent cetyl trimethylammonium bromide (CTAB); (f) a zwitterionic detergent.

Detergents

Detergents are amphiphilic molecules; that is, they possess both hydrophilic and hydrophobic regions. In solution, they form micelles, with the hydrophilic region exposed, and the hydrophobic region protected from interaction with the solvent. Detergents disrupt membranes, forming mixed micelles with lipids, and solubilizing membrane proteins. The structures of some experimentally used detergents are shown in Fig. 10.4. A similar mechanism of action is involved in the emulsification of fats by such naturally-produced detergents as taurocholic acid (p. 114). Some cationic detergents are used as antibacterial agents, being incorporated into ointments, handcreams, etc. They act by disrupting bacterial cell walls.

Membrane asymmetry

The two faces of biological membranes are exposed to different intracellular compartments, or to the cytoplasm and the extracellular space; the structural components may therefore be different in each half of the bilayer. This has been shown by studying the interaction of membrane components with reagents which are im-penetrant, and which therefore attack exposed lipids, proteins or carbohydrates on one face of the bilayer only. Reagents used in this way include impenetrant alkylating or acylating agents, specific antibodies, lectins (proteins, usually derived from plants, which bind specifically to certain carbohydrate residues) and enzymes (such as proteases, phospholipases, neuraminidase, and lactoperoxidase, which, in the presence of I^- and H_2O_2, can catalyse iodination of protein). In a few cases it has been possible to investigate separately modification of each side of a membrane, for example in erythrocytes, which can be lysed and resealed with the membrane in the usual orientation, or everted, with the cytoplasmic face exposed.

Partial lipid asymmetry occurs in the membranes of erythrocytes, with phosphatyl serine and phosphatidyl ethanolamine located predominantly in the internal face, and phosphatidyl choline and sphingomyelin largely in the external face. The membranes of bacteria and the enveloped viruses are also asymmetric, but have a different lipid distribution. The significance of lipid asymmetry is unknown; nor is it known whether this asymmetry occurs in all membranes. Since different lipids have different charges, and tend to contain different types of acyl chain, the two halves of the bilayer could have different fluidities and surface charges.

In contrast to the partial asymmetry exhibited by lipids, that of carbohydrate and proteins is total. The carbohydrate moieties of glycolipids and glycoproteins are found only on the extracytoplasmic face (i.e. the *external* face of plasma membranes and the *luminal* sides of intracellular membranes), and each protein has only one possible orientation in a membrane. Although membrane proteins have lateral mobility, they cannot rotate about an axis parallel to the bilayer, since this would involve passage of hydrophilic parts of the protein through the hydrophobic centre of the bilayer, and would therefore have an enormous activation energy.

Membrane biogenesis

With the exception of mitochondrial membranes, all the membranes of animal cells are derived from the endoplasmic reticulum. As one type of membrane matures to form another, the lipid composition is changed, existing membrane proteins are segregated out and new ones are inserted. This process of segregation occurs in the Golgi apparatus; the mechanism by which particular proteins are directed to the appropriate membrane is not understood.

The biosynthesis of phospholipids is described elsewhere (chapter 7); many of the enzymes required (e.g. those involved in the methylation of phosphatidyl ethanolamine) are themselves membrane-bound, so that synthesis and insertion take place concomitantly, but it is not known how lipids are translocated from one side of the bilayer to the other. This translocation is not measurable in liposomes and in viral membranes, but occurs rapidly in the membranes of growing cells: it must therefore be catalysed. Experiments designed to reveal points or areas of growth in bacterial membranes have not been successful, but such growth points could well be obscured by the lateral mobility of membrane components. It appears that concomitant insertion of proteins and lipids is not essential, these components being inserted separately and turned over at different rates.

Some of the principles underlying the synthesis and insertion of intrinsic membrane proteins have been worked out from studies of the replication of vesicular stomatitis virus, cultured in mammalian cells. G-protein is an intrinsic protein of the viral membrane (mol. wt 69 000), the N-terminus and two carbohydrate chains being external, and the C-terminus internal. It is synthesized on membrane-bound ribosomes, and passed into the lumen of the endoplasmic reticulum. Like most secreted proteins (p. 220) the nascent G-protein has an N-terminal signal of about 20 hydrophobic amino acids, which permits it to pass through the membrane, and which is removed on the

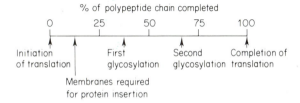

Fig. 10.5 Sequence of events in the synthesis of vesicular stomatitis virus G-protein.

luminal side of the endoplasmic reticulum. Synthesis, translocation and glycosylation are strictly synchronized (Fig. 10.5); if the protein is synthesized in vitro on free ribosomes, and membranes added after it is completed, it is neither inserted nor glycosylated. From the lumen of the endoplasmic reticulum, G-protein is transported to the Golgi apparatus (probably in 'coated vesicles', membrane fragments stabilized by complexing with the protein clathrin). Following modification of the carbohydrate chains in the Golgi, the protein travels to the plasma membrane, eventually becoming incorporated into the viral membrane when the virus buds out of the cell.

Although the 'signal' mechanism of concerted synthesis and insertion may apply to many membrane proteins, incorporation of some proteins into membranes can occur after synthesis is completed, rather than being obligatorily coupled to translation: examples of this are

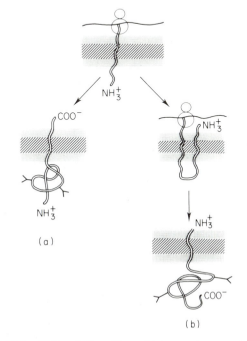

Fig. 10.6 Different dispositions of transmembrane proteins produced by coupled synthesis and insertion.

the coat protein of bacteriophage M13, and the bee venom protein mellitin.

The asymmetry of orientation of membrane proteins is decided by the process of insertion and is maintained because once inserted, proteins are unable to rotate through the bilayer or flip to the other side. Since most proteins are synthesized in the cytoplasm, it follows that some are inserted directly into the cytoplasmic face, others being fed through during translation and remaining exposed on the extracytoplasmic face (see Fig. 10.6).

Many proteins destined for incorporation into organelles are synthesized as precursors which appear as completed proteins in the cytoplasm, before uptake occurs. This applies to proteins of the mitochondrial matrix (e.g. carbamyl phosphate synthetase, p. 129), of the mitochondrial inner membrane (cytochrome c_1 and subunits IV–VII of cytochrome oxidase, p. 186), at least one inter-membrane protein (cytochrome c peroxidase, in yeast), and peroxisomal proteins. Uptake of these proteins appears to require energy. The extra polypeptide sequences found in the precursor forms of these proteins are necessary for uptake and insertion into membranes; but those proteins which are synthesized within mitochondria (p. 218) probably do not appear as precursors.

Glycosylation

Linkage of oligosaccharide chains to proteins takes place in the endoplasmic reticulum, as the nascent protein is extruded through the membrane and into the lumen. Oligosaccharides may be linked to asparagine or serine (Fig. 10.1): those destined for attachment to asparagine are built up whilst attached to dolichol, a mixture of long-chain isoprenyl alcohols of the general formula

$$\text{H---[CH}_2\text{---}\overset{\overset{\textstyle CH_3}{\textstyle |}}{C}\text{=CH---CH}_2]_n\text{---CH}_2\text{---}\overset{\overset{\textstyle CH_3}{\textstyle |}}{CH}\text{---CH}_2\text{---CH}_2\text{OH}$$

where n is 18–20. Individual monosaccharides are transferred from nucleoside-diphosphate sugars to dolichol phosphate, forming a dolichol pyrophosphate-oligosaccharide from which the carbohydrate is eventu-

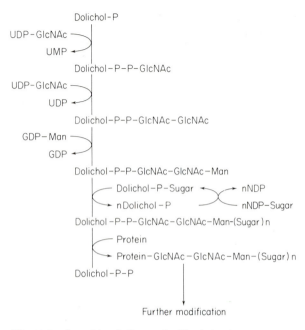

Fig. 10.7 Assembly of oligosaccharide chains for linkage to asparagine in glycoproteins. GlcNAc = *N*-acetyl glucosamine. Man = mannose.

ally transferred to the nascent polypeptide (Fig. 10.7). This mechanism of glycosylation probably applies to the asparagine-linked chains of both membrane proteins and secretory proteins, though not to serine-linked oligosaccharides, which are built up by direct transfer of monosaccharides from nucleoside-diphosphate sugars to protein. The glycosyl transferases which catalyse this process are located in the Golgi apparatus, and post-synthetic modification of asparagine-linked carbohydrate chains can also occur in the Golgi. Thus after transfer of the 'core' polysaccharide (the structure of which is shown in Fig. 10.1) from dolichol phosphate to VSV-G protein, three glucose and four $\alpha(1\text{-}2)$-linked mannose residues are removed by specific glycosidases, and fucose, galactose, and sialic acid residues added in the Golgi.

Membrane-related Processes

Membrane transport

Uptake of substrates by cells and communication between subcellular compartments are essential for metabolic activity. Transfer of molecules across mem-

branes by passive diffusion obeys Fick's law of diffusion:

$$\mathcal{J} = -D\frac{dc}{dx}$$

where $\mathcal{J}$ is the flux, D the diffusion constant (which depends on the type of membrane and the diffusing solute) and dc/dx the concentration gradient of the solute across the membrane. Thus the flux is directly proportional to the solute concentration difference across the membrane. Uptake of some uncharged molecules such as urea, methanol, formamide and ethylene glycol by liposomes, erythrocytes and other vesicles appears to be by diffusion, although the diffusion constants are larger than would be expected if these molecules simply dissolved in the hydrophobic core of the bilayer—possibly they pass through partially hydrated regions of the membrane. Some ions also cross membranes by passive diffusion, those with the largest diffusion constants being those with large hydration shells: relative rates of diffusion in liposomes and erythrocytes are

$$CNS^- > I^- > Br^- > Cl^- > F^-$$

Weak acids such as acetic and pripionic acids, and bases such as ammonia and methylamine, also permeate readily. There is a correlation between the rate of permeation and the concentration of uncharged species in solution—the weaker the acid or base, the more rapidly it permeates. The relative rates of diffusion of monocarboxylic acids are

$$CH_3CH_2COOH > CH_3COOH > HCOOH >$$
$$ClCH_2COOH > Cl_2CHCOOH$$

suggesting that the protonated, uncharged acids, rather than their anions, are the species which diffuse most rapidly. Most metabolites are ionized at intracellular pH-values, and do not undergo uncatalysed translocation through membranes; pyruvate and the di- and tricarboxylates of the tricarboxylic acid cycle, for example, are strong acids with negligible rates of passive diffusion, and are taken up by mitochondria via permeases in the inner mitochondrial membrane. However, uncatalysed permeation of membranes by carbon dioxide and urea, for which there are no known permeases, is obviously of physiological importance.

Facilitated diffusion

Translocation of metabolites through membranes is brought about by permeases, intrinsic membrane proteins which bind a substrate on one side of the membrane, conduct it through the bilayer, and release it on the other side. They are analogous to enzymes, the reaction catalysed being the equilibration of substrate concentrations across a membrane. Like enzymes they show saturation kinetics, the rate equation describing transport being analogous to the Michaelis equation (p. 27):

$$\mathcal{J} = \frac{\mathcal{J}_{max}}{1 + (K_m/[S])}$$

where [S] is the substrate concentration, and K_m a constant, analogous to the Michaelis constant, which is related to the affinity of the permease for its substrate—it may approximate to the dissociation constant of the substrate–permease complex, and has the units of concentration (e.g. mM). $\mathcal{J}_{max}$ is the maximal flux, achieved at infinite substrate concentration; it is related to the molecular activity or turnover number of the permease, and has the units of reaction velocity (e.g. μmol substrate transported/mg membrane protein/min).

Also like enzymes, permeases show specificity for their substrates, and are subject to inhibition by specific inhibitors. The sugar permease of the erythrocyte membrane shows strong selectivity for D-glucose ($K_m = 8$ mM): it transports other D-sugars, such as D-mannose, D-galactose and D-xylose, but has lower affinities for them (K_m values 25 mM, 40 mM and 60 mM respectively), while L-sugars are transported only slowly, the K_m for L-glucose being > 3 M. It is inhibited by reagents, such as N-ethyl maleimide, which react with thiol groups, by cytochalasin B, and by the glycoside phlorhizin. Other specific transport systems also exist within the erythrocyte membrane: one carries various nucleosides, such as adenosine and uridine, and is inhibited by the substrate analogue nitrobenzyl-thioinosine. Another erythrocyte permease is an anion carrier, which catalyses the obligatory exchange of two anions, e.g. Cl^- for HCO_3^-. Permeases of this type, which bring about the coupled transport of two substrates in opposite directions, are called antiporters. The exchange of Cl^- for HCO_3^- across the erythrocyte membrane (the 'chloride shift') is of importance in the transport of oxygen and carbon dioxide in the blood (p. 251).

Ionophores

A number of fungal antibiotics specifically increase the permeability of biological membranes to certain cations; since they have similar effects on liposomes, these antibiotics do not alter the specificity of a permease which is present in the membranes, but by themselves bring about ion transport: they are therefore known as ionophores. These compounds are of interest in investigating the roles of ions in cellular processes (e.g. the ionophore A23187 specifically induces calcium permeability, and has been used to study the role of Ca^{2+} in exocytosis) and also in studying mechanisms of membrane transport. Valinomycin forms a complex with K^+, allowing this ion to cross membranes; the rate of transport of Na^+ is about 10 000-fold lower. Nigericin, which has similar specificity, has an ionizable carboxyl, and may exist either as the K^+-complex or the protonated form: it catalyses exchange of K^+ for H^+. Ion transport catalysed by valinomycin and nigericin is greatly reduced at low temperatures, when the fluidity of the bilayer decreases and the diffusion of the ion–antibiotic complex becomes slower; in contrast, ion transport by gramicidin A is much less temperature-dependent. This ionophore shows rather low ion specificity, and is thought to dimerize within the membrane, forming a rigid pore which spans the bilayer and through which ions may diffuse.

Active transport

Solute diffusion, whether uncatalysed or brought about by permeases, dissipates concentration gradients, the solute moving *down* the concentration gradient until the concentration of free solute is the same on both sides of the membrane. Thus the concentration of free glucose within the erythrocyte cannot exceed the plasma glucose concentration. The creation of high solute concentrations, by transport *against* the concentration gradient, requires the input of energy, and is known as active transport. Active transport across biological membranes is brought about by permeases, and, like facilitated diffusion, shows saturation kinetics, substrate specificity, and susceptibility to specific inhibitors. In addition, active transport systems need a source of energy, which may be ATP or some other high-energy intermediate, or it may be electrical, in the form of a membrane potential (p. 185).

Na$^+$, K$^+$-ATPase

One of the best studied active transport systems is the Na$^+$, K$^+$-dependent ATPase, which is found in the plasma membranes of many cells, including the erythrocyte. This enzyme, of molecular weight 300 000, contains two identical subunits of approximate molecular weight 100 000, and two of 50 000. It catalyses the outward translocation of 3 Na$^+$, and the inward translocation of 2 K$^+$: both ions are transported against the concentration gradient, the concentrations of Na$^+$ in the plasma and erythrocyte cytoplasm being 150 mM and 5 mM respectively, and of K$^+$, 5 mM and 60 mM. Energy is provided by the hydrolysis of ATP to ADP and inorganic phosphate. This take place on the cytoplasmic side of the membrane (see Fig. 10.8), requires Mg^{2+} (which is not transported) and the presence of both K$^+$ and Na$^+$, and occurs in two stages: in the first, the carboxyl group of an aspartate in the large subunit becomes phosphorylated, and in the second, this mixed anhydride is hydrolysed with the release of inorganic phosphate:

$$\text{E—COOH} + \text{ATP}^{4-} \xrightarrow{\text{Mg}^{2+},\ \text{Na}^+} \text{E—CO—OPO}_3^{2-}$$
$$+ \text{ADP}^{3-} + \text{H}^+$$

$$\text{E—CO—OPO}_3^{2-} + \text{H}_2\text{O} \xrightarrow{\text{K}^+} \text{E—COOH} + \text{HPO}_4^{2-}$$

There is some evidence that phosphorylation of the enzyme in the first stage of hydrolysis results in a conformation change which expels sodium ions from within the cell, while dephosphorylation in the second stage is linked to the uptake of potassium ions. Ion transport and ATP hydrolysis are inhibited by the cardiac glycoside ouabain.

Ca^{2+}-ATPase

Another active transport system pumps calcium ions into the specialized endoplasmic reticulum of muscle cells. Contraction of skeletal muscle is initiated by release of Ca^{2+} from the sarcoplasmic reticulum in response to a nerve impulse, and is terminated when Ca^{2+} is reaccumulated by the active transport process. The pump, a protein of molecular weight about 100 000, hydrolyses ATP to ADP and inorganic phosphate, with the coupled translocation of 2 Ca^{2+}. Like the Na$^+$, K$^+$-ATPase of plasma membrane, this Ca^{2+}-ATPase undergoes reversible phosphorylation during its catalytic cycle. Since this enzyme comprises about 70% of the protein of sarcoplasmic reticulum, it can easily be isolated. It is inactive when completely depleted of lipid, but is reactivated by the addition of phospholipid, allowing investigation of the dependence of activity on specific phospholipids. Maximal activity is achieved when phosphatidyl choline is added, in lipid:protein ratios greater than or equal to 35 mol/mol.

Active accumulation of glucose

The Na$^+$, K$^+$-ATPase and Ca^{2+}-ATPase use the energy derived from hydrolysis of ATP in their active translocation of ions, but not all active transport systems hydrolyse ATP; some derive the energy required for active transport of one solute by simultaneously transporting a different solute *down* its concentration gradient. In the small intestine, active uptake of glucose from the lumen by intestinal mucosa is catalysed by a Na$^+$-glucose symporter, which transports both glucose and Na$^+$ into the mucosal cell; active transport of glucose thus collapses the concentration gradient of Na$^+$ which is established by the Na$^+$, K$^+$-ATPase. Thus hydrolysis of ATP indirectly provides the energy required for glucose uptake, being coupled to it by the electrochemical gradient of Na$^+$ (Fig. 10.8).

Chemiosmotic transport mechanisms

In principle, the ion-gradient established by a single ion-translocating ATPase can provide the energy utilized by several different active uptake systems, if they are all linked to inward movement of the same ion. Another example of indirectly-coupled active transport is found in chromaffin granules, membrane-bounded secretory vesicles of the adrenal medulla, which contain very high concentrations of ATP (0.15 M) and the

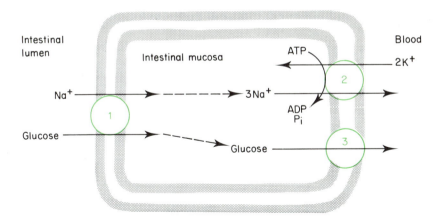

Fig. 10.8 Active transport of glucose from intestine to blood, by intestinal mucosal cells. 1: Na$^+$-glucose symporter; 2: Na$^+$, K$^+$-ATPase; 3: glucose uniporter.

catecholamine hormones adrenaline and noradrenaline (0.7 M). These vesicles are very similar to the catecholamine-containing secretory vesicles of sympathetic nerve terminals. The granule membrane contains an ATPase which translocates protons into the granule matrix, creating a transmembrane pH gradient (acid inside); since this ATPase transports a charged species (H$^+$), it is said to be electrogenic. The inward translocation of protons results in the protonation of basic groups (such as the amino acid side chains of proteins) within the granule matrix, and hence an imbalance of charge across the granule membrane; this can be expressed as an electrical potential, known as the

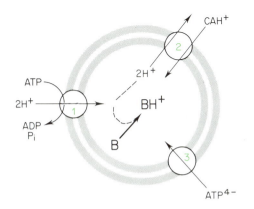

Fig. 10.9 Accumulation of catecholamines and ATP by the secretory granules of the adrenal medulla. 1: H$^+$-translocating ATPase; 2: catecholamine-proton antiporter; 3: ATP uniporter. B represents the internal buffers of the granule, such as proteins.

membrane potential ($\Delta\psi$). The transmembrane pH gradient and membrane potential provide the energy for active uptake of catecholamines by a specific carrier, which catalyses exchange of one positively charged catecholamine molecule for two protons (Fig. 10.9). Although the ATPase translocates 2 H$^+$ per ATP hydrolysed, the overall stoichiometry of uptake is much less than 1 catecholamine/ATP, because the pH gradient is partially dissipated by proton leakage from the granules, or by the action of other permeases in the membrane.

A system of interacting metabolite antiporters is found in the inner membrane of mitochondria, and is discussed in chapter 9.

Structure of permeases

Only a few permeases have been isolated and studied, and in no case has the molecular mechanism been elucidated in any detail. Permeases operate by binding the substrate on one site of the membrane, and providing a channel for its passage; in the case of active transport, passage of the substrate involves an energy-dependent conformation change in the permease, so that it binds the substrate tightly on one side of the membrane, and weakly on the other. All permeases are transmembrane proteins: movement of the permease from one side of the membrane to the other, or rotation within the membrane, would be very difficult because of the interaction of different parts of the protein with the membrane and with the aqueous phases on either side, and is contrary to experimental observations of the asymmetric orientation of all membrane proteins. No permease has been yet studied in enough detail for the structure of the conducting channel to be defined, but it seems likely that such channels are formed by bundles of α-helices, such as occur in those intrinsic proteins in which the polypeptide chain traverses the membrane a number of times (see Fig. 10.3). Since the diameter of an α-helix is 1.5 Å (p. 18), the interior of a single helix would be too small to accommodate the substrate, but a bundle of several helices could form a large pore within the membrane. Hydrophobic amino acid side chains on the outside of such a structure would interact with the lipids of the bilayer, other side chains providing lateral interaction between the helices, and others binding groups for substrates with the pore. Since the pitch of the helix is 5.4 Å, and there are 3.6 amino acids per turn, some 20–25 amino acids (6 helical turns) would be needed for an α-helical protein chain to span the hydrophobic part of the membrane; a pore formed from a bundle of 6 helices, as shown in Fig. 10.3, would probably require a protein of molecular weight 25 000 or more.

Membrane-bound enzymes

Some enzymes are membrane-bound even though the reactions they catalyse do not involve translocation of the substrate across the membrane. One such enzyme is glucose-6-phosphatase, which is located on the cytoplasmic face of the endoplasmic reticulum: the reason for this location is unknown. The endoplasmic reticulum is also the site of important electron-transport systems, concerned with the hydroxylation of drugs and sterols, and with production of unsaturated fatty acids. These reactions involve cytochromes b_5 and P_{450}, and their associated reductases (chapter 9).

Cytochrome b_5

Cytochrome b_5 is an amphiphilic membrane protein of molecular weight 16 000; it can be solubilized by disruption of the membrane with nonionic detergents such as triton X-100, or by treatment with proteases such as trypsin. The detergent-solubilized protein (mol. wt 16 000) has a tendency to associate into octomers, whereas the protease-solubilized cytochrome has a lower molecular weight (11 000, corresponding to amino acids 1–88 of the cytochrome) and does not oligomerize. The protein has a hydrophilic, haem-containing domain which is anchored to the membrane by a hydrophobic domain of about 31 amino acids at the C-terminal end of the polypeptide chain. This hydrophobic domain can interact with the acyl chains of phospholipids in the membrane bilayer, or with other hydrophobic proteins; it causes oligomerization of the detergent-solubilized form, and is removed by trypsin (Fig. 10.10). Cytochrome b_5 reductase and cytochrome P_{450} reductase have essentially similar structures, a globular region of

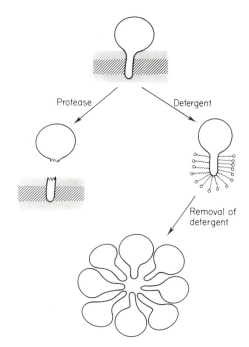

Fig. 10.10 Solubilization of the amphiphilic protein cytochrome b_5, by proteolysis or detergent treatment.

each protein containing the prosthetic group, and a hydrophobic 'tail' at the C-terminus anchoring this to the membrane.

Sucrase–isomaltase

The sucrase–isomaltase complex of the brush border membrane of small intestine has two homologous glycoprotein subunits, the larger catalysing the hydrolysis of isomaltose, the smaller, hydrolysis of sucrose. The entire amphiphilic enzyme complex is synthesized as a single polypeptide which, after insertion into the endoplasmic reticulum, is cleaved into the two subunits, one (isomaltase, mol. wt 140 000) anchored to the membrane by a hydrophobic sequence at its N-terminal, the other (sucrase, mol. wt 120 000) held to isomaltase by non-covalent interaction. The complex is exposed at the extracytoplasmic face of the plasma membrane after fusion of a vesicle derived from the endoplasmic reticulum (Fig. 10.11).

Oxidative phosphorylation

Mitochondrial oxidative phosphorylation consists of two separate but coupled processes: the oxidation of substrates such as NADH, succinate, reduced electron-transferring flavoprotein (p. 97) or glycerol-1-phosphate by the mitochondrial electron-transport chain, with the reduction of oxygen to water, and the synthesis

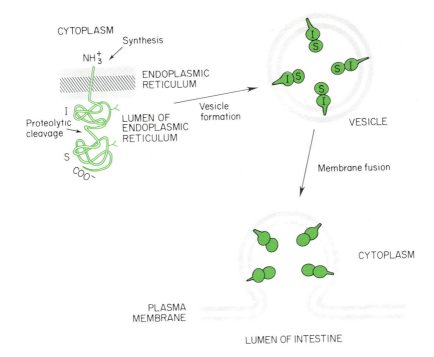

Fig. 10.11 Probable mechanism by which the sucrase–isomaltase complex is synthesized and inserted into the plasma membrane of intestinal cells.

of ATP from ADP and inorganic phosphate by the mitochondrial ATPase complex. All of the enzyme systems responsible are located in the mitochondrial inner membrane. The energy available from substrate oxidation is used to drive ATP synthesis, an endergonic reaction, and the two processes are efficiently coupled only in intact mitochondria. Mechanical or chemical disruption of the mitochondrial inner membrane causes uncoupling—the wasteful oxidation of substrates without concomitant ATP synthesis—and energy derived from oxidation appears as heat. This is because the link between oxidation and phosphorylation is a proton-motive force ($\Delta\mu_{H^+}$, the sum of a transmembrane proton-gradient and membrane potential) which is created by the electron-transport chain, and utilized by the ATPase as it synthesizes ATP.

The proton-motive force

Passage of electrons along the electron-transport chain results in translocation of protons across the mito-chondrial inner membrane. The mechanism of proton ejection is still unknown, but experiments with respiratory chain complexes I, III and IV (see Table 10.2), purified and incorporated into liposomes, have shown that each complex catalyses active, electrogenic proton transport. Presumably the passage of electrons

between the redox components of these complexes produces conformation changes which, by alteration of the pK_a of ionizing groups, results in asymmetric proton uptake and release: protons are taken up from the matrix, and released into the intermembrane (cristae) space. Whatever the mechanism, electron transport results in a rise in the matrix pH, relative to the cristae space (which is in communication with the cytosol). This proton concentration gradient (ΔpH) is partially converted into a membrane potential ($\Delta\psi$, negative inside) by the exchange of protons for cations, and hydroxyl ions for anions. K^+/H^+ and Na^+/H^+ anti-porters have been identified in the inner membrane of mitochondria, and a whole series of anion permeases also exist (see chapter 9).

Coupled rat-liver mitochondria have ΔpH $= -1.4$, and $\Delta\psi = 150$ mV: these give a proton-motive force of 230 mV, according to the expression

$$\Delta\mu_{H^+} = \Delta\psi - \frac{RT}{F}\Delta pH$$

where R is the gas constant (8.314 J/mol.K), T the temperature (310 K), and F the Faraday constant. This represents the energy available for ATP synthesis, or for other endergonic reactions. The proton-motive force is maintained only if protons do not leak back into the

185

Table 10.2 Mitochondrial respiratory-chain complexes

Complex	Polypeptides	Prosthetic groups	Inhibitors
I. NADH-ubiquinone reductase	25	FMN Non-haem iron (5 types)	Amytal Rotenone Piericidin
II. Succinate-ubiquinone reductase	2	FAD Non-haem iron (3 types) ? Cytochrome b	Thenoyl trifluoroacetone
III. Ubiquinone-cytochrome c reductase	6 (1*)	Cytochrome b (2 types) Non-haem iron Cytochrome c_1	Antimycin 8-hydroxyquinoline-N-oxide
IV. Cytochrome c oxidase	7 (3*)	Cytochrome a Cytochrome a_3 Cu (2 mols/mol)	Cyanide Azide Carbon monoxide

* Indicates the number of mitochondrially-synthesized polypeptides.

matrix. The mitochondrial inner membrane is an excellent insulator: despite the high voltage gradient (approximately 3×10^5 V/cm) proton leakage is slow, and the energy released by oxidation is conserved with high efficiency.

Synthesis of ATP

Isolated mitochondrial membranes hydrolyse ATP rapidly. This ATPase activity is catalysed by the enzyme which *synthesizes* ATP in intact mitochondria, and is simply the reverse of the reaction which occurs in vivo:

$$ADP^{3-} + HPO_4^{2-} + H^+ + energy \rightleftharpoons ATP^{4-} + H_2O$$

The amount of energy required ($\Delta G'_{ATP}$) depends on the intramitochondrial concentrations of ATP, ADP and phosphate:

$$\Delta G'_{ATP} = \Delta G'_0 + RT \ln \frac{[ADP][P_i]}{[ATP]}$$

Since $\Delta G'_0$ (the standard free energy of hydrolysis of ATP), is approximately -30 kJ/mol, $\Delta G'_{ATP}$ is probably about 55 kJ/mol (see p. 156). The immediate source of this energy is the proton-motive force: like the electron transport chain, the ATPase translocates protons, ATP synthesis resulting in the re-entry of protons into the matrix. When hydrolysing ATP, the enzyme translocates protons in the opposite direction. Electron transport coupled to ATP synthesis produces a proton circuit through the mitochondrial inner membrane: this is summarized in Fig. 10.12.

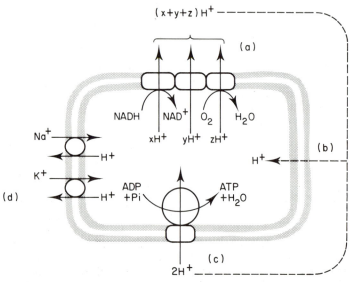

Fig. 10.12 Ion movements across the mitochondrial inner membrane. (a) H^+-translocating electron-transport chain: the quotients x, y and z are not accurately known; (b) proton leakage; (c) proton flow through the H^+-translocating ATPase, resulting in ATP synthesis; (d) Na^+/H^+ and K^+/H^+ antiport, which interconverts the transmembrane pH-gradient (ΔpH) and membrane potential ($\Delta\psi$). Translocation of ADP, ATP and phosphate (p. 160) are not shown.

Respiratory control

The tight coupling of phosphorylation to substrate oxidation results in respiratory control, the dependence of substrate oxidation on the presence of ADP and inorganic phosphate. Isolated mitochondria, supplied with phosphate and saturating concentrations of substrates, consume oxygen only slowly, but when ADP is added the rate of oxygen uptake increases some 6–8 fold. Respiratory control is very important, as it ensures that substrate is not needlessly oxidized if the ratio $\frac{[ATP]}{[ADP][P_i]}$ is high; when ATP is consumed in the cell, this ratio falls, and oxidative phosphorylation is stimulated.

Respiratory control is exercised by the proton-motive force. As respiration proceeds, proton translocation by the respiratory chain increases both ΔpH and $\Delta\psi$, and $\Delta\mu_{H^+}$ rises. Phosphorylation of ADP occurs as long as

$$\Delta G'_{ATP} \leqslant nF\Delta\mu_{H^+} \quad \text{(where } n = 2\text{; see below)}$$

but as ADP is phosphorylated, the ratio $\frac{[ATP]}{[ADP][P_i]}$ increases, $\Delta G'_{ATP}$ rises, and phosphorylation slows down and eventually stops. Since the electron-transport chain has to translocate protons *against* the proton-motive force which, because it is no longer being dissipated by the action of the ATPase, has risen to its limiting value of about 230 mV, respiration also slows down and stops. Utilization of ATP by endergonic processes within the cell decreases the ratio $\frac{[ATP]}{[ADP][P_i]}$, resulting in a stimulation of ATP synthesis by the ATPase; proton translocation by this enzyme lowers $\Delta\mu_{H^+}$, and respiration is stimulated. By this means, cellular ATP consumption controls the activity of the respiratory chain; this in turn affects the ratio $[NAD^+]/[NADH]$ within mitochondria, and the ratio of oxidized to reduced flavin in flavoprotein dehydrogenases. These ratios exert a controlling effect on the tricarboxylic acid cycle, and on β-oxidation of fatty acids.

Uncouplers

In damaged mitochondria, respiratory control is reduced or absent, because proton leakage collapses the proton-motive force, short-circuiting the flow of protons through the ATPase. Some chemical agents produce a similar short-circuit, by conducting protons through the inner membrane; such compounds are called uncouplers. The structures of some uncouplers are shown in Fig. 10.13; all are lipid-soluble weak acids, which are thought to act as protonophores by the mechanism shown in Fig. 10.13. In the presence of uncouplers, mitochondrial electron transport proceeds

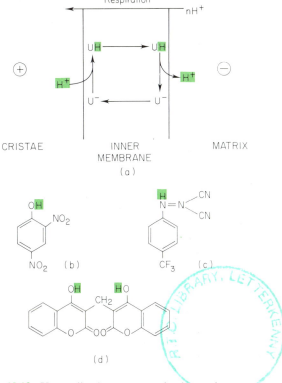

Fig. 10.13 Uncoupling by proton conduction, and structures of some uncouplers. (a) Mechanism of H$^+$-transport by uncouplers; (b) 2,4-dinitrophenol; (c) *p*-trifluoromethoxycarbonylcyanide phenylhydrazone (FCCP); (d) dicoumarol. Dissociable protons are shown in green.

at its maximal rate, irrespective of the ratio $\frac{[ATP]}{[ADP][P_i]}$, and energy is wasted. The toxic effects of uncouplers have been seen in factory workers exposed to 2,4-dinitrophenol, which was once manufactured as an explosive.

Brown fat

Brown fat is thermogenic tissue, and is plentiful in hibernating and in cold-adapted animals. Mitochondria in brown fat have little or no ATPase, the energy from electron transport being used for heat production rather than ATP synthesis. This occurs because the proton-motive force is dissipated by the action of a specific ion-conducting channel, which permits hydroxyl, chloride and other anions to cross the membrane. This channel is controlled by fatty acids and various nucleotides. Brown fat mitochondria are therefore physiologically uncoupled, the uncoupling being hormonally regulated in response to the need for heat production. The quantitative significance of brown fat in the metabolism of adult humans is not known.

The stoichiometry of oxidative phosphorylation

In experiments with isolated mitochondria, measurement of the oxygen consumed during phosphorylation of known amounts of ADP gives the P/O ratio (moles phosphate esterified/atoms oxygen consumed). In well-coupled mitochondria this ratio is about 3 for oxidation of NADH, 2 for succinate and glycerophosphate and 1 for electron donors, such as ascorbate, which reduce cytochrome c. Experimental estimates of the number of protons ejected during passage of two electrons along the electron-transport chain (the H$^+$/2e ratio) vary between 6 and 12 for NADH oxidation. Since the ATPase almost certainly translocates 2 protons in the phosphorylation of ADP (H$^+$/P = 2), the P/O ratio for NADH might under some circumstances be greater than 3. Only 6 H$^+$ are translocated back into the matrix in the production of 3 ATP, but extra protons may be required for substrate uptake by mitochondria, or may simply be lost by leakage across the inner mitochondrial membrane; in any case, since the coupling between oxidation and phosphorylation is not chemical but electrical, there is no reason why P/O ratios should always have integral values.

Low concentrations of uncouplers lower P/O ratios, as respiratory control is reduced; higher concentrations completely abolish respiratory control and phosphorylation.

Inhibitors of oxidative phosphorylation

Three distinct types of inhibitor act on phosphorylation. *Respiratory inhibitors* bind to components of the electron-transport chain, preventing electron flow and proton translocation and therefore inhibiting both oxygen consumption and phosphorylation. Examples are antimycin, which binds to cytochrome b, and cyanide, which inhibits cytochrome c oxidase. *Uncouplers*, as discussed above, inhibit phosphorylation, but stimulate respiration and ATP hydrolysis, which proceed at their maximal rates in uncoupled mitochondria. *Phosphorylation inhibitors*, such as oligomycin, inhibit the H$^+$-translocating ATPase or the uptake of ADP (p. 160). This inhibits phosphorylation of ADP, and in coupled mitochondria also inhibits oxygen uptake, as there is no means of dissipating the proton-motive force when the ATPase is inhibited. The inhibitory effects of these compounds on electron transport are overcome by uncouplers, which allow a distinction to be made between respiratory inhibitors and phosphorylation inhibitors.

Structures of the mitochondrial respiratory chain complexes

When mitochondrial membranes are treated with detergents such as cholate, the components of the oxidative phosphorylation are solubilized, and can be separated from each other by conventional techniques of protein purification. Fractionation of this type yields four respiratory chain complexes, each of which contains a number of proteins and electron-transfer components such as iron–sulphur centres, flavin and haem: the composition of each is given in Table 10.2. Two of these complexes contain proteins which are synthesized within the mitochondria, the remaining protein components of these and the other complexes being imported from the cytoplasm. Together with ubiquinone and cytochrome c, the four complexes form the complete respiratory chain, which catalyses the transfer of electrons from NADH or succinate to oxygen. Such electron transport can be reconstituted by mixing the complexes; in order to reconstitute oxidative phosphorylation the ATPase complex (see below) must also be added, together with phospholipid. Under carefully controlled conditions respiratory complexes and ATPase then reconstitute into sealed phospholipid vesicles which are capable of oxidative phosphorylation.

Cytochrome c oxidase

Each complex couples electron transport to proton translocation, creating the proton-motive force which is the energy source for ATP synthesis. The most stable and best-studied is cytochrome c oxidase (complex IV). The probable arrangement of the seven subunits of this complex within the mitochondrial membrane is shown in Fig. 10.14. It catalyses the reaction

$$2 \text{ cyt.c}^{++} + 2\text{H}^+ + \tfrac{1}{2}\text{O}_2 \rightarrow 2 \text{ cyt.c}^{+++} + \text{H}_2\text{O}$$

Oxygen is consumed on the inner (matrix) side of the membrane, and cytochrome c^{++} oxidized on the outer (cristae) side; in addition to the two protons consumed by the reaction, two more are translocated from matrix to cristae, with each atom of oxygen reduced. It is not known which subunits bind the prosthetic groups haem and copper, and which are involved in H$^+$-translocation, but all of these functions may reside in subunits I and II, as some bacteria have simple cytochrome oxidases containing only two subunits, which are similar to those of eukaryotic enzymes.

The H$^+$-translocating ATPase

The mitochondrial ATPase complex (mol. wt about 400 000) has two segments; one, known as F$_1$, contains the binding-site for ATP, and the other, F$_0$, forms a proton-conducting channel through the inner mitochondrial membrane. In electron-microscopy of mitochondria which have been negatively-stained with reagents such as phosphotungstate, F$_1$ molecules appear as spherical particles, of diameter about 9 nm, on the inner surface of the membrane. These are easily

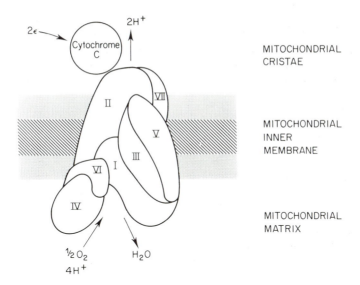

Fig. 10.14 Probable arrangement of cytochrome oxidase subunits in the mitochondrial inner membrane, determined by protein labelling studies.

detached, and have ATPase activity which is insensitive to oligomycin, although this compound inhibits oxidative phosphorylation and ATP hydrolysis in intact mitochondria. This is because oligomycin binds to a protein in F_o, the membrane segment of the ATPase. Attachment of F_1 to F_o involves 'oligomycin-sensitivity conferring protein', which does not itself bind oligomycin, by may be part of the 'stalk' linking F_1 to F_o.

H$^+$-translocating ATPases in the mitochondria of all species, and in bacteria, are basically similar (Fig. 10.15). There are five types of subunit in F_1, in the stoichiometry $\alpha_3\beta_3\gamma\delta\epsilon$: β contains the site of ATP hydrolysis. F_o contains about four subunits, some of which are synthesized in the mitochondria. One is a low molecular-weight, hydrophobic protein which can be reconstituted in liposomes to form an oligomycin-sensitive, proton-conducting channel.

Inhibitors of the ATPase may act either on F_1, blocking hydrolysis or synthesis of ATP, or on F_o, blocking H$^+$-translocation. These processes are tightly coupled in the F_1–F_o complex, and are then blocked by all types of inhibitor.

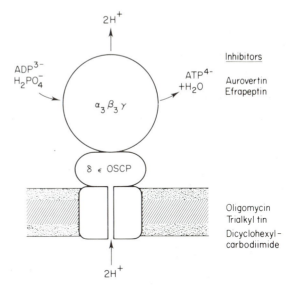

Fig. 10.15 Structure of the H$^+$-translocating ATPase of mitochondria, showing the 'head' and 'stalk' (F_1) and the membrane segment (F_o).

Cell-surface Components

Cells are recognized by the surface components of their plasma membranes: such recognition occurs in endocytosis, in the rejection of tissues after transplants, in adhesion between cells, in attack by viruses, and in the binding of hormones. It is thought that the carbohydrate parts of glycoproteins and glycolipids are particularly important in cellular recognition processes.

Blood group substances and histocompatibility antigens

Surface components which are specific to the cells of an individual are called histocompatibility antigens. Injected or grafted cells from another individual with a different antigenic pattern are recognized as foreign by the immune system of the recipient, and are destroyed. The best-characterized cell-surface antigens are those determining the blood group (A, B or O), which occur on erythrocytes, on epithelial and endothelial cells, and in soluble form in some secretions. The differences between the blood groups are due to various modifications of a particular glycolipid (the H antigen), by addition of *N*-acetyl galactosamine (in individuals of blood group A), galactose (blood group B) or both (blood group AB). In individuals of blood group O, neither transferase enzyme is expressed, and the H antigen is not modified (see p. 67).

These are not the only antigens on the erythrocyte surface; the Rhesus system is of clinical importance, as Rh$^-$ individuals often have antibodies against the Rh antigen, whereas another system (MN) causes few problems in blood transfusion, as antibodies against these determinants are not usually present in human serum. MN secificity is associated with the erythrocyte membrane protein glycophorin, and appears to reside both in the polypeptide chain and in the attached oligosaccharide. The chemical nature of the Rh antigen is not yet known.

Histocompatibility antigens present on other cell types, such as leucocytes and fibroblasts, have also been characterized. The HLA and related HLB and HLC antigens are transmembrane glycoproteins of molecular weight about 45 000, carrying a single asparagine-linked oligosaccharide chain. They have some sequence homology with IgG, and are non-covalently associated with a non-glycosylated polypeptide of molecular weight 12 000, known as β_2-microglobulin (Fig. 10.16).

During the development of a cell, certain cell-surface antigens may be lost and others appear. These differentiation antigens have been isolated from lymphocytes, and appear to be similar in general structure to the histocompatibility antigens discussed above.

Viruses that attack eukaryotic cells recognize specific receptors on the cell surface: the first stage of attack by Semliki Forest Virus or Vesicular Stomatitis Virus has been shown to be binding of the virus to cell-surface glycoproteins, which interact with glycoprotein 'spikes' on the viral envelope. Some bacteria also attach themselves to eukaryotic cell surfaces, for example cells of *Vibrio cholerae* become bound to the brush-border of the intestine during infection, although they do not invade the cells.

Certain mammalian cells have the ability to recognize and bind other types of cell from the same organism: e.g.

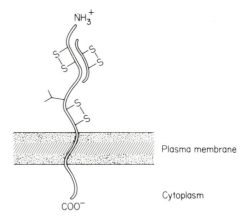

Fig. 10.16   Structure of histocompatibility antigen.

Kupffer cells of the liver bind asialoerythrocytes (red blood cells from which cell-surface sialic acid has been removed), a process which probably occurs during the removal of aged erythrocytes from circulation—there also the recognition process clearly involves cell-surface glycolipid or glycoprotein.

Cell–cell adhesion

Morphogenesis requires adhesion between cells of the same type, a process mediated by fibronectin, a glycoprotein of molecular weight about 200 000, which is also involved in the binding of cells to collagen. Fibronectin is not itself an intrinsic membrane protein, but it has been shown to interact specifically with a glycoprotein in the plasma membrane of fibroblasts. Many transformed cells, which undergo uncontrolled growth and division, appear to lack fibronectin, and it appears that alteration in cell-surface glycoproteins, with consequent disturbance of a cell–cell adhesion, may be important in the generation of tumours.

Membrane receptors for hormones and neurotransmitters

The mechanisms by which metabolic pathways are controlled by signals arriving from outside the cell are discussed in chapter 13. Cells respond to a wide range of extracellular ligands, varying from small neurotransmitters, such as acetylcholine and noradrenaline, to large polypeptide hormones such as insulin and glucagon; these exert their effects by interacting with specific membrane receptors. (Steroid hormones are an exception to this, in that they enter the target cell, interact with cytoplasmic, rather than membrane-bound, receptors, and then proceed to the nucleus.)

Occupation of a receptor results in an increase in the

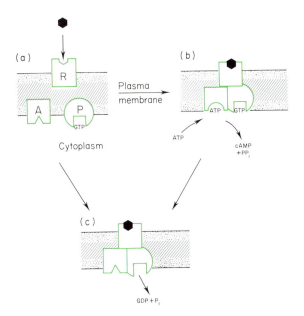

Fig. 10.17 The β-adrenergic receptor. (a) Binding of the hormone to R, the receptor, leads to association of adenyl cyclase (A) and a GTP-binding regulatory protein (P), with stimulation of cAMP synthesis (b). The complex becomes inactive when GTP is hydrolysed (c).

cytoplasmic concentration of some other molecule, known as the 'second messenger'. Cyclic-AMP and calcium ions each function as second messengers for several different hormones; the effects of other hormones, such as insulin, may be mediated by other, so far unrecognized, compounds. β-Adrenergic receptors, which respond to adrenaline, and noradrenaline, function by activation of adenyl cyclase; their mode of action is shown diagrammatically in Fig. 10.17. The receptor is a separate entity from the cyclase; in turkey erythrocytes, there are 600–1000 receptor molecules per cell, with a moderate affinity for catecholamines ($K_d = 6 - 10 \times 10^{-6}$ M). Adenyl cyclase is probably a free-floating, membrane-bound enzyme, which is not permanently associated with any receptor; when it collides with an occupied receptor it is stimulated to produce cyclic AMP:

$$ATP \rightarrow cAMP + P\text{-}P_i$$

Cells frequently contain several types of hormone receptor, each of which can activate adenyl cyclase. A third component of the system is a separate GTP-binding protein, which probably has a regulatory function. Hormone activation of adenyl cyclase requires GTP, which is itself hydrolysed; an analogue of GTP which cannot be hydrolysed still stimulates adenyl cyclase, and indeed produces chronic, rather than transient, activation of the cyclase by hormones. Hydrolysis of GTP by the regulatory protein therefore seems to terminate the activation of the cyclase.

Cholera toxin exerts its effects by stimulation of adenyl cyclase. The toxin has two types of subunit: B, which binds to the toxin receptor (the membrane ganglioside G_{M1}) and A, which catalyses the ADP-ribosylation of the GTP-binding protein:

$$NAD^+ + protein \rightarrow protein\text{—}ADPR + nicotinamide$$

Modification of the regulatory protein abolishes its GTPase activity, causing it to irreversibly activate adenyl cyclase. The increased concentration of cAMP in the intestinal epithelium produces loss of salt and water to the intestinal lumen, resulting in massive diarrhoea.

11

Nucleic Acid Synthesis: Transcription

Genetic Function of Nucleic Acids

All organisms possess genetic information which, by specifying the amino acid sequences of the proteins synthesized by the organism, programmes their structure and metabolic activity. This information is, in both prokaryotes and eukaryotes, contained in one or more chromosomes, the genetically essential component of which is nucleic acid: usually deoxyribonucleic acid (DNA), but in a few cases ribonucleic acid (RNA). Nucleic acids are linear polymers of nucleotides, and most chromosomes, including those of all higher organisms, contain double-stranded DNA; however, the genomes of some of the simplest known bacterial viruses (bacteriophages), as well as those of many plant and animal viruses, contain single-stranded RNA, whilst other phages contain single-stranded DNA. Up to 65% of the chromosomes of higher organisms is protein, and although DNA was recognized as a constituent of chromosomes many years ago, it was only comparatively recently shown that amino acid sequence information is carried by nucleic acids.

Bacterial transformation

This was shown by experiments in which non-pathogenic, uncapsulated strains of pneumococci (pneumonia bacteria) were genetically transformed into pathogenic, capsulated forms by the addition of DNA which had been isolated from heat-killed cells of a capsulated strain. The transforming activity of the extract was associated with DNA, being destroyed by deoxyribonuclease, but not by ribonuclease, nor by proteases. The transformed strains remained capsulated through many cell divisions, suggesting that DNA had been incorporated into the chromosome. Other genetic transformations of bacteria, such as resistance to penicillin, streptomycin and other antibiotics, are also brought about by acquisition from resistant strains of DNA which specifies the production of particular proteins.

Viruses

Our acceptance of the general applicability of these results owes much to the study of the structure of viruses, and their replication in infected cells. Viruses are complexes of nucleic acid, protein, and, sometimes, carbohydrate and lipid. They have no metabolism of their own, and can only replicate within a host cell. They may parasitize bacteria, in which case they are called bacteriophages, or the cells of plants or animals, and they are responsible for many diseases, such as influenza, polio, psittacosis and rabies. Some viruses cause tumours, and viruses may well be responsible for at least some forms of human cancer. The properties of some viruses are summarized in Table 11.1.

Infection of bacterial cells occurs by attachment of the virus to specific receptors in the bacterial cell wall, followed by passage of viral nucleic acid into the cell; in the case of plant and animal viruses, the entire virus particle enters the cell. The viral genome is replicated, and also translated into protein, by the enzyme systems of the host cell, which may themselves be modified during infection, in order to make them synthesize viral material more efficiently. Newly formed viral proteins and nucleic acids then assemble to form virus particles; the host cell lyses, releasing the phage particles, and another cycle of infection begins. If phage T2, a virus parasitic on *E. coli*, is labelled in its DNA with radioactive phosphorus (^{32}P), and in its protein coat with radioactive sulphur (^{35}S), then allowed to infect *E. coli* and multiply, the progeny phage are found to contain the parental ^{32}P-DNA, but no radioactive protein. This is because the protein of the infecting phage does not enter the cells whereas the DNA enters, is replicated and translated into new coat protein, and used again when new virus particles are assembled. Indeed, it is in some cases possible to infect cells with viral DNA from which the coat protein has been removed, and produce the same cycle of viral replication and host cell lysis. Thus the viral protein contains no genetic information: it serves only to protect the nucleic acid core, and act as a vehicle for infection.

Table 11.1   Characteristics of some viruses

	Molecular weight	Nucleic acid (%)	Diameter (nm)	Shape
Single-stranded DNA viruses				
Bacteriophage φX174	6.2×10^6	29	25	Polyhedral
Double-stranded DNA viruses				
Bacteriophage T7	3.8×10^7	41	6	Tadpole
Papilloma (human wart)	5.3×10^6	8	50	Polyhedral
Adenovirus	2.0×10^8	5	70	Polyhedral
Herpes	10^9		150	Polyhedral
Single-stranded RNA viruses				
Bacteriophage Q β	4×10^6	38	25	Polyhedral
Poliomyelitis	6.7×10^6	28	30	Polyhedral
Rabies	1.5×10^8	2	70	Bullet
Double-stranded RNA viruses				
Reoviruses	7×10^7	14	60	Polyhedral

Chromosomes

In the same way, the protein components of the chromosomes of higher organisms contain no genetic information. Eukaryotic DNA is located mainly in the nucleus (a small fraction being found in mitochondria, where it also has a genetic function), is metabolically very stable, and for each organism, is present in a fixed amount in all cells containing a set of diploid chromosomes, half as much being present in haploid (sperm and ovum) cells. Each of these properties is consistent with the role of DNA as a carrier of genetic information, an idea which ultimately received confirmation when the sequences of nucleotides in chromosomes were determined, and compared with the amino acid sequences of the proteins translated from them.

Classical genetic studies suggested the gene as the fundamental unit of inheritance: the appearance of a particular characteristic in progeny is determined by the production or non-production of particular enzymes, each enzyme being encoded by one or more genes, the expression of which is predictable by Mendelian laws. The study of bacterial conjugation, in which DNA is passed from one bacterial cell to another, in a sexual process, established the chemical nature of the gene: it is a linear stretch of nucleic acid in the chromosome. The enterobacterium *E. coli* has a single chromosome, in the form of a closed loop of double-stranded DNA. Male cells possess a fertility factor (F) which can exist as a small, free piece of DNA, known as an episome. This episome can integrate into the main chromosome of the cell when the cell becomes Hfr (high frequency of recombination)—it will now undergo conjugation with a female cell, which lacks the F factor. During conjugation, the male chromosome breaks (at the point of integration of the F factor), and begins to replicate, the progeny entering the female cell: the chromosomes

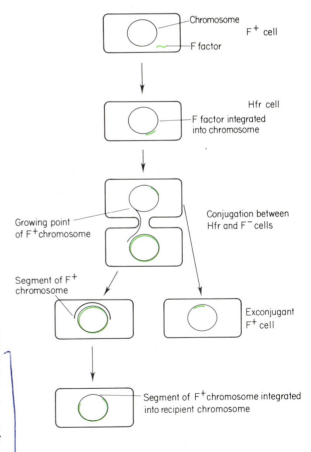

Fig. 11.1   Bacterial conjugation.

193

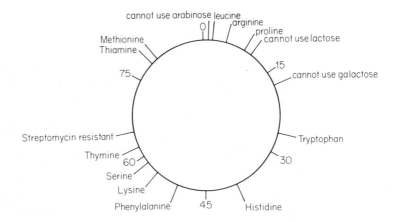

Fig. 11.2 Genetic map of the circular chromosome of *E. coli*, showing the positions of mutations conferring amino acid requirement, inability to use sugars, or antibiotic resistance.

become joined by a process known as 'crossing over', to form a recombinant chromosome. Thus genetic information is transferred from the Hfr cell to the recipient, and appears in the progeny. This process is shown schematically in Fig. 11.1. Complete transfer of the chromosome takes about 90 minutes at 37°C, and can be interrupted by mechanically separating the cells during conjugation (e.g. by use of a high-speed blender). By interrupting conjugation at various times, and determining which genes had been transferred from male to female cells, the positions of the genes on the circular chromosome of *E. coli* were determined. Fig. 11.2 shows some of the features of the genetic 'map' of *E. coli*; the gene for a particular enzyme is found at a particular point on the map, genes specifying enzymes with related metabolic functions often being clustered together. Simple circular chromosomes are found in mitochondria, and many viruses, as well as in bacteria; eukaryotic chromosomes differ in being linear, and associated with a large number of proteins, in a complex known as chromatin. Furthermore, eukaryotic cells tend to contain many, non-identical chromosomes—there are 23 pairs in man. These are segregated in the nucleus, unlike prokaryotic chromosomes, which are exposed to the cytoplasm.

Information flow in protein synthesis

Despite the nuclear location of the chromosomes in eukaryotes, protein synthesis occurs very largely in the cytoplasm, where there is no DNA; this suggests that the genetic information stored in DNA must be transferred to the cytoplasm, where it is translated into the amino acid sequence of proteins, by some coding material. Although protein synthesis in prokaryotes is not spatially separated from the chromosome, bacterial extracts can carry out protein synthesis in vitro in the absence of DNA, so in these organisms too, DNA does not participate directly in protein synthesis. The intermediate between DNA and protein is RNA. In eukaryotes, RNA is synthesized almost exclusively in the nucleus; protein synthesis in cell-free extracts is sensitive to ribonuclease, and the content of RNA is greatest in those organs with a high protein synthetic activity, such as liver, pancreas, brain, small intestine and the reticulo-endothelial system: it is low in kidney, heart and skeletal muscle.

The flow of information, whereby DNA specifies the amino acid sequence of protein, is thus from DNA to RNA, and RNA to protein: this is sometimes called the 'central dogma of molecular biology'. Some revision of it was necessary to account for events in the replication of RNA-containing animal tumour viruses, where the genome (RNA) directs the synthesis of DNA, which in turn directs the synthesis of multiple copies of the viral RNA. The process of DNA-directed RNA synthesis, which occurs in all cells synthesizing protein, is called transcription, whereas RNA-directed DNA synthesis is called reverse transcription. The synthesis of protein, in accordance with an amino acid sequence specified by RNA, is translation.

$$\text{DNA} \xrightarrow[\text{Reverse transcription}]{\text{Transcription}} \text{RNA} \xrightarrow{\text{Translation}} \text{Protein}$$

The Chemical Structure of Nucleic Acids

Covalent structure

The covalent structures of DNA and RNA are shown in Fig. 11.3: they have considerable similarities. Both are linear, unbranched polymers of four types of nucleotide, the nucleotides being linked by 5′-3′ phosphodiester bonds; that is, the 5′ and 3′ hydroxyls of neighbouring ribose residues are covalently joined through phosphate. There are two major differences of chemical structure between DNA and RNA: first, DNA contains 2′-deoxyribose, which has no hydroxyl at the 2′-position, whereas the pentose in RNA is D-ribose; and second, while both DNA and RNA contain the purine bases adenine and guanine and the pyrimidine cytosine, the fourth base is thymine (5-methyl uracil) in DNA, and uracil in RNA. Some bases in DNA are sometimes found to be modified, e.g. by methylation, and some forms of RNA contain 'unusual' bases—e.g. transfer RNA contains pseudouracil, dihydrouracil, inosine and many others.

Since each ribose or deoxyribose is asymmetrically linked to its neighbour, the polynucleotide strand has a *polarity*—the end in which the 5′-hydroxyl of a ribose is not attached to another nucleotide is the 5′-end, and that with a free 3′-hydroxyl (or phosphate), the 3′-end. It is important to consider this polarity when discussing the direction in which a nucleic acid is replicated, transcribed or translated.

The phosphate residues of the polynucleotide 'backbone' are strongly acidic, whereas the attached purines and pyrimidines are only weak bases (see Table 5.1, p. 52) so that nucleic acids carry multiple negative charges even at quite low pH values. They also have strong affinity for cations such as Mg^{2+}, and for basic proteins such as histones and protamines.

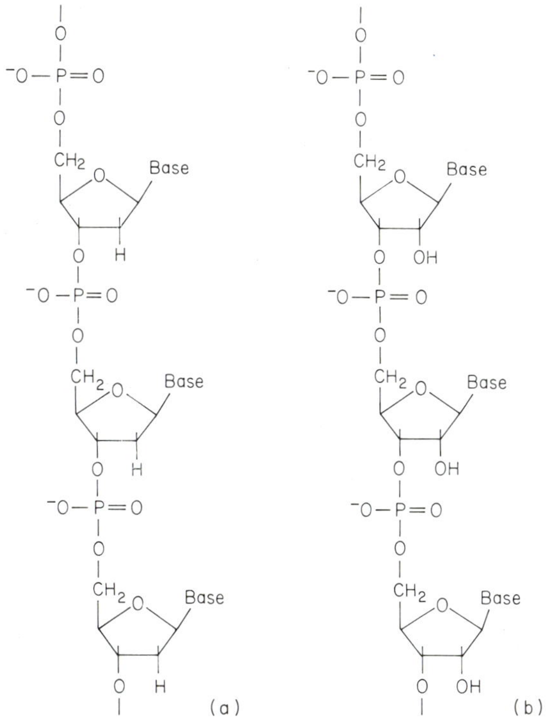

Hydrolysis of nucleic acids

Nucleic acids are chemically quite stable, over a wide range of pH values. Under acidic conditions, loss of purines from DNA may occur by hydrolysis of the glycosidic bonds between C^1 of 2-deoxyribose and N^9 of purines—the 'backbone' is left intact, and the product is called 'apurinic acid'. The phosphodiester bonds may be hydrolysed under alkaline conditions, and RNA is much more susceptible than DNA to alkaline hydrolysis: this is because the 2′-hydroxyls of RNA can become deprotonated and attack the neighbouring 3′-phosphate, breaking the 'backbone' and producing cyclic nucleoside 2′,3′-phosphates, which are hydrolysed to a mixture of nucleoside 2′- and 3′-phosphates. In addition to these hydrolyses, some cross-linking reactions between bases are important in the production of mutations by ultraviolet light (p. 201), and base-specific chemical cleavage procedures are used in the determination of the base-sequence of DNA.

The destruction of unwanted (often exogenous) genetic material is important to all organisms, and there are many types of enzymes (nucleases) which attack nucleic acids. They invariably hydrolyse phosphodiester bonds, breaking the link between phosphate and either 5′-hydroxyl or 3′-hydroxyl. Particularly active nucleases have been obtained from pancreas, spleen, and snake venom. Exonucleases are enzymes which sequentially remove nucleotides from one or other end of the chain; endonucleases make internal cuts. Many of these enzymes show little base specificity, although some

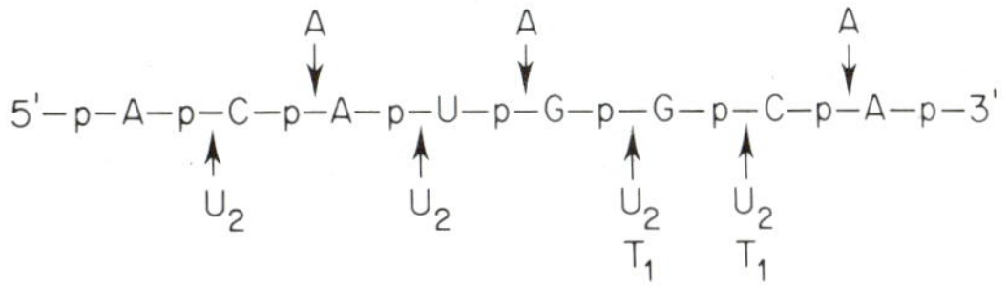

Fig. 11.3 General structure of (a) DNA and (b) RNA backbone.

Fig. 11.4 Specificity of ribonucleases. Ribonuclease A cleaves on the 3′-side of pyrimidines, U_2 on the 3′-side of purines, and T_1 on the 3′-side of guanosine.

DNases are specific for either single-stranded or double-stranded substrates. The discovery of a new class of microbial DNases, known as restriction endonucleases, has been of paramount importance in DNA cloning. These enzymes, specific for double-stranded DNA, recognize sequences of up to six bases; their experimental use is discussed later. Ribonucleases hydrolyse the bond between phosphate and 5'-hydroxyl, and show some base specificity (Fig. 11.4).

The secondary structure of DNA

For DNA derived from all organisms except some simple viruses, analysis of the base content has shown that the number of adenine and thymine bases is equal, as is the number of guanines and cytosines: $A/T = G/C = 1$. However, the ratio of $(A + T)$ to $(G + C)$ varies widely between organisms, although for a given organism, it is the same for DNA derived from all tissues, and independent of age. No such relationship is found for RNA. This finding, which was for long a puzzling one, was explained by the elucidation of the double-helical structure of DNA, following study of the X-ray diffraction patterns of DNA fibres. The structure is shown in Fig. 11.5: its principal features are:

1. Two separate DNA chains are wound together in a right-handed double helix, the strands running in opposite directions.
2. The helix is maintained by 'base-pairing'—hydrogen bonding between adenine and thymine, or guanine and cytosine bases. In each base-pair, the hydrogen-bonding partners are in different chains.
3. The planar base-pairs are stacked one above the other, on the inside of the helix, and perpendicular to its axis, while the polar 'backbone', of alternating deoxyribose and phosphate residues, is external, and interacts with the solvent.
4. The helix is 2.0 nm in diameter, and makes a complete turn every 3.4 nm; each turn contains 10 base-pairs.

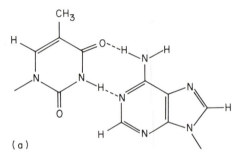

(a)

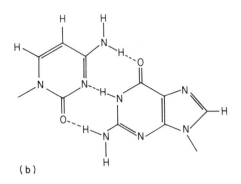

(b)

Fig. 11.6 Base-pairs found in DNA: (a) adenine–thymine; (b) guanine–cytosine.

The structures of the base-pairs are shown in Fig. 11.6. The strictness of the purine–pyrimidine base-pairing rules is due to the geometry of the double helix: there is not enough room for two purines to base-pair, and two pyrimidines would be too far apart. Since each base is paired with its hydrogen-bonding partner on the other chain, the sequences of bases in the two chains are not

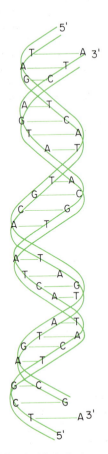

Fig. 11.5 The double helical structure of DNA.

independent—they are *complementary*. The only forms of DNA which do not obey the base-pairing rules are those from viruses in which the genome is single-stranded. The base-paired, double helical model of DNA structure accounts for the physical properties of DNA, described below, and also suggests how the molecule may be replicated, or transcribed into RNA. Since RNA is almost always single-stranded, there is usually no restriction on its base ratios.

Physical properties of DNA

DNA can be isolated from most sources by a sequence of steps: treatment of the cells with a detergent, such as deoxycholate, to extract water-soluble components, followed by shaking with chloroform or phenol to denature proteins, and addition of ethanol, to precipitate the DNA from aqueous solution. It can then be wound out onto a glass rod. DNA is readily soluble in dilute salt: its solution is highly viscous, because of the great length of the DNA chains. The *E. coli* chromosome contains about 3×10^6 base-pairs, has a molecular weight of 2×10^9, and an extended length of almost 1 mm; a single eukaryotic chromosome may be 20–40 times longer. However, DNA is easily broken into shorter lengths by mechanical stirring, and purified DNA is invariably sheared, unless isolated very carefully.

Buoyant density

The density of DNA reflects its secondary structure and base content. It is measured by isopycnic (equilibrium density) gradient centrifugation: the DNA is mixed with a high concentration (about 6 M) of caesium chloride and subjected to prolonged high-speed centrifugation (typically, 65000 rpm, for three days). A density gradient, in which sedimentation of the heavy CsCl molecules is balanced by diffusion, is established in the centrifuge tube, and the DNA migrates to its

equilibrium position in the gradient, where it forms a band. The position is determined solely by the density of the DNA, and not, for example, by the length of the strands—the experiment is usually performed with sheared DNA. In this way, double-stranded DNA can be separated from single-stranded, and from RNA. Furthermore, the buoyant density of the double-stranded DNA can be directly correlated with its base composition. This is because base-pairing between guanine and cytosine involves three hydrogen bonds, but there are only two hydrogen-bonds between adenine and thymine (see Fig. 11.6). The higher the content of G and C, the more compact the structure of the DNA, and the higher the density (Fig. 11.7).

Strand-separation

For the same reason, a higher GC content also makes the DNA strands more difficult to separate. If double-stranded DNA is heated, denaturation and strand-separation occur: this is a cooperative process, which takes place over a rather narrow temperature range, the midpoint being known as the melting temperature (T_m). Denaturation can be followed by measurement of viscosity, which decreases on 'melting', or spectrophotometrically. Separation of the strands breaks the hydrogen bonds, with consequent redistribution of electrons in the bases: this leads to an increase of about 40% in the absorbance of the DNA solution, measured at 260 nm, a phenomenon known as *hyperchromicity* (Fig. 11.8). The melting temperature can be directly correlated with the base composition (Fig. 11.7), T_m always increasing with an increasing GC-content.

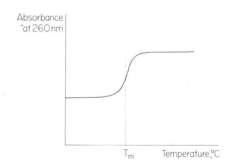

Fig. 11.8 'Melting' of DNA, shown by an increase in its ultraviolet absorbance (hyperchromicity) above the 'melting temperature', T_m.

Hydridization

If double-stranded DNA is first heated to separate the strands completely, then slowly cooled, the base-pairs re-form and double-stranded molecules reappear. This process is called annealing, and has several experimental applications. First, it can be used to assess the

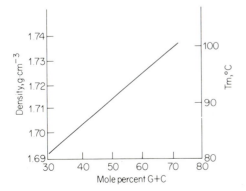

Fig. 11.7 Relationship between physical properties and base content in DNA.

repetitiveness of base sequences in DNA: if the same sequence occurs many times within the genome, a piece of single-stranded DNA will readily find a complementary partner, and annealing is rapid; conversely, if the strands do not contain reiterated sequences, they can base-pair in one way only, and annealing is slow. The degree of annealing can be determined either by use of a single-strand specific nuclease, to digest away un-annealed DNA, or by passage of the solution through hydroxyl-apatite, a crystalline form of calcium phosphate, which retains double-stranded DNA, but not single-stranded. Second, hybrid duplexes can

be formed between DNAs derived from different organisms: the degree of hybridization indicates the genetic similarity between them (e.g. only 25% hybridization occurs between human and mouse DNA). Third, DNA–RNA hybrids can be formed, in which a strand of RNA base-pairs with a strand of DNA. In this way one can determine the region of the genome from which the RNA has been transcribed (see below). Inspection of such DNA–RNA hybrids by electron microscopy has provided direct evidence for the existence of non-informational interruptions, or introns, in eukaryotic genes.

DNA Replication

When a cell divides, each chromosome has to be exactly duplicated, and a copy given to each of the daughter cells. The double-helical model of DNA structure suggests that this could occur by separation of the two

DNA strands, followed by synthesis of two new strands, each one complementary to a parental strand. This scheme of DNA replication is called 'semiconservative', and evidence in support of it came from experiments

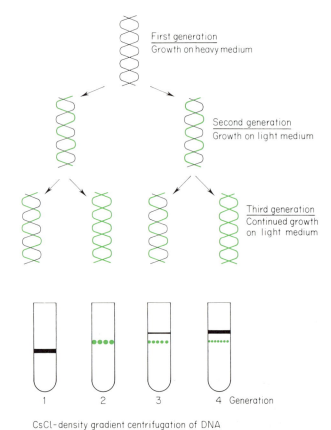

Fig. 11.9 Semiconservative replication of DNA. In the experiment of Meselsohn and Stahl, centrifugation on CsCl gradients separates DNA of different densities after each replication.

with *E. coli* (Fig. 11.9). The bacteria were grown on a nitrogen source containing a stable, heavy isotope (^{15}N), which became incorporated into the DNA. The cells were then transferred to a medium containing ^{14}N, and growth allowed to continue: preparations of DNA were made after each round of cell division, and analysed by caesium chloride density gradient centrifugation (p. 197). After growth on ^{15}N, the DNA was a duplex of two heavy strands; on transfer to ^{14}N, followed by one cell division, the density of the DNA decreased, through formation of a duplex of light and heavy strands; with further cell divisions, a new species (a duplex of two light strands) appeared, along with duplexes containing one heavy strand. These experiments were performed with sheared DNA, and did not indicate whether separation of the parental DNA strands occurred before or during replication. Electron micrographs of intact, circular bacterial chromosomes show that replication is initiated at one point, and proceeds around the circle until complete, while genetic experiments show that replication proceeds in both directions simultaneously: that is, that there are two replication forks, moving away from the common origin, toward each other. In the linear chromosomes of eukaryotes, replication is initiated at multiple sites (up to several thousand in each chromosome) and proceeds bidirectionally, until the replicated regions meet (Fig. 11.10).

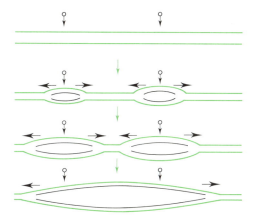

Fig. 11.10 Bidirectional replication of DNA in eukaryotes. Replication begins at several origins (o), and proceeds until the replicating forks meet.

Semiconservative replication of DNA poses several topological problems: the strands, which are covalently closed circles in bacteria, mitochondria and some viruses, must be unwound (at the rate of several thousand revolutions per minute) and separated; since the strands run in opposite directions, synthesis must proceed in different directions on the new strands: from 5′ to 3′ on one, and from 3′ to 5′ on the other.

Enzymes of DNA replication

E. coli contains at least three enzymes capable of synthesizing DNA, known as DNA polymerases I, II and III. All catalyse Mg^{2+}-dependent polymerization of deoxyribonucleoside triphosphates, and require both a primer (a stretch of nucleotide with free 3′-hydroxyl, to which the next nucleotide is added) and a template (to specify the sequence of bases in the newly synthesized, complementary strand). All of the properties of DNA polymerase III are consistent with its being the major enzyme of DNA replication. Mutants lacking DNA polymerase I can synthesize DNA but are unusually sensitive to ultraviolet light, which therefore suggests that one of the functions of this polymerase is repair of damaged DNA. DNA polymerase II also seems to be dispensable, and its function is unknown.

DNA chains synthesized by DNA polymerases always grow by addition of nucleotides to the 3′-end,

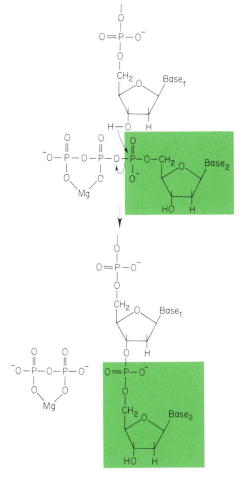

Fig. 11.11 Growth of a DNA primer from the 3′-end, in the reaction catalysed by DNA polymerase.

the template being copied from its 3′- to its 5′-end (Fig. 11.11): no polymerase is known which adds nucleotides to 5′ ends.

Priming of DNA synthesis

Nucleotides newly incorporated into DNA first appear in short fragments, which are later joined, so that the strand with overall 3′- to 5′- growth is extended discontinuously, by retrograde synthesis of these fragments (i.e. *toward* the replicating fork). In principle, the other strand might grow continuously, but it, too, appears to be synthesized in short stretches, at least in prokaryotes. The newly-synthesized fragments are called 'Okazaki pieces', after their discoverer, and are about 1000 nucleotides long in bacteria, but not more than 200 nucleotides long in eukaryotes. Synthesis of each fragment is initiated with a stretch of RNA, about five nucleotides long, laid down by an RNA polymerase ('primase') which, unlike DNA polymerases, does not require a primer; this RNA primer is then extended by DNA polymerase III. Finally, the RNA primer is

removed, and replaced by DNA: DNA polymerase I is able to do this, since, although it has only a single polypeptide chain, it has three separate enzymic activities: polymerase, 5′ → 3′ nucleotidase and 3′ → 5′ nucleotidase. The enzyme can digest an RNA primer from its 5′-end, while extending the preceding Okazaki piece from its 3′-end. An endonuclease, ribonuclease H, which can remove RNA sequences from mixed RNA/DNA polymers, may also function in DNA replication. The action of these enzymes may produce a newly synthesized chain which has no missing nucleotides, but which contains some breaks, in which a 5′-phosphate is next to a 3′-hydroxyl: such breaks can be closed by the action of DNA ligase, an enzyme which uses the energy derived from the hydrolysis of ATP to $AMP + P\text{-}P_i$ or, in some bacteria, NAD^+ to $NMN + AMP$. The events of DNA replication are shown schematically in Fig. 11.12.

Unwinding proteins

In addition to these enzymes, several other proteins are required for DNA replication in *E. coli*: most were discovered by genetic methods, and not all have been characterized biochemically. Furthermore, some of these proteins appear to be dispensable for the replication of some, but not all, viral DNAs. Of the proteins of known function, two are concerned with unwinding DNA. 'Unwinding protein' unwinds the double helix by binding tightly to single-stranded DNA ahead of the replicating fork. In closed circular double stranded DNA, unwinding causes supercoiling in other parts of the genome: these supercoils are removed by 'topoisomerases', which introduce cuts in one strand, then rejoin the ends after rotation of the strands about each other.

DNA replication in eukaryotes

These proteins of DNA replication have been identified in bacteria: far less is known about the enzymology of DNA replication in eukaryotes, although it is presumably similar. Eukaryotic nuclei contain at least 2 DNA polymerases, one of which functions in DNA repair; a different DNA polymerase is found in mitochondria. Bacterial replication is faster than that in eukaryotes (about 10^5 nucleotides per minute, compared to about 3×10^3 nucleotides per minute at each eukaryotic replicating fork), but the replication of eukaryotic chromosomes involves the synthesis and attachment of many proteins, such as histones: this seems to be closely coupled to DNA synthesis.

Viral genome replication

Since the chromosomes of many viruses contain single-stranded DNA or even RNA, the mechanism of

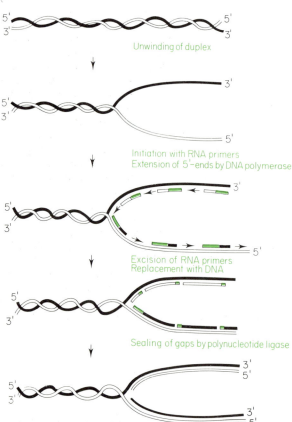

Fig. 11.12 DNA replication: the sequence of events at the replicating fork.

Unwinding of duplex

Initiation with RNA primers
Extension of 5′-ends by DNA polymerase

Excision of RNA primers
Replacement with DNA

Sealing of gaps by polynucleotide ligase

genome replication must be somewhat different from the replication of double-stranded DNA already discussed. In those viruses containing single-stranded DNA (e.g. bacteriophages ϕX174 and M13) a complementary DNA strand is synthesized immediately after infection; the resultant double-stranded DNA is then replicated in the normal way, and the new copies used as templates for the synthesis of new single-stranded progeny DNA. On the other hand, replication of the single-stranded RNA which constitutes the genome of the simple bacteriophages R17, MS2 and Qβ does not involve DNA at all. On infection, an RNA polymerase ('replicase'), translated directly from the viral genome, synthesizes a complementary RNA strand, which then serves as a template for new rounds of viral RNA synthesis in a similar way to the DNA-directed synthesis of DNA (although RNA duplexes are unstable, and only occur as transient intermediates in RNA replication).

Tumour viruses

Genome replication in RNA tumour viruses (such as Rous sarcoma virus and avian myeloblastosis virus) operates differently: immediately following infection, single-stranded DNA, complementary to the RNA template, is synthesized by a 'reverse transcriptase' which has been translated from the viral genome. This single-stranded DNA is then made double-stranded by the action of DNA polymerase; the double-stranded DNA, known as a provirus, integrates into the host chromosome. It can then be transcribed back into RNA, which codes for viral proteins. Viruses which employ this mode of replication are called 'retroviruses'.

Although they differ in detail, the mechanisms by which these viral genomes are replicated are based on the same principles as DNA replication: each new polynucleotide strand is synthesized, from 5′- to 3′-end, from nucleoside triphosphates, the base sequence being dictated by that in the complementary template, and accuracy being maintained by extremely specific base-pairing.

Mutations

A mutation is a structural alteration in the genome of an organism, which can lead to an alteration in the sequence of a protein translated from it. In many bacteria, the mutation rate is about one detectable mutation in 10^6 gene duplications: if a gene is about 10^3 nucleotides long, and a change in any of these is detectable, this suggests that errors in DNA replication are made only once in 10^9 base-pairs, a value much lower than expected from considerations of the chemistry of base-pairing, and the likelihood of mismatching. It is probable that errors in DNA synthesis are relatively

frequent, but are corrected by the 3′ → 5′ exonuclease activity which is a feature of most DNA polymerases; mutant DNA polymerases with low 3′ → 5′ exonuclease activities are less accurate, and the reverse transcriptase of RNA tumour viruses, which has no exonuclease activity, has an error frequency of 1 in 600.

Substituted bases, such as 5-bromouracil, a thymine analogue, may be incorporated into DNA and bring about errors of base-pairing during replication. Mutations may also be caused by chemical agents such as nitrous acid, which deaminates bases: adenine is converted to hypoxanthine, which forms H-bonds with cytosine, rather than uracil, and adenine is deaminated to uracil. Nitrogen mustards, such as $CH_3N(CH_2CH_2Cl)_2$, form covalent bonds between neighbouring guanines, and ultraviolet light can also cause cross-linking, neighbouring thymidine bases reacting to form 'thymidine dimers'. Mutagens are useful experimentally, and they occur widely: about 85% of substances known to be carcinogens have been shown to be mutagens.

Most cells are to some extent able to recognize and excise damaged regions of DNA, and to repair the strand; a specific endonuclease makes a cut in one strand near, for example, a thymidine dimer, an exonuclease (which may be one of the DNA polymerases) excises the damaged region, DNA polymerase replaces the missing nucleotides, and DNA ligase joins the break in the strand. In a rare inborn error of metabolism, xeroderma pigmentosa, one of the repair enzymes is missing, and there is a greatly increased sensitivity of the skin to sunlight, and a high incidence of skin cancer.

The mutagens discussed above tend to cause 'point mutations', alterations in a single base resulting in a single amino acid replacement in the translated protein; other mutagens, such as the acridine dyes, become inserted between the stacked bases of DNA and cause mismatching. This results in insertions or deletions of nucleotides during replication, and errors in translation which affect long amino acid sequences (p. 208). These are called 'frameshift mutations', and they can also arise as a result of 'crossing over', in which two homologous double-stranded DNA molecules break, and reunion takes place between the strands of different molecules. Such crossing over usually takes place with fidelity, but occasional mismatching can result in the insertion or deletion of nucleotides.

Selective inhibition of viral DNA replication

In the replication of many viruses, synthesis of viral nucleic acid requires enzymes which are themselves virally encoded. This offers the possibility of combating viral infections with drugs which inhibit viral enzymes, but which do not affect enzymes of the host cell. One of the most successful examples is acyclovir, which is active against herpes, though not against vaccinia, adenoviruses or RNA viruses. Acyclovir is an analogue of guanosine, in which ribose is replaced by a hydroxymethoxymethyl group:

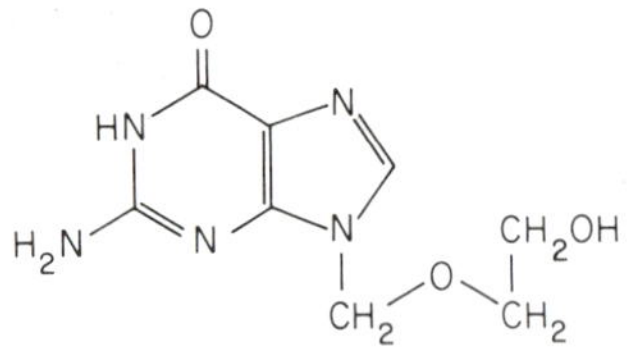

It is taken up by cells, and phosphorylated by thymidine kinase to the monophosphate (an analogue of dGMP). This is further phosphorylated to the triphosphate, which inhibits DNA replication.

The specific effects of acyclovir on virus-infected cells derive from the fact that it is a very poor substrate for mammalian thymidine kinase, while the viral enzyme has a 200-fold lower K_m for acyclovir, and a much higher V_{max} (p. 27). Furthermore the triphosphate is a good substrate for viral DNA polymerase, but not for the mammalian polymerase. It is incorporated into nascent viral DNA, and since it lacks the 2'-hydroxyl of dGTP, it causes chain termination. The viral DNA polymerase remains bound to the terminated DNA, and is thereby inactivated.

Transcription of DNA into RNA

Types of transcript

Before it is used in protein synthesis, the sequence information in DNA is always transcribed into RNA. There are three types of transcript, only one of which is translated into protein. *Ribosomal RNA* (rRNA) is a structural component of *ribosomes*, complexes of RNA and protein which are the sites of protein synthesis in the cytoplasm. Up to 80% of the RNA in a cell is rRNA, but only a small fraction of the DNA codes for it. Bacteria contain only a few rRNA genes, and eukaryotes have 50–1000 identical copies of the gene, located in a special structure called the nucleolus. Ribosomal RNA does not contain any protein sequence information, and ribosomes must therefore be programmed with this information before they can polymerize amino acids. This function is performed by *messenger RNA* (mRNA): it accounts for only 1–5% of the cellular RNA, but, as is shown by RNA–DNA hybridization experiments (p. 198), about 90% of the genome codes for mRNA, since a different message is required for each type of protein synthesized. The third form of RNA, *transfer RNA* (tRNA) acts as an adaptor for amino acids during protein synthesis: it comprises some 10–15% of total cellular RNA and, like rRNA, is encoded by reiterated genes—there are about 8000, for some 60 different types of tRNA, in the toad *Xenopus*, for example.

In the chromosomes of prokaryotes, the genes for rRNA, mRNA and tRNA are separated by short non-informational sequences only; but eukaryotic chromosomes contain, in addition to multiple copies of some genes, long non-coding regions, and many eukaryotic mRNA genes are interrupted by untranslated regions (introns—see p. 204).

Mechanism of transcription

All forms of RNA are synthesized by the same mechanism, although in eukaryotes different enzymes are responsible for making rRNA, mRNA and tRNA. DNA-directed RNA synthesis is catalysed by a ubiquitous enzyme, RNA polymerase, which requires all four nucleoside triphosphates (ATP, GTP, CTP and UTP), a divalent cation (Mg^{2+} or Mn^{2+}) and a DNA template. It is able to initiate the synthesis of new chains of RNA without a primer, and, like DNA polymerase, adds nucleotides to the free 3'-hydroxyl. Thus the nascent RNA chain grows at its 3'-end, whilst the 5'-end carries the triphosphate of the initiating nucleotide. The product is complementary to the DNA template, with uridine base-pairing to adenine, and cytosine to guanosine. In vitro, particularly if partially denatured DNA is used as the template, both strands may be transcribed. However, since the complementary strands of DNA have different sequences, if both were transcribed and translated in vivo, two different proteins might be made. Genetic evidence shows clearly that each gene codes for only one protein, so only one strand of DNA should be transcribed. This was confirmed with bacteriophages in which the base compositions of the complementary DNA strands are different enough for the strands to be separated by caesium chloride density gradient centrifugation; hybridization with RNA transcribed from the original duplex showed that only one DNA strand had been transcribed. However, not all RNA is transcribed from the same DNA strand. In some viruses (e.g. bacteriophage λ) and in *E. coli*, different genes lie on different DNA strands. Since the DNA strands are antiparallel, transcription of genes on different strands must be in different directions.

RNA polymerase

Bacterial RNA polymerase has a complicated structure, involving three types of subunit, known as α, β and β' (mol. wt 40 000, 150 000 and 160 000, respectively) in the stoichiometry $\alpha_2\beta\beta'$. This complex is called the 'core enzyme'; a third type of subunit, σ (mol. wt 90 000) is required for specific RNA chain initiation, $\alpha_2\beta\beta'\sigma$ being called the 'holoenzyme'.

Synthesis of RNA begins with attachment of RNA polymerase to the DNA duplex at the beginning of the gene to be transcribed. The attachment site is called the promoter, and is about seven nucleotides long. Promoters from many bacterial and viral genes have been found to have related sequences. Only the holoenzyme ($\alpha_2\beta\beta'\sigma$) attaches specifically to promoters, although the core enzyme can initiate non-specifically by random attachment to DNA. The promoter sequence is not itself transcribed into RNA: transcription begins 6–7 bases further on, the first nucleotide in the RNA always being a purine, so that new RNA chains, unless they are modified after transcription, begin pppA... or pppG.... Elongation of the nascent RNA chain occurs without the σ subunit, which dissociates from the RNA polymerase complex. Termination and release of the completed RNA chain may require another subunit, ρ, which also attaches transiently to RNA polymerase.

Bacterial RNA polymerases are inhibited by the rifamycin class of antibiotics, of which rifampicin is a member. The mushroom *Amanita phalloides* contains the poison α-amanitin, which inhibits only eukaryotic RNA polymerases and has been useful in proving the existence of three separate enzymes. Polymerase 1, located in nucleoli and responsible for the synthesis of rRNA, is insensitive to α-amanitin; polymerase 2 (sensitive to low concentrations of α-amanitin), and polymerase 3 (sensitive only to high concentrations) are located in the nucleoplasm, and synthesize mRNA and tRNA, respectively.

In prokaryotes, most controls of protein synthesis operate on transcription, genes being switched on and off by specific proteins which bind near the promoter and which affect the initiation of mRNA synthesis. In multicellular organisms cell differentiation involves, in addition, different rates of transcription of different parts of the chromosome. These controls are discussed in chapter 13.

Actinomycin D, an antibiotic from *Streptomyces*, binds tightly to DNA and inhibits transcription in both prokaryotes and eukaryotes.

Determination of nucleotide sequences in nucleic acids

Sequence determination in nucleic acids is an apparently formidable task: since there are usually only four types of constituent nucleotide, a random sequence would, on average, contain the same triplet coding every 64 nucleotides. The problem of sequencing has, however, been solved, first for tRNA, then for viral mRNAs, and more recently for DNA.

RNA sequences

The strategy first used to sequence RNA was similar to that developed for proteins; the molecule was cleaved with specific endonucleases, the fragments separated by electrophoresis, and individual fragments sequenced using base analysis and endo- or exonuclease treatment. The order of fragments was determined from overlaps derived from treatment of the original RNA with a different endonuclease. The specificities of some ribonucleases are shown in Fig. 11.4.

In sequencing transfer RNAs the matching of overlapping fragments was made easier by the frequent occurrence of unusual bases.

DNA sequences

Although there are no simple base-specific DNases, the availability of restriction endonucleases made possible the determination of nucleotide sequences in DNA. These enzymes are produced by many bacteria, and degrade exogenous DNA, such as viral DNA. They make double-stranded cuts within specific sequences of

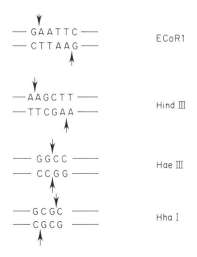

Fig. 11.13 Nucleotide sequence specificity of restriction endonucleases. Note that in most cases the cuts on complementary DNA strands are not directly opposite each other, and that fragments therefore have 'sticky' ends.

4–6 nucleotides; since these sequences do not occur within the chromosome of the producer cell, the bacterial genome is not attacked, but the invasive DNA may be cut in a few places. The sequence specificities of a few restriction endonucleases are shown in Fig. 11.13; many enzymes, of differing specificity, are now available. In most cases the cuts are staggered, so that the ends of the fragments have short, single-stranded regions, and will stick to each other by base-pairing. This 'stickiness' is important in cloning DNA.

Restriction endonucleases are used to cleave DNA molecules into large but manageable fragments; a number of techniques for sequencing these have been developed. All involve radioactive labelling of the fragment with ^{32}P, and generation of a set of sub-fragments extending from the endonuclease cut to one of the points of occurrence of a specific base. These are separated electrophoretically according to size, and detected by autoradiography. The nucleotide sequence can be obtained directly from the autoradiograph, and the sequence of over 200 nucleotides can be obtained from a single experiment.

Sequences in viral genomes

The nucleotide sequences of many viral, bacterial and eukaryotic genes are now known; some of the results were initially very surprising. For instance, phage ϕX174 has a single-stranded, closed circular DNA chromosome of 5375 nucleotides, which codes for nine proteins. The total molecular weight of these proteins suggests that at least 6500 nucleotides would be needed to specify them. When the nucleotide sequence of the viral genome was obtained, the genes for proteins D and J were found to be adjacent, although genetic evidence suggested the order D–E–J. The nucleotide sequence for protein E was found to be within the D gene, but read in a different frame. Similarly, the nucleotide sequence for protein B was continued within the gene for protein A, again read in a different frame. The device of overlapping genes may be confined to viruses, which, because of their small size, must minimize the amount of nucleic acid that they contain. It obviously places tight constraints on sequences, since a mutation in a region of the genome where two genes overlap may affect two proteins simultaneously.

Interrupted genes

Comparison of the nucleotide sequences of eukaryotic genes with those of the corresponding RNA transcript

reveals that genes are frequently made up of translatable sequences (exons) interrupted by untranslated regions (introns), for which complementary sequences are not found in the mRNA. This is also demonstrable by electron microscopy of DNA–RNA hybrids: the introns are seen as looped out, nonhybridized regions of DNA,

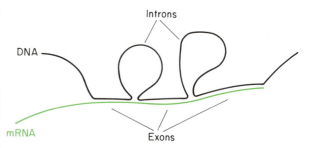

Fig. 11.14 Appearance of a DNA–mRNA hybrid in which the DNA contains introns. Sequences corresponding to the introns are not present in mRNA, and the introns therefore appear as single-stranded loops in the hybrid.

between the exon-RNA hybrid regions (Fig. 11.14). The genes for *Drosophila* rRNA, yeast tRNA^{tyr} and rabbit globin all contain a single non-translatable insertion; the gene for human preproinsulin contains two introns (Fig. 11.15) and that for chicken ovalbumin, six. Introns have also been found in viral genes (SV40 and adenovirus) and in some yeast mitochondrial genes, but not in prokaryotes. Their function is unknown. It appears that each gene is transcribed in its entirety, the introns subsequently being excised during processing of the RNA. Since most if not all eukaryotic messenger RNA first appears as heterogeneous nuclear RNA, extensive post-transcriptional processing must always occur (see p. 208).

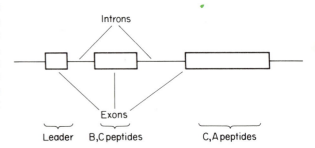

Fig. 11.15 Arrangement of exons and introns in the human preproinsulin gene. The three exons, shown as boxes, are translated into consecutive sequences in the polypeptide.

Cloning

Cloning, or 'genetic engineering', is the introduction of genetic material from one organism into another, where it may be expressed. This may be done for experimental purposes, in order to study the expression of a particular gene in a convenient system; or in order to transform the recipient organism genetically, so that it acquires a new biological activity through synthesis of a new protein. This offers the possibility of 'repairing' genetically defective cells, or of obtaining strains of microorganisms which synthesize and secrete proteins of clinical use, such as insulin or interferon.

Transfer of genetic material between eukaryotes and prokaryotes has been performed in both directions: mouse DNA has been introduced into *E. coli* using a bacteriophage or bacterial plasmid as vector, and *E. coli* DNA has been put into cultured mouse cells using polyoma virus as a vector. The vector is a small piece of DNA which may be isolated, covalently combined with the DNA from the donor cell, and then introduced into the recipient; it must have the property of being readily taken up and expressed by the recipient cell. Suitable vectors are viral DNA or, in the case of bacteria, plasmids. Plasmids are small, extrachromosomal, closed-circular DNAs, which exist within the

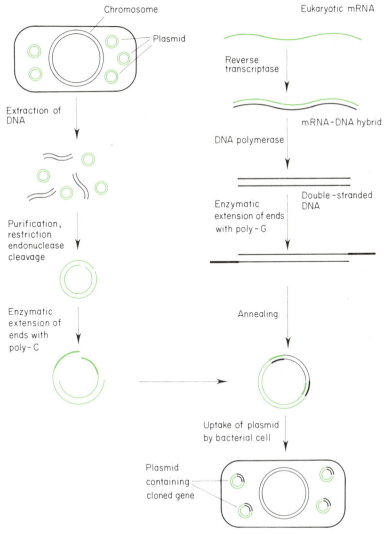

Fig. 11.16 Scheme for cloning of a eukaryotic gene in a bacterium. The mRNA for the protein required is transcribed into complementary DNA, which is introduced into the bacterium by means of a plasmid vector.

bacterial cell, and are replicated independently of the main chromosome. They frequently carry antibiotic resistance genes, a property which aids the selection of transformed cells. Thus the plasmid ColE1 contains the gene for the antibiotic protein colicin E1. It exists in about 20 copies per cell, although the number may be artificially increased to about 2000, and it can be easily isolated. This plasmid contains about 6500 nucleotides, and a single site where cleavage by the restriction endonuclease E CoR I can occur: it has proved useful in the cloning of several different eukaryotic genes in *E. coli*. A simplified scheme for the cloning of a mammalian gene in a microorganism, using a plasmid as a vector, is shown in Fig. 11.16.

Isolation of genes

The first problem to be solved is identification and isolation of the DNA fragment required for cloning. Although it must exist somewhere within the chromosome (or one of the many chromosomes, in higher organisms) it will be only a tiny fraction of the total complement of DNA within the cell, and furthermore, in eukaryotes the sequence information may not be continuous, but interrupted by untranslated regions. If the messenger RNA for a particular protein can be isolated (which is often possible, since mRNA can be assayed in vitro using a cell-free protein synthesizing system), it can be used to select the correct DNA fragment by hybridization, or it may be transcribed by reverse transcriptase into complementary DNA, which can be converted enzymatically to its double-stranded form, and then cloned.

Combination with the vector

The selected DNA fragment is combined with the vector, which has itself been cleaved with a restriction endonuclease: this in vitro recombination requires that the ends of the fragments are 'sticky'. The ends of restriction fragments are normally of this type (Fig. 11.13), or the fragments to be joined may be made 'sticky' by other means, e.g. the enzymatic attachment of complementary nucleotide sequences (Fig. 11.16). The recombined strands are covalently joined, and the vector introduced into the host; the transformed host cells must now be selected. This may be done by checking whether DNA derived from individual clones can hybridize with the appropriate radioactive (^{32}P) mRNA, or by looking for production of the required protein, e.g. by the ability to react with specific antibodies, raised against that protein.

Expression of cloned DNA

A number of problems may be associated with expression of the cloned gene by the recipient cell. In particular, the gene must possess recognizable regulatory signals: to start synthesis of mRNA, and to terminate transcription, and, within the mRNA itself, to initiate and terminate translation. All bacterial genes contain such signals, but those associated with eukaryotic genes are not recognized in bacteria. One solution is to insert the cloned DNA into the middle of a bacterial gene, thus using existing regulatory signals; but the product of translation will then be a hybrid protein, with additional amino acid sequences (derived from the bacterial gene) at either end. These sequences may be removed after the protein is isolated. This approach was used successfully in cloning the DNA sequence for somatostatin, a peptide hormone containing only fourteen amino acids. Since the amino acid sequence was known, the corresponding DNA was synthesized chemically, rather than isolated, and this DNA sequence, followed by a termination codon, inserted near the end of the gene for a large bacterial protein. The hybrid protein synthesized by the bacterium was isolated, and the somatostatin sequence cleaved off chemically.

Preproinsulin has been produced in a similar way, by cloning complementary DNA produced by reverse transcription of pancreatic β-cell mRNA; this sequence was inserted into a bacterial gene for penicillinase, carried on a plasmid. Penicillinase is secreted by the bacterium, so the hybrid protein synthesized by the clone carried the penicillinase 'signal' sequence (p. 220) at its N-terminus, and appeared in the culture medium. The unwanted bacterial sequences were removed from the penicillinase/preproinsulin hybrid by treatment with trypsin.

Applications of cloning

Cloning techniques may also be applied to the production of viral coat proteins, which could be used for immunization of humans against, for example, rabies or hepatitis; and to the production of interferon (p. 218), which may be of value in the treatment of viral infections and possibly cancer. It appears that the 'signal' sequences present in some eukaryotic secretory proteins can be recognized by the bacterial membrane enzymes responsible for processing pre-proteins, so that even the job of removing the leader sequence may be performed by the bacteria. Unfortuately many potentially useful proteins are glycosylated, and attachment of carbohydrate moieties (a process which occurs in the endoplasmic reticulum and Golgi apparatus of eukaryotic cells) is unlikely to occur in bacteria. Although interferon is a glycoprotein, its biological activity does not depend upon its being glycosylated; possibly the carbohydrate is necessary to ensure the long survival of interferon within the body.

12

Protein Synthesis: Translation

The Structure of Ribosomes

Ribosomes are ribonucleoprotein complexes which sequentially polymerize amino acids to make protein. They are found in the cytoplasm, within mitochondria and, in higher plants, chloroplasts, but not within other organelles (including the nucleus, where most DNA is situated). If cells are exposed to a short pulse of radioactive amino acid and then fractionated, incomplete radioactive polypeptides are found associated with ribosomes, which appear in the electron microscope as approximately spherical particles, 22 nm across, and are very plentiful—there are about 15 000 in an *E. coli* cell, making up 20% of the cell mass.

Single ribosomes are dimers of a large and a small ribosomal subunit, each subunit containing a different rRNA, and many proteins. The complement of proteins, sizes of rRNA molecules and molecular weights of the ribosomal subunits are different in prokaryotes and eukaryotes (Table 12.1), eukaryotic ribosomes being the more complex. Ribosomes, ribosomal subunits and rRNA molecules are often referred to by their sedimentation constants (expressed in svedbergs, abbreviated to S) which are related, though not proportional, to their molecular weights.

Bacterial ribosomes

In bacteria, the large (50 S) ribosomal subunit contains one molecule of 23 S rRNA, plus one other much smaller rRNA (5 S) and about 34 different proteins,

each probably being present in only one copy; the small (30 S) subunit contains 16 S rRNA and about 21 proteins. The function of the rRNA is presumably to orientate the proteins and maintain their conformation, and probably also to bind mRNA and tRNA (see below). Studies of the physical properties of rRNA suggests that it has a complicated secondary structure, containing many base-paired 'hairpins' of unequal size, and inspection of its nucleotide sequence reveals many regions where such hairpins are possible. Very few of the ribosomal proteins have been assigned functions although one, in the large ribosomal subunit, is the enzyme catalysing peptide bond formation during protein synthesis, and genetic studies have identified others as the sites of binding of certain antibiotics, resistance to a given antibiotic being associated with an amino acid sequence change in a particular ribosomal protein.

Ribosome synthesis

The ribosomal RNAs of bacterial and eukaryotic cytoplasm (but not mitochondria) are transcribed as a single precursor (pre-rRNA) of sedimentation constant 30 S in bacteria, 45 S in eukaryotes, from which the two large rRNA molecules are produced by nuclease digestion. The 5.8 S rRNA present in eukaryotic ribosomes is also derived from the precursor, but 5 S rRNA is transcribed from separate genes.

Table 12.1 Properties of ribosomes and ribosomal subunits

		Bacteria	Eukaryotes	Mitochondria
Ribosome	Sedimentation constant	70 S	80 S	55 S
	Molecular weight	2.8×10^6	4.5×10^6	2.8×10^6
Large ribosomal subunit	Sedimentation constant	50 S	60 S	39 S
	Molecular weight	1.8×10^6	3.0×10^6	1.7×10^6
	Proteins	34	?45	?50
	rRNA	23 S, 5 S	28 S, 5.8 S, 5 S	16 S
Small ribosomal subunit	Sedimentation constant	30 S	40 S	30 S
	Molecular weight	1.0×10^6	1.5×10^6	1.1×10^6
	Proteins	21	33	?25
	rRNA	16 S	18 S	12 S

Ribosomes, like many multisubunit proteins and viruses, assemble spontaneously from their constituents. The assembly of the ribosomal subunits has been studied in vitro by mixing purified rRNA, and purified ribosomal proteins: the final, completely assembled particles behave exactly like those isolated from bacteria. This is another approach to the problem of assigning functions to particular ribosomal proteins, although omission of one protein from the mixture may cause several others to be missing from the assembled particle, so interpretation of the results can be complicated. Ribosome assembly follows a defined pathway, with proteins added in a particular order: it even seems to begin before processing of the pre-rRNA is complete.

Messenger RNA

The cytoplasmic ribosomes within a particular organism are probably identical (although they differ from those in mitochondria) and contain no protein sequence information; before synthesizing a protein, they must be programmed for its amino acid sequence. Following phage infection of bacterial cells there is no synthesis of new ribosomal RNA, but immediate synthesis of viral proteins occurs on existing ribosomes. Obviously the virus supplies the programme to subvert the protein synthesis machinery of the cell. The new programme is encoded in another form of RNA, messenger RNA (mRNA), the base sequence of which codes for the sequence of amino acids in the protein, and is translated by the ribosome. Messenger RNAs are quantitatively a minor component of the cell, and tend to be short-lived (chapter 13); their discovery and purification was therefore difficult, the first to be isolated being stable mRNAs from cells producing mainly one protein (e.g. globin mRNA, from reticulocytes). Using purified or even synthetic mRNAs, ribosomes can be re-programmed to make protein in vitro; such experiments were crucial in the solving of the 'genetic code'.

Base sequences in mRNA

The nucleotide sequences of many mRNAs have now been determined. Messenger RNAs vary greatly in size, reflecting the different lengths of the polypeptides for which they code. In addition to the region containing amino acid sequence information, virtually all messenger RNAs contain untranslated regions at the 5'- and 3'-ends. Most eukaryotic mRNAs are found to contain 7-methyl guanosine, linked through a triphosphate, at the 5'-end. This 'cap' is not coded by DNA, but is added after transcription of the gene, by the action of specific enzymes. At the 3'-end there is usually a tail of up to 150 adenylyl residues (poly A), also added after transcription. The functions of the 'cap' and the poly A 'tail' are still obscure; neither is found in bacterial mRNAs, while some eukaryotic mRNAs lack one or the other, and yet are apparently translated normally. The mRNA of poliovirus has no cap, but does have poly A at its 3'-end; since this mRNA is made in the cytoplasm, the addition of poly A cannot be connected with export of mRNA from the nucleus. Poly A has been suggested to stabilize the mRNA, because there is some correlation between the length of the poly A tail and the lifetime of mRNA. Histone mRNA is an example of a message which has no poly A at all.

Messenger RNA, like ribosomal RNA, is transcribed in eukaryotes as a precursor which first appears as heterogeneous nuclear RNA (HnRNA), a major fraction of RNA in the nucleus. This has molecular weight between 10^5 and 10^7, and a short lifetime. It contains poly A, and appears to be processed into mRNA; however, a large fraction (up to 90%) of HnRNA never leaves the nucleus.

The Genetic Code

Number of bases in a codon

The amino acid sequences of proteins are encoded in the nucleotide sequences of the genes from which they are translated. Obviously since there are only four types of nucleotide in DNA and RNA, there must be at least three bases in a codon (the set of nucleotides in mRNA which specifies one amino acid): this gives 64 (4^3) different codons for the 20 amino acids.

The triplet nature of the code was confirmed by genetic experiments, in particular by the use of 'frame-shift' acridine mutagens (p. 201). Insertion or deletion of a base invariably results in a totally inactive gene product: this suggests that codons are read sequentially along the mRNA—a point mutation affects only one codon, but insertion or deletion of a base alters the reading frame, affecting the translation of every

AUG Met	GUC Val	G**G**A Gly	ACU Thr	GAU Asp	AUA Ile	GUC Val	AAU Asn	UCC Ser	GUC Val	AAG Lys	CCG Pro
AUG Met	GUC Val	GAA Glu	CUG Leu	**A**UA Ile	UAG Stop	UCA Ser	AUU Ile	CCG Pro	UCA Ser	AGC Ser	CG
AUG Met	GUC Val	GAA Glu	CUG Leu	UAU Tyr	AGU Ser	CA**A** Gln	UUC Phe	CGU Arg	CAA Gln	GCC Ala	G
AUG Met	GUC Val	GAA Glu	CUG Leu	UAU Tyr	AGU Ser	CAU His	UCC Ser	GUC Val	AAG lys	CCG Pro	

Fig. 12.1 Frame-shift mutations caused by deletions from a gene. Deletion of one nucleotide (G, from the third codon in line 1) gives nonsense: in this case the misread message contains a stop codon. Deletion of a second nucleotide also gives nonsense, but a third deletion restores the original protein sequence, apart from a short mis-sense region (underlined).

codon after the mutation. Partially-functional genes can sometimes result from an insertion near to a deletion, and a combination of three (but not two or four) deletions *or* three insertions can also produce an active gene product. This is shown in Fig. 12.1, where deletion of three bases results in the subtraction of one codon: the resultant protein has a limited region of misreading, which may not affect its activity.

Identities of codons

The codon assignments were made using synthetic polynucleotides as messengers. An RNA-like polymer containing only uridine (poly U) stimulates ribosomes to polymerize phenylalanine: the codon for phenylalanine is therefore UUU. Other assignments were made using heteropolymers, with known or statistically predictable codon contents. Even sequences of just three bases (single codons, in other words) stimulate the appropriate aminoacyl-tRNA to bind to ribosomes (see below), allowing unambiguous identification of the amino acid corresponding to each codon (Table 12.2).

The translatable codons specify the 20 amino acids commonly found in proteins. Asparagine and glutamine have different codons from aspartate and glutamate, but there is no codon for cystine, which is formed by oxidation of a pair of cysteines in the completed protein. Any unusual amino acid must be made by post-translational modification of the protein.

Only 61 of the 64 codons specify amino acids; the remaining three, UAA, UAG and UGA, are termination codons, which cause release of the nascent polypeptide chain.

Table 12.2 The genetic code: identity of codons in mRNA (read in the direction $5' \rightarrow 3'$)

First base	Second base				Third base
	U	C	A	G	
U	Phe	Ser	Tyr	Cys	U
	Phe	Ser	Tyr	Cys	C
	Leu	Ser	Stop	Stop[2]	A
	Leu	Ser	Stop	Trp	G
C	Leu	Pro	His	Arg	U
	Leu	Pro	His	Arg	C
	Leu[3]	Pro	Gln	Arg	A
	Leu	Pro	Gln	Arg	G
A	Ile[4]	Thr	Asn	Ser	U
	Ile	Thr	Asn	Ser	C
	Ile[4]	Thr	Lys	Arg[5]	A
	Met[1]	Thr	Lys	Arg[5]	G
G	Val	Ala	Asp	Gly	U
	Val	Ala	Asp	Gly	C
	Val	Ala	Glu	Gly	A
	Val	Ala	Glu	Gly	G

[1] May act as initiator codon, specifying met-tRNA$_{met}^{F}$.
[2] Read as trp in all mitochondria, except those of plants.
[3] Read as thr in yeast mitochondria, but leu in human mitochondria.
[4] Read as met in human mitochondria.
[5] Termination codons in human mitochondria.

If one examines the amino acid sequences of mutant forms of a protein, such as the abnormal haemoglobins, the amino acid replacements are usually seen to have arisen from single nucleotide changes: e.g. in sickle-cell haemoglobin (HbS), glutamic acid at position 6 of the β-chain is replaced by valine, the result of the codon change GAA → GUA. However, in 'nonsense' mutants, where an amino acid codon is altered to a 'stop' codon, a shortened polypeptide is produced: this is almost always inactive.

Direction of reading

The direction in which mRNA is read was also established using synthetic polynucleotides such as AUUUUUUUU..., which has the codon AUU (isoleucine) at its 5'-end, followed by repeated UUU (phenylalanine) codons. The message is read from its 5'-end, the first codon translated specifying the N-terminal amino acid of the protein (see p. 213).

Nature of the genetic code

The genetic code is 'commaless', the message being a series of contiguous codons without spacers or punctua-

tion; it is unambiguous, as there is only one amino acid corresponding to each codon; and it is degenerate, some amino acids being specified by as many as six codons. In many cases degeneracy arises through the equivalence of G and C, and of A and U, in the third nucleotide of the codon. Such equivalent codons can usually be read by the same transfer RNA, a phenomenon known as wobble (p. 212).

Although the code appears to be universally valid in prokaryotes and eukaryotic cytoplasm, there are some differences in the genetic code used in mitochondria: e.g. UGA, which is a 'stop' codon in cytoplasmic protein synthesis, codes for tryptophan in mitochondria. Many of the other codons, including the 'stop' codon UAA, have their usual identities, but there are some differences between the codes used in mitochondria of different species: these are listed in Table 12.2.

Transfer RNA

Although the sequence of nucleotides in messenger RNA determines the amino acid sequence of the protein synthesized by the ribosome, amino acids cannot interact directly with the messenger; codons are recognized by base-pairing to transfer RNAs (tRNAs), which carry covalently-attached amino acids. Transfer RNAs thus act as adaptors or 'labels' for amino acids, enabling them to be recognized correctly by the protein-synthesizing machinery; there is at least one tRNA for each amino acid, though not as many as one for each codon. The reason for this discrepancy is explained on p. 000. Once an amino acid has become attached to its tRNA, its identity is specified only by the tRNA. This was shown by attaching cysteine to its appropriate tRNA (written $tRNA^{cys}$) then reductively removing the sulphydryl group of the amino acid, to give alanine:

$$cys + tRNA^{cys} \longrightarrow cys - tRNA^{cys} \xrightarrow{\text{Ni/H}_2}$$
$$ala - tRNA^{cys} + H_2S$$

The reduction had no effect on the tRNA, but when the modified aminoacyl-tRNA was used by an in vitro system synthesizing haemoglobin, alanine was inserted into positions normally occupied by cysteine, having been recognized only by the $tRNA^{cys}$ to which it was attached.

Folding of transfer RNA

Transfer RNAs are relatively short (75–80 bases, giving a molecular weight of about 25 000). Studies of their physical properties suggest that tRNA molecules have

a complex secondary structure, with extensive internal base-pairing. It is possible to account for the base-pairing by a number of structures, including the 'cloverleaf' shown in Fig. 12.2. This structure maximizes the base-pairing, and similar structures can be written for all of the tRNAs so far sequenced; it is thought that they all have similar three-dimensional structures since mixtures of tRNAs co-crystallize, and tRNAs from one organism can be 'activated' (i.e. have the correct amino acid attached) by enzymes from other species.

The following elements are common to all known tRNA structures:

1. Guanosine at the 5'-terminal, and the sequence -CCA at the 3'-terminal. The amino acid becomes attached through its carboxyl, by an ester link to a ribose hydroxyl of the 3'-terminal adenosine.
2. A 'DHU loop' of 8–12 unpaired bases, always containing the unusual base dihydrouracil.
3. A loop of 7 unpaired bases, containing the sequence TψCG, where ψ is pseudouridine.
4. A loop of 7 unpaired bases containing the anticodon, a sequence of three bases complementary to the codon in the mRNA which specifies the amino acid. The anticodon always has uridine on its 5'-side, and a purine, often methylated, on its 3'-side.
5. A smaller loop, of variable size, between the TψCG and anticodon loops.

The unpaired loops are separated by base-paired 'stems'. The function of the anticodon loop is to base-pair with the appropriate complementary codon in mRNA; the other loops are presumably concerned with

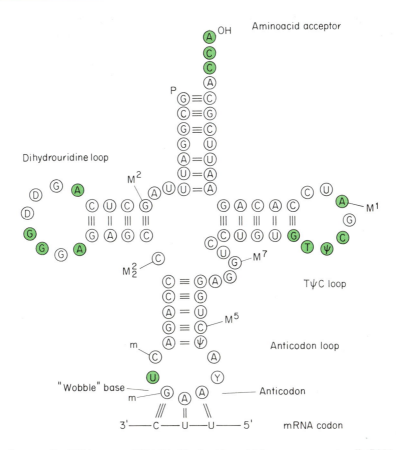

Fig. 12.2 Structure of a transfer RNA, yeast tRNA^phe. Nucleotides which are common to all tRNAs are shaded. ψ, pseudouridine; D, dihydrouridine; T, ribothymidine; Y, a 'hyper-modified' purine; M, base methylation; m, ribose methylation.

binding of tRNA to ribosomes, and with recognition by the amino acid activating enzymes which covalently attack the correct amino acid. The cloverleaf structure has been confirmed by X-ray diffraction studies of tRNA crystals: it is more compact than depicted in Fig. 12.2, since the DHU and TψCG loops interact with each other through a number of hydrogen bonds. The anticodon loop and —CCA terminus protrude at right-angles to each other, the whole molecule being rigid and roughly L-shaped, as shown in Fig. 12.3.

Recognition of codons by tRNA

Since there are 61 different amino acid codons, there could be as many different tRNAs, bearing the appropriate anticodons: for some amino acids, as many as six isoacceptor tRNAs. However, some purified tRNAs can recognize several different codons; this can only be explained if base-pairing between the third base of the codon (at its 3′-end) and the first of the anticodon (at its 5′-end) is less rigid than usual, perhaps because

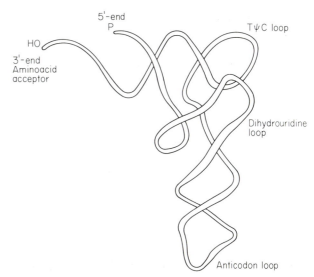

Fig. 12.3 Three-dimensional arrangement of the tRNA chain, determined by X-ray crystallography.

of the distorted structure of the anticodon. This is known as 'wobble' base-pairing: the following base pairs have been found in codon:anticodon binding:

Anticodon	Codon
A	U
C	G
G	U or C
U	A or G
I	A, U or C

'Wobble' base-pairing reduces the number of tRNAs necessary to match all 61 codons, although in some cases strict discrimination in the third base of the codon is necessary (e.g. UGA = stop, UGC = trp; $tRNA^{trp}$ obviously must not recognize UGA). The base inosine (I) occurs frequently in anticodons, permitting the tRNA to recognize up to three alternative codons. In mitochondria, where there are some differences in the genetic code, some tRNAs are able to recognize up to four codons, and the total number of tRNAs required is further reduced, to 23–26. Despite this economy, there is sometimes more than one cytoplasmic tRNA for a given codon—these may be present in different concentrations, and the reason for their existence is unclear. There are no tRNAs corresponding to the termination codons, which are recognized instead by protein 'releasing factors' (p. 216).

Transfer RNAs contain a high proportion of unusual nucleotides, such as inosine, pseudouridine, dehydro-uridine, ribothymidine and various methylated derivatives: about 60 different ones are known, and all are formed by enzymic modification of the tRNA, after transcription. Like ribosomal RNA, tRNAs are synthesized as longer precursors, which are shortened by nuclease digestion to the mature form; e.g. the tyrosine tRNA of *E. coli* is formed from its precursor (pre-$tRNA^{tyr}$) by removal of 39 nucleotides from the 5'-end, and 3 from the 3'-end. Some precursors are even more elaborate, and contain two different tRNA sequences within the same transcript.

Amino Acid Activation

Amino acids can only be recognized and used for protein synthesis after they have been 'activated', i.e. covalently joined to the appropriate tRNA. Attachment is catalysed by amino acid activating enzymes (aminoacyl tRNA synthetases) which are specific for both amino acid and tRNA. The energy for formation of the ester bond is provided by the hydrolysis of ATP to AMP and pyrophosphate. This is a typical ligase reaction:

$$RCHCOOH + ATP \rightleftharpoons RCHCO—AMP + P-P_i$$
$$| \qquad\qquad\qquad |$$
$$NH_2 \qquad\qquad\qquad NH_2$$

$$RCHCO—AMP + tRNA \rightleftharpoons RCHCO—tRNA + AMP$$
$$| \qquad\qquad\qquad\qquad |$$
$$NH_2 \qquad\qquad\qquad\qquad NH_2$$

The reaction proceeds through an enzyme-bound aminoacyl adenylate intermediate, which is a mixed anhydride of AMP and amino acid: the acyl group is then transferred to tRNA, forming an ester link with the 3'-terminal adenosine. The amino acid may be attached to either the 2'- or the 3'-hydroxyl, depending on which enzyme activates the tRNA. There is at least one, and sometimes more than one, activating enzyme for each amino acid, and a given enzyme often recognizes several 'isoacceptor' tRNAs. Thus the enzyme does not primarily recognize the anticodon.

Aminoacyl tRNA synthetases

Aminoacyl-tRNA synthetases are the only components of the protein-synthesizing system that recognize amino acids directly; after activation, there is no way of recognizing and discarding amino acids attached to the wrong tRNA. These enzymes must therefore be able to discriminate accurately between different amino acids, and between tRNAs. Some amino acids are quite similar to each other in structure: e.g. valine and isoleucine differ by a methylene group only. The strength of binding to enzymes of these two amino acids can only differ by the binding energy contributed by this group (8–12 kJ mol^{-1}), which would account for a 100–200-fold preference for isoleucine, relative to valine, by isoleucyl-tRNA synthetase. However this enzyme only makes the mistake of synthesizing val-$tRNA^{ile}$ 1 in 3000 times. It appears to possess an 'editing' activity, which results in hydrolysis of incorrectly activated aminoacyl tRNAs, presumably at a separate binding site: although enzyme-bound valyl-AMP may be formed, it is hydrolysed if $tRNA^{ile}$ is bound to the enzyme. Although this process wastefully hydrolyses ATP, it is essential in preserving the fidelity of amino acid activation. In a similar way, valyl-tRNA synthetase has been found to activate threonine, giving thr-$tRNA^{val}$, which it also hydrolyses. The error frequency in activating isoleucine is higher than with most amino acids, and the overall rate of occurrence of errors in translating mRNA is less than

1 in 10^4. One case in which an aminoacyl-tRNA synthetase can be 'fooled' is in the substitution of p-fluorophenylalanine for phenylalanine: this analogue becomes attached to tRNA^{phe}. Many amino acid analogues inhibit aminoacyl tRNA synthetases by competing with the amino acid substrate, but only rarely is an analogue activated and incorporated into protein.

Translation

Direction of protein synthesis

The direction of translation was determined by exposing growing cells (reticulocytes) for short periods to radioactive amino acids. The intensity of labelling in the completed globin chains was found to increase toward the C-terminus, showing that chain initiation had occurred at the N-terminus. This is a universal rule. Since translation of mRNA is known to be in the direction $5' \rightarrow 3'$, the first translated codon from the 5'-end corresponds to the N-terminal amino acid of the protein.

Sequence of events

The synthesis of proteins on ribosomes can be divided into three phases: first, *initiation*, the formation of a complex between mRNA, two ribosomal subunits, and the tRNA carrying the initiating (N-terminal) amino acid; second, *elongation*, a repeated cycle of aminoacyl-tRNA binding, addition of an amino acid to the C-terminus of the nascent chain, and release of tRNA; and third, *termination*, the release of the completed polypeptide from the ribosome and, often, of the ribosome from the messenger. The details of the events occurring in each of these phases have been worked out largely from in vitro experiments with cell-free translation systems, and the functions of particular soluble or ribosomal proteins have been investigated using antibiotics which block specific steps in protein synthesis, antibodies raised against the purified proteins, or, in prokaryotes, mutations affecting the production or function of the protein.

Initiation

The two non-identical subunits which form a functional ribosome are only transiently associated, and become separated when translation of each polypeptide is complete. The holo-ribosome (70 S in prokaryotes, 80 S in eukaryotes) is unable to bind mRNA, and initiation of new peptide chains occurs with the sequential addition of subunits to mRNA and the initiator aminoacyl-tRNA. A protein (initiation factor 3) has the function of preventing the association of ribosomal subunits in the absence of mRNA.

Newly synthesized proteins always have methionine as the N-terminal amino acid, because translation in eukaryotes is always initiated with methionyl-tRNA. The single methionine codon, AUG, can code for either N-terminal or internal methionine, but there are two tRNAs for methionine, one of which (tRNA$_F^{met}$) is used in initiation, the other (tRNA$_M^{met}$) for methionines within the protein sequence. Both are charged by the same methionyl-tRNA synthetase. During initiation, met-tRNA$_F^{met}$ binds to the 40 S ribosomal subunit before mRNA, so its binding cannot at first be specified by a codon.

Initiation requires, in addition to the components already mentioned, GTP, and a number of soluble proteins called initiation factors. In eukaryotes, the sequence of events is:

1. Formation of a complex between the small (40 S) ribosomal subunit, met-tRNA$_F^{met}$, GTP and the initiation factors eIF-2 and eIF-3.
2. Addition of mRNA and eIF-4, in a reaction which may require ATP.
3. Binding of the large (60 S) ribosomal subunit, with initiation factor eIF-5.
4. Hydrolysis of the bound GTP to GDP + P_i, and release of all of the initiation factors.

Ribosomes have two binding sites for aminoacyl-tRNA, called the P (peptidyl) and A (aminoacyl) sites; the initial binding of met-tRNA$_F^{met}$ is unusual in occurring at the P site—during elongation, binding of aminoacyl-tRNA is always at the A site. The role of the initiation factor eIF-1 is not clearly understood—possibly it may help to dissociate free ribosomes into their constituent subunits.

Since the codon AUG can specify both initial and internal methionines, and mRNAs have untranslated sequences at their 5'-ends, the ribosome must have some way of recognizing the start of the message. This may be provided by the tertiary structure of the mRNA, or the 5'-cap of methylguanine may be important, since 7-methyl GTP is an effective inhibitor of initiation; however, some capless mRNAs can be efficiently translated.

The complete sequence of events in initiation (Fig. 12.4) leads to formation of the 80 S 'initiation complex', and elongation of the peptide chain can now begin.

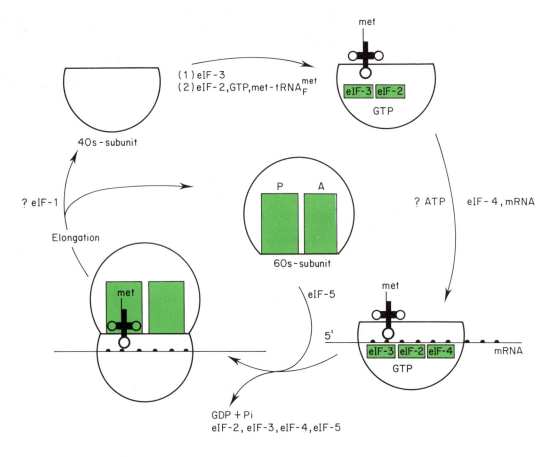

Fig. 12.4 The initiation of translation: formation of the 'initiation complex', by the sequential addition of large and small ribosomal subunits and methonyl-tRNA to mRNA, with the aid of protein initiation factors.

Initiation of translation in prokaryotes

In prokaryotes, there are some differences. Translation is initiated with *N*-formyl methionyl tRNA, which is formed by transfer of a formyl group from N^{10}-tetrahydrofolate to met-$tRNA_F^{met}$. Methionyl-$tRNA_M^{met}$, which recognizes internal methionine codons, is not a substrate for the formyl transferase. The complex between mRNA and the prokaryotic 30 S ribosomal subunit forms *before* binding of met-$tRNA_F^{met}$ (contrast Fig. 12.4), and a series of prokaryotic initiation factors, apparently equivalent to the eukaryotic factors, is required. It has been shown that prokaryotic ribosomes bind to a recognition sequence in the mRNA, some 3–7 nucleotides long and situated 10–14 nucleotides before the initiation codon, in the untranslated 5′-leader. This binding occurs by base-pairing to a sequence at the 3′-end of the 16 S ribosomal RNA. The 5′-end of the messenger is unimportant, as even endless, circular mRNAs can be translated in prokaryotic systems.

Elongation

Three separate steps occur with each amino acid added to the peptide chain; together they constitute the 'elongation cycle'. They are:

1. *Binding of aminoacyl-tRNA to the A site of the ribosome, as dictated by the next codon of the message.* This requires a soluble protein called elongation factor 1 (EF-1) and GTP, which is hydrolysed to GDP and inorganic phosphate.

EF-1 is a mixed dimer of EF-1α and EF-1β: EF-1α forms a complex with GTP and aminoacyl-tRNA, which can then bind to ribosomes. Following hydrolysis of GTP, a complex of EF-1α with GDP dissociates from the ribosome, and EF-1β displaces GDP from EF-1α.

2. *Formation of the peptide bond.* This is catalysed by an enzyme, peptidyl transferase, which is a constituent protein of the large ribosomal subunit, and involves hydrolysis of the peptidyl-tRNA occupying the P-site (or, in the first round of elongation, methionyl-$tRNA_F^{met}$) and transfer of the acyl residue to form a peptide bond

with the α-amino group of the aminoacyl-tRNA in the A-site. There is no requirement for any soluble protein or for GTP. The discharged tRNA dissociates from the P-site, leaving the nascent polypeptide linked to the tRNA in the A-site.

3. *Translocation of the peptidyl-tRNA from the A- to the P-site, and of the ribosome along the mRNA, by one codon, toward the 3'-end.* This process requires another soluble elongation factor (EF-2), and involves the hydrolysis of GTP. The P-site is now occupied by peptidyl-tRNA, and the A-site vacant, ready to receive the next aminoacyl-tRNA.

In prokaryotes, a very similar sequence of events occurs during elongation, the elongation factors corresponding to EF-1α, EF-1β and EF-2 being EF-Tu, EF-Ts and EF-G. The sequence of aminoacyl-tRNA binding, peptide bond formation and translocation of peptidyl-tRNA is repeated until the polypeptide is complete, the nascent chain always being attached through its carboxy-terminus to the tRNA corresponding to the last-added amino acid. As translation proceeds, the ribosome moves along the mRNA, and a new codon specifies the binding of each incoming aminoacyl-tRNA to the A-site. These events are shown diagrammatically in Fig. 12.5, and the factors required summarized in Table 12.3.

Termination

There is no tRNA with anticodon complementary to the terminator codons; instead, a protein-releasing factor binds to the large ribosomal subunit, and causes peptidyl transferase to release the completed chain from its C-terminal linkage to tRNA. In eukaryotes, a single releasing factor (RF) recognizes any of the terminator

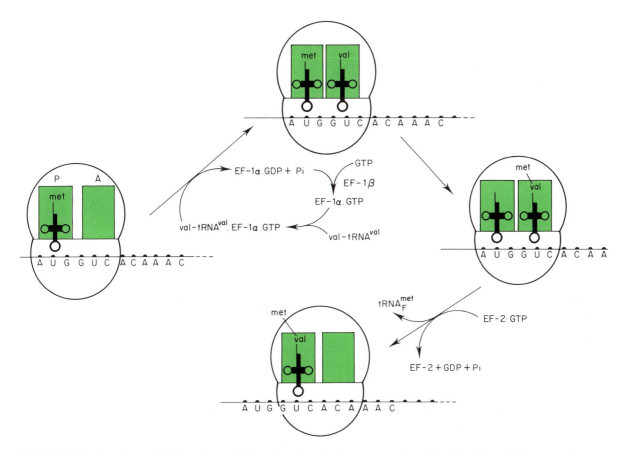

Fig. 12.5 The elongation cycle of protein translation: in each cycle the nascent polypeptide chain is extended by one amino acid, with the consumption of 2 mols of GTP.

Table 12.3 Soluble proteins required in translation of mRNA

Function	Eukaryotic factor	Prokaryotic equivalent
Initiation		
Release of ribosome from mRNA	eIF-1	IF-1
Binding of initiator tRNA	eIF-2	IF-2
Dissociation of ribosomal subunits	eIF-3	IF-3
Binding of small ribosomal subunit	eIF-4 (a, b, c, d)	
Binding of large ribosomal subunit	eIF-5	
Elongation		
Binding of aminoacyl tRNA	EF-1α	EF-Tu
Binding of GTP to EF-1α (EF-Tu)	EF-1β	EF-Ts
?	EF-1γ	
Translocation of peptidyl-tRNA	EF-2	EF-G
Termination		
Release of polypeptide	RF	RF-1, RF-2 RF-3

codons, but in bacteria two releasing factors are known, with different codon specificities: RF_1, which recognizes UAA or UAG, and RF_2 which recognizes UAA or UGA. A third factor, RF_3, is also necessary: its function seems to be to bind GTP. After release of the protein the deacylated tRNA is released, and the ribosomal subunits dissociated from the messenger RNA and from each other. Antibiotics which inhibit peptidyl transferase also block chain termination.

Energy requirements

Each step of elongation requires the hydrolysis of 2 molecules of GTP to GDP + P_i; the energy derived from this hydrolysis is used to produce the conformation changes required to translocate tRNA and mRNA on the ribosome. Peptide bond formation does not require hydrolysis of GTP, and appears to be driven by hydrolysis of the bond between the nascent peptide chain and tRNA, which has a large negative free energy change ($\Delta G^0 = -28$ kJ mol^{-1}). This was originally provided during amino acid activation, by hydrolysis of ATP to AMP + P-P_i. Thus there is a formal requirement for 4 nucleoside triphosphates per amino acid (2 GTP, plus 2 ATP, since 2 ATP are required to rephosphorylate AMP to ATP). Another molecule of GTP is consumed during initiation of each peptide chain.

Bacterial ribosomes also catalyse an 'idling' reaction, in which GTP is converted by ATP to an unusual nucleotide, guanosine with pyrophosphate attached to the 3'- and 5'-hydroxyls (5'ppG-3'pp). This reaction is promoted by uncharged tRNA and occurs under conditions of amino acid starvation; ppGpp accumulates in the cell, inhibiting synthesis of rRNA and tRNA (and also inhibiting some other enzymes). This is known as the 'stringent response'.

Polysomes

As translation of a message proceeds, additional ribosomes can, if the message is long enough, become attached at the initiation site on the mRNA, and follow each other toward the 3'-end; those near the 5'-end carry short nascent polypeptide chains which have just been begun, those near the 3'-end longer ones which are near completion. Aggregates of ribosomes, joined by mRNA, are called polyribosomes or polysomes; they are disrupted by RNase treatment, or by lowering the Mg^{2+} concentration (e.g. by chelation with EDTA). The ribosomes, which are about 22 nm in diameter, may be separated by as little as 3 nm, giving one ribosome for every 80 nucleotides in the mRNA, which is therefore very efficiently translated. The message for a globin molecule, coding for about 146 amino acids, is often associated with 4–6 ribosomes, and longer messages produce even larger polysomes. Polysomes of different sizes have different sedimentation constants, and are readily separated by centrifugation through a sucrose density gradient. The density of all the polysomes is about the same, and is greater than that of the sucrose solution at the bottom of the gradient. In this type of centrifugation, separation of the polysomes depends solely on the sedimentation constant; the larger the polysome, the faster it moves down the gradient.

In bacteria, the DNA is exposed to the cytoplasm, and translation of mRNA may begin even before its transcription from DNA is complete. Bacterial messenger RNAs are frequently polycistronic: that is, they contain the nucleotide sequences specifying more than one protein. Often these proteins have metabolically related functions; e.g. one single mRNA contains ten cistrons, coding for ten different enzymes involved in histidine biosynthesis. It is about 12 000 nucleotides long. Polycistronic mRNAs usually contain untranslated regions at each end and between each protein message: initiation of translation may occur not only at the 5'-end, before the first cistron, but at each initiator codon, which is preceded by the ribosomal binding sequence. No polycistronic mRNAs have been found in eukaryotes.

Inhibitors of translation

Although the mechanism of translation appears to be basically the same in prokaryotes and eukaryotes, there are enough differences in structure of the ribosomes and the soluble protein factors for many agents to selectively inhibit either prokaryotic or eukaryotic protein synthesis. This is fortunate, because many antibiotics which are very potent inhibitors of bacterial protein synthesis have no effect on protein synthesis in the mammalian cytoplasm. Some antibiotics even discriminate between different prokaryotes, e.g. lincomycin and the streptogramins are more active against gram-positive bacteria than gram-negative. Since synthesis of viral proteins takes place on ribosomes made by the host cell, there is little possibility of inhibiting it specifically; but some viral gene product which is involved in viral replication, such as DNA-polymerase or RNA-polymerase, might be selectively inhibited (see p. 201).

Antibiotic resistance

Although the chemical structures of many, though not all, antibiotics are known, this does not usually give much idea of their mechanism of action. In microorganisms, antibiotic-resistant mutants frequently show alterations in the sequence of particular ribosomal proteins, which may tentatively identify the binding site of the antibiotic; but often the ribosomal components involved in binding the drug are not the same as those altered in the resistant mutant (thus, erythromycin is known to bind to proteins L15 and L16 in the 50 S ribosomal subunit, but resistance can be conferred by alterations in L4, L22 or in the 23 S rRNA). This reflects the interactions of the different proteins within the ribosome structure. Resistance to antibiotics can also arise through alteration of the bacterial membrane, so that the drug no longer enters the cell, or through production of an enzyme which destroys the activity of the antibiotic, by chemically modifying it. Such enzymes are usually specified on a piece of extrachromosomal DNA (plasmid): see p. 205.

Puromycin

The structure of puromycin, an antibiotic which inhibits protein synthesis in both prokaryotes and eukaryotes, is shown in Fig. 12.6. It resembles the 3'-terminal of tyrosyl-tRNA, binds to the A-site of the ribosome and accepts transfer of the nascent peptide chain from the P-site onto its free amino group. However, it is unable to undergo translocation, and in any case, the peptide chain cannot be removed from the sugar moiety of puromycin by peptidyl transferase, since it is held by an amide, rather than an ester, linkage. Peptidyl puromycin dissociates from the ribosome; the effect of the antibiotic is therefore to cause premature chain termination, and production of short polypeptides attached to puromycin at the C-terminus. Even formyl-methionyl-tRNA$_F^{met}$ undergoes reaction with puromycin, suggesting that it is bound directly to the P-site during initiation of translation. Because of its lack of specificity, puromycin has no clinical use, but this and other antibiotics have been useful in dissecting the various stages of chain elongation in prokaryotes. A summary of the effects of some common antibiotics is given in Table 12.4.

Inhibitory proteins

The proteins colicin E3 and cloacin are produced by certain strains of *E. coli* (which are themselves resistant to the effects of these antibiotics). Colicin E3 acts only on 70 S ribosomes: it cleaves the 16 S rRNA near the 3'-end, and although no rRNA is lost from the ribosomes, they become much less efficient in protein synthesis.

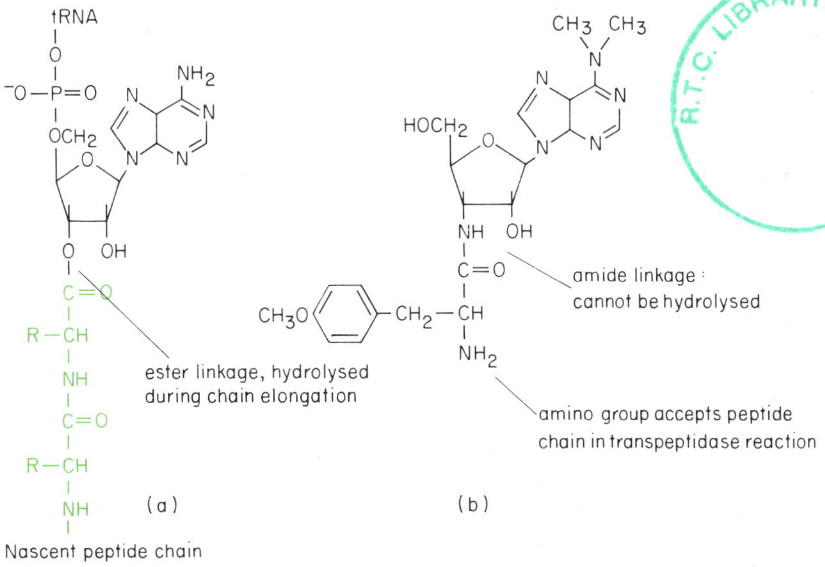

Fig. 12.6 Comparison of the structures of (a) peptidyl tRNA and (b) puromycin.

Table 12.4 Antibiotics which inhibit protein synthesis

Antibiotic	Reaction inhibited	Binding site	Specificity*
Edeine	mRNA binding	Small subunit	P, E
Kasugamycin	Formyl-met-		
Neomycin	tRNA$_{met}^F$	Small subunit	P
Negamycin	binding		
Tetracyclines	Aminoacyl	70 S ribosomes	P
Streptogramins	tRNA binding	Large subunit	P
Puromycin	Chain elongation	Large subunit	E, P
Chlorampheni- col			P
Lincomycin	Trans-		P
Sparsomycin	peptidation	Large subunit	E, P
Kanamycin			P
Cycloheximide			E
Erythromycin	Translocation	Large subunit	P
Fusidic acid	Dissociation of EF-G	EF-G, EF-2	P, E

* Specificity: P—prokaryotic ribosomes; E—eukaryotic ribosomes.

Inhibition of eukaryotic protein synthesis

Inhibitors of many of the steps of eukaryotic protein synthesis also exist. One of the most potent, cycloheximide, inhibits the peptidyl transferase of 80 S ribosomes. The plant proteins abrin and ricin, which are among the most toxic substances known, also act on 60 S subunits; they apparently catalyse some modification of a ribosomal protein, but the chemical nature of the alteration is unknown. Diphtheria toxin is a protein which inhibits eukaryotic protein synthesis, by catalysing the modification of the protein elongation factor EF-2. In this reaction, ADP-ribose, derived from NAD$^+$, is covalently attached to the protein

$$\text{NAD}^+ + \text{EF-2} \xrightarrow{\text{Diphtheria toxin}} \text{Nicotinamide} + \text{ADP-ribose—EF-2}$$

Since diphtheria toxin is a catalyst in the inactivation of EF-2, a single molecule of toxin is capable of completely inactivating protein synthesis in a cell.

Interferon

Interferons are glycoproteins which are produced by some eukaryotic cells on exposure to viral infection, and which have the effect of protecting other cells in the same organism against viral infection. They are very active inhibitors of protein synthesis, only a few molecules of interferon per cell being needed to greatly reduce virus production on infection. Interferons have

two separate effects: an increased rate of destruction of viral messenger RNA, and a decrease in the rate of translation of viral messengers. Interferons induce production of a number of enzymes, including a protein kinase which catalyses the ATP-dependent phosphorylation of the initiation factor eIF-2, thus inhibiting translation, and oligoisoadenylate synthetase, an enzyme which catalyses the formation of the unusual nucleotide ppApApA. The activity of both of these enzymes is dependent upon the presence of double-stranded RNA, which appears during viral infection. Viral RNA is degraded by ribonuclease F, an enzyme which is present in cells even in the absence of interferon, but which is activated by oligoisoadenylate. A phosphatase, which dephosphorylates eIF-2, and a phosphodiesterase which degrades oligoisoadenylate, are also induced. Thus cells which produce interferon 'warn' other cells, which are then competent to inhibit viral protein synthesis, should infection occur. Recently it has been suggested that interferons may also act against some types of cancer, which has led to an interest in the production of pure interferon by cloning techniques (p. 205). There are several immunologically distinct forms of interferon in humans.

Protein synthesis in organelles

Mitochondria contain multiple copies of a closed-circular, double-stranded DNA chromosome. The size of this varies from one species to another: animal mitochondrial DNA has a molecular weight of about 10^7, and encodes the three rRNAs, a set of tRNAs, and a few proteins. These proteins are numerically a small fraction of the many proteins in the inner mitochondrial membrane, but they are nevertheless essential to oxidative phosphorylation. They include subunits I, II and III of cytochrome oxidase (the three largest and most hydrophobic subunits of this enzyme, which also contains four subunits synthesized in the cytoplasm) and cytochrome b; one, or in some species two subunits of the proton-translocating ATPase, and, in yeasts, one protein of the large subunit of mitochondrial ribosomes, are also synthesized in mitochondria. Protein synthesis in mitochondria has many of the characteristics of that in prokaryotes, a fact which has been taken as evidence for the origin of mitochondria as prokaryotic symbionts. It is initiated with formyl-methionyl-tRNA$_F^{met}$, takes place on ribosomes which are smaller than those in the cytoplasm and, in particular, is insensitive to cycloheximide, but sensitive to antibiotics, such as chloramphenicol, which inhibit prokaryotic protein synthesis. Antibiotic treatment of bacterial infections could therefore interfere with protein synthesis in the mitochondria of the patient, although in practice the antibiotic often fails to reach the mitochondrial matrix.

There is no evidence for import of mRNA from the cytoplasm for translation on mitochondrial ribosomes; indeed, this seems very unlikely, since there are differences in the genetic codes used in the cytoplasm and in mitochondria.

Post-translational events in protein synthesis

Upon completion and release from the ribosome, proteins must assume the correct three-dimensional structure for biological activity; in many cases, this occurs only after covalent modification of the polypeptide, known as post-translational processing. Most evidence indicates that protein folding occurs spontaneously; denatured proteins refold in vitro, the refolding apparently following a defined pathway, to a final conformation which is a state of minimum free energy. Folding probably starts even before chain termination, as nascent proteins, still attached to polysomes, are able to bind coenzymes, and may even show enzymic activity. A post-translation modification which almost all proteins undergo is removal of the initiator amino acid. In prokaryotes the N-terminus is deformylated, and in some cases the methionine itself removed, after completion of the chain. In eukaryotes the N-terminal methionine is removed while the nascent chain is still attached to the ribosome, and well before it is completed.

Precursor forms of proteins

Many proteins are synthesized as precursors, which are processed to the mature form when they arrive at their site of action. The pancreatic proteases chymotrypsin, trypsin, elastase and carboxypeptidase are all synthesized as inactive zymogens, which are activated, by proteolytic cleavage, in the duodenum: this prevents their exhibiting unwanted protease activity before they reach the digestive tract (see p. 123). Insulin first appears in a form, proinsulin, with very little biological activity: it is converted to insulin by removal of an internal

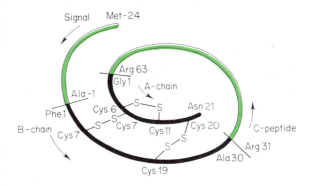

Fig. 12.7 Structure of preproinsulin. The 'signal' sequence (24 amino acids at the N-terminus) and the C-peptide (33 amino acids) are removed proteolytically, leaving the A and B chains joined by two disulphide bridges.

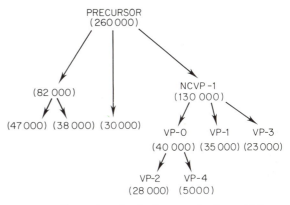

Fig. 12.8 Generation of poliovirus proteins from a high molecular weight precursor, by post-translational processing.

sequence of 30 amino acids, the C-peptide, leaving the A and B chains linked by disulphide bridges (Fig. 12.7). This seems to be a device to ensure correct folding and formation of the right disulphide bonds (separated A and B chains do not refold together correctly, or form the right disulphide bonds on reoxidation) rather than

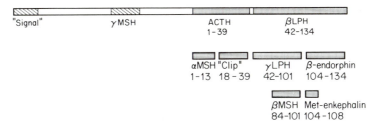

Fig. 12.9 Maturation of bovine pituitary ACTH–βLPH precursor. ACTH, adrenocorticotrophic hormone; LPH, lipotropins; MSH, melanocyte-stimulating hormones. 'Signal' is a putative sequence necessary for secretion of the precursor, 'clip' a sequence of unknown function produced by proteolytic processing of ACTH. Numbers refer to amino acids within the ACTH–LPH ('pro-opiocortin') sequence.

a control of insulin activity, since even when stored before release, in the secretory granules of the pancreatic β-cell, insulin is largely in its mature form.

Two further important examples of post-translational proteolytic processing are poliovirus, in which the single primary translation product is a precursor of molecular weight 260000, which is finally processed into seven proteins; and the enkephalins, small peptides which bind to opiate receptors and which are derived from a large polypeptide which also contains the sequences of several pituitary hormones. The processing of these precursors is shown schematically in Figs. 12.8 and 12.9.

Other examples of post-translational modification include glycosylation, which is particularly important for extracellular and membrane proteins (see p. 179), and addition of covalently-linked prosthetic groups such as biotin, lipoate or, in the case of C-type cytochromes, haem. In collagen, the structural protein of much connective tissue, a series of modifications takes place, including hydroxylation of proline, oxidation of lysine, glycosylation, and the formation of interchain and interhelix bonds (p. 221).

Synthesis of proteins for secretion

Electron microscopy reveals the existence of polysomes attached to the cytoplasmic face of the endoplasmic reticulum. Membrane-bound polysomes can be isolated from cell homogenates, and are found to be engaged in the synthesis of proteins destined for secretion, such as serum albumin, the immunoglobulins or, in the pancreas, protease zymogens and hormones such as insulin and glucagon. Conversely, free ribosomes synthesize intracellular proteins. There appears to be no structural difference between membrane-bound and free ribosomes; in vitro, synthesis of secretory proteins can be initiated on free ribosomes, which become bound when membrane vesicles are added. The attachment of ribosomes to the endoplasmic reticulum takes place only when the nascent protein has a 'signal' sequence of

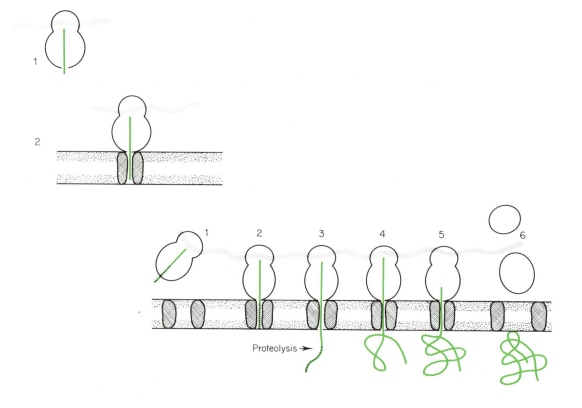

Fig. 12.10 Coupled synthesis and secretion of a protein. (1) Initiation of translation on free ribosomes: the N-terminus of the protein contains a signal sequence which (2) interacts with proteins in the endoplasmic reticulum. These form a pore through which the nascent protein passes. The signal is removed by a membrane-associated protease (3), and the nascent protein folds up (4, 5). It is finally discharged into the lumen of the endoplasmic reticulum (6).

hydrophobic amino acids, which anchor the ribosome to the membrane, probably by interacting with intrinsic membrane proteins. As protein synthesis proceeds, the nascent polypeptide chain passes through the membrane and into the lumen of the endoplasmic reticulum, where it may undergo glycosylation, and is eventually exported from the cell when membrane vesicles derived from the endoplasmic reticulum fuse with the plasma membrane and release their contents. The sequence of events during the synthesis of secretory proteins (shown in Fig. 12.10) is believed to be as follows:

1. Translation of the messenger RNA is initiated on free ribosomes, in the usual way.
2. As elongation of the polypeptide chain proceeds, part of the polypeptide, known as the 'signal sequence', interacts with the membrane.
3. The signal binds to intrinsic membrane proteins, which form a pore, through which the protein moves as it is synthesized.
4. On termination of translation, the ribosomal subunits dissociate from the membrane; the completed protein is left on the opposite side.
5. In most cases the signal sequence is removed proteolytically.

The 'signal' is usually a sequence of some 15–30 hydrophobic amino acids at the N-terminus, which is removed shortly after or even during synthesis, and is not found in the mature protein. Proteins bearing this sequence are called 'pre-proteins', and are rapidly processed to the mature form by proteases in the endoplasmic reticulum: indeed, pre-proteins were first discovered following translation of mRNA in vitro, under conditions which did not lead to removal of the 'signal'.

Proinsulin first appears as pre-proinsulin (Fig. 12.7) with an N-terminal extension sequence of 21 amino acids, most of which are hydrophobic:

H$_2$N-TrpMetArgPheLeuProLeuLeuAlaLeuLeuVal
LeuTrpGluProLysPheAlaGluAla-

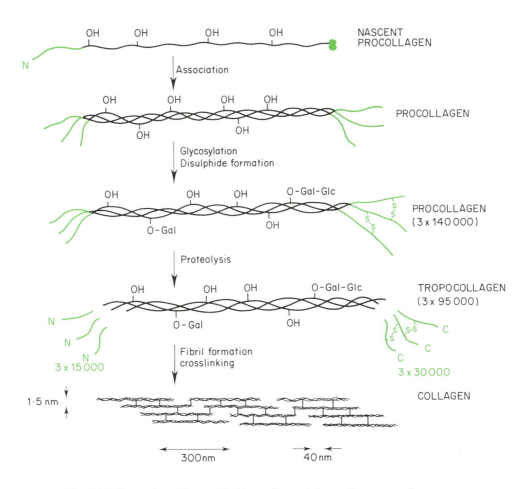

Fig. 12.11 Processing and assembly of procollagen chains, to form tropocollagen.

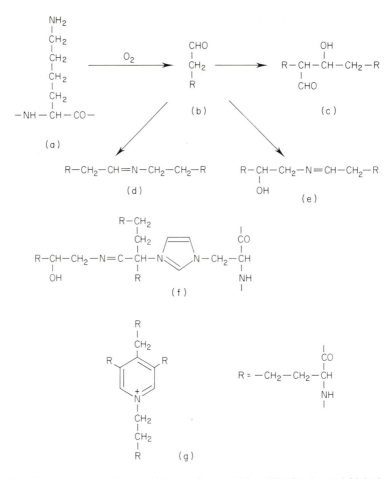

Fig. 12.12 Formation of cross-links in collagen and elastin. Lysine (a) is oxidized to its semialdehyde (b) which may dimerize (c) or react with lysine or hydroxylysine (d, e). Other cross-links are through the formation of histidino-hydroxymerodesmosine (f, in collagen only) and desmosine (g, in elastin only).

Other pancreatic pre-proteins, such as pre-trypsinogen and pre-procarboxypeptidase have N-terminal 'signals' with similar sequences. However, the 'signal' need not always be at the N-terminus; in ovalbumin, a hydrophobic *internal* sequence (amino acids 234–253, from a total of 385) fulfils the signal function. It is exposed during synthesis but becomes buried when the protein folds after secretion, and it is not, of course, removed.

The 'signal' mechanism of coupled synthesis and transmembrane secretion also applies to extracellular bacterial proteins, and to some proteins which remain inserted in the membrane. Other forms of post-translational processing, such as glycosylation, are coupled to translation of the nascent protein; this is discussed in chapter 10.

Post-translational modification of collagen

The collagen class of fibrous proteins are subject to extensive post-translational cleavage, glycosylation and cross-linking. This is quantitatively very important, as collagen constitutes more than 30% of the body's protein, and defects of collagen synthesis and processing are associated with many painful diseases.

Collagen precursors are synthesized on the rough endoplasmic reticulum, pass to the Golgi, and are eventually secreted. The stages in processing are:

1. *Hydroxylation.* Three different hydroxylation reactions occur: conversion of proline to 3-hydroxy-proline and 4-hydroxyproline, and of lysine to 5-hydroxylysine. These reactions are catalysed by specific mixed function oxidases, which also have a

requirement for 2-oxoglutarate, and are begun before translation is complete.

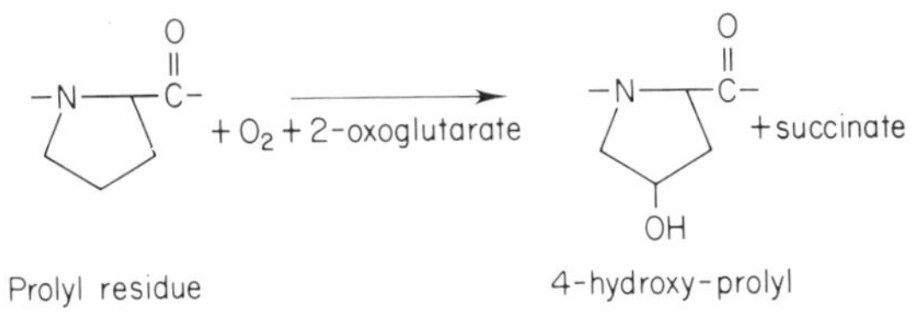

$$\text{Prolyl residue} + O_2 + 2\text{-oxoglutarate} \longrightarrow 4\text{-hydroxy-prolyl} + \text{succinate}$$

Prolyl residue   4-hydroxy-prolyl

Proline hydroxylase has an iron-containing prosthetic group, and requires ascorbate to maintain it in the reduced (ferrous) form; vitamin C deficiency leads to the symptoms of scurvy, as a result of inactivation of this enzyme (see also p. 279).

2. *Glycosylation.* Collagen is unique in having mono- and disaccharides of glucose and galactose covalently attached to the 5-hydroxyl of 5-hydroxylysine. Collagens from different tissues are glycosylated to different extents, by specific glycosyltransferases in the Golgi.

3. *Proteolytic cleavage.* Collagen chains are synthesized as precursors of molecular weight about 140 000; three such precollagen chains associate to form a triple helix (Fig. 2.7) with interchain disulphide bonds. Proteolytic removal of peptides from the N-terminal and C-terminal of each chain leaves tropocollagen, in which each chain has a molecular weight of about 95 000 (Fig. 12.11).

4. *Fibril formation.* Tropocollagen spontaneously associates to form collagen fibrils: each tropocollagen molecule is about 300 nm long, and overlaps its neighbour by 260 nm.

5. *Cross-linking.* Fibrils are strengthened by covalent intra- and inter-chain cross-links: these are formed by oxidation of lysine to its semialdehyde, which then undergoes further reactions (Fig. 12.12). Sweet peas contain the compound β-(α-glutamyl)aminopropionitrile, which inhibits lysine oxidase and thereby reduces cross-link formation, leading to collagen fragility.

13

Control Mechanisms

Control of Enzyme Activity

The factors which control the rates of enzyme-catalysed reactions have been discussed in chapter 3. In mammals, the pH is usually close to 7.4 (apart from some special cases, such as the digestive tract and some intracellular compartments) and the temperature virtually invariant; these are therefore not important controlling influences. The major long-term control of enzyme activity is through the concentration of enzymes, by varying the rate of synthesis and/or degradation: more rapid controls are exercised through changes in concentrations of substrates or inhibitors, and by covalent modifications of enzymes.

Substrate concentration

The activity of simple single-substrate enzymes is described by the Michaelis equation:

$$v_0 = \frac{k_{cat}[e]}{1 + \frac{K_m}{[S]}}$$

This predicts that the rate of the reaction is directly proportional to enzyme concentration, and that the effect of a change in substrate concentration depends upon the degree of saturation of the enzyme with substrate: if $[S] \gg K_m$, the rate is virtually independent of substrate concentration, but if $[S] \ll K_m$ the equation approximates to $v_0 = \frac{k_{cat}}{K_m}[e][S]$, so the rate becomes directly proportional to substrate concentration. The Michaelis equation is, however, only applicable when $[S] \gg [e]$; within the cell, enzyme concentrations may be very large and may even exceed the concentration of substrate, so that the effects of a change in substrate concentration may be hard to predict.

Cooperativity in substrate binding

For all enzymes of the Michaelis type, the change in substrate concentration needed to produce a given change in activity is the same; for example, to increase

v_0 from $0.1\,V_{max}$ to $0.9\,V_{max}$ requires an 81-fold increase in substrate concentration. This ratio is independent of enzyme concentration and is unaffected by simple inhibitors. However, not all enzymes obey the Michaelis equation: regulatory enzymes frequently show velocity-substrate curves of the type shown in Fig. 13.1, which deviate from Michaelis (hyperbolic) behaviour in one of two ways: *positive cooperativity*, in which v_0 increases sharply over a narrow range of substrate concentration, making v_0 very sensitive to changes in [S] in this range; or *negative cooperativity*, in which increase in [S] produces a more gradual increase in v_0 than is predicted by the Michaelis equation. Both types of curve are approximately described by the empirical Hill equation:

$$v_0 = \frac{V_{max}}{1 + \frac{K}{[S]^{n_h}}}$$

in which n_h is the Hill coefficient; with $n_h > 1$ there is positive cooperativity in substrate binding, while negative cooperativity occurs when $n_h < 1$. From this equation is derived the expression

$$\log \frac{v_0}{V_{max} - v_0} = n_h \log [S] - \log K$$

allowing evaluation of n_h from the slope of a Hill plot (Fig. 13.2). The values of r, the ratio of substrate concentrations needed to increase v_0 from $0.1\,V_{max}$ to $0.9\,V_{max}$ with various values of n_h, are shown in Table 13.1.

Mechanism of positive cooperativity

Positive cooperativity derives from an *increase* in the apparent affinity of an enzyme for its substrate, as the substrate concentration and degree of saturation increases; this is usually because each molecule of enzyme has more than one substrate binding site, and binding of substrate to one site increases the affinity of the remaining sites for substrate. As a rule, no more than one substrate binding site occurs on each polypeptide

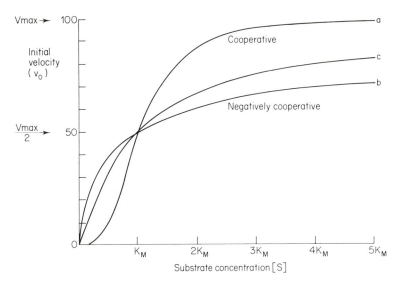

Fig. 13.1 Effect of (a) cooperativity and (b) negative cooperativity in substrate binding, on the Michaelis plot for an enzyme. Curve (c) is for an enzyme exhibiting normal substrate binding, which obeys the Michaelis equation (see Fig. 3.4).

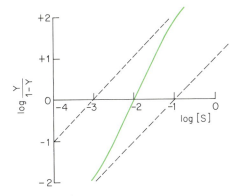

Fig. 13.2 Hill plot for an enzyme showing positive cooperativity in substrate binding. Between $Y = 0.1$ and $Y = 0.9$ the plot is a straight line, the slope of which equals the Hill coefficient ($n_h = 2$, in this case). It is asymptotic to lines of unit slope.

Table 13.1 Values of the Hill coefficient (n_h) and r (r = the ratio of substrate concentrations necessary to produce $v_0 = 0.9\ V_{max}$ and $v_0 = 0.1\ V_{max}$)

Binding	n_h	r
Cooperative	4	3.00
	3	4.33
	2	9.00
Hyperbolic	1	81
Negative-cooperative	0.8	243
	0.6	1520
	0.5	6560

chain, and enzymes of this type are composed of several subunits which are often, though not always, identical. Binding of the first substrate molecule produces a protein conformation change which increases the affinity for subsequent substrate molecules.

The 'concerted-transition' model

The simplest theory to account for cooperative substrate binding assumes that the protein consists of identical subunits, which are always arranged symmetrically with respect to each other. These subunits are known as protomers (usually protomers correspond to individual polypeptide chains, although in some cases each protomer might itself consist of two or more polypeptides), and the protein can exist in at least two states, which differ in the conformation of the protomers, and in their affinity for substrate. The enzyme state with low (or zero) affinity for the ligand is known as T (taut), that of higher affinity the R (relaxed) state; the unliganded enzyme is a mixture of the two forms. Binding of the ligand is predominantly to the R state, because of its higher affinity, and displaces the T/R equilibrium toward R; since each protein molecule has several ligand binding sites this increases the total number of empty, high-affinity binding sites so that the next ligand molecule is bound more readily. This sequence produces cooperativity in ligand binding, and is shown in Fig. 13.3.

The degree of cooperativity exhibited depends on three parameters; the conformational equilibrium constant [T]/[R], the ratio of ligand affinity constants of the two states K_R/K_T, and the number of protomers (n).

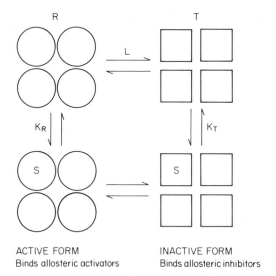

R T

K_R K_T

ACTIVE FORM
Binds allosteric activators

INACTIVE FORM
Binds allosteric inhibitors

Fig. 13.3 The 'concerted' model of cooperativity, in which an oligomeric enzyme (shown here as a tetramer) can exist in two symmetrical conformations, R (relaxed) and T (taut), with different affinities for the substrate S, denoted by K_R and K_T. L is the equilibrium constant for the conformation change: $L = [T]/[R]$.

The larger any of these, the greater the degree of cooperativity. For yeast glyceraldehyde 3-phosphate dehydrogenase, which shows cooperativity in the binding of its substrate NAD^+, $[T]/[R] = 60$, $K_R/K_T = 25$, and $n = 4$. This enzyme has four identical subunits, and a number of physical studies have shown that a conformation change follows the binding of NAD^+.

Cooperative oxygen binding by haemoglobin

The best-studied example of a protein which shows cooperative ligand binding is haemoglobin. Adult haemoglobin has four similar polypeptide subunits, in the ratio $\alpha_2\beta_2$, each with an oxygen-binding haem prosthetic group. There is strong cooperativity in oxygen binding, the Hill coefficient being 2.7–2.9. This is a result of the conformation change which follows the binding of an oxygen molecule to deoxyhaemoglobin, and is triggered by a movement of the iron atom to which O_2 binds. In deoxy Hb this atom has five nitrogen ligands (four from pyrroles in the porphyrin ring system, one from histidine $\alpha87$ or $\beta92$) and is in the 'high spin' state with four unpaired electrons. It lies slightly outside the plane of the porphyrin. Binding of O_2, the sixth ligand, alters the iron (still in the ferrous state) to the 'low spin' form, resulting in a decrease in the size of the atom, and its movement into the plane of the porphyrin ring by about 0.06 nm (Fig. 13.4). This in turn moves the proximal histidine, and produces a conformation change in that subunit: this is transmitted to other subunits, with the breaking of some non-covalent intersubunit bonds, and the formation of new ones. The binding of O_2 to one subunit of deoxy Hb therefore shifts the conformation of *all* subunits to that of oxy Hb. This concerted transition from the T (deoxy) to the R (oxy) state increases the affinity of haemoglobin for oxygen about 200-fold, resulting in a sigmoid binding curve for oxygen (see p. 249).

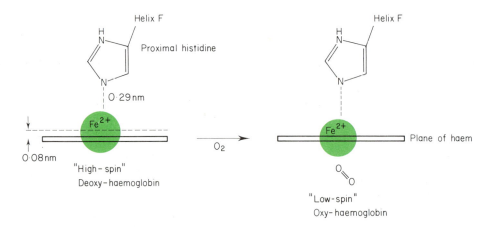

Fig. 13.4 The 'trigger' for the conformation change in haemoglobin. Binding of O_2 alters the spin state and reduces the diameter of an iron atom; this moves into the plane of the haem, producing a conformation change which is transmitted to neighbouring subunits.

Negative cooperativity

Negative cooperativity also occurs in multimeric enzymes, as a result of stepwise, rather than concerted, conformation changes, binding of each substrate molecule altering the conformation of a subunit and producing a decrease in substrate binding affinity in the unliganded subunits. Binding of NAD^+ by rabbit muscle glyceraldehyde 3-phosphate dehydrogenase shows negative cooperativity, the successive dissociation constants being $K_2 < 10^{-9}$ M, $K_3 = 2 \times 10^{-7}$ M, $K_4 = 2 \times 10^{-5}$ M; the first dissociation constant, K_1, is too small to be measured accurately.

Control by inhibitors

Most enzymes can be shown to be inhibited by substrate analogues, which compete with the substrate and form 'dead-end' complexes with the enzyme. Products of the reaction, which are of course analogues of the substrate, inhibit in a similar way, by forming enzyme–product complexes. Even high concentrations of the substrate can cause inhibition, for example by binding to the enzyme at the wrong point in the reaction sequence (p. 32). Although product inhibition, in particular, is probably important in controlling the activity of many enzymes in vivo, simple types of inhibition do not provide very efficient metabolic control, since the inhibitor binds at the active site, and must therefore be structurally related to the substrate: a metabolite which acts as a simple inhibitor can only affect the enzyme which produces it. More sophisticated control can be exerted when the enzyme has a *regulatory site*, distinct from the active site, to which an inhibitor or activator may bind, altering the affinity of the enzyme for its substrate, and/or the rate of catalysis. Since these modulators of activity need not be structurally related to the substrate, such enzymes are called allosteric ('other shape'); allosteric control of enzymes, by effectors which may be structurally and metabolically distant from the substrates of the enzyme at the control point, is important in the regulation of many metabolic pathways.

Mechanism of allosteric effects

The best-studied examples of allosteric enzymes are those involved in biosynthetic pathways in bacteria; the first committed step in the pathway is often inhibited by the product of that pathway which, because the pathway contains very many steps, is structurally quite unlike the substrate. This is an efficient form of control, since there is no wasteful production of intermediates between the first step and the final product.

Allosteric effects could occur in a monomeric enzyme, providing it had both a regulatory and a catalytic site, but examples of this are rare; more often, allosteric enzymes are multimeric, and show cooperative binding either of substrates or of effectors. For this reason allosteric behaviour is frequently confused with co-operative effects, which are often but not *necessarily* found in allosteric enzymes. An enzyme of this type is bacterial aspartate transcarbamylase, which catalyses the synthesis of carbamyl aspartate, the first intermediate in the biosynthesis of pyrimidines (p. 54). In *E. coli*, ATCase is inhibited by CTP, and activated by ATP; it shows cooperativity in the binding of one of its substrates, aspartate, and this cooperativity is increased by CTP, resulting in inhibition of activity, and decreased by ATP, resulting in activation. The catalytic and regulatory sites are actually on different subunits of the enzyme, which has six catalytic and six regulatory polypeptides; these subunits can be separated without loss of catalytic activity, which is then unaltered by the allosteric effectors. Mammalian ATCase (p. 54) is rather different, being part of a multienzymic complex, of which a different enzyme activity is inhibited by UTP (like CTP, a product of the biosynthetic pathway) and activated by ATP. The existence of separate catalytic and regulatory subunits in *E. coli* ATCase is exceptional:

Table 13.2 Some enzymes showing allosteric control features

Enzyme	Cooperatively-bound substrate	Positive effector (activator)	Negative effector (inhibitor)
Acetyl-CoA carboxylase	—	Citrate	Acyl CoA
Citrate synthase	—	AMP, ADP	ATP
Fructose bis-phosphatase	Fructose bis-phosphate		AMP
Glutamate dehydrogenase	—	ADP	GTP
Glycogen phosphorylase	—	AMP, Glucose-1-phosphate	ATP Glucose-6-phosphate
Glycogen synthetase	—	Glucose-6-phosphate	
Isocitrate dehydrogenase	—	ADP, Ca^{2+}	ATP
Phosphofructokinase	Fructose-6-phosphate	AMP, ADP Phosphate, NH_4^+	ATP Citrate
Pyruvate kinase (liver)	PEP	Fructose bis-phosphate	Alanine

most allosteric enzymes are polymers of identical subunits, each of which contains a catalytic and a regulatory site, or even several regulatory sites, each of which binds a different effector. This permits control by several different metabolites simultaneously. Table 13.2 lists some mammalian enzymes which have allosteric control sites.

K and V systems

Mechanistically, allosteric effects can be explained by concerted conformational changes (Fig. 13.3), allosteric inhibitors binding more strongly to the T than to the R state, and activators binding more strongly to R than to T. The R state has either a greater affinity for substrate than does the T state (the allosteric 'K system') or catalyses the reaction at a greater rate (the 'V system'). Displacement of the equilibrium toward R, by a substance which binds more strongly to R than to T, activates the enzyme; conversely, displacement toward T causes inhibition.

The cooperative binding of oxygen by haemoglobin is affected in vivo by 2,3-diphosphoglycerate, which is present in erythrocytes at a concentration similar to that of haemoglobin (4–5 mM). This substance binds strongly to deoxy-haemoglobin (the T state), and less strongly to oxyhaemoglobin (the R state). It therefore acts as an inhibitor of oxygen binding, increasing cooperativity so that there is an approximately 500-fold difference in oxygen affinity between the two forms. In some metabolic diseases such as hexokinase deficiency, which result in decreased concentration of 2,3-diphosphoglycerate in erythrocytes, the affinity of haemoglobin for oxygen is increased; conversely, an increase in the concentration of diphosphoglycerate (which occurs in pyruvate kinase deficiency), decreases the affinity of haemoglobin for oxygen. The concentration of 2,3-diphosphoglycerate within red blood cells increases to 7–10 mM during adaptation to high altitude.

This inhibitor is unusual in binding *between* the subunits of haemoglobin, with a stoichiometry of 1 molecule per molecule protein. Most allosteric inhibitors have a binding site on each subunit.

The allosteric T/R equilibrium of haemoglobin is also pH-dependent; the pK_a values of a number of groups (such as the N-terminal valines of the α-chains, and C-terminal histidines of the β-chains) decrease on oxygenation of haemoglobin, so that a decrease in pH favours deoxygenation:

$$HbO_2 + nH^+ \rightleftharpoons HbH_n^+ + O_2$$

At physiological pH, the value of n is about 0.7; as haemoglobin is transferred from the lungs to tissues, P_{O_2} falls from 100 mmHg to about 40 mmHg, and the pH from 7.4 to 7.2 or less.

In allosteric K systems, the binding of both substrates and effectors is cooperative; an example is liver pyruvate kinase, which has a sigmoid Michaelis plot for the substrate PEP, cooperativity being increased by the inhibitor alanine, and decreased by the activator fructose bis-phosphate. In V systems, substrate binding is not cooperative; for example pyruvate carboxylase shows no cooperativity in the binding of its substrates pyruvate, ATP and CO_2, but V_{max} is increased by the allosteric activator acetyl-CoA, which binds cooperatively.

Control of Enzyme Activity by Reversible Covalent Modification

Allosteric control operates through conformation changes brought about by non-covalent binding of metabolites, acting as intracellular signals; conformation changes can also result from reversible *covalent* modification, through the attachment of some residue, such as phosphate: this is generally controlled by extracellular signals (hormones).

The irreversible activation of enzymes by proteolysis has already been described (zymogens, p. 33). Where covalent modification is used as a means of metabolic control, the elements of the control system are (1) an extracellular signal; (2) a receptor on the cell surface (see p. 190); (3) a transducer which, interacting with the receptor, produces an intracellular signal; (4) a modifying enzyme, whose activity is altered by the intracellular signal; (5) a target enzyme, which catalyses the reaction under control, and is a substrate for the modifying enzyme, undergoing covalent modification with alteration of activity; and (6) an enzyme which reverses the modification of the target enzyme.

Control of glycogen metabolism in muscle

Glycogen synthesis and degradation are controlled by the hormones adrenaline and insulin; only the mechanism of action of adrenaline is understood at present, and it exemplifies the control system outlined above. Adrenaline does not enter the cell; occupancy of its receptor (see p. 191) activates a membrane-bound adenylate cyclase, which increases the intracellular concentration of 3′,5′-cyclic AMP (cAMP), the second messenger for adrenaline. This initiates a cascade of reactions, resulting in accelerated glycogenolysis and reduced glycogen synthesis; the participants are

(i) cAMP-dependent protein kinase

Protein kinases catalyse the transfer of phosphate from ATP to serine or threonine hydroxyls in proteins:

$$\text{Protein} + \text{ATP} \rightarrow \text{Protein}-\text{PO}_4^{2-} + \text{ADP}$$

Protein kinases are ubiquitous, and many are only active in the presence of cAMP. Cyclic-AMP-dependent protein kinases have two catalytic and two regulatory subunits, and dissociate when cAMP binds to the regulatory subunits, with a large increase in the rate of phosphorylation of the protein substrate by the catalytic dimers.

(ii) Phosphorylase kinase

This enzyme, itself a protein kinase, is the substrate of cAMP-dependent protein kinase. Its regulation is complex, in that its activity is controlled by phosphorylation at two distinct sites, and by the concentration of Ca^{2+}.

The subunit structure is $(\alpha\beta\gamma\delta)_4$, α and β being regulatory, phosphorylatible subunits, γ the catalytic subunit and δ the regulatory Ca^{2+}-binding subunit. The dephospho-form of phosphorylase kinase has a high activation constant for Ca^{2+} ($K_A = 23$ μM), which is reduced to 1.5 μM on phosphorylation of the β subunit; in other words, phosphorylation makes the enzyme more sensitive to activation by Ca^{2+}. The increase in activity is totally dependent upon Ca^{2+}, and follows phosphorylation of a single serine in β; phosphorylation of α increases the rate at which β is dephosphorylated by phosphatases (see below). Subunit δ is identical with the regulatory Ca^{2+}-binding protein calmodulin (p. 230), which is itself similar to troponin C, the Ca^{2+}-dependent regulator of muscle contraction, and phosphorylase kinase can also be activated by troponin C.

(iii) Glycogen phosphorylase

This enzyme is the substrate of phosphorylase kinase. Dephosphophosphorylase is a dimer of identical subunits, each of which can be phosphorylated at a single serine; on phosphorylation the dimers associate to tetramers, with an increased rate of glycogen breakdown (see p. 80):

$$\text{Glc}_n + \text{P}_i \rightarrow \text{Glc-1-P} + \text{Glc}_{n-1}$$

Activation can also be produced by the allosteric effector AMP, and this effect is very important; indeed, a strain of mice which lacks phosphorylase kinase shows unimpaired control of glycogen degradation, and in these animals the activatory effect of AMP on glycogen phosphorylase is presumably the only means of accelerating glycogen breakdown.

(iv) Glycogen synthetase

This enzyme catalyses the elongation of glycogen chains (p. 80), and is a tetramer of identical subunits, each of which can be phosphorylated at three distinct sites, by separate protein kinases. These are cAMP-dependent protein kinase, phosphorylase kinase, and a kinase which is probably specific for glycogen synthetase and is known as 'glycogen synthetase kinase 3'. There is some evidence that insulin lowers the activity of this third kinase; at any rate it lowers the degree of phosphorylation of glycogen synthetase.

Phosphorylation of glycogen synthetase increases the K_A for the allosteric activator G6P; with each phosphorylation there is a further increase in K_A^{G6P}, and the phospho- and dephospho-forms are sometimes called D (dependent on G6P) and I (independent).

(v) Cyclic nucleotide phosphodiesterases

Cyclic AMP, which is continuously produced by adenyl cyclase when the adrenaline receptor is occupied, is hydrolysed by phosphodiesterases:

$$3',5'\text{-cAMP} + \text{H}_2\text{O} \rightarrow 5'\text{ AMP}$$

There are several types of phosphodiesterase, one of which, active on both cAMP and cGMP, is controlled by Ca^{2+}. These enzymes are inhibited by caffeine and theophylline, and these drugs therefore enhance the stimulatory effects of adrenaline.

(vi) Protein phosphatases

Phosphorylation of proteins is reversed by phosphomonoesterases, of which there are several in muscle. One acts only upon the α-subunit of phosphorylase kinase; another acts on phosphorylase kinase (β subunit), glycogen phosphorylase and glycogen synthetase. Its activity is controlled by a specific inhibitor protein, which is itself phosphorylated by cAMP-dependent protein kinase, and is inhibitory only in the phosphorylated form.

The hormonal control of glycogen levels in muscle (and with minor differences, liver) is summarized in Fig. 13.5: even this complex scheme is somewhat simplified. Cyclic AMP is the second messenger not only for adrenaline in muscle, but for many other hormones with various actions on other tissues; specificity is achieved through the presence of specific receptors. Activation of glycogenolysis by adrenaline

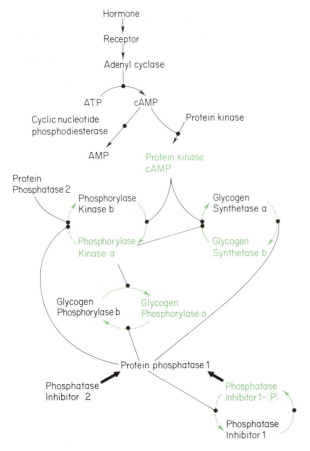

Fig. 13.5 Control of glycogen metabolism in muscle, by a hormone (adrenaline). The activated forms of enzymes and inhibitors are those shown in green. → phosphorylation steps; --→ dephosphorylation.

Table 13.3 **Some enzymes which are controlled by phosphorylation/dephosphorylation**

Activated by phosphorylation	Inhibited by phosphorylation
Cholesterol esterase	Acetyl-CoA carboxylase
Glycogen phosphorylase	Citrate lyase
Phosphorylase kinase	Glycogen synthetase
Phosphofructokinase	HMG CoA reductase
Triglyceride lipase	Pyruvate dehydrogenase
	Pyruvate kinase (liver)

pyruvate dehydrogenase activity, activating the phosphatase and inactivating the kinase by an unknown mechanism.

Phosphorylation of enzymes is a form of allosteric regulation, and many of the enzymes in Table 13.3 are also controlled by non-covalent allosteric effects (Table 13.2). The two forms of regulation act together, phosphorylation being largely controlled by extracellular signals, and modifying the response to allosteric effectors. For example, phosphorylation of liver pyruvate kinase lowers its activity by increasing K_m for the substrate PEP, decreasing K_I for the allosteric inhibitor alanine, and increasing K_A for the activator FDP. Phosphorylation of acetyl CoA carboxylase also inhibits, by decreasing K_I for acyl CoA, and increasing K_A for citrate.

Regulation by other forms of reversible covalent modification occurs in bacteria; phosphorylation is the most common process in mammals, though other modifications, such as ADP-ribosylation (p. 232) are also found.

occurs in a few seconds, each step in the enzyme cascade producing a large amplification of the signal. Adrenaline also inhibits glycogen synthesis; insulin promotes glycogen synthesis, by increasing the proportion of the dephospho (I) form of glycogen synthetase. The mechanism of this effect is not clear, but it is not by lowering the concentration of cAMP; more likely, insulin has its own second messenger.

Control of other enzymes by phosphorylation

Several other enzymes are controlled by hormones which affect their phosphorylation; some are listed in Table 13.3. In several cases phosphorylation is by cAMP-dependent protein kinase, but pyruvate dehydrogenase is controlled by a different protein kinase, and by a Ca^{2+}-dependent phosphatase; insulin increases

Control by Ca^{2+} levels

A rise in cytoplasmic Ca^{2+} levels occurs on hormonal or neural stimulation of many types of cell, Ca^{2+} acting as the second messenger in the coupling of specialized cellular processes (muscular contraction, exocytosis, etc.) to an external stimulus. Occupation of specific receptors by neurotransmitters leads to movement of Na^+ into, and K^+ out of, the cell; the resultant change in the membrane potential opens voltage-dependent Ca^{2+} channels in the membrane. In striated muscle, calcium initiates contraction by binding to troponin C; in smooth muscle it binds to calmodulin, an acidic protein (mol. wt 17 000) with four Ca^{2+} binding sites per molecule. In the presence of Ca^{2+} calmodulin activates myosin light-chain kinase; phosphorylation of one of the light chains of myosin produces contraction of smooth muscle.

Calmodulin is present in many types of cell, and mediates the activatory effects of Ca^{2+} on a number of enzymes, including phosphorylase kinase and cyclic nucleotide phosphodiesterase (p. 229). The antipsychotic phenothiazine drugs show Ca^{2+}-dependent binding to calmodulin, and inhibit its action. This may have effects on concentrations of cAMP or cGMP, or on the phosphorylation of specific neural proteins.

When the extracellular stimulus is removed, cytoplasmic Ca^{2+} levels fall as Ca^{2+} is taken up by the endoplasmic reticulum (p. 182) or mitochondria

(p. 161), or is expelled through the plasma membrane by ATP-dependent Ca^{2+} pumps which are themselves under the control of calmodulin.

There is considerable evidence that increased intracellular Ca^{2+} is connected with accelerated turnover of phosphatidyl inositol phosphates in the plasma membrane, but the significance of this is unknown.

Control of Enzyme Levels

The controls discussed so far vary the *activity* of enzymes without altering the total amount. A different series of controls determines the *amount* of a protein that is present in a cell: these are summarized in Fig. 13.6.

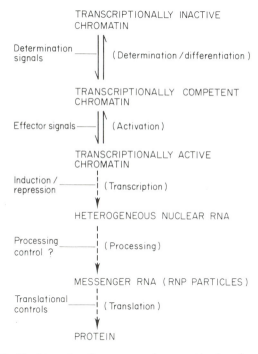

Fig. 13.6 The hierarchy of gene expression operating in eukaryotes.

Chromatin structure

Eukaryotic DNA exists as a complex with nuclear proteins; since the DNA of a single chromosome may be several centimetres long, this complex is highly organized. The basic unit of chromatin structure is the nucleosome, a segment of DNA about 200 nucleotide pairs long associated with two molecules each of the histones H2A, H2B, H3 and H4, and one of H1. The arrangement of these is shown in Fig. 13.7: DNA is wound around the protein octomer (H2A, H2B, H3, H4)$_2$ to form a 'nucleosome core particle', with H1 molecules separating each of these coils. The core particle contains two supercoils of DNA, the double strand being in the usual double-helical conformation (p. 196). Histones are small, very basic proteins (Table 13.4), the amino acid sequences of which are highly conserved. Also present in chromatin, although at a

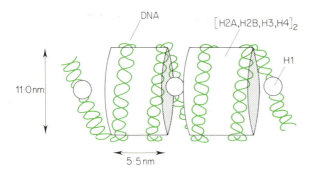

DNA

$[H2A,H2B,H3,H4]_2$

H1

11·0 nm

5·5 nm

Fig. 13.7 The microscopic structure of chromatin. Each unit, known as a nucleosome, consists of two turns of supercoiled, double helical DNA, containing 200 base pairs, and wound on an octameric histone core.

fraction of the levels of histones, are many different non-histone proteins, some basic, some acidic.

In most cells, at least 80% of the DNA is in a transcriptionally inactive state, and is never expressed; during cell differentiation conversion of DNA to a transcriptionally active form may involve replacement or modification of chromatin proteins. Histones undergo many types of reversible covalent modification: phosphorylation, acetylation, methylation, and attachment of poly-ADP-ribose (derived from NAD^+). In addition, histone H2A can become covalently attached to a protein known as ubiquitin. Such modification presumably alters the interaction of histones with DNA. Histone phosphorylation and dephosphorylation occurs at different stages of cell division, and is also affected by some hormones.

Table 13.4 Histones in eukaryotic chromatin

Histone	Molecular weight	Number of amino acids	Lys/Arg ratio
H1	21 000	215	20
H2A	14 000	129	1.25
H2B	13 800	125	2.5
H3	15 300	135	0.72
H4	11 200	102	0.79

Control of transcription

In bacteria, synthesis of some proteins can be started and stopped very quickly, in response to a change in the need for them. Synthesis proceeds either at the maximal rate (providing the amino acid supply is not limiting) or not at all, and bacterial enzymes may be constitutive (synthesized all the time), inducible (synthesized only when required) or repressible (synthesized unless not required). When inducible or repressible enzymes are

not made, it is because synthesis of the corresponding mRNA has been blocked: induction of the enzyme occurs through the initiation of messenger synthesis.

Mechanism of enzyme induction

Transcriptional control, as it operates in prokaryotes, is summarized in Fig. 13.8. To transcribe a gene into mRNA, RNA polymerase binds at a site (the promoter site) close to the beginning (5'-end) of the message. Next to the promoter is the operator region, a sequence of about 20 base-pairs, which can bind a specific repressor protein. When the repressor protein is bound, transcription of the gene cannot occur, but the repressor is an allosteric protein which can also bind a metabolite (the inducer), which lowers its affinity for the operator. In the presence of this metabolite, therefore, the repressor leaves the operator, and transcription of the gene begins.

Bacterial mRNAs are often polycistronic (p. 216); a single operator may therefore regulate the synthesis of several proteins simultaneously, and this is known as coordinate control. A set of proteins where synthesis is controlled by the same operator is called an operon; the *lac* operon, for example, consists of three proteins needed for the metabolism of lactose: lactose permease, β-galactosidase and thiogalactoside transacetylase. When the organism is grown in the absence of lactose, the *lac* operon is repressed; in the presence of lactose all three proteins in the operon are coordinately induced.

The *lac* repressor has been isolated: it is a tetramer of identical polypeptides of molecular weight 37 000 which binds tightly to DNA ($K_d = 10^{-13}$ M), and this binding is prevented by the inducer, a metabolite of lactose called allolactose. The repressor protein is present in very limited quantities (10–20 molecules per cell), its synthesis being constitutive, but slow, through very infrequent initiation of transcription.

Mechanism of enzyme repression

A modified form of this control mechanism exists for repressible enzymes. In the *his* operon ten enzymes involved in bacterial histidine biosynthesis are translated from a single mRNA, the synthesis of which is blocked in the presence of his. The *his* repressor protein binds to DNA *only* in the presence of a co-repressor, which is formed when histidine is present in the cell in sufficient quantities: this co-repressor is histidyl-tRNA.

Catabolite repression

As well as this system of negative control of transcription by repressor proteins, there is sometimes positive control of bacterial operons by gene activator proteins; for example, rapid transcription of the *lac* operon requires the presence of 'catabolite gene activator protein' (CAP). This protein forms a complex

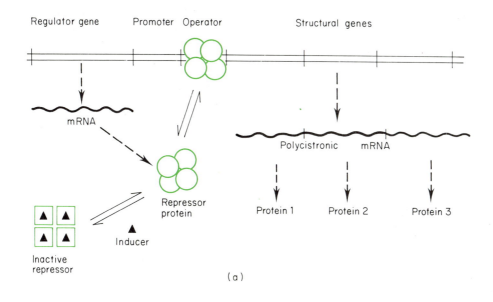

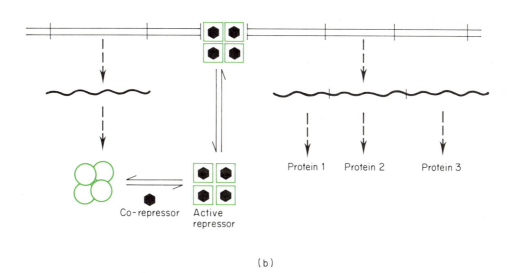

Fig. 13.8 Control of transcription in prokaryotes. A repressor protein (shown as a tetramer) is produced by a regulator gene, and binds to the operator, which controls the transcription of several genes (an 'operon'). In inducible systems (a) the inducer prevents binding of the repressor; in repressible systems (b) a co-repressor is necessary for repressor binding to occur.

with cAMP and binds to the *lac* region of the DNA, near the promoter, producing a 50-fold increase in the rate of transcription. The cytoplasmic concentration of cAMP is high in the absence of glucose so that, providing lactose is available to effect induction, the *lac* operon is transcribed. Glucose lowers the concentration of cAMP, preventing the binding of CAP and inhibiting transcription even in the presence of the inducer. This effect is known as catabolite repression.

Transcriptional control in eukaryotes

Higher organisms synthesize some enzymes which are induced by, for example, steroid hormones, but it is not known whether the mechanism of gene control is similar to that operating in bacteria. Examples are tyrosine aminotransferase and tryptophan oxygenase, which are induced by glucocorticoids. The steroids enter the cell, becoming bound to specific receptor proteins which are then transported into the nucleus; there they may affect transcription directly, or alter the rate of one of the post-transcriptional steps.

In some cases, control of transcription by extracellular signals is mediated by cAMP. For instance, synthesis of the pituitary hormone prolactin is inhibited by dopamine, which reduces the concentration of cAMP by inhibiting adenyl cyclase. This inhibits synthesis of prolactin mRNA, and the effects of dopamine are overcome by cAMP analogues.

Polyamines

The diamines putrescine and cadaverine, and the polyamines spermidine and spermine (derived from L-methionine) have stimulatory effects on the synthesis of proteins and nucleic acids and on cell proliferation, the mechanisms of which are not understood. Ornithine decarboxylase, which is involved in putrescine biosynthesis, is an inducible enzyme with a very rapid turnover time, and shows several peaks of activity during the cell cycle.

$$H_2NCH_2CH_2CH_2CH_2NH_2$$
Putrescine

$$H_2NCH_2CH_2CH_2CH_2CH_2NH_2$$
Cadaverine

$$H_2NCH_2CH_2CH_2NHCH_2CH_2CH_2CH_2NH_2$$
Spermidine

$$H_2NCH_2CH_2CH_2NHCH_2CH_2CH_2CH_2NHCH_2CH_2CH_2NH_2$$
Spermine

Post-transcriptional control

The cytoplasmic mRNA of eukaryotic cells is derived from the heterogeneous nuclear RNA which is the immediate product of transcription; a number of post-transcriptional processing events occur, including excision of introns and joining the coding portions of the message, 'capping' of the 5'-end, internal methylation and attachment of a poly-A 'tail' at the 3'-end; it is quite likely that some control of gene expression is exerted here. In liver, about 80% of hnRNA is not processed, and is degraded without ever leaving the nucleus. Very soon after transcription, hnRNA becomes associated with proteins, forming ribonucleoprotein particles (RNPs) which have a protein:RNA ratio of 4:1. These invariably contain several 'core' proteins and, in addition, enzymes such as poly-A synthetase, ribonuclease and protein kinase. RNPs which are not associated with ribosomes, and therefore probably not translatable, are found in the cytoplasm of many cells; it appears that some protein constituent of the particles prevents translation, since RNPs associated with polysomes show some differences in protein composition,

and the mRNA which is recovered from 'untranslatable' particles, after protein is removed, is readily translated.

Degradation of mRNA

Studies of protein synthesis at different stages of the cell cycle show that histone synthesis ceases abruptly when DNA replication is complete, because there is rapid breakdown of histone mRNA. Messenger degradation requires protein synthesis, and is itself blocked by cycloheximide and actinomycin D, inhibitors of translation and transcription (p. 218). However, the means of its control is not known.

Control of translation

The multiplicity of factors involved in the translation of mRNA (p. 216) suggests that control of protein synthesis might be exercised at this level, and there is some evidence that protein synthesis during developmental changes in cells is controlled through the availability of isoacceptor tRNAs (p. 211) or protein initiation factors.

The best-studied example of translational control is in mammalian reticulocytes, in which globin synthesis is controlled by the availability of haemin, which is necessary for conversion of the apoprotein to haemoglobin. Initiation of translation is blocked in the absence of haemin; this occurs through activation of a protein kinase which phosphorylates the initiation factor eIF-2, inhibiting formation of the complex between met-$tRNA_f^{met}$ and the 40 S-ribosomal subunit (p. 213). The mechanism of protein kinase activation is not clear, but several protein kinases, both cAMP-dependent and -independent, are found in reticulocytes; some of these also phosphorylate ribosomal proteins.

Environmental factors, such as the levels of glucose and amino acids, also affect translation rates; in bacteria, amino acid starvation leads to accumulation of the nucleotide ppGpp (p. 216) but the mechanism of control in eukaryotes is unknown. Viral infection often leads to inhibition of protein synthesis in the host cell, probably through effects on initiation. Several bacterial toxins inhibit translation of eukaryotic mRNA by specific mechanisms, and interferon selectively inhibits the translation of viral mRNAs (p. 218).

Degradation of enzymes

In rapidly-growing bacteria, enzymes whose synthesis is repressed are not destroyed; their concentration within the culture is simply diluted by cell division. In animal cells, degradation of enzymes is much more important. A simple mathematical analysis shows that if the rate of synthesis of an enzyme is k_1, and the first-order rate-constant for degradation k_2, then the concentration of enzyme in the steady state is k_1/k_2, but the time taken to reach the steady state depends only on k_2; thus even if complete repression reduces k_1 to zero, the enzyme will only disappear slowly unless k_2 is large.

The measured half-lives ($t_\frac{1}{2}$) of different enzymes vary from about 11 min (ornithine decarboxylase) to over 100 days (muscle glyceraldehyde 3-phosphate dehydrogenase) and there is considerable evidence that these breakdown rates can be altered, with consequent variation in the turnover rate. Sometimes the substrate of an enzyme protects it against degradation: for example, tryptophan protects tryptophan oxygenase,

causing an apparent induction by increasing the half-life about 7-fold from its normal value of 2 h. This enzyme is also induced by hormones such as cortisone, and this induction is blocked by inhibitors of transcription or translation.

Similar considerations apply to the concentrations of mRNA, hormones, intracellular 'second messengers' and allosteric effectors: unless the rate of removal of the substance is large, in relation to its concentration at the control site, the rate at which its concentration changes will be slow even if the rate of synthesis changes greatly, and the usefulness of the substance as an effector of the process that it is supposed to control will be small.

Appendix: Definitions

Cooperative binding. Binding in which the affinity for the substrate increases with substrate concentration: it is characterized by values of the Hill coefficient (n_h) greater than 1.

Negatively cooperative binding. Binding in which the affinity for the substrate decreases with substrate concentration; it is characterized by the Hill coefficient being less than 1.

Allosteric enzyme. An enzyme which is regulated by the binding of an allosteric effector, at a site distinct from the substrate binding site.

Allosteric K-system. An enzyme in which an allosteric effector alters the affinity for the substrate. Such enzymes usually show cooperativity in substrate binding, and the effector alters the Hill coefficient.

Allosteric V-system. An enzyme in which an allosteric effector alters the maximum velocity of the reaction, but not the affinity for the substrate. Such enzymes may show cooperativity in binding of the allosteric effector, but not the substrate.

Half-life ($t_\frac{1}{2}$). The time taken for the concentration of a substance to fall to half of its original value, if it is removed by an irreversible, first-order process (such as radioactive decay). It is not strictly applicable to recycling systems. It is related to the rate-constant of the process by $t_\frac{1}{2} = \dfrac{2.3 \ln 2}{k} = \dfrac{0.693}{k}$.

Turnover rate (R). The total quantity of a compound moving through a pool (Q) in unit time, the pool being in steady state.

Fractional turnover rate. The same as the rate-constant of removal, since $R = kQ$. If there are several exits (including reversible ones) k is the sum of all the individual values.

Turnover time. The time taken for a quantity equal to the pool size (Q) to move in and out. It is numerically equal to $1/k$. For an irreversible first-order system, $1/k$ is the time taken to fall to $1/e$ of its original value, but this cannot be said to be a *turnover* time.

14

The Chemistry of Blood

In this chapter we consider blood as if it were a tissue—that is, we describe some of the important components, and their immediate function, without necessarily considering their relation to other tissues and organs. This has the advantage that we are able to describe some of the interesting new information about major systems in blood, in particular *complement* (p. 246). It has the disadvantage that information about haemoglobin is dispersed throughout the book. The structure of the subunits and the way that they interact with each other and with haem is described in chapter 13; genetic variants of the primary sequence that are of clinical importance are discussed in this chapter, together with the maintenance of haem structure by red cell metabolism. In chapter 15 the work that haemoglobin does in both O_2 and CO_2 transport, and its relation with internal organs, principally lung and kidney, is described. This is so important that to have included it in this chapter would have overshadowed the other topics.

Because of the way in which biochemistry is normally defined, as if it had no relation to cell biology, we have seriously under-stressed the importance of leukocytes, both of lymphocytes as the source of specific antibodies, and the role of 'killer' cells; and also the importance of macrophages in phagocytosis, and their cooperation with both the immunoglobin and the complement systems. This is a serious omission which will not be completely reversed until medical science is more logically ordered. A brief reference to the 'neutrophil burst' has been made in chapter 9 (pp. 166–168).

Haemoglobin

Normal haemoglobins

The relation between haem and the protein structure of haemoglobin has been outlined in chapters 2 and 13. Here we are concerned to document the normal haemoglobin occurring in blood, and some of the abnormal variants. The work of haemoglobin in red cells as it transports oxygen from the lungs, and helps

Table 14.1 Normal haemoglobins

	Chains	
Embryonic haemoglobins e.g. Hb Gower 2	$\alpha_2 \epsilon_2$	Normally present only during the first trimester of fetal life
Fetal haemoglobin HbF	$\alpha_2 \gamma_2$	Predominates after first trimester and at birth: largely replaced by HbA by 3–6 months
Adult haemoglobin HbA HbA$_2$	$\alpha_2 \beta_2$ $\alpha_2 \delta_2$	Principal adult haemoglobin Normally 2–3% of total

to transport CO_2 from peripheral tissue, is considered in chapter 15.

There are two embryonic haemoglobins which do not contain α chains, but otherwise the α chain is always present, and γ, δ and ϵ chains may be regarded as substitutes for β. Note that, although α and β chains are both normal in themselves, α_4 does not occur, and β_4 is not stable and is discussed under abnormal haemoglobins. The association of α and β subunits to form the normal haemoglobin molecule is discussed in chapter 13.

Abnormal haemoglobins

Genetic considerations

Humans each possess a pair of genes for making β chains. Structural abnormalities of β chains behave as alleles of the normal β-chain genes, as does β-thalassaemia (to be discussed in a moment).

Caucasians and Orientals have two pairs of α-chain genes, while Negroes have three genes, and Melanesians a single pair. The important point is that all these genes are located on a different chromosome from the β-chain genes; thus there is not, as might be expected, a single operon to control the synthesis of both types of peptide chain. The experimental evidence is that, although the two chains are synthesized at slightly different rates, the production of finished chains is very highly coordinated in normal bone marrow.

Inherited abnormalities exist in which mRNA production for one or the other chain is defective or non-existent. There is then a relative surplus of the other chain. These are the *thalassaemias*. Curiously, although the defect is in the rate of production of a normal chain, rather than production at the normal rate of a defective chain, inheritance of the thalassaemias behaves as though they were allelic with point mutations.

There are 141 residues in the α chain, and 146 in the β chain. Nine residues (only) in each chain may be regarded as invariant, so that there are $132 + 137 = 269$ possible point mutations in either α or β chain. Almost all of these have been detected, and most are quite harmless. In addition, for some variants that may be considered to have a protective effect, and are relatively common, individuals may be heterozygous for two abnormalities in combination (e.g. HbSC—see sickle cell haemoglobin, below). It is possible only to mention a few of these abnormalities, which have a clinical or a theoretical interest.

The thalassaemias

α-thalassaemias. Total failure of α-chain formation prevents the formation of an HbF (Table 14.1), as well as of adult haemoglobin. It is fatal in utero (hydrops fetalis).

Incomplete failure of α-chain formation leads to an excess of non-α chains. The tetramer γ_4 is known as Hb Bart's. Although it is unstable, and has a higher affinity for O_2 than does normal haemoglobin, sufferers only have a mild anaemia, as γ_4 disappears soon after birth. A separate disease in which HbH (β_4) accumulates, and causes premature destruction of the red cells, is more serious.

β-thalassaemias are more widespread, not only in the Mediterranean (from which sea they were first named), but in a belt right across the Middle East to Thailand. They seem to offer some protection against malaria, but the abnormality is not found in Africans. Complete suppression of β-chain synthesis leads to severe anaemia (thalassaemia major or Cooley's anaemia) with some persistence of HbF, but a much reduced total haemoglobin. Defects with partial loss of β chain (or β and δ chains) are quite common, but the anaemia is usually only mild.

Hereditary persistence of fetal haemoglobin

This defect, HbFH or 'high-F gene', occurs when the normal changeover, just after birth, from synthesis of γ chains to β chains fails to take place. It is distinguished from β-thalassaemia by the fact that there is no excess of α chains, and the blood haemoglobin level is, in fact, normal. HbF has a higher affinity for O_2 than HbF, so that the tissues may be somewhat anoxic, but the clinical symptoms are mild.

Point mutations

Haemoglobins M

These are a group of rare defects which have a strong theoretical interest. The 'pocket' in which the haem residue sits in haemoglobin contains two His residues, one at 87 in the α chain or at 92 in the β chain, which coordinately bonds the Fe atom (see p. 226), and another at 58 (α chain) or 63 (β chain). This latter does not coordinate with haem, but is very close to the O_2 binding site.

Four variants of HbM are known in which one or other of these histidines, on either α or β chain, has been replaced by tyrosine. Tyr has a phenolic —OH, sufficiently acidic to ionize and stabilize the Fe of the haem in the *ferric* form. The met-haemoglobin reductase of red cells (see p. 239) is not sufficiently powerful to keep the Fe^{3+} reduced. Thus in principle 50% of the total Hb in the affected blood of homozygotes would be MetHb and unable to carry oxygen; moreover, the normal subunit interactions could not take place, and the binding curve would be abnormal.

In a fifth HbM defect, Val 67 on the β chain is replaced by Glu, whose terminal —COO$^-$ is near enough to the haem also to stabilize the ferric form.

In the event, these defects are all known only as heterozygotes. The blood is dark in colour, the affected individuals are cyanosed in appearance, but they survive into old age without difficulty.

Sickle cell haemoglobin

This is the only common single residue mutation which causes severe clinical disability in homozygotes. The latter, indeed, may expect a short and painful life, terminating before adulthood. The persistence of the allele over a wide area of central Africa, from coast to coast, where 16% of children may be heterozygotes, is almost certainly because it confers protection in early life against malaria, especially the highly fatal cerebral malaria. The reasons for this are not clear cut, and only heterozygotes are protected. Sickle cell anaemia is also common in other populations of African origin, particularly in the United States.

The abnormality is codified as HbS ($\beta^{6\text{Glu}\rightarrow\text{Val}}$). The replaced residue is thus not near the haem, but on the outside of the folded peptide chain. The outstanding characteristic of HbS is that the deoxygenated form is markedly less soluble than normal deoxyhaemoglobin. It crystallizes out within the red cell to form, ultimately, long rods or fibres which distort the red cells into a sickle shape (hence the name of the disease). Sickle cells have a much shorter lifetime than normal cells—about 17 days. In consequence, the total blood haemoglobin is much reduced. Symptoms are severe pain, necrosis, and spleen infarction, among others.

Since oxygenated HbS is just as soluble as other haemoglobins, precipitation and sickling can be rapidly reversed if the red cells return quickly to the lungs. Thus heterozygotes (sickle cell trait) suffer only a mild anaemia unless exposed to unfavourable conditions— vigorous exercise, low external oxygen concentrations, etc.

Although the defect is in the β chain, the formation of HbS fibres requires the α chains for stacking; β^{S}_{4} tetramers are not insoluble. Many other heterozygote abnormalities coexist with sickle cell trait; among them may be mentioned HbC$^{(6\text{Glu}\rightarrow\text{Lys})}$. β-Chains of HbC may be found included in the insoluble deoxyhaemoglobin of HbSC patients. On the other hand, the chains of HbF do not participate in stacking, so that the double heterozygotes HbS/HbFH have milder anaemia than if they had HbS alone.

Erythrocytes

Mature mammalian erythrocytes contain no nucleus and no mitochondria. They metabolize only glucose, which they do only very slowly. One litre of packed human red cells consumes about 2 mmoles glucose/hr at 37°C; red cells from several other mammalian species use glucose even more slowly than this. There is therefore a tendency to regard erythrocytes merely as inert bags, packaging haemoglobin molecules so tightly as to be almost at the limit of their solubility (the concentration *within* red cells is about 46 g/litre cell water, or 7 mM), as a means of avoiding the osmotic and viscosity problems that so much extra protein would produce, if it were simply dissolved in plasma.

The metabolism of red cells, however slow, does nevertheless have two important functions. The first is to keep its own membrane intact. There is a large number of inherent conditions in which the red cell membrane shows abnormal fragility, which may be manifested in abnormal shape (spherocytoma), shortened red cell lifetime (often accompanied by reticulocytosis), or in impaired response to a challenge by a drug or other chemical. Here we have space only to mention two such genetic defects.

1. *Favism (primaquine sensitivity)*

In this condition the glucose-6-phosphate dehydrogenase (see p. 78, and Fig. 14.1 below) of red cells is much reduced in activity; typically it is about 15% of normal. In most cases the bearer of this defect is unaware of it until he ingests one of a range of drugs, either anti-malarials such as primaquine, or anti-bacterial drugs such as sulphonamides. A component of the fava bean (a common item of the diet in the Mediterranean) is another precipitating agent. The response is an acute haemolytic crisis. The biochemical reason is that the drugs cause oxidation of lipid components of the cell membrane and also oxidation of haemoglobin to methaemoglobin. The former is normally countered by glutathione peroxidase (see chapter 9, p. 167, and also below). NADPH is necessary to keep glutathione reduced, but the capacity of the red cells to do this is not great enough if G6PDH is defective.

2. *Pyruvate kinase deficiency*

Again there are many genetic variants of this defect. The red cell is peculiar (see p. 239) in that its Embden–Meyerhof pathway has a bypass, which means that it does not necessarily produce more ATP than it consumes (cf. the overall balance of the 'normal' pathway, p. 72). In these circumstances partial loss of activity of one of the two enzymes producing ATP may mean that the Embden–Meyerhof pathway is no longer even self-sufficient in ATP, and it may not be able to maintain the flux of glucose to pyruvate which is necessary to provide reducing equivalents for glutathione and methaemoglobin reductases. Here again the deficiency, although ostensibly in energy metabolism, is fundamentally concerned with repairing damage to the membrane and to the cell contents resulting from the presence of O_2.

Methaemoglobinaemia

The second function of red cell metabolism is to prevent the accumulation of metHb. Not all methaemoglobinaemias involve red cell fragility; we have already seen that heterozygotes for HbM deficiencies (p. 237) have a normal life expectancy. Any reduction of the oxygen-carrying capacity of the blood must, however,

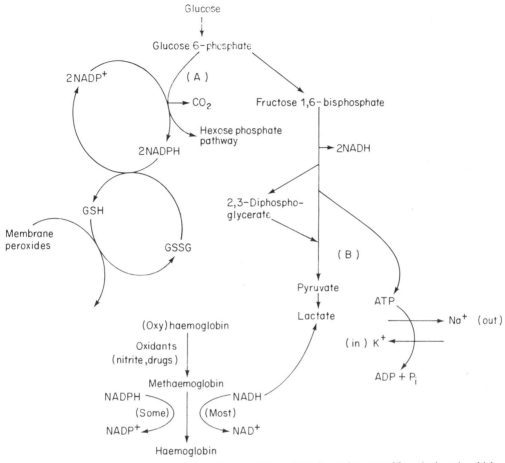

Fig. 14.1 Summary of glucose metabolism, and associated redox reactions, in erythrocytes. Note the loop in which 2,3-diphosphoglycerate (see this page) is formed. Glucose-6-phosphate dehydrogenase defects act at point (A). Pyruvate kinase defects act at point (B).

be potentially serious, particularly to tissues with a heavy O_2 demand, such as brain.

Red cells contain a flavoprotein, *methaemoglobin reductase*, which continually reduces the metHb formed in small amounts as Hb transports O_2 around the circulation. It can use both NADH and NADPH as substrates, but the former is preferred. The red cells have a certain capacity to increase NADH/NADPH production, but even in the absence of enzymic defect, this capacity can be swamped by agents that promote the oxidation of the Fe^{2+} in haemoglobin. Among many such substances, nitrite may be specifically mentioned. Nitrate can accumulate in drinking water, both from overuse of agricultural fertilisers, and as an end-product of sewage treatment. Though in itself harmless, it is easily reduced to nitrite in the intestinal tract. The latter is a mild oxidant which oxidizes Hb to metHb, at a rate which may be beyond the capacity of the cells to counteract. As it happens, this only occurs in infants, because HbF is more easily oxidized than HbA, but the

resulting methaemoglobinaemia is serious enough for brain damage to be caused. In locations in which high NO_3^- concentrations in drinking water may at any time be expected, precautions are taken to keep a stock of NO_3^--free water for making up artificial milk preparations.

The diphosphoglycerate shunt

The preceding discussion has underlined the fact that the demand on red cell metabolism for reducing equivalents is greater than the demand for energy as ATP. Nevertheless, control of the Embden–Meyerhof pathway is always organized in such a way that the rate of ATP demand decides the rate of flux through the pathway. In order to counter this, a shunt system has evolved in erythrocytes. It is shown in Fig. 14.1. Instead of 1,3-diphosphoglycerate reacting with ADP to give 3-phosphoglycerate and ATP, it can also be transformed by a mutase to 2,3-diphosphoglycerate. Since this compound cannot transfer either phosphate group to ADP, the result is loss of 1 ATP (per triose) in the energy balance of the complete pathway.

2,3-Diphosphoglycerate is hydrolysed by a specific phosphatase to 3-phosphoglycerate and inorganic phosphate, but as the activity of the phosphatase is low, 2,3-DPG accumulates in the red cells. Its concentration in normal erythrocytes is about 5 mM, while that of ATP is 2 mM. Both these compounds are effector ligands for haemoglobin, and the consequences for O_2 transport in blood are discussed in the next chapter. Here it is only necessary to point out that, although the concentration of 2,3-DPG does vary with O_2 demand, the changes must necessarily be slow, because of the very slow throughput of glucose in the system. For example, 2,3-DPG doubles in response to living at lower O_2 tension (high altitude), but $t_{\frac{1}{2}}$ for the change is 15 hr. Changes in 2,3-DPG in response to instantaneous changes in oxygen load are not to be expected.

Blood Clotting

The major processes involved are shown in Fig. 14.2. They are the following:

(a) Factor XII (Hageman factor) circulating in plasma is activated on contact with collagen. This can only occur if the blood vessels are damaged, and is the start of the *intrinsic pathway* of activation of coagulation. Blood collected in siliconed vessels without platelet damage, from which the platelets are then removed by centrifugation, will not clot for long periods. The platelets possibly activate Factor IX.

(b) There follows a succession of steps in which proteolytic enzymes of the serine esterase type are successively set free by the removal of a masking peptide. Factor X (Stuart factor) can be activated not only by Factor IXa, but also by Factor VIIa. The latter is activated by extracts of several tissues (especially brain). This mode of activating Factor X is known as the *extrinsic pathway* (see Fig. 14.2). It is less well understood than the intrinsic.

In the activation of Factor X, not only IXa, but also the active form of a non-enzymic protein, Factor VIIIa, has to be involved, as well as phospholipids and Ca^{2+} ions. Factor VIII is a very large protein (mol. wt 1.1×10^6) normally present in very low concentrations in plasma. Its absence causes classical haemophilia. The gene coding for it occurs on the X chromosome. It is therefore transmitted by females, but the defect occurs only in males (about 1 in 10000 male infants).

(c) Factor Xa converts prothrombin (Factor II) into thrombin (Factor IIa), again by selective removal of a (large) peptide-masking fragment. This reaction requires Ca^{2+} ions, phosphatidyl-serine or -ethanolamine, and the active form of Factor V (accelerin) as well as Xa.

(d) Thrombin hydrolyses Arg—Gly bonds near the ends of the two α and β chains of fibrinogen, which is a long fibrous molecule (subunit structure $\alpha_2\beta_2\gamma_2$), with globular end domains. The peptides released by this hydrolysis are small, but rich in acidic amino acids. Their removal is thought to alter the charge density of the molecules to such an extent that the fibrin associates (predominantly end to end) to form an open lattice or *soft clot*.

(e) Inter-chain peptide bonds are formed between the fibrin molecules by Factor XIIIa, by the following mechanism:

This produces a rigid lattice or *hard clot*.

Thrombin activates Factor XIII (the transglutaminase) by hydrolysing bonds in the two α chains, which with two β chains make up the inactive Factor XIII in plasma. After the loss of two peptide fragments, the β chains which are regulatory or masking, diffuse away.

$$\alpha_2\beta_2 \xrightarrow{\text{thrombin}} \alpha_2'\beta_2 + 2 \text{ peptides}$$

$$\textit{Factor XIIIa}$$

$$\alpha_2'\beta_2 \longrightarrow \underset{\substack{\textit{active}\\ \textit{transglutaminase}}}{\alpha_2' + 2\beta}$$

Platelets contain an α_2 dimer which on release is more rapidly converted to α_2' than is the tetramer in plasma.

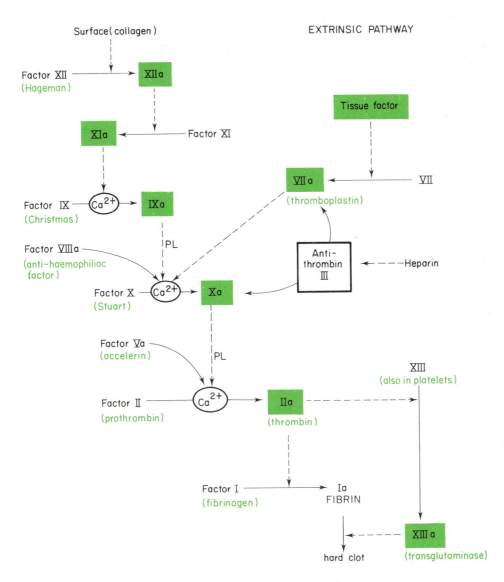

Fig. 14.2 Major stages in blood clotting.

(f) After a few days, clots are lysed by plasmin (fibrinolysin), formed from a plasma precursor (plasminogen) by proteolytic hydrolysis. The bacterial proteolytic enzyme streptokinase is also used clinically to activate plasminogen. Plasmin itself preferentially hydrolyses arginine—lysine bonds in fibrin.

The role of platelets

The summary above describes the process of coagulation as it would occur in a silicone-treated vessel. In vivo, however, step (a) would be preceded by the formation of a platelet plug.

Platelets are disc-shaped cells that have lost their nuclei, but contain many dense granules and also mitochondria. They adhere tightly to collagen (see also p. 240), which is not exposed unless the vascular lining is damaged. Upon adherence they become spherical, with spiny extrusions, and ADP and serotonin (p. 148) are released from the granules into the surrounding plasma. The ADP causes the platelets to become sticky; this causes more platelets to adhere and release their ADP. The result is the formation of a plug which will seal the rupture in a small blood vessel until the formation of a fibrin clot begins (6–12 min).

Platelets have at least two other functions in clotting: they act as a store both of the phospholipids for the activation of Factors X and II (Fig. 14.2), and of the fast-activated form of Factor XIII. Secondly, they provide the precursor (arachidonic acid) and the enzymes which catalyse the formation of thromboxane, the powerful vasoconstrictor which is related to the prostaglandins. This is described in more detail on p. 120 (see also p. 248).

Inhibitors of clotting

Blood coagulation is an irreversible cascade process. The later stages will therefore occur much more rapidly than the early stages, because there is a higher concentration of the catalysts. This can be seen by comparing the standard clotting time, 6–12 min, with the time of clotting after adding Ca^{2+} to decalcified plasma (effectively initiating conversion of IX to IXa),

which is 3 min, and the time taken to clot when thromboplastin is added to recalcified blood (15 sec). In order to prevent the appearance of thrombin in plasma, in the absence of injury, blood contains many clotting inhibitors, usually acting on enzymes early in the cascade. One of the most studied is antithrombin III, which forms inactive complexes with Factors VIIa and Xa, and also (slowly) with thrombin itself. A hereditary defect leading to absence of antithrombin III from plasma gives rise to severe thrombosis.

Heparin is an activator of antithrombin III, and can prevent clotting indefinitely, both in vivo and in vitro. In vitro, the Ca^{2+}-complexing agents oxalate, citrate or ethylene diamine tetra-acetic acid (EDTA) are used to inhibit clotting. They would have the same effect in vivo, so care has to be taken with their use (e.g. as an antidote in heavy metal poisoning).

Plasma Proteins

About 80 proteins can be visualized in plasma by sophisticated two-dimensional analytical techniques, e.g. electrophoresis and immunodiffusion. There are probably many more than this which can be measured by specialized techniques such as enzyme assay, radio-immunoassay or immunoradiometric assay (cf. alpha fetoprotein, below). A one-dimensional separation method will visualize about 30 components (see Fig. 14.3(b)). Only a few of these can be mentioned here. Table 14.2 lists some of the major components in descending order of average concentration.

Plasma is noticeably more dense than water (S.G. ∼ 1.025). The original classification into albumin and globulin depended on solubility in water or dilute salt solution; the first electrophoresis revealed three globulin fractions (α, β and γ) with different mobilities. Fig. 14.3(b) shows that more modern techniques do not necessarily separate components in this order, although the terms 'γ-globulin', 'pre-beta' or 'pre-albumin' continue to reflect the use of electrophoresis for separations.

Albumin

This is both the smallest in size of the major proteins of plasma, and the one present in highest concentration. The concentration falls noticeably in under-nourishment. Surprisingly, in view of the complex polymorphism of haemoglobin and of the haptoglobins, only one variant is known, which gives rise to a double band on electrophoresis (bisalbuminaemia).

It is difficult to describe a precise function for

albumin. It is easily shown that in the normal person, reduction in plasma albumin concentration is associated with oedema, so that the protein plays an important part in the maintenance of colloid osmotic pressure. Similarly, albumin transports free fatty acids, most of the plasma bilirubin, and several other small lipophilic molecules.

However, in the rare inherited defect *analbuminaemia*, albumin is virtually absent from plasma, and the consequences are minimal. The colloid osmotic pressure is regulated partly by increases in the concentration of the globulin fractions, and partly by the establishment of a lower arterial blood pressure.

Albumin is synthesized in liver, at the rate of 3–4 g/day in an adult.

Alpha fetoprotein

Fetal plasma contains a protein very similar to albumin and coded for on a neighbouring gene. It contains, however, about 5% carbohydrate. The concentration falls to an extremely low level (10 ng/100 ml), soon after birth. Alpha fetoprotein is of importance for two reasons:

(a) In spina bifida and other neural tube defects, the protein 'spills over' into amniotic fluid. Measurement of the concentration provides a very reliable way of diagnosing spina bifida before birth. It is also possible to screen mothers during pregnancy, with less reliability, since a fraction of the protein crosses the placental membrane. Very sensitive analytical methods are needed for this.

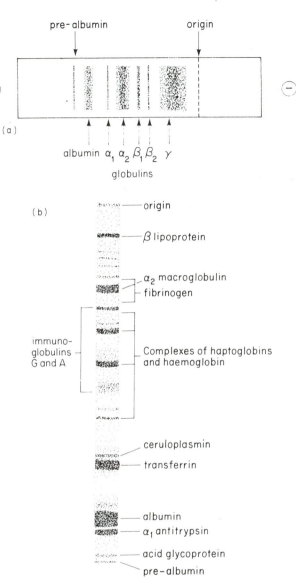

Fig. 14.3 Electrophoretic separation of serum proteins. Serum, rather than plasma, was used in these separations because fibrinogen obscures other proteins when plasma is used. (a) Simple separation on cellulose acetate, to show the rapidly moving albumin (and pre-albumin) and the five major groups of more slowly moving globulins. (b) Two-stage separation on polyacrylamide gel. Although this shows many more proteins as discrete bands, it still does not separate well the haptoglobins or the immunoglobulins. About 80 proteins have been identified by a combination of separation techniques.

Table 14.2

Plasma proteins	Normal values (g/100 ml)
Total	6.3–7.8
Albumins	3.2–5.1
α_1-Globulins	0.06–0.39
α_2-Globulins	0.28–0.74
β-Globulins	0.69–1.25
Immunoglobulins (or γ-globulins)	0.8–2.0
IgA	0.15–0.35
IgG	0.8–1.8
IgM	0.08–0.18
IgD	approx. 0.003
Fibrinogen	0.2–0.4
Mucoprotein	approx. 0.135
Haptoglobins	0.03–0.19

They have great potential value for the diagnosis of cancer, particularly for early screening, but at the moment the process of establishing the reliability and precision of the sensitive assays required, together with the expense of the sensitive assays required, means that they are not in routine clinical use.

Only 7 or 8 of the proteins in plasma are regularly measured in clinical practice, and not all these measurements are of great value. The summary below does not include enzymes, whether normally present in plasma, such as the acid and alkaline phosphatases, or those which leak out from damaged cells (creatine kinase, lactate dehydrogenase).

Albumin. The importance of albumin as a measure of nutritional state or in relation to oedema should be self-evident. Low also in kidney failure.

Alpha fetoprotein. Not routine. Useful for diagnosis of fetal neural tube defects, or primary hepatoma.

α_1-Antitrypsin (better called α_1 protease inhibitor). This protein inhibits leucocyte proteases, which in its absence attack the connective tissue support of lung and liver.

Antithrombin III (see p. 242). Reduced in heparin therapy, and in about 6% of women taking a contraceptive pill regularly.

Ceruloplasmin (ferroxidase, p. 273). For estimating copper status.

Clotting factors   Diagnosis of abnormalities of clot-
Complement   ting and of resistance to bacterial infections.

Haptoglobin. These proteins bind the α chain of

(b) In adult life, a raised alpha fetoprotein concentration is diagnostic for primary hepatoma, and is of possible value in one or two other cancerous conditions.

There are several such *oncoproteins* which are found in plasma when tumour cells have become established.

haemoglobin if it escapes from the red cell. Not very useful for diagnosis.

Lipoproteins (see chapter 7). Abnormalities in lipid metabolism; diagnosing hypercholesterolaemia.

Thyroxin-binding globulin. For estimating thyroxin status. Not very useful as free thyroxin in plasma may be the same in the presence or absence of TBG.

Transferrin. For estimating iron status.

Defence Mechanisms

Immunoglobulins

These rather large proteins move most slowly on the simplified electrophoresis strips (see Fig. 14.3(a)), and therefore came to be known as the γ or 'gamma' globulins. They are complex proteins whose general structure is shown in Fig. 14.4. There are two types of chain, the *light* and the *heavy*, synthesized separately, but connected to each other by one —S—S— bond. There are two heavy and two light chains in each immunoglobulin molecule, but in IgM, five such dimers associate together, and IgA has a tendency to dimerize (see molecular weights, Table 14.3). The light chains (mol. wt 22 000) can be of two classes, κ and λ; they are overproduced in so-called immunocytoma tumours (mostly myelomatosis) and are lost in the urine as

Bence-Jones proteins. The heavy chains exist in five classes (see Table 14.3), which are used to subdivide the immunoglobulins themselves.

The unique property of the immunoglobin peptide chains is that at the N-terminal end of each heavy or light chain there is a region (or 'domain') which does not have a constant primary sequence. These regions are known as the variable domains, V_L and V_H respectively, and within them there are three or four *hypervariable sequences*, where most of the variations occur. It is beyond the scope of this book to discuss how a clone of lymphocyte cells (mostly β cells) is stimulated to produce a particular antibody, of defined sequence in the variable domains, against a single antigen, but the result is that the Y-shape of the complete immunoglobulin creates a pocket, of individual three-dimensional

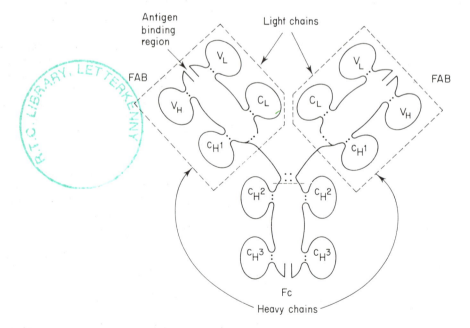

Fig. 14.4 Diagrammatic view of an immunoglobulin molecule. Papain splits the heavy chains below the —S—S— bonds (indicated by ..) which link the heavy and light chains, releasing two FAB (fragment antigen binding) peptides, which can still bind to an antigen, but which do not have the other biological properties of the complete Ig (see text). The latter properties (e.g. complement binding) reside on the Fc region of the heavy chains. Pepsin splits below the intra-chain —S—S— links, releasing two Fc fragments per Ig molecule. The diagram shows that each light chain has one variable domain (V_L) and one constant domain C_L, while each heavy chain has one variable domain (V_H) and three constant domains (C_H^{1-3}). Some Ig molecules have four or more constant domains (see text).

Table 14.3 Physical properties of immunoglobulins

	IgA	IgD	IgE	IgG	IgM
Light-chain	κ or λ	κ or λ	κ or λ	κ or λ	κ or λ
Mol. wt	22 K	22 K	22 K	22 K	22 K
No. of constant domains	1	1	1	1	1
Heavy chain	α	δ	ϵ	γ	μ
Mol. wt (including CHO)	55 K	62 K	70 K	50 K	70 K
No. of constant domains	3	3	4	3	4
Extra components	SC, J				J
Total mol. wt	162 K (390 K)*	178 K	188 K	146 K	880 K
Concentration in serum (mg/100 ml)	210	3	0.01	1200	190

* Dimer, including secretory piece (SC).

internal form, into which a part of the antigen molecule can fit closely, and bind tightly, with rather little cross-reaction between different antigens. This is visualized in Fig. 14.5.

The part of the antigen which reacts with the immunoglobulin can be quite small. In Fig. 14.5 the reacting molecule is phosphoryl choline (see p. 108), which is a *hapten*, i.e. it will not itself stimulate antibody production, but will confer antigenic properties on a neutral polymer if it is attached covalently to it. Each immunoglobin molecule has at least two antigen binding sites. It can either bind two molecules of

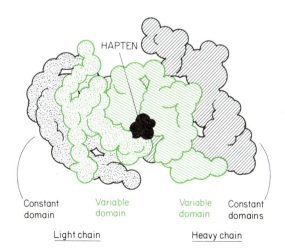

Fig. 14.5 Three-dimensional representation of an immuno-globulin binding to a hapten. The latter (phosphoryl choline) is much smaller than a complete antigen with which the antibody molecules would co-precipitate, but the sketch shows clearly the deep pocket in which the variable domains enfold the reactive molecular shape.

antigen, or if the antigen is a cell, it can form multiple attachments to a single cell surface. IgM, with 10 binding sites, is particularly effective in the classical antibody function of co-precipitating with an antigen.

The non-variable parts of the Ig molecules, called Fc in Fig. 14.4, do not just act as a convenient handle for the variable domains. In fact the five Ig classes have quite different secondary properties, and these are conferred by the C_H2 and C_H3 domains (also C_H4 in IgM) of the five types of heavy chain. The relative concentrations in serum do not reflect these properties. For example, IgA dimers bind a separate polypeptide chain, a secretory component (SC), which seems to facilitate secretion through mucous membranes. On the other hand, IgG is the only immunoglobulin class which can pass through placental membranes, and thus confer some of the mother's immunity to the fetus.

Table 14.4 Secondary biological properties of immunoglobulins

Secretion through mucous membranes	IgA
Movement through placenta	IgG
Binding to mast cells	IgE
Complement binding	IgM(+ +), IgG
Binding to phagocytic cells (opsonization)	IgG1, IgG3
Antibody-dependent cytotoxicity (killer-cell activation)	IgG

Although IgE is almost absent from serum, it is present in extravascular tissue, bound through domains C_H2 and C_H3 to the plasma membranes of mast cells, particularly in the lung. Attachment of antigens to these cell-bound IgE molecules triggers off a series of reactions at the cell receptor which result in degranulation of the cells and the release of histamine and other substances (*immediate allergic reaction*). It can be shown that the concentration of cell-bound IgE is higher in allergy sufferers than in the normal population.

IgG antibodies bind to phagocytosing cells (monocytes and neutrophils) through a specific receptor, thus making a bifunctional link between the antibody (or foreign cell) and the phagocyte (*opsonization*). In addition, foreign cells to which IgG has been bound are more susceptible to attack by a non-phagocytosing mechanism involving so-called 'killer' cells (K-cells). This is *antibody-mediated cellular cytotoxicity*. In each case a specific domain in the Fc region of the antibody region is involved.

In activation of the *complement* system (see below), it is the C_H2 domains of IgG and IgM that bind C1q, the first component of the classical complement pathway.

At the present moment, least is known about the secondary biological functions of IgD. For the other classes, integrated biochemical and immunological study has shown how several mechanisms, each concerned with defence against foreign polymers or cells, have come to be associated with a single polypeptide chain (the heavy chain). These are summarized in Table 14.4. The C_H1 domain appears to be concerned mainly with the non-covalent tertiary structure between the two prongs of the Y-shape of the intact molecule, so that the secondary properties summarized in Table 14.4 are almost entirely associated with the C_H2 and C_H3 domains. Little is known at present about the functions of the C_H4 domains of IgM and IgE.

Complement

It has long been known that antibodies alone do not protect animals from bacterial infection. Antibodies agglutinate bacteria which have infected a host, but do not lyse them. For lysis a second defence mechanism is required, known as *complement*. This consists of a series of labile proteases, each of which activates the next, until a lytic complex is formed which attacks the bacterial cell walls. As Fig. 14.6 shows, there are two methods of initiation of the cascade, one (the classical pathway) which involves binding to an aggregated antibody, and one (the alternative pathway) which is less well understood, but which probably involves direct recognition of antigenic polysaccharides of bacterial cell walls.

The complete protein chemistry of factor C5 and the way in which it combines with factors C6–C9 have not yet been worked out, but some features of the earlier

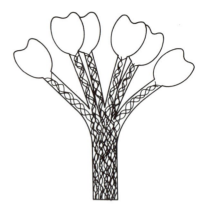

Fig. 14.7 The structure of complement factor C1q, showing the six antibody-binding heads, and the six connecting strands whose structure is very similar to that of collagen. This protein is known as 'the bunch of tulips'.

stages of the classical pathway may be briefly described by reference to Fig. 14.6.

The initial phase is the most striking part of the system. It depends on a large protein (mol. wt 400 000) called C1q which has six sets of three protein chains (see Fig. 14.7). Each of the sets has a globular head and a long collagen-like stem; the 6 stems are firmly bound to each other at the N-terminal end. The heads bind strongly to the second constant domain, C_H2, of antibody molecules, but only if the latter have been aggregated by antibody. Thus each molecule of C1q probably binds six to twelve antibody molecules in all.

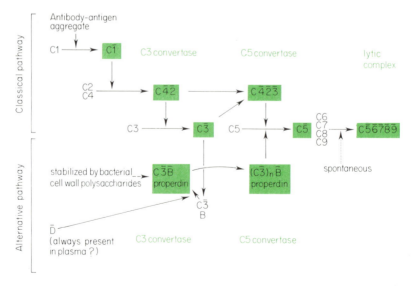

Fig. 14.6 Pathways of complement activation. The diagram shows both the classical (upper) and the alternative (lower) pathways. The factors shaded in green are the active forms, indicated in print by a bar over the identifying number, thus: C3 (inactive), C̄3 (active).

246

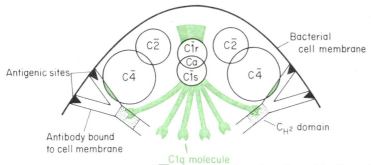

Fig. 14.8 Sequential binding of C1q, C1r, C1s, and C$\overline{42}$ to antibodies bound to a bacterial cell membrane. The cell membrane has been made concave purely for the sake of clarity. Note how the 'bunch of tulips' shape of C1q enables it to bind to the C$_H$2 domains of several antibody molecules, giving a very firm binding. Ca^{2+} is also involved in binding. The later stage in the process is the arrival and conversion of C5, to give a complex that can lyse the cell membrane.

Two smaller proteins, C1r and C1s, can then bind to the end of the C1q molecule, as shown in Fig. 14.8. C$\overline{1}$r is a specific protease which cleaves a small peptide from C1s (its only substrate); this unmasks protease activity in C1s.

C$\overline{1}$s is less specific and will split a terminal peptide from both C4 and C2. Both these proteins are originally in solution, but a small fraction of activated C4 (C$\overline{4}$) can bind covalently (the nature of the bond is still uncertain) to the Fab region of the antibody molecules. C2 then binds to C4, and the two together form another protease, the 'C3 convertase' or C$\overline{42}$, which is very unstable, with a life of less than a minute. The reason for this, and for the binding to the cell–antibody–C1 complex, appears to be to prevent the appearance of uncontrolled protease activity free in plasma.

C3 convertase, as the name suggests, activates C3, probably also by proteolysis, but the detailed structure of C3 has not yet been worked out. Enough has been said, however, to indicate many similarities with the blood clotting cascade (p. 241). However, the complement system has some unique features.

1. It is not an amplification cascade, since—unusually—the first enzyme (the protease C$\overline{1}$r) and its substrate C1s are present in equimolar concentrations.
2. The complex binding of the components to antibody, and ultimately to the bacterial cell surface, ensures both a high local concentration of active molecules and avoidance of indiscriminate proteolysis.

There are a number of inhibitors and activators, particularly of C3, that will not be discussed here.

Fig. 14.6 shows that the 'alternative pathway' interacts with the classical pathway at the level of C3. It is not known precisely how initiation takes place, but it is thought that a low level of the first protease (D) is present in plasma all the time, but that the system is greatly speeded up if polysaccharides (such as bacterial cell wall material) bind to the convertase enzyme C3B,

and protect it from the inhibitors mentioned in the previous paragraph.

Finally, many of the small peptides which are split off during the complement activation process are themselves biologically active. C3a and C5a, for example, are active on the walls of blood vessels. C5a, as well as the complex C$\overline{5}$C$\overline{6}$C$\overline{7}$, are leucotactic substances.

Leucotactic substances

Histological observation shows that leucocytes of various types are found in high concentration near the site of a tissue injury or bacterial infection. This is partly because they can change shape and move along a chemical gradient of an attractor substance.

Many such attractants are known. Complement fragments, especially C5a, have already been referred to; another group are small peptide fragments possessing an *N*-formyl-methionine residue at the N-terminal end. FMet-Leu-Phe is the most active of these. Since the initiating Met in eukaryote protein synthesis is not formylated, these fragments must be a result of prokaryote (i.e. bacterial) protein synthesis. Both C5a and the FMet peptides bind to specific receptor sites on neutrophils.

A third group of substances are derivatives of arachidonic acid. The most powerful attractant currently known is 5,12-dihydroxyeicosatetraenoic acid (5,12-diHETE) (note the shifts in double bonds by comparison with arachidonate):

247

This is formed in a series of reactions whose beginning is indicated in Fig. 14.9. Glutathionyl or cysteinyl derivatives of the HETEs are the slow-acting substances called the *leukotrienes*.

The sequence of events leading to the release of HETEs from neutrophils is shown in Fig. 14.9. It is typical of several events, e.g. release of thromboxane, p. 242, which are initiated by receptor binding, followed by activation of a Ca^{2+}-dependent phospholipase A_2 which sets free arachidonate. Note that all the reactions take place within the membrane.

A second point of interest shown in Fig. 14.9 is that glucocorticoids bind to a separate receptor on the same membrane and release a lipoxygenase inhibitor (called lipomodulin) which diminishes HETE secretion. This complex picture provides some molecular explanation of one way in which steroid hormones and drugs reduce inflammation. The penalty for this relief is that they may also weaken the leukocyte response to a bacterial infection by suppressing the formation of a gradient of attractant substances.

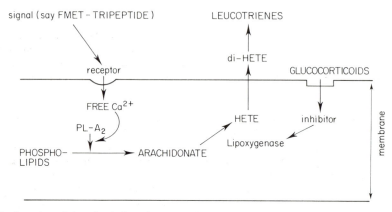

Fig. 14.9 Schematized version of the stimulation of a neutrophil to produce a leucocyte attractant (di-HETE). Notice that all the reactions shown take place within the cell membrane. Condensation of HETE with glutathione to form a leucotriene probably also takes place within the cell. Inhibition of arachidonate oxidation by lipomodulin, as a result of glucocorticoid binding to a receptor, is also indicated.

15

Fluid, Electrolyte and Acid–Base Balance

Gas Transport

Oxygen

Haemoglobin

The structure of haemoglobin and its relation to function have been described in chapters 2 and 13. In this chapter its physiological role in transport of both O_2 and CO_2 will be discussed together with the way in which its behaviour is modified by its localization in erythrocytes. Some abnormal haemoglobins, regarded purely as genetic variants of protein structure, were mentioned in the previous chapter. The only one which it is necessary to refer to here is fetal haemoglobin (HbF).

Figs 15.1 and 15.2 show the relationship between the O_2 and saturation curves of pure adult Hb(HbA) at pH 7.4, and related saturation curves of importance. It will be seen that although they all (except myoglobin, Fig. 15.1, included for comparison) have the same characteristic sigmoid shape, the steeply rising portion

of the curve can be shifted to the right or the left in various circumstances. This is equivalent to saying that the P_{O_2} of half-saturation (the P_{50}) can be greater or smaller than the value characteristic of pure HbA in solution, which is about 11 mmHg (1470 pascals).

Diphosphoglycerate binding. This is discussed as an example of an allosteric heterotropic effector, in chapter 13. Here we are more concerned to stress that it is the presence of up to 5 mM diphosphoglycerate in red cells that is primarily responsible for the difference between the saturation curve of pure HbA (P_{50}, 11 mm) and of Hb in lysed red blood cells (P_{50}, 26.5 mmHg). This difference had been known for many years before the effect of DPG binding on the quaternary structure of haemoglobin was discovered. One molecule of DPG binds per molecule of haemoglobin (not per haem unit).

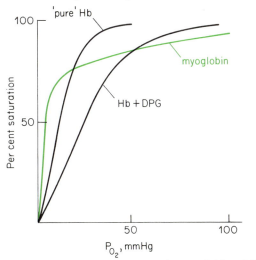

Fig. 15.1 Oxygen saturation curves for myoglobin, adult haemoglobin (HbA) and HbA in the presence of 5 mM 2,3-diphosphoglycerate. Note the hyperbolic shape of the myoglobin saturation curve, and the marked reduction of O_2 affinity in the presence of DPG.

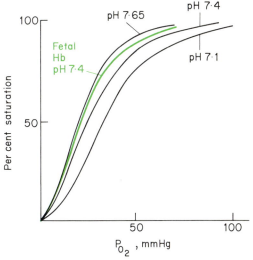

Fig. 15.2 The Bohr effect. The saturation curve for pH 7.4 is very similar to the 'HbA + DPG' curve in Fig. 15.1. Note how increased acidity (pH 7.1) diminishes and decreased acidity (pH 7.65) increases the affinity of Hb for O_2. The saturation curve for fetal Hb (HbF) has been included for comparison with HbA. It shows a slight but significant increase in O_2 affinity compared with the latter.

As emphasized in the previous chapter, DPG changes in concentration only very slowly. Changes in DPG, and thus of effective oxygen affinity, are observed in chronic hypoxia, such as is produced by moving to high altitudes, and occur within 24 hr. (Note that another physiological response to this stress is increased erythropoiesis.)

Fetal haemoglobin. Pure fetal haemoglobin (HbF, chapter 14) has a lower affinity for O_2 than adult haemoglobin, but it binds diphosphoglycerate less readily than does HbA, so that *in the red cell*, HbF has a higher affinity for O_2 (P_{50}, 22 mmHg). This is indicated in Fig. 15.2.

Although the difference in P_{50} between HbF and HbA in red cells is small, at moderately low O_2 tensions the difference in saturation will be quite large. For example, at a P_{O_2} of 26 mmHg, HbA will be 50% saturated, but HbF about 70% saturated. This explains how O_2 is transferred from maternal to fetal blood across the placenta. At low O_2 tensions the curve for HbF is very much like that of HbA so that unloading in fetal tissues to supply their O_2 requirements is adequate.

pH: the Bohr effect

This effect, which depends on changes in the environment of a particular histidyl residue, His[146], on deoxygenation, altering its effective pK (making it a weaker acid), can be most easily understood from Fig. 15.2. Increase in [H^+]—*decrease* in pH—shifts the saturation curve to the *right*, so that at any pH more acid than that of normal blood, haemoglobin has *less* affinity for O_2 than it would have at pH 7.4. Conversely, a fall in [H^+]—*increase* in pH—shifts the curve to the *left*, giving a *greater* affinity than at pH 7.4. The effect is continuous, so that the magnitude of the difference in affinity depends on the extent of the pH shift.

The converse of the Bohr effect is also important: because HbO_2 is a stronger acid than Hb, the protein releases protons into the medium as it takes up O_2. Unlike the DPG effect, the Bohr effect is shown by each β chain so that the maximum gain or loss of protons per molecule should be 2. In fact, because at pH 7 the histidyl residues are already partly ionized, the net production of H^+ per mole of Hb on complete oxygenation is 0.7 mole. (Haemoglobin has other ionizable side chains, and is a net anion at pH 7, but these groups are not involved with loss or gain of O_2.) The proton uptake and release described in this paragraph is important in balancing the changes in pH that follow changes in blood P_{CO_2}.

Carbon dioxide

CO_2 diffuses from tissues into blood as the latter is passing through the capillary system. The evidence is

that it passes across cell membranes as dissolved and thus electrically neutral gas. The hydration of CO_2 to H_2CO_3 is rather slow, and no doubt much of the gas would be transported to the lungs in simple solution if it were not for the presence in red cells of a Zn-containing enzyme, *carbonic anhydrase*, which speeds up the reaction about 1000-fold

$$H_2O + CO_2 \rightarrow H_2CO_3 \quad K_{eq}\ 0.003 \qquad (1)$$

The ionization of H_2CO_3 is almost instantaneous:

$$H_2CO_3 \rightarrow H^+ + HCO_3^- \quad pK = 3.8 \qquad (2)$$

The second ionization of HCO_3^- to CO_3^{2-} and H^+ can be neglected in this context, because it only begins to be significant at pH 8.

The 'true' ionization constant quoted in equation (2) shows carbonic acid to be a relatively strong acid (as strong as lactic acid). However, because it is technically very difficult to determine [H_2CO_3] separately from [CO_2], it is usual to determine the sum of the concentration of H_2CO_3 and dissolved CO_2. From this is derived a secondary equilibrium constant,

$$K_1' = \frac{[H^+][HCO_3^-]}{[CO_2 + H_2CO_3]} \equiv \frac{[H^+][HCO_3^-]}{qP_{CO_2}} \qquad (3)$$

with a derived pK_1' which has the more familiar value of 6.1. The normal pH of blood is thus very much to the alkaline side of the pK, but the discussion of the $CO_2/H_2CO_3/HCO_3^-$ buffer system in chapter 1 makes it clear that carbonic acid is almost unique in providing effective buffering, in physiological conditions, even at pH 7.4.

Note that as P_{CO_2} in alveolar air varies, both [CO_2] and [H_2CO_3] in blood will vary simultaneously. In equation (3), q is a proportionality constant.

If other things were equal, the increased concentration of [$CO_2 + H_2CO_3$] in venous blood would lead only to a trifling increase in [HCO_3^-] in red cells, but at the same time as CO_2 is entering plasma, O_2 is leaving it, to support tissue oxidations. As we have seen in the previous section, de-oxygenation of haemoglobin makes it a less strong acid, and it takes up protons. This has the effect of pulling the reaction of equation (2) to the right, increasing substantially the amount of HCO_3^- in the red cells. At the same time, although Hb remains a net anion, the ImH^+ (protonated histidinyl) residues formed when haemoglobin loses its oxygen, act as cations to balance the increase in bicarbonate anion. There is thus no need for any cation movement across the red cell membrane.

The extent to which the effective reaction

$$H^+ + HbO_2 \rightarrow HHb + O_2 \qquad (4)$$

comes into play depends on many factors, not least the Respiratory Quotient (chapter 16). In general, however, the buffering power of the red cell haemoglobin is great enough to prevent a drop of more than 0.03 pH unit *in plasma*.

It is common to say that the acid produced by metabolism in man, as CO_2, is equivalent to 25 Eq H^+, or about 2 litres of concentrated HCl, per 24 hr. This statement is somewhat misleading; it represents the *potential* yield of H^+ if reactions (1) and (2) went to completion for every molecule of CO_2 released from tissues. In the absence of carbonic anhydrase and of haemoglobin, most of the CO_2 would stay in simple solution, as indicated in the discussion of gas transport by perfluoro compounds (p. 252). In addition, the release of CO_2 is continuous with time. The calculation nevertheless shows how potentially serious a derangement of gas transport could be, in terms of the body's acid–base balance. This is discussed in a later section.

Carbamino compounds. CO_2 reacts with primary amino groups to give carbamino compounds (substituted carbamates):

$$-NH_3^+ + CO_2 \rightleftharpoons -NH \cdot COO^- + 2H^+ \qquad (5)$$

The reaction is rapid, and reversible when the P_{CO_2} drops. As equation (5) shows, the process does nothing to buffer acid resulting from CO_2 liberation, indeed the reverse.

In the conditions pertaining to blood, the terminal $-NH_2$ groups of haemoglobin chains are the only important candidates for carbamino compound formation. There is some uncertainty about the extent to which it occurs, since diphosphoglycerate binds to the same charged groups, and the two ligands compete with one another. Most authorities suggest that slightly less than 30 % of the *extra* CO_2 in venous blood is in the form of carbamino compounds, with 65 % as HCO_3^-, and only 8 % as dissolved $CO_2 + H_2CO_3$.

The process in the lungs

In the lungs, the events take place in the reverse order. The changes have to be very rapid, because it is estimated that a bolus of blood spends only about 1 second in close proximity to the alveolar membrane, the only place in the lungs where gas exchange can take place.

The chloride shift

Almost all the HCO_3^- produced when CO_2 diffuses into capillary blood is formed in erythrocytes. This produces an osmotic problem: the number of osmotically active particles inside the cells increases. Red cell membranes do possess a Na^+/K^+ pump, but it is not very active, so that for all practical purposes the red cell is impermeable to cations (including H^+). Any adjustment of osmotic pressure must therefore come about by movement of anions— Cl^-, HCO_3^- and OH^-.

Applying the Gibbs–Donnan equilibrium to this system, we have

$$\frac{[HCO_3^-]_c}{[HCO_3^-]_p} = \frac{[Cl^-]_c}{[Cl^-]_p} = \frac{[OH^-]_c}{[OH^-]_p} = \frac{[H^+]_p}{[H^+]_c} \qquad (6)$$

It is also true that the total concentration of impermeable anions (proteins) inside the red cell is greater than that in plasma. It follows that the total concentration of diffusible anions inside the cell is less than that outside, and in particular

$$[Cl^-]_p > [Cl^-]_c \quad \text{and} \quad [HCO_3^-]_p > [HCO_3^-]_c$$

The effect of a rise in $[HCO_3^-]_c$ in capillary blood, caused by the hydration of CO_2 and subsequent ionisation, is that HCO_3^- moves out of the cells while Cl^- moves in. This is called the *chloride shift*; note that it is a consequence, not a cause, of carbonic anhydrase activity and buffering by haemoglobin.

The result of the shift is, however, that although almost all the HCO_3^- in venous blood was formed inside the cells, the greater part of the extra bicarbonate (almost 90 % of it) is to be found in the plasma. Thus the 'standard bicarbonate' of acid–base measurements is in effect the *plasma* bicarbonate.

Some water moves into the cells with Cl^-, so that the haematocrit of venous blood is distinctly greater than that of arterial blood, the chloride shift reversing itself, in line with all the other changes in solute concentration, as the blood passes through the lungs.

The exchange of gases is summarized in Fig. 15.3.

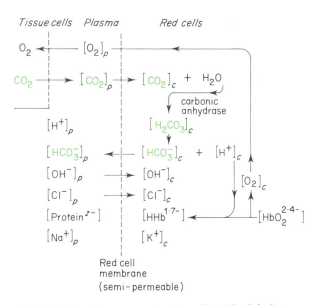

Fig. 15.3 The changes occurring in capillary blood during the liberation of oxygen to, and acquisition of carbon dioxide from, the tissues. In the lungs the changes are reversed. 'Protein^{z-}' refers to plasma proteins, of varying net charge.

The effect of P_{CO_2} on oxygen transport

When CO_2 dissolves in water, the solution becomes more acid (equations (1) and (2), p. 250). For a long time there was discussion whether there is also a direct effect of CO_2 on O_2 transport by haemoglobin, i.e. one independent of any change in pH. It is now agreed that, because CO_2 combines with terminal NH_2 groups of haemoglobin chains as does DPG, it can act in much the same manner to decrease O_2 affinity. The effect is not regarded as important as that of DPG or of pH.

Gas transport by perfluoro compounds

Many organic liquids in which all the H atoms have been replaced by F, have a tremendous capacity for dissolving gases, including both O_2 and CO_2. A few of them, including perfluorodecalin,

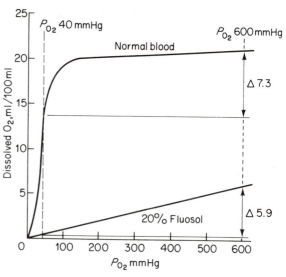

Perfluorodecalin

are sufficiently non-toxic to be injected in emulsified form, into the bloodstream. They have been used in a few cases as partial or complete replacements for red blood cells in patients for whom this is advisable. Effects on the kidney and reticuloendothelial system prevent such substitutes from being used for more than a few days, but they provide a completely adequate gas transport for both gases for a short period. Fig. 15.4 shows that the relation between the oxygen concentration of the blood substitute and P_{O_2} is linear, instead of

Fig. 15.4 Oxygen dissociation curve of a perfluoro blood substitute compared with that of normal blood. 'Fluosol' is a mixture of perfluoro compounds, predominantly perfluorodecalin. Note that although Fluosol carries very little O_2 at the normal P_{O_2} in the lungs (~ 100 mmHg), the solubility curve is linear, so that on increasing P_{O_2} (by breathing pure O_2), the ΔO_2 on moving from lungs to tissue capillaries can be made almost as great for Fluosol 'blood' as it is for normal blood.

sigmoid, and that consequently a wider range than normal between the loading and unloading pressures of O_2 is required, but this can be achieved by arranging for the patient to breathe pure O_2.

A second interesting point is that CO_2 is dissolved in the tiny droplets of perfluoro compound, and not in plasma, so it is not hydrated, and H^+ ions are not produced. This differs from the chain of events that occurs when CO_2 dissolves in normal blood, as we have already seen (p. 250), but is nevertheless not harmful.

Electrolyte Status of the Blood and Tissues

The last part of this chapter is devoted to a discussion of the ways in which the body attempts to compensate for changes in pH or in gas transport. Before starting this it is desirable briefly to summarize the ways in which the body regulates its electrolyte concentrations, because these limit very severely the responses to disturbances of acid–base balance.

The ionic compositions of plasma, extracellular fluid and intracellular fluid are compared in Table 15.1. These figures are in mEq, so that the balance between total cations and total anions (required for electro-neutrality) can be assessed. The value for phosphate is rather lower than its molecular weight would suggest, because its activity coefficient is appreciably < 1.0. In Table 15.2, on the other hand, the *total amount* of the ions in the various fluid compartments are given, together with the estimated values for the total skeleton. The values in this table are in moles, because the activity coefficient of Ca and PO_4 in bone cannot be calculated.

The values for the multivalent ions Mg^{2+}, Ca^{2+}, SO_4^{2-}

Table 15.1 Composition of plasma, extracellular fluid, and intracellular fluid (muscle). Values are in mEq/litre

	Plasma	Extracellular fluid	Intracellular fluid
Cations			
Na^+	145	139	10
K^+	4.5	4	158
Mg^{2+}	1.6	1	31
Ca^{2+}	5	4	6
Sum	156	148	205
Anions			
Cl^-	103	112	1
HCO_3^-	27	25	10
HPO_4^{2-}	2	2	24
SO_4^{2-}	1	1	19
Protein	18	2	65
Organic acid	5	6	16
Organic phosphate	—	—	70
Sum	156	148	205

Table 15.2 Distribution of electrolytes in various compartments of the body. Values are totals and are given in millimoles.

	Plasma	ECF	ICF	Bone
Volume (litres)	3	12	25	—
Na^+	435	1670	250	1180
K^+	13.5	48	3950	80
Mg^{2+}	2.4	6	390	800
Ca^{2+}	7.5	24	75	33000
Cl^-	310	1345	25	—
HCO_3^-	80	300	250	4100
HPO_4^{2-}	3	12	300	19300

These figures are very approximate, and should be used only as a guide to relative distributions.

and HPO_4^{2-} have been included for completeness. They may be neglected for the present discussion, although phosphate becomes important in a later section (p. 256). We can see that although the reservoir of Na^+ in bone is not negligible (and is readily available if needed, because it is in the outer hydration shell (cf. chapter 17)), nevertheless the major portion of the body's Na^+ is in the extracellular fluid. The Na^+ in intracellular fluid is mostly associated with the nucleus, and is probably not readily available for acid–base balance.

On the other hand, a much greater proportion of the total body K^+ (over 90%) is in the intracellular fluid. Although K^+ loss from cells, with replacement by Na^+ or H^+, can form an important part of the body's defences against cation loss, there cannot be a simple replacement of K^+ for Na^+ in extracellular fluid, because high extracellular $[K^+]$ (> 8 mM) will cause cardiac arrest. A negative K^+ balance is therefore a slow and chronic response to derangement of acid–base balance.

The foregoing analysis assumes that the cations Na^+ and K^+ alone determine the body's electrolyte balance, with the anions responding passively (in the absence of pH change). Although this is true at the level of cellular membranes, Cl^- is deterministic in one respect: there is no reservoir of the ion; almost all of it is in the plasma and extracellular fluid. It is in fact difficult to determine

whether there is truly any intracellular chloride, other than in red cells. Most authorities assume that Cl^- is distributed according to the Donnan equilibrium, which would make intracellular $[Cl^-]$ about 1 mM (see Table 15.1), and then use this calculation to make a correction for other ions in extracellular fluid. There is thus no independent estimate of intracellular chloride, but the concentration must be very low. It follows that any loss of Cl^-, as for example in vomiting, leads automatically to a reduction in total extracellular Cl^-, and thus to a reduction in extracellular fluid volume, unless the Cl^- is replaced by HCO_3^-. This, however, would introduce a change in extracellular pH which would have to be corrected. Thus chloride ion balances cannot be completely neglected in assessing electrolyte balance.

The foregoing analysis takes as its fundamental premise the proposition that the control of ionic strength (tonicity) is more important even than control of pH. Some degree of extracellular dilution—leading to oedema—can be tolerated, but a dilution of intracellular fluid relative to the extracellular medium leads to swelling of the cells themselves, which can very easily lead to occlusion of minor blood vessels, and so to tissue death.

The fact that the kidneys are regulating two separate entities, fluid and electrolyte balance, and pH, which at times have different and indeed opposing requirements, makes renal function, especially that part of it under hormonal control, far from simple to understand.

Control of Acid–Base Balance

Acidosis

The formation of acid, i.e. proton-donating, groups is more common in the body than is the formation of bases, i.e. proton acceptors. The body's defences against acidosis, by the bicarbonate/CO_2 system, and by other mechanisms, are more efficient than the defences against alkalosis. A mild degree of acidosis is almost a normal condition; the oxidation of the —SH of cysteine to sulphate (p. 139), together with the hydrolysis of phospho-diesters, e.g. nucleic acids, to inorganic phosphate, are a continuous source of non-volatile acid. On a largely vegetarian diet the oxidation of salts of acids such as citrate and malate, effectively to $KHCO_3$, may on the other hand induce a mild alkalosis.

These disequilibria of dietary origin are not serious enough to cause concern. There are processes which release acid into the blood (acidaemia) in sufficient amounts to activate the physiological defence mechanisms; they can be either acute or chronic. The most common are lactic acidosis, usually from intense exercise, and keto-acidosis (p. 107) which is usually only serious if it arises from poorly-controlled diabetes. Other chronic acidoses can arise from inborn errors of metabolism, such as phenylketonuria (p. 149) or maple syrup urine disease (p. 145), but in these the acidosis is usually less important than the other consequences.

(a) *The respiratory response*

When acid metabolites enter the blood the pH falls, and the ratio $[HCO_3^-]/[CO_2 + H_2CO_3]$ (see chapter 1) also falls. Stated in another way, the reactions

$$HA \rightleftharpoons H^+ + A^-$$
$$H^+ + HCO_3^- \rightleftharpoons H_2CO_3 \qquad (7)$$

occur. Their equilibria are very far to the right. H_2CO_3 dissociates almost completely into H_2O and CO_2 (p. 250). and P_{CO_2} increases. This increases the amount of CO_2 blown off in the lungs. This brings the ratio $[HCO_3^-]/[CO_2 + H_2CO_3]$, which is equivalent to the ratio $[HCO_3^-]/qP_{CO_2}$ (equation (3)) back to normal, and the pH returns to normal.

The acidosis has been *compensated* by a respiratory response, but the composition of the extracellular fluids is not normal because the absolute concentration of HCO_3^- has been reduced, and the anion A^- of the acid metabolite is still present. If the acidosis is only temporary, the metabolite may be metabolized (as after exercise) or else excreted. The acid–base status then slowly returns to normal with less CO_2 being blown off in the lungs, and the $[HCO_3^-]$ in extracellular fluids rises to its normal level.

This description has concentrated on the response of the bicarbonate/CO_2 system. There are other buffers in blood and extracellular fluid which increase the resistance of the body to changes in $[H^+]$. The most important of these is haemoglobin. Each molecule contains 36 histidyl residues, many of which have pK's close to 7.4; only 4 of them are involved in the change of acidity of Hb on oxygenation (p. 250). The buffering power of the others is quite independent of the state of oxygenation of the red cells. Plasma proteins also contain histidyl residues, but they are not quantitatively so important as those of haemoglobin.

From the Henderson–Hasselbalch equation (equation (3)) it is possible to determine the acid–base status of an individual by measuring any two of the variables pH, total CO_2 and P_{CO_2}. A complete description of this is beyond the scope of this book. In interpreting the results, for example by the Siggaard Andersen nomogram, it will be observed that the haemoglobin content of the blood is taken into account. This is in order to allow for the extra buffering power of the histidyl residues, as described in the previous paragraph.

Apart from the pH itself (values outside the range 7.0–7.6 are barely compatible with life), the amount of Brønsted base (proton acceptors) still remaining in the blood and extracellular fluid is the most important quantity, becase this determines the remaining capacity of the respiratory acid–base system, including the blood proteins, to respond to further acidotic episodes. It is usual to specify an HCO_3^- concentration of 24 mEq/litre at pH 7.4, or a P_{CO_2} of 40 mmHg, as normal. A concentration of HCO_3^- below this level, when corrected to pH 7.4, is characterized as a *base deficit*. The base deficit read from the nomogram will be somewhat larger than the HCO_3^- deficit, because of the haemoglobin (and plasma protein) supplementation.

This explanation of response to acidosis has been written in terms of an acid metabolite arising in tissues and accumulating in blood. There can also be a *respiratory acidosis*, in which CO_2 is not removed at the normal rate from the lungs. This may be due to acute respiratory failure, drug-induced respiratory paralysis, or lung congestion. In these situations P_{CO_2} will be high with pH low, but $[HCO_3^-]$ will be normal or even slightly high.

(b) *The renal response*

It is clear that in respiratory acidosis the acid–base status of the blood cannot be returned to normal by a physiological response of lung function. If the kidneys are functioning normally, it is they which compensate for the disturbance. The kidneys also play a part in compensating acidoses of non-pulmonary origin, unless the latter are very short-lived. A brief outline of the mechanism of kidney function is included at the end of this chapter. Here the chemical events will be summarized, with the kidneys treated as a 'black box'.

The response to respiratory acidosis provides the simplest picture. From the Henderson–Hasselbalch equation (equation (3)), one may infer that the pH will be restored to normal if the concentration of HCO_3^- increases, so that the ratio $[HCO_3^-]/[CO_2 + H_2CO_3]$ returns to normal. In ordinary circumstances, HCO_3^- is completely absorbed from the glomerular filtrate (for a mechanism see Fig. 15.5). Any increase in plasma $[HCO_3^-]$ therefore depends on two factors. The first is exchange of H^+ for Na^+, so that in effect the tubule cells are manufacturing H_2CO_3 from endogenous CO_2 (see Fig. 15.5), but are secreting $NaHCO_3$ into the plasma. The second factor (usually neglected) is the constancy of the osmotic pressure of plasma and ECF; this demands that an increase of $[HCO_3^-]$ must be balanced by a decrease of $[Cl^-]$. There must therefore be an increase in the net excretion of Cl^- into urine.

Metabolic acidosis

As we have seen, the respiratory response to a mild metabolic acidosis may leave the blood pH completely normal, but there will be a base deficit, a lowered P_{CO_2}, and an excess of the corresponding anion in the plasma. If this anion is lactate, it will be removed from plasma, chiefly by liver. Other anions must be excreted, but excretion is not normally a problem—the ion is simply not reabsorbed in the tubular system—but maintenance of electrical neutrality is a problem with which the body may find it very difficult to cope. In effect, unless the metabolic acid is undissociated at the pH of the urine, an equivalent cation has to be excreted for every anion. This cannot be H^+, because of the limited capacity of the kidneys to excrete an acid urine. The point may be made explicit by reference to sulphate, a normal constituent of urine. Normal urine contains about 50 mEq/litre of SO_4^{2-}, but one does not suppose that urine is 0.05 N in H_2SO_4. The sulphate ions are balanced by cations, either Na^+ or K^+. This daily occurrence is not a problem, particularly as the intake of both Na^+ and K^+ are usually greater than minimal requirements

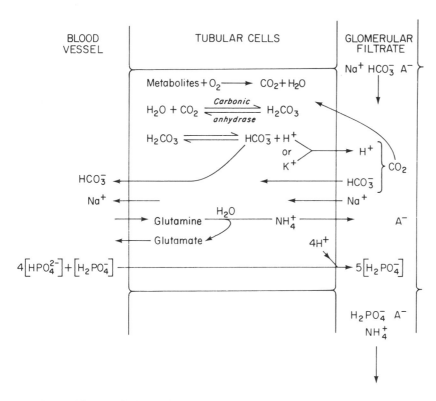

Fig. 15.5 The acidification of urine and ammonium ion excretion by the kidney. The diagram shows reabsorption of HCO_3^- both by diffusion into the tubular cells and by conversion to H_2CO_3 ($\equiv H_2O + CO_2$). NH_4^+ is shown exchanging with Na^+; the latter also exchanges with H^+ or K^+ (see Fig. 15.7). The diagram also shows buffering of H^+ by excretion of HPO_4^{2-} and its conversion to HPO_4^{2-}. The permanent anion A^- is not reabsorbed, and leaves the kidney largely neutralized by NH_4^+.

(chapter 17). However, in serious metabolic acidosis the provision of cations, whether Na^+, NH_4^+, intracellular K^+, or Ca^{2+} from bone, may become limiting so that renal compensation of acidosis is incomplete. Before outlining this important topic of cations in urine, it is convenient to discuss the capacity of the kidneys to acidify urine.

The excretion of H^+ in urine

The minimum pH of urine is about 4.5; this is equivalent to 0.03 mEq/litre, a trivial amount. The only way for excretion of H^+ to be increased is by the simultaneous excretion of a Brønsted base—a proton acceptor with a pK of 4.5 or greater. In normal urine the only such base of importance is HPO_4^{2-}. Daily excretion of phosphate is about 50 mEq. The amount of H^+ removed from solution by transfer of phosphate ion (pK 6.8) from an environment of pH 7.4 to one of pH 4.5 is about 80% of the phosphate excreted, i.e. about 40 mEq in a normal person. This is a maximum; if the urine pH is 6 the amount of proton removed is slightly less. All other acid excretion has to be balanced by 'fixed' cations, or by proton uptake by the acid anion itself.

The acids to be found in urine have a wide range of pK's. As already mentioned, H_2SO_4 will always be fully ionized. Lactic acid (pK 3.9) will also be fully ionized, but the second dissociation constant of malic and citric acids (~ 5) ensures that one —COO^- of each of these acids can act as a proton acceptor in urine. These acids are metabolically unimportant, but acetoacetic acid (pK 3.6) and 3-hydroxybutyric acid (pK 4.4) can be excreted in considerable amounts in ketosis. Acetoacetic acid will always be fully ionized, but 3-OH butyrate can accept about 0.5 H^+ equivalent per equivalent of acid, at pH 4.5. There is usually more 3-OH butyric than acetoacetic acid in ketotic blood and urine.

There are two Brønsted bases that can appear in urine, which seem at first sight to be suitable acceptors for a significant amount of H^+. These are bicarbonate and ammonia:

(a) $\qquad H^+ + HCO_3^- \rightarrow H_2CO_3 \qquad pK'$ 6.1

and $\qquad\qquad$ base

(b) $\qquad H^+ + NH_3 \rightarrow NH_4^+ \qquad pK$ 9.2

In neither case, however, do these reactions lead to a net loss of H^+ from the kidney. As Fig. 15.6(a) shows, the H_2CO_3 formed in (a) returns (as CO_2) to the tubule cells and re-dissociates into H^+ and HCO_3^- there. The bicarbonate is transferred to the blood, but the H^+ remains in the cells.

With (b), the proton is retained in the cells before the uncharged ammonia diffuses into the lumen (Fig. 15.6(b)). Thus for every proton bound in the

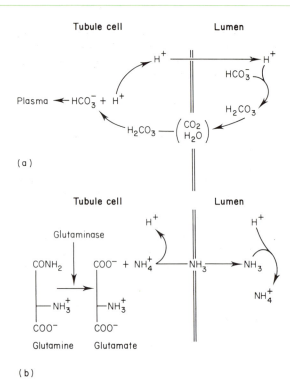

Fig. 15.6 (a) Failure of HCO_3^- to act as a proton acceptor in urine. The H_2CO_3 that is formed diffuses back into the tubular cells and re-dissociates, forming the same amount of H^+ there as before. (b) Failure of NH_3 to buffer protons in urine. The hydrolysis of glutamine releases NH_4^+ in the tubule cells. For this to pass through the wall of the lumen as NH_3, it must leave a proton behind, which is exactly equivalent to the proton that is taken up in the luminal fluid.

urine, an equivalent proton is gained by the tubule cells.

It is nevertheless true that, taking protein catabolism as a whole, the transition from the neutral peptide bond —CO—NH— to the neutral molecule area H_2N—CO—NH_2 involves no net proton changes, while the transition from the peptide bond to the formation (and excretion) of an ammonium ion NH_4^+ results in the formal loss of one H^+ from the system. This approach is unrewarding, however, because it implies that urea and ammonia are alternatives which replace each other according to the acid–base status of the individual. In fact, NH_4^+ formation and excretion is entirely under the control of the kidney, while urea synthesis occurs quite independently in liver, in response to the nitrogen balance of the organism, and can be manipulated independently of the acid–base state. It is more useful to consider NH_4^+ as a slow, but very important, mechanism for conserving cation, confined entirely to the kidneys. This is considered in more detail in the next section.

It must be remembered that we are here considering only the *net* loss to the body of H⁺; the total rate of secretion of H⁺ into the lumen is very much greater than the net loss (see bottom of page).

Alkali cations in acidosis

Normal urine contains about 150 mEq/litre Na^+ (and Cl^-) and about 60 mEq/litre K^+. It also contains about 40 mEq/litre NH_4^+, and is slightly acid. The amount of Na^+ and K^+ excreted is very variable, because it depends on the intake (usually in excess of requirements, see chapter 17), and on other circumstances, e.g. loss through sweating. With a normal acid–base status, the loss of Na^+ (and Cl^-) can be reduced almost to zero to conserve cation, although there is always a slight loss of K^+ (because there is no plasma-directed K^+ pump).

When the urine becomes more acid than pH 6, the compensatory mechanisms described above come into play. Buffering by phosphate actually conserves Na^+, because of the effective replacement of Na_2HPO_4 (in plasma) by NaH_2PO_4 (in urine), but the excretion of organic anions in metabolic acidosis, or Cl^- in respiratory acidosis, necessarily implies the loss of 'fixed' base, which is very largely Na^+. As Table 15.2 shows, there is only a limited reservoir of Na^+ within the body, and an acute massive acidosis can lead to immediate problems of reduced extracellular fluid volume, increased haematocrit, and so on. In chronic severe acidosis, there is some supplementation by mobilization of Ca^{2+} from bone, K^+ from cells (by replacement with H⁺), and even of Mg^{2+}. However, the major relief of Na^+ loss comes from quite a different mechanism.

This involves replacement of Na^+ by NH_4^+ (produced by the hydrolysis of glutamine in the kidney). In general, the mechanism is slow-acting; it takes about 5 days for a significant change to become established, although increased NH_4^+ output is seen after short-term lactic acidosis produced by exercise, or respiratory acidosis induced by re-breathing CO_2. However, the final response, which comes partly from activation of glutaminase and partly from synthesis of new glutaminase protein, can be very large. 500–600 mEq NH_4^+ can be secreted per day, over an indefinite period, and Na^+ excretion can fall well below the normal level. The glutamine required for this mechanism is synthesized in liver, but as mentioned above, the liver does not regulate amino acid group catabolism. A negative N balance is quite common in acidosis.

It should be pointed out that the continued excretion of large quantities of anions, whether neutralized by NH_4^+ or not, does produce a marked diuresis. If the acidosis is a result of uncontrolled diabetes, this is likely to be made worse by glucosuria.

Alkalosis

The body's defences against alkalosis are less effective than against acidosis. This is because the H_2CO_3/HCO_3^- buffer system has less capacity at pH's above 7.4 (chapter 1, Fig. 1.3), and also because the urine has very little capacity to buffer OH^- ions, and indeed no known mechanism for secreting these ions directly. It is fortunate that alkalosis is not so common as acidosis: quite often it is a response to hypokalaemia.

(a) *The respiratory response*

The respiratory centre is sensitive both to P_{CO_2} and to pH. A rise in pH will cause under-breathing, which raises the blood P_{CO_2} and so brings the ratio $[HCO_3^-]/[CO_2 + H_2CO_3]$ back to normal (equation (3)). However, under-breathing leads to O_2 deficit and thus cannot be continued indefinitely. The respiratory response to alkalosis is consequently limited.

(b) *The renal response*

Even if there were to be direct secretion of OH^- into urine (e.g. in exchange for Cl^-), the reaction

$$OH^- + CO_2 \rightleftharpoons HCO_3^-$$

would ensure that bicarbonate ion would be the apparent product. Thus the response to alkalosis resolves itself into the secretion of an alkaline urine containing HCO_3^-. This has immediate consequences for electrolyte balance, because a cation (usually Na^+) must also be excreted. The excretion of NH_4^+, to conserve Na^+, is not possible because the hydrolysis of glutamine in the kidney is inhibited almost completely by a rise in pH.

Bicarbonate in urine

The glomerular filtrate contains *ca.* 25 mM HCO_3^-, but normal urine contains little or no bicarbonate. As the filtrate volume is approximately 200 litres/24 hr, some 5000 mEq of HCO_3^- are removed from urine every day. Most people think that the mechanism is by secretion of H⁺, which converts the HCO_3^- to H_2CO_3 and subsequently to CO_2, which diffuses out from the lumen (Fig. 15.6(a)). This would imply that the *total* excretion of H⁺ into the lumen of the tubules would also be of this order of magnitude (cf. p. 258). The mechanism is for the moment irrelevant; the question is, what controls the appearance of the ion in the final urine?

Recent research has shown that reabsorption of HCO_3^- is inversely related to the arterial blood volume. If the volume is expanded above normal, reabsorption of HCO_3^- (and Na^+) is depressed, HCO_3^- begins to appear in urine, and a threshold maximum concentration of 28 mEq/l in plasma has been observed experimentally.

However, if the blood volume is below normal, no plateau for HCO_3^- reabsorption is seen. The view that there is no absolute plasma maximum value is strengthened by the fact that in respiratory acidosis (p. 254), the plasma $[HCO_3^-]$ may rise to 40 mEq/l, yet no bicarbonate appears in the urine.

If the effects of changes in blood volume are allowed for, it seems that HCO_3^- begins to appear in urine when $[H^+]$ secretion falls off, as a result of rising pH. Although this agrees in a general way with the mechanism depicted in Fig. 15.6(a), the precise relation between pH and H^+ secretion is not known (cf. p. 260).

Vomiting. Loss of any fluid with a ratio of $[Cl^-]/[HCO_3^-]$ greater than that found in plasma—typically gastric juice—will lead to alkalosis. Control by excretion of an alkaline urine may be difficult because the fluid loss will reduce the blood volume (see above).

Hypokalaemia. In part of the renal tubules there is a secretory mechanism in which K^+ and H^+ ions compete with each other. Consequently if the plasma $[K^+]$ is low, more H^+ ions are secreted into the lumen, and the plasma becomes alkaline. It used to be thought that this alkalosis could only be released by administration of K^+, but it has been found that expansion of the extracellular fluid volume with NaCl solution will depress HCO_3^- reabsorption sufficiently (see above) to correct the alkalosis.

A frequent cause of potassium alkalosis is increased aldosterone in the blood, which increases the reabsorption of Na^+ and the excretion of K^+.

How the Kidney Functions

The kidney is a device for actively reabsorbing, from a protein-free filtrate, ions whose conservation is important to the body—Na^+ and other cations, Cl^- and to some extent HPO_4^{2-}, and also glucose and amino acids. Secondly, it reabsorbs almost all the water presented to it, a vital function for a terrestrial animal. Thirdly, it secretes a number of substances unwanted by the body, notably H^+ and NH_4^+, and in doing so helps regulate the acid–base balance of the body fluids.

These functions are highly energy-intensive; at rest the kidneys consume about 7% of the total O_2 used by the body. The only primary coupling mechanism between metabolic energy and reabsorptive processes so far discovered is the Na^+/K^+ pump (chapter 10). This is very concentrated in the kidney, especially in the cortex (ascending limb of loop of Henle). The pump appears to be located entirely on the basement membrane; it is absent from the luminal side of the cells (apical membrane). This is important as it allows Na^+ entry from the lumen to be coupled in a number of ways (Fig. 15.7).

There is a discrepancy between estimates of the activity of the Na^+/K^+ pump and the turnover of ATP, calculated from the Q_{O_2}. Present estimates are that about 50% of the Na^+ entry is linked directly to ATP hydrolysis; perhaps 25% could be exchanged for H^+ (Fig. 15.5), but the remaining 25% is unaccounted for, energetically speaking. There is good evidence that in the thick ascending loop of Henle, the apical membrane contains an active Cl^- pump, which brings Na^+ into the cells with it. In view of the absence of a Cl^- reservoir in the body (Table 15.2), an active reabsorption of Cl^- is to be expected. However, even at this point the overall pumping of NaCl probably depends on the extrusion of Na^+ through the basement membrane.

It is known that the formation of H^+ ions in the proximal convoluted tubules and elsewhere in the kidney depends on carbonic anhydrase activity (Fig. 15.5), since inhibitors of this enzyme prevent acidification of the urine and lead to the appearance in it of HCO_3^- ions. However, nothing is known of the mechanism of H^+ transport across the apical membrane, or whether a simple Na^+/H^+ exchange would be energetically favourable.

In the proximal tubules, glucose and amino acids are actively absorbed by co-transport with Na^+, as in the small intestine (pp. 70 and 124). Possibly amino acids are also absorbed by the γ-glutamyl transferase mechanism described in chapter 8 (p. 124). Phosphate and Ca^{2+} are also actively reabsorbed. Most of the K^+ in the filtrate also appears to be reabsorbed in these tubules, so that the K^+ present in urine largely arrives by secretion (in competition with H^+) in the distal tubules and collecting ducts.

The countercurrent mechanism

About 80% of the solids in the glomerular filtrate are absorbed in the proximal tubule, and about 80% of the H_2O follows the solutes, leaving a fluid roughly isotonic with plasma. Urine is however hypertonic. The final concentration of the tubular fluid is carried out by a

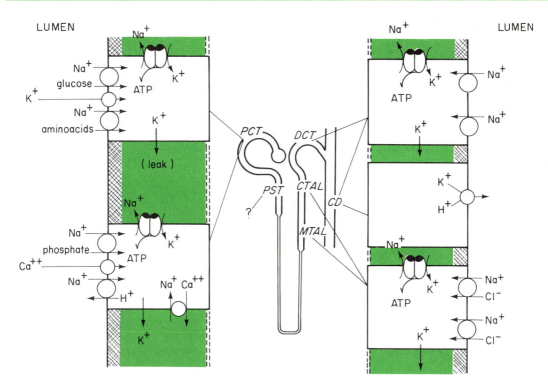

Fig. 15.7 Reabsorption and exchange mechanisms in the kidney, and their probable relation to the Na^+/K^+ pump. PCT = proximal convoluted tubule; PST = proximal straight tubule; MTAL, CTAL = medullary and cortical thick ascending limb; DCT = distal collecting tubule; CD = collcting ducts. The substance reabsorbed in the thick ascending limb is shown as NaCl. However, it is agreed that the primary process here (see Fig. 15.8(a)) is a chloride pump, although it is powered by a Na^+/K^+ ATPase, and Na^+ secondarily enters the tubule cells at this point.

countercurrent mechanism, illustrated in Fig. 15.8. The mechanism depends on the establishment of a solute concentration gradient, running from cortex to medulla, maintained by differences in permeability of the tubules at various points.

The active part in this process is played by the ascending tubule, particularly the thick-walled part, and by the collecting tubule. The whole of the ascending tubule is almost impermeable to H_2O, and the upper (thick-walled) part possesses the Cl^- and Na^+ pump referred to earlier (see Fig. 15.7). This has the effect of forming in the distal tubule a rather dilute urine, in which the major solute is urea.

By contrast, the collecting tubule is permeable to H_2O, and, progressively towards the collecting duct, to urea also (but not to NaCl). The effect is to produce an increasingly hypertonic urine, in which the urea concentration rises eventually to about 750 mM. The matrix surrounding the tubules is hypertonic; at the upper (cortical) end because of NaCl (concentration 4–600 milliosmoles), and in the lower (medullary) end because of a gradient of urea rising to 500 milliosmoles, superimposed on the hypertonic NaCl solution. The

total osmolarity at the bottom of the loop of Henle is thus about 1000 mOsm.

We may now return to the thin descending loop of Henle, in which the fluid is originally isotonic, but loses water as it flows down into the cortex, so that at the bottom of the loop its concentration is also 1000 mOsm. As it enters the ascending limb, it loses some NaCl and gains some urea by diffusion, but as already mentioned, does not gain H_2O.

The channels for water and for urea in the distal tubules and collecting ducts are controlled by vaso-pressin, which opens them so that water moves out of the ducts. Vasopressin is known to increase adenyl cyclase activity in the membrane and hence the concentration of cAMP. There is increasing evidence that cAMP activates the phosphorylation of one or more membrane proteins, which presumably alters the configuration of the membrane, but the precise details of the way in which permeability is changed are not known at the present time.

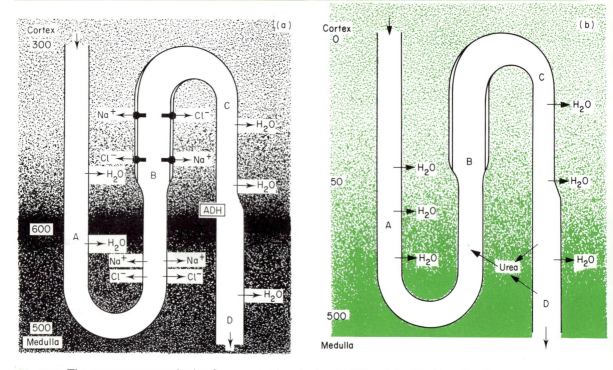

Fig. 15.8 The countercurrent mechanism for concentration of urine. (a) Effect of the chloride and sodium pumps on the external osmolarity around the loop of Henle and the collecting ducts. Note that the descending limb of the loop (A) and the collecting tubule (C–D) are permeable only to H_2O, which moves out of the lumen to the higher osmolarity in the surrounding tissue. At the beginning of the ascending loop, the filtrate has become concentrated, so that Na^+ and Cl^- can move outward by diffusion, but in the thick ascending limb (B) the remaining Na^+ and Cl^- ions are pumped out against a concentration gradient. The figures in the left-hand margin refer to the osmolarity with respect to NaCl.

(b) Effect of urea permeability on the external osmolarity around the loop of Henle and the collecting ducts. It is mainly the collecting ducts (D) that are permeable to urea, which flows out from the urine, in which it has become concentrated. The extratubular fluid thus becomes very hyperosmolar in the inner cortex and medulla of the kidney, which assists water loss from the tubular fluid as it flows from A to B. The figures in the left-hand margin refer to the osmolarity with respect to urea.

Control of acid–base balance

It is not known how H^+ secretion is accomplished, nor how HCO_3^- is reabsorbed, or NH_4^+ secreted in exchange for Na^+. There are great technical difficulties in increasing our knowledge of these processes, particularly since HCO_3^- is such an unstable ion. It is therefore perhaps more helpful at the present time to adhere to the 'black box' description of renal acid–base control described earlier (pp. 255 to 258), than to try to explain renal control of acid–base balance in terms of the molecular biology of the membrane.

The kidney has several other functions, such as the primary production of a protein-free glomerular filtrate, effects on blood pressure and renal blood flow through the renin–angiotensin system, and modification of 25-OH-cholecalciferol (chapter 17). These have had to be omitted in this necessarily restricted survey of kidney function.

16

Energy Requirements and Food Intake

The energy which the human body expends may be divided, from the point of view of an outside observer, into two parts. The first relates to the energy required to move the body or any object in its surroundings. The energy equivalent of the actual movement could, if necessary, be calculated by classical physical methods, but it would not be equal to the total energy expended. The efficiency of the body as a physical mechanism for some kinds of straightforward movement has been estimated to be about 25%, but for more complicated operations (e.g. lifting, bending) it is very difficult to calculate.

The second part of the energy expenditure is that which is required to keep the organism intact. From the viewpoint of physical chemistry, the body is an *unlikely* conglomeration of essentially unstable compounds, dissolved in or surrounded by a very precise but unusual salt solution, and maintained, as a rule, above the temperature of its surroundings. A good deal of chemical energy must be expended merely to preserve such a combination intact. Some of the energy may be used in physiological ways—in the osmotic work of the kidneys, in the pumping of the blood to transport oxygen and remove wastes, in the contractions of the pulmonary muscles. But it must be remembered that each cell throughout its life must expend energy to maintain its integrity, whether or not it is taking part in the activity of the whole organism. As a striking example, the energy expenditure of the central nervous system has been estimated to be 10% of that of the whole body at rest, but there is very little variation in this energy utilization between directed thought, unconsciousness, or complete imbecility.

All the energy expended to keep the organism intact appears as heat, except for the not inconsiderable amount which is represented by water vapour in expired air or in insensible perspiration. This energy is expressed in joules (J). Most of the energy used in physical work also appears as heat. Nevertheless, much of the chemical energy of the food, although it is expressed in joules, has performed a number of vital biological functions in the tissues before it reappears in the environment as heat.

The Measurement of Energy Expenditure

As all the body's energy expenditure appears as, or can be converted into, heat units the simplest way of measuring energy expenditure is by using a calorimeter. The calorimeters which were constructed many years ago for this purpose need not be described in detail. They consisted in essence of a chamber, well insulated, from which the heat was removed by piped water, whose flow rate and temperature rise were accurately measured. Arrangements had to be made for gas exchange, and the water formed by the subject was collected and weighed.

Such an apparatus is impossible for routine use, or for measuring energy expenditure in physical work, and *indirect methods* are now used. The guiding principle is the chemical theorem (Hess' Law) that *the energy liberated in a chemical reaction is the same, no matter by which intermediate steps it is carried out*. Thus the energy liberated (as heat) in the reaction

$$C_6H_{12}O_6 + 6O_2 \rightarrow 6CO_2 + 6H_2O$$

is exactly the same whether it is carried out in the laboratory or in the body, by way of pyruvic acid and the tricarboxylic acid cycle, so long as the end products in each case are CO_2 and H_2O. Thus by measuring the heat given off in a calorimeter on burning a given quantity of fat, carbohydrate, or protein, it is possible to say what must be the energy liberated by a body consuming known quantities of the three foodstuffs. The calorimeter usually used is the so-called bomb calorimeter which can be filled with oxygen under pressure and ignited electrically. A correction has to be made for protein, as one of the biological end products, urea, is different from that obtained in a calorimeter. The figures obtained are 37.6 kJ/g for fat, 17.2 kJ/g for carbohydrate (mixed), and 18 kJ/g for protein.

This method of assessing energy expenditure is only applicable to long-term studies; it cannot be used for measuring the energy used in specific physical tasks. For this, use is made of the fact that the combustion, for example, of 1 gram-molecule of $C_2H_{12}O_6$ always produces 180×17.2 kJ, and it always requires 6 moles of O_2. Thus one can state the energy produced when 1 mole, or, more usefully, 1 litre of oxygen is used to combust carbohydrate (or fat, or protein). This comes to 19.7 kJ/litre for fat, 17 kJ/litre for carbohydrate, and 18.6 kJ/litre for protein. The only problem now is to decide the proportions of the three foodstuffs actually being oxidized in any given period. The amount of protein can be estimated by finding the amount of urea excreted, but since the proportion of protein oxidized is not usually more than 10%, it is often ignored completely in the cruder investigations. For the estimation of the proportions of fat and carbohydrate being combusted, the *respiratory quotient* (strictly the non-protein RQ) is used.

$$RQ = \frac{\text{Volume of } CO_2 \text{ produced}}{\text{Volume of } O_2 \text{ consumed}}$$

The RQ of

$$C_6H_{12}O_6 + 6O_2 \rightarrow 6CO_2 + 6H_2O$$

is 6/6 or 1.0. The RQ of fat oxidation is 0.71. Tables exist showing the joules liberated per litre of oxygen for RQs between 0.71 (19.2 kJ/litre) and 1.00 (21.1 kJ/litre). Since the difference is small, the RQ, to a first approximation, may be neglected. Any of the recognized apparatus for measuring gas exchange may be used for estimating energy expenditure by the indirect method. There are limitations. The subject should not be changing in weight, or synthesizing fat from carbohydrate (the RQ for this > 1), which means that measurements should be made in the post-absorptive period. The subject should not be in oxygen debt, or developing acidosis or alkalosis.

The basal metabolic rate

The energy expended in maintaining the integrity of the organism (see p. 261) may be tentatively identified with the *basal metabolic rate*. The latter is strictly the heat loss when voluntary activity is at a minimum, and is carefully defined. The subject must be at mental and physical rest, 12 hours after the last meal, and in an equable temperature (about 18°C). Even a small departure from these conditions increases the observed heat output considerably. The energy expenditure for adults, in the conditions laid down, is found to depend on sex and surface area. It is 167 kJ/m²/hour for males and 157 kJ/m²/hour for females, and remarkably constant throughout the normal population. The rate is much higher in infants and children, and begins to decrease again after 40 years of age. This decrease, which is not usually accompanied by changes in eating habits, is partly the reason for overweight, and also hypothermia, in old age.

The BMR is not fundamentally concerned with maintaining the body temperature, since it is unchanged in subjects living in climates with temperatures above 37°C. The fetus also has a BMR in utero. The basal heat output has, however, been adapted, under the control of the thyroid gland, to be part of the temperature-regulating mechanism. It alters markedly in thyroid dysfunction. Reduction in BMR by 20% or more indicates myxoedema (hypothyroidism); an increase of 20% or more is characteristic of thyrotoxicosis (hyperthroidism). Measurements of BMR were formerly used in the assessment of thyroid function, but they were time-consuming to make. Moreover, there is a considerable variation in BMR between individuals, so that only rather large changes from the average values could be accepted as indicative of thyroid dysfunction. More sophisticated clinical tests are now standard practice.

Total energy expenditure

The usual way of assessing total energy expenditure is to calculate the BMR, which is usually about 6.7×10^6 J/day for an adult male, and to add to it figures representing the energy used in the various activities of the day. Ten per cent of the total thus obtained is then added to allow for Specific Dynamic Action (see p. 268). Many textbooks have tables of energy expenditure per hour for various activities; they are not very reliable for various reasons. Extended work includes numerous short rest periods which are not allowed for by the figures, and leisure activities account for a great deal of the total energy expenditure. Finally, the efficiency of food utilization varies enormously between individuals; it is therefore pointless to calculate food requirements from an estimation of the activity of a single person. In heavy labour the energy expenditure varies to some extent with the weight of the subject.

A table of daily energy expenditure, obtained from groups of subjects, and checked against food intake, is given in Table 16.1. The figures are only indicative, and they do not allow for biological variation and differences in leisure activity.

Stress must be laid on the large energy expenditure at 10–13 years, corresponding with the pubertal growth spurt. This is only partly accounted for by the great

Table 16.1 Daily energy expenditure in kilojoules

	Males	*Females*
Infants < 1 year	460 per kg	
Children, 4–9 years	6300–8400	
Children, 10–15 years	up to 21 000	up to 17 000
Clerical worker or student, 24–25 years	11 700–12 500	10 000
Housewife, 30 years		9 200
Miner at coal face	17 000	
Very arduous labour or athletic training	up to 21 000	
Housewife, 50 years		6 300 +
Sedentary worker, 50 years	10 000	

physical activity of children of this age. The synthesis of new tissue requires much energy, partly because the compounds to be synthesized have chemical energy locked up in them, partly because, in accordance with physicochemical principles, a good deal of the energy supplied is wasted, and appears as heat. Even in rapid synthesis, such as the production of milk, the elaboration of 1 pint of milk (containing 1550 kJ) requires at least 3100 kJ of energy. With slower growth, when the formed tissue has to be warmed, oxygenated, and transported, it has been estimated that the laying down of 1 kg of tissue in a year requires 340 kJ/day extra energy. This principle applies not only to *growth in childhood* but also to *convalescence*. Surgery or feverish infections are almost always accompanied by severe protein loss, and loss of weight. It is possible to calculate the energy content of the lost tissue, but in order that it may be replaced, *at least* twice the amount of joules so calculated must be included in the diet, suitably spread out over the recovery period. The loss is largely due to slowing-down of synthesis, not an increased rate of breakdown.

The other point to notice from Table 16.1 is that rather a small fraction of adults has an *average* energy expenditure > 10 000 kJ/day. In Western countries, however, food intake (including alcohol) commonly supplies about 12 000 kJ/day. The discrepancy is the source of the obesity which is a pervasive clinical problem in those countries.

Food Intake

Food as an energy supply. From the point of view of useful energy, i.e. that convertible into ATP or otherwise available for biological reactions, it does not matter greatly whether the food intake is carbohydrate, fat, protein, or alcohol. Present information suggests that about 65% of the energy combined in fat and carbohydrate can be trapped as ATP. The conversion factor for protein is undoubtedly considerably lower, but the problem with most diets is to raise the protein intake above the minimum requirement, so that its inefficiency as a source of useful energy is unimportant. The human body can convert glucose into any other carbohydrate or lipid required by the cells, so the chief limitations on the proportions of foodstuffs eaten are those outlined below:

1. A minimum of 2500 kJ/day of carbohydrate ($\equiv$ 150 g/day) is required to avoid the ketosis which would otherwise arise from fat catabolism (chapter 7), and to give a protein-sparing effect (chapter 8).

2. A diet containing no fat is intolerably bulky; 35% fat is recommended by many authorities. However, current recommendations are to make this figure the *upper* limit (see below).

3. Enough fat must be taken to provide the fat-soluble vitamins particularly vitamin A (see chapter 17).

4. The minimum protein intake for body maintenance is 40 g/day (see below).

5. The contribution made by alcohol is irregular, but should not be underestimated. A pint of beer of moderate strength will provide 10% of the daily energy requirement of most adults. Alcoholic drinks are very poor in vitamins, and vitamin deficiencies (particularly of thiamine) are a problem in chronic alcoholism.

Table 16.2 shows the calorific value, per pound, of some common foodstuffs.

The high energy value of such fatty foods as pork products, chocolate, and nuts should be remarked.

Table 16.2 Calorific values, per pound, of some foods

	Kilojoules		*Kilojoules*
Bread	5000	Eggs (10–16)	3200
Cakes	7500	Milk (per pint)	1715
Potatoes (uncooked)	1840	Whisky (per pint)	5860
Rice (cooked)	540	(24 nips)	
Butter	15000	Beer (per pint)	1050
Ham	up to 12550	Chocolate	12000
Beef	5400	Nuts	10700
Cheese	9000	Sugar	7800

Carbohydrate

The limitation on the fat content of the diet, discussed below, and the unlikelihood that the protein content will rise much above 10% for most people, implies that a higher proportion of the diet will be carbohydrate than would have been recommended some years ago. Present emphasis is on ingesting this as starch, which is relatively slowly broken down in the gut, rather than as sucrose, which is quickly absorbed, and in addition gives rise to dental plaque. The possible dangers of feeding diets containing lactose to malnourished children have been discussed in chapter 6 (p. 70).

'Roughage', as explained in chapter 6, is also predominantly carbohydrate, but is by definition not digested, and therefore does not contribute to the body's energy balance.

Fats

It appears now to be established that man, like many other animals, cannot synthesize linoleic acid, but this seems to be the only essential fatty acid. Only in very young children, and possibly in sufferers from the various sprue syndromes, may a deficiency state arise, and there is no precise minimum requirement for adults. Current recommendations, which are aimed primarily at influencing cholesterol metabolism, are a maximum of 35% of the energy intake as fats; 30 g/day as linoleate or other poly-unsaturated acids; and only 10% of the total energy intake as saturated fatty acids. These last two recommendations would be very difficult to achieve in the U.K. without a very comprehensive change in food habits.

Proteins

The problem of recommended requirements for protein nutrition has been unexpectedly difficult to solve. This is partly because there is no single criterion by which optimal intake may be judged, and partly because the optimum depends on a number of variables: the type of protein being fed, the energy equivalent of the non-protein part of the diet, the pattern of protein intake in relation to other nutrients, and whether the recommendation is for a growing person or an adult.

We may first briefly discuss the possibility that there is an optimum protein intake; this implies that too much protein would be harmful. There is some evidence from animal experimentation that very high protein diets are correlated with a shorter life span. Eskimos, probably the best-nourished segregated group with a very high protein intake, do have a somewhat shorter life expectation, on average, than Europeans, but the correlation with dietary protein is at best obscure. It is probable that other aspects of high-protein diets, e.g. cholesterol intake or general overfeeding, are more important determinants of life span than protein itself, and we may therefore concentrate on the definition of a minimum satisfactory protein intake.

The fact that there has to be a minimum intake is determined by the experimental observation that humans, like other animals, continue to excrete nitrogenous material when they are placed on a protein-free diet. They are then in *negative nitrogen balance*. So far as adults are concerned, the question to be answered would appear to be quite simple: what is the minimum amount of protein that has to be supplied in order to re-establish nitrogen balance? (Note that anything above the minimum intake will not, in long-term experiments, cause a positive balance to arise; the N output continues to balance the intake, and the carbon skeletons of the amino acids replace some of the molecules of lipid or carbohydrate that would otherwise be metabolized to provide energy.)

Since adaptation to a change in protein composition of the diet is very slow, this can only be satisfactorily answered by long-term experiments. In practice, an answer has been obtained in two ways: by observing the state of health of volunteers living on restricted protein diets for long periods, or by estimating nitrogen loss from subjects on high-calorie protein-free diets. The former method suggests a minimum intake of about 40 g/day, which is supported by observations of the minimal satisfactory protein intake for patients with renal failure who are on long-term dialysis.

The alternative approach, i.e. N output on a protein-free diet, was first used many years ago, before it was realized that daily protein turnover, as shown in Fig. 16.1, is so large. It was then called 'wear-and-tear' protein metabolism. It is now called 'maintenance' protein requirement, and is measured by the N excretion after 6–7 days on a protein-free diet. The value

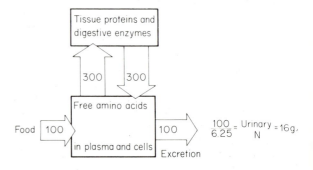

Fig. 16.1 Protein balance in an adult male on a high protein diet (100 g/day). Total flux through the amino acid pool is 400 g/day = 5.7 g/kg/day. Note the relatively small part played in this turnover by amino acids absorbed from the gut.

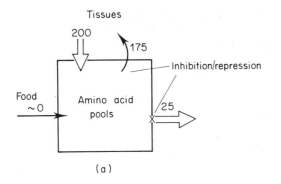

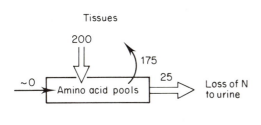

Fig. 16.2 Two alternative models of protein metabolism on a protein-free diet. (a) In this model the size of the amino acid pools is normal, but both protein synthesis and irreversible catabolism are depressed. (b) In this model the amino acid pools are reduced in size, and irreversible loss is reduced in proportion. In neither case is the rate of irreversible loss necessarily a good guide to the *optimal* rate of protein synthesis (cf. Fig. 16.1). This means that the rate of irreversible loss may not be a good guide to a minimal protein intake for nitrogen balance.

for a 70 kg man is about 3.8 g/day, equivalent to about 24 g protein/day.

This type of measurement is theoretically not very satisfactory. It assumes that as the intake (Fig. 16.2) rises from zero, the size of the amino acid pool will not change, and the rate of deamination and urea synthesis will not increase, whereas it is known that if more protein is fed than is needed for N balance, deamination and urea synthesis certainly do increase.

A better guide is the overall rate of tissue protein synthesis, which is found not to increase, in adults, if the protein intake is above 0.4 g/kg, equivalent to 28 g for a 70 kg man. The difference may seem trifling, but it must be remembered that whole adult populations live on the verge of protein malnutrition, and in famine or other catastrophe, the question of how much daily protein to supply for 10^6 people, given that most foods only contain about 10 % protein, may pose a considerable transport problem.

The minimum adult protein requirement derived above presupposes that the protein given will be utilized

Table 16.3 Protein synthesis rates at various ages

	Total protein synthesis (g/kg)	(g/kg^0.75)	Net deposition (g/kg)	'Renewal' (turnover)
Premature infants	10.9	11.6	1.6	10.0
Infants, 1 year	6.3	10.9	0.15	10.75
Adult males	3.0	8.7	—	8.7
Elderly males	3.2	9.0	—	9.0

All values are rates per 24 hr period.

with maximal efficiency. Many proteins, including those of cereals and of meat, are only about 70 % efficient for replacement purposes (cf. p. 266), so that the 28 g/day minimal requirement becomes approx. 40 g/day of average dietary proteins (0.55 g/kg). This is in agreement with long-term observations on individual subjects. Taking into account their smaller stature, it is also in agreement with the estimated daily protein intake of much of the population of S.E. Asia (about 30 g/day).

There is general agreement that the protein requirement of children is larger than that of adults, and ought to be set at about 0.6 g/kg/day nominal, equivalent to about 0.85 g/kg/day of average protein. This is partly due to the net accretion of protein in growth (see Table 16.3), and also to the fact that for each gram of protein laid down, about 1.4 g is actually synthesized. This is shown diagrammatically in Fig. 16.3. Very often the 'extra' protein is degraded and the N excreted. In some instances this can be shown conclusively; for example, during the synthesis of collagen, as in wound repair, a good deal of collagen is degraded, so that the increase in synthesis is accompanied by increased urinary excretion of hydroxyproline. The ratio of 1:1.4 for

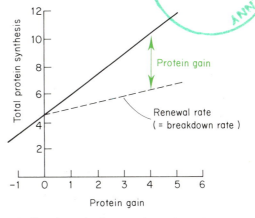

Fig. 16.3 Protein retained vs protein synthesized in growing children. The graph shows that as the protein gain per day increases, the breakdown rate also increases, so that rapid protein gain involves more total protein synthesis (and hence more energy expenditure) than would be expected from the simple energy cost of the protein laid down. The units are g protein/kg body weight/day.

protein retained/protein synthesized during growth is very favourable; pigs, which are reckoned to be very efficient converters of nutrient intake among farm animals, have a ratio of 1:2.2.

Fig. 16.3 shows that it is very difficult to specify an optimal protein intake for children, since one cannot say dogmatically that 5 g net protein synthesis/day is better than 1 g. Problems of protein malnutrition in childhood are considered in a later section.

The effectiveness of different proteins in the diet

To state the amount of protein in the diet means nothing if the type of protein is not specified. This is because the body requires certain amino acids, which it cannot make itself, to be supplied in at least minimal amounts. These are the *essential amino acids* (see chapter 8). Proteins really deficient in one or more amino acids, such as gelatin or zein, will not enable nitrogen equilibrium to be reached if they are the sole dietary protein, no matter how much of them is included in the food intake. An empirical measure of the efficiency of a protein as a supply of essential amino acids is the *biological value*; this is

$$\frac{\text{g protein retained}}{\text{g protein fed}} \times 100$$

The diet must of course be free from proteins other than that under test. Estimates of biological value range from 97 for egg protein to about 63 for gelatin.

Amino acid analysis has shown that most proteins are seriously deficient in only one essential amino acid; usually the deficiency is in *tryptophan*, *lysine*, or *methionine*. The biological value of a protein tells us that it is deficient in essential amino acids, but does not tell us which is missing. The Food and Agriculture Organization of UNO has worked out a 'protein score' based on the essential amino acid content of proteins. Although the score was worked out a number of years ago, more recent research has not invalidated it, despite the fact that one point (below) is still a matter of controversy. The score is estimated on the following lines. Investigations by various workers have given a tentative picture of the absolute requirements of essential amino acids per kg body weight, which are shown in Table 16.4.

This table shows that tryptophan is the amino acid needed in least amount, and the 'pattern' in the last column of the table is arrived at by expressing all the other amino acid requirements as fractions of that of tryptophan. This is the pattern of the composition of an ideal food protein. Milk protein, of high biological value, contains 1.4 g tryptophan per 100 g protein. We can now say that the ideal protein should contain, per 100 g, 1.4 g tryptophan, 2.8 g threonine, 4.2 g lysine, etc. This step in the argument is important, because the

Table 16.4 Minimal requirements for essential amino acids

	Infants (mg/kg body wt/day)	Adult men (mg/kg body wt/day)	'Pattern' (tryptophan = 1)
Histidine*	30	—	—
Isoleucine	70	10	3.0
Leucine	160 (135)	14	4.0
Lysine	160 (100)	12	3.5
Methionine + cysteine	60 (50)	13	3.7
Phenylalanine + tyrosine	125	14	4.0
Threonine	115 (70)	7	2.0
Tryptophan	17	3.5	1.0
Valine	90	10	3.0

* Not certainly essential for adults.
For the figures in parentheses in column 2, see explanation on p. 267.

Table 16.5 Essential amino acid content of foods in g/100 g protein

	Pattern	Egg	Beef	White flour	Beans
Isoleucine	4.2	6.7	5.2	4.1	5.6
Leucine	5.6	8.8	8.7	6.9	8.4
Lysine	4.9	6.2	8.4	1.95	7.15
Phenylalanine + tyrosine	5.6	5.7	4.0	5.0	6.4
Total sulphur amino acids	5.2	5.3	3.7	3.0	1.95
Threonine	2.8	4.8	4.3	2.7	4.3
Tryptophan	1.4	1.6	1.2	1.1	0.9
Valine	4.2	7.2	5.4	4.1	5.9
Protein score	100	100	$\frac{1.2}{1.4} \times 100$ = 86	$\frac{1.95}{4.9} \times 100$ = 40	$\frac{1.95}{5.2} \times 100$ = 38
Biological value		96	77	53	46

concentration of essential amino acids in the diet is significant, as well as the pattern. The proportion of essential amino acids in the overall composition can nevertheless vary with age: this point is discussed below. Food proteins are given a score by dividing the concentration of the amino acid in which they are most deficient by the concentration in the ideal protein (Table 16.5).

This technique makes it possible to provide a proper complement of essential amino acids by mixing incomplete proteins. In general all animal proteins (except gelatin) are *complete*; cereal proteins are *incomplete*, lacking *lysine*, and leguminous proteins are incomplete, lacking *methionine*. A suitable mixture of cereals and legumes can therefore provide a full complement of essential amino acids. (The terms *first class* and *second class* proteins should be avoided, as they give no indication of the possibility of mixing incomplete proteins in the diet.)

The only real point of controversy about this technique is the requirements for infants. The nutritional experiments on which these tables are based were carried out on infants, using artificial diets supplemented with amino acids. This is not entirely satisfactory, and American experts prefer to estimate infant requirements by calculating the amounts of amino acids in the amount of human milk protein that gives optimal growth. This produces the figures quoted in parentheses in Table 16.4. There are discrepancies for some amino acids; these are only very large for lysine and threonine. The discrepancy for lysine is particularly disturbing, first because lysine is most likely to be deficient in cereal protein, and secondly because the absorbed lysine from the diet may be significantly lower than the analytical figures suggest. Lysine and reducing sugars form inactive condensation products during cooking.

Babies may thrive best on human milk, whose proteins contain 37% 'essential' amino N, but adults need less N in total, and also proportionately less of the essential amino acids. For them, if 15% of the total amino-N is made up by the amino groups of essential amino acids (with the rest made up by, say, glycine or glutamate) this is marginally adequate. Children are more like infants: about 32% of their total amino-N intake, at least, should be from essential amino acids.

One other facet of protein nutrition must be discussed. There is *no storage* of protein or of amino acids in the body. Therefore all amino acids required for synthesis of replacement protein must be absorbed at the same time. Most work has been done on the omission of tryptophan. If this amino acid is injected into rats fed on a diet which is otherwise complete, the amount of protein synthesized in the liver depends on the interval separating the injection from the food intake. Even a 4-hour interval reduces the rate of protein synthesis very sharply. It is found that injection simultaneously with food not only prevents the accumulation of incomplete peptide chains, but also in some way activates the ribosomal protein-synthesizing machinery. This latter effect appears to be specific, at least in liver, to tryptophan; leucine stimulates protein synthesis in muscle for reasons at present not at all understood. Thus, if incomplete proteins of different types form part of the diet, they should be mixed at each meal for maximum effectiveness. Similarly, since protein synthesis requires a good deal of energy, carbohydrate or fat should be taken with protein for maximum replacement of tissue protein. If this is not done, part of the absorbed amino acids will be catabolized to supply energy.

Protein–calorie malnutrition

Protein deficiency is particularly widespread among children. Two syndromes are prevalent in the literature, *kwashiorkor* (a West African name) and *marasmus* (a word derived from Greek). These are extreme states, and most cases are not so clear-cut, so that the term *protein–calorie malnutrition* (PCM) is more generally useful.

Typically, kwashiorkor attacks children some 3–6 months after the belated weaning customary in many African societies. Apart from failure to gain weight and to thrive, the abdomen is distended, with an enlarged liver, and there may be oedema in the legs, but the child may appear well nourished. The oedema may be due to a lowered concentration of plasma albumin, or to Na retention, as a result of reduced activity of the Na pump in the kidney. In marasmus, which is the problem most frequently encountered, there is generalized wasting, so that the flesh appears to shrivel and the skin to hang loose. In both diseases there tends to be an intolerance to the disaccharides sucrose and lactose. These are fermented by bacteria in the gut, so that the stools are frothy, copious and fetid.

In kwashiorkor, the diet contains 'excess' energy (i.e. an excess relative to the possible growth rate), and fat synthesis is high. The plasma insulin level is also high. The enlarged liver is fatty and this is not easy to explain; it may be due to choline or phospholipid deficiency. The disaccharide intolerance is caused by a deficiency of sucrase and lactase in the mucosa of the small intestine (chapter 6), attributable to diminished protein synthesis. It may persist for months after other symptoms have been relieved. Note that the lactose intolerance makes it undesirable to use whole milk solids as a source of protein in treatment of either disease.

Marasmus arises when a child has neither enough protein nor enough carbohydrates, so that its tissue proteins are slowly used for gluconeogenesis, and there is generalized wasting. The plasma insulin level may be low. It is clear that in any given situation, either protein or energy deficiency may predominate. In addition, there are likely to be vitamin and mineral deficiencies, particularly if there is diarrhoea, and there will be increased suspectibility to infection.

Much effort has been spent in looking for simple tests which will diagnose these conditions, or better the pre-disease states. None of them have proved satisfactory in field conditions, and clinical definitions are now being used again. Children of 60–80% expected weight-for-age are said to be suffering from kwashiorkor if oedema is present. If they are below 60% of expected weight, they are called marasmic if oedema is absent. The distinction is not entirely pedantic; protein supplementation will not relieve marasmus unless the energy intake is also increased.

Specific dynamic action

When a subject is kept in such conditions that his metabolic rate is perfectly steady (not necessarily in BMR conditions), and then allowed to eat, his metabolic rate, measured either as heat output or oxygen consumption, goes up. The increase is proportional to the calorie value of the food taken in, but is not the same for all foodstuffs. For carbohydrate and fat the increase is about 5%, but for protein almost 30%. This rise in heat output which follows eating is known as the *specific dynamic action* (SDA) of foodstuffs. It is of course waste heat from the point of view of cellular energetics. Partly it is due to the chemical energy lost in hydrolysis of the food, and the physical energy lost in intestinal movements, but a considerable SDA has been shown to be produced by the injection of amino acids directly into the bloodstream. It seems likely that the large SDA of protein is a consequence of the facts that amino acid oxidation is not always so efficient a source of useful cellular energy as carbohydrate or fat oxidation, and that urea synthesis from amino acid nitrogen requires considerable energy expenditure (chapter 8). This view is strengthened by the fact that in periods of rapid growth when amino acid oxidation is suppressed, the SDA of protein may also disappear.

The SDA of a typical meal of a European diet is about 10%. It is usual in calculating energy requirements to find the total energy output of the various daily activities and to add on 10% of this to allow for the SDA. The grand total then represents the joules which must be supplied in the daily diet.

The necessity for an adequate supply of minerals and vitamins (see chapter 17) must be remembered in preparing diets. The considerations outlined in this chapter are only the basis of proper nutrition.

17

Accessory Food Factors

By accessory food factors is meant, in general, the vitamins and minerals which are present in the diet only in traces, but which are essential for the normal function of the body. However, two of the minerals, calcium and phosphorus, are required in more than trace amounts, while a number of the amino acids and one unsaturated fatty acid must be provided in some quantity since the body cannot synthesize them. They are not usually classified as *accessory* food factors.

Minerals

A requirement for *chromium* has been reported, and in animals *selenium* is both required for satisfactory reproduction and toxic in excess. Thus it is impossible at the present time to give a list of inorganic materials which will include all the elements required in traces. The most important inorganic requirements are for *Na*, *K*, *Ca*, *Mg*, *Fe*, *Cu*, *Zn*, *Mo*, *Cl*, *I*, and *P*. *S* as sulphate is essential for the formation of certain body constituents, but it is always organic sulphur, in the form of the sulphur-containing amino acids, which is oxidized to sulphate in vivo. *Co*, as far as is known at present, is only necessary in the form of vitamin B_{12} (p. 279). *Se* is a constituent of glutathione peroxidase (p. 140). The involvement of *Cr* and *V* in normal human physiology is not yet proven. *F* does not form part of the normal body requirements, but it must be mentioned since it may be added regularly in traces to drinking water as a preventative of dental caries.

Potassium is the main inorganic cation present in intracellular fluid, as distinct from plasma and extra-cellular water. Its concentration is about 3.2 g (80 mEq) per kg (muscle). It also occurs to the extent of about 4 mEq per litre in plasma. Since it is universally abundant in foodstuffs, whether of plant or animal origin, there is usually no difficulty in supplying the daily requirements, probably of the order of 1 g per day (20–30 mEq). Disturbances of potassium metabolism may occur, particularly in derangements of the adrenal cortical gland.

Sodium. The average adult excretes about 4 g Na^+ per day (150–200 mEq), of which about 1 g (50 mEq) is in foodstuffs before flavouring is added. Intakes lower than 50 mEq may be tolerated with difficulty, but the requirement may be greatly increased in many circumstances, particularly when a hot climate or prolonged exercise have caused a considerable loss of Na^+ in perspiration. Insufficient Na^+ intake can give rise to heat exhaustion or heat stroke, characterized by muscular weakness, nausea, and fever. The excretion of sodium is carefully regulated by renal mechanisms under the control chiefly of aldosterone, and may be abnormal in diseases of the adrenal gland.

Chloride is the chief anion present in the body, particularly in the extracellular fluid. The excretion in urine is slightly higher than that of Na^+, which would put the average daily intake at about 7 g Cl^- (200 mEq). However, the kidney is capable of secreting a urine almost free of Na^+ and Cl^-, although this is often an indication of incipient heat exhaustion.

The requirements for *Ca*, *Mg*, and *Fe* are very different, but the absorption of each from the food is complicated by the fact that these three metals form insoluble hydroxides in alkaline conditions. All three also form salts which are insoluble, or in which the metal is poorly ionized. For these reasons a great deal of Ca, Mg, or Fe which is actually ingested with the food may be excreted in the faeces unless the conditions in the intestine are favourable.

Calcium and phosphate metabolism

The structure of bone

Bones and teeth are distinguished from all other tissues in the body by their high mineral content. The basic formula of this material is that of *hydroxyl-apatite*:

269

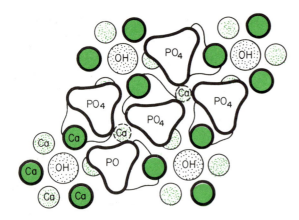

Fig. 17.1 Simplified representation of the hydroxylapatite lattice. The figure shows the crystallographic unit cell, which is repeated in all three dimensions. The ions and inter-ionic distances are roughly to scale, but the individual atoms of the phosphate groups have been replaced by triangular shapes for simplicity. These ions are actually tetrahedral. There are two layers of ions in the unit cell, in the direction at right angles to the paper. One layer (heavy colour and thick outlines) is at three-quarters of the unit height, the other (shaded colour and thin outlines) is at one-quarter unit height, except for the two Ca^{2+} ions with dashed outlines (half unit height). Two hydroxyl ions are superimposed on each other, in the 'tunnel' formed by the Ca^{2+} ions, at one-quarter and three-quarter unit height.

$Ca_{10}(PO_4)_6 \cdot (OH)_2$, although bone salt as prepared contains considerable amounts of other ions, chiefly Na^+, Mg^{2+}, CO_3^{2-}, and organic anions. However, like most inorganic molecules hydroxylapatite in the solid state does not exist as single molecules, but as a regular crystal lattice (Fig. 17.1). Ions of similar charge and shape can replace Ca^{2+} and PO_4^{3-} and OH^- in the lattice; thus OH^- can be replaced by F^-, PO_4^{3-} by CO_3^{2-}, and Ca^{2+} by Mg^{2+}, Sr^{2+}, Ba^{2+}, and Ra^{2+}, among others. For this reason the formula of any particular sample of bone salt cannot be precisely specified. Cations which become part of the lattice structure can remain in bone for extended periods, often with serious consequences (see below). The crystals of hydroxylapatite are very small—never more than 0.1 mm long—and they are surrounded by a hydration shell of water molecules, which is very extensive. For example, 46 % of the body's Na^+ is in this shell, but outside the crystal structure proper.

The crystals are so small that they would never form a rigid structure, able to withstand tension, if they were not held together by the matrix. This consists of fibrils of collagen regularly arranged, to which the crystals are attached in a regular fashion. Between the fibrils is the ground substance, a mixture of hyaluronic acid and chondroitin sulphates attached to a carrier protein, see Fig. 6.11, chapter 6. In the development of bone, the matrix is formed first, in the cartilage, and when bone

is resorbed, the matrix disappears together with the bone salt. The proper formation of the matrix, especially of the collagen fibrils, is essential to the formation of tough bone.

In serum the Ca and P concentrations often vary inversely (see p. 282), and it is often said that this is a result of the *solubility product* relationship. This states that for any solution of an ionized salt in contact with the solid (i.e. a saturated solution), the product of the activities of the ions is a constant, equal to the solubility product $K_{s.p.}$. In the simplest case, if the solution is merely saturated with respect to a single univalent salt in it, the concentrations of both ions will be equal, but this need not necessarily be so. In the present case the solubility product would be

$$K_{s.p.} = [Ca^{2+}][PO_4^{3-}]f_{Ca} \cdot f_{PO_4},$$
(f_{Ca} and f_{PO_4} are the activity coefficients)

The fact that the serum inorganic Ca and P concentrations appear to obey some such relationship has been used as evidence that these ions are in equilibrium with the solid bone salt, implying that plasma is a saturated solution of bone salt. This, however, is not so, for the following reasons.

The simple salt which precipitates from solutions containing Ca^{2+} and phosphate at neutral pH is $CaHPO_4 \cdot 2H_2O$. At the physiological plasma concentrations: 1.3 millimolar (ionized) Ca^{2+}, 1.3 millimolar phosphate, the solubility product for this compound is not reached, i.e. it will not precipitate from normal plasma. However, $CaHPO_4 \cdot 2H_2O$ is unstable at neutral pH and rapidly transforms into hydroxylapatite, which is much less soluble. For various reasons this latter compound does not have a true $K_{s.p.}$, but it has been shown by experiment that if bone salt is added to serum, or imitations of serum, the concentrations of dissolved Ca and P will fall, i.e. they will precipitate on the crystals.

We thus have the situation that plasma, and the extracellular fluid, are *supersaturated* with respect to bone salt, but are not saturated with respect to $CaHPO_4 \cdot 2H_2O$. This approach explains one fact— that calcium phosphate does not precipitate all over the body. It does not normally do so because $K_{s.p.(CaHPO_4 \cdot 2H_2O)}$ is not exceeded. In hypercalcaemia (plasma Ca > 15 mg per cent) it can and does so; there is calcification of soft tissues and the formation of renal stones.

It remains to explain why the plasma Ca and P do not precipitate on the already formed bone salt, i.e. why the normal Ca and P concentrations are not much lower than they are. In hypercalcaemia, this precipitation does, in fact, take place, with the formation of 'marble bone'. In normal bone it seems likely that the apatite crystals are surrounded by a medium in which the activity, but not necessarily the concentration, of Ca has been reduced by comparison with that of plasma. Complexing with multivalent organic acids would achieve this; it is known that the concentration of citrate is higher in bone than in plasma, and there are good reasons for suspecting that osteoclasts are a good source of citrate.

The final problem is how the bone salt ever precipitates at all. Phosphate supplementation in vitamin D deficiency (cf. p. 280) leads to diffuse calcium phosphate precipitation all over the collagenous matrix. Normal mineralization starts with hydroxylapatite precipitation at nucleation centres. These have recently been identified with 'matrix vesicles', small vesicles which bud off from osteoblasts, and which contain a non-specific phosphatase, a Ca pump and also membrane phospholipids that are known to bind Ca^{2+}. The vesicles appear first to accumulate Ca^{2+} from the matrix fluid, and then at a later stage to raise the inorganic phosphate concentration by hydrolysis of organic phosphates in the vicinity. This recalls the original Robison theory of phosphate accumulation. When the local concentrations of Ca and P become high enough, crystals of hydroxylapatite form, and rapidly grow because the overall concentration of Ca and P_i in the matrix is supersaturated with respect to hydroxylapatite, as discussed above.

The properties of matrix vesicles are not yet properly understood since they contain no enzymes capable of metabolizing glucose, and therefore no continuous source of ATP or of other organic phosphates; these could only be produced extracellularly by lysis of osteoblasts or other connective tissue cells. Nevertheless, the consistent appearance of these vesicles in mineralization zones provides the explanation, long lacking, for nucleation centres in bone. Similar vesicles have been seen in calcifying aorta. It is noteworthy that bone growth is associated with high concentrations of plasma alkaline phosphatase. Furthermore, the rare inborn error hypophosphatasia, in which there is a partial absence of alkaline phosphatase, is accompanied by the excretion of ethanolamine phosphate (of unknown origin), and is very often associated with defects in bone formation.

In spite of its insolubility, hydroxylapatite is not metabolically stable. Calcium ions, and presumably phosphate ions also, rapidly exchange with the plasma from the surface of the crystals. The turnover rate is not very large when compared with the total amount of Ca in the skeleton, but it represents complete exchange of the plasma Ca every 2–3 minutes. Even in adult bone there is extensive remodelling of the Haversian canal system by osteocytic activity, which means that apatite crystals may be redissolved, and also that crystals previously adjacent to a canal may be locked away in the interior of the bone structure where the circulation of extracellular fluid is very poor. As a result of this,

cations which have replaced Ca ions on the surface of a crystal may remain in the bones more or less permanently. If these ions are long-lived radioactive isotopes, this can be very serious because of their proximity to the bone marrow and the likelihood of blood cell disorders. Bone is also a reservoir for carbonate, which is released in acidosis.

The average adult body contains 1.5 kg of calcium, 99 % of which is in the skeleton. The concentration in plasma is normally about 10 mg per 100 ml, and is kept very constant by the regulating action of the parathyroid gland. Thus if the absorption of Ca from the intestine is not sufficient to replace that excreted, it may be taken from the bones. This process if continued may lead to serious fragility and weakness of the skeleton, but as a 30 % deficiency of bone salt is barely detectable radiologically, Ca imbalance must continue for several years before the effects become serious. Resorption of skeletal Ca is exacerbated by pregnancy and lactation.

The daily excretion of Ca is about 1 g, of which 200 mg is found in the urine. It is not easy to tell how much of the 800 mg found in the faeces has been excreted in the digestive juices, and how much has never been absorbed at all. Absorption of Ca is known to be inefficient if the pH of the duodenum is at all alkaline, e.g. in achlorhydria or emotional disturbances inhibiting gastric secretion; also in the presence of oxalate, citrate, benzoate, or particularly of phytate (a polyphosphoric ester of inositol, occurring in cereal products). The presence in the diet of fat or galactose is said to improve the efficiency of Ca absorption. The most important factor, however, is vitamin D (see p. 279). The importance of phosphate in relation to Ca absorption is less regarded than formerly because, although the efficiency of absorption ought to be proportional to the Ca/P ratio in the diet, it is known that children on low-calcium diets absorb Ca much more efficiently than would be possible if this ratio were a major controlling factor.

Ca requirements are difficult to state exactly as there may be great day-to-day variations in balance. The adult probably requires about 1 g per day. Children require *relatively* more of both because of increasing skeletal size and because the bones are poorly calcified at birth. Pregnant and nursing mothers also require a good deal of extra Ca for the requirements of the fetus and later of the milk. Ca absorption is often poor in the aged, who may require more than the adult in middle life.

There are few good sources of Ca; milk and milk products are by far the most important.

In the same way as calcium, *phosphate* is chiefly required by the body for the formation of bone salt. Unlike calcium, however, it is found in every cell in many different compounds, nucleic acids, coenzymes, phosphate esters of sugars, etc., which are vital for cellular processes. This wide distribution means that it is readily available in the diet, and phosphate deficiency is rarely met with. In agriculture, on the other hand, adequate phosphate in the soil is of prime importance for the growth of plants.

A primary deficiency of *magnesium* probably never occurs because, like K^+, it is so universally distributed in (cellular) foodstuffs. In an average diet 2–400 mg may be present, but as with Ca^{2+}, much of this may be excreted in the faeces unabsorbed. Mg^{2+} deficiency, with symptoms of depression and weakness, with reduced intracellular Mg^{2+} concentrations, may occur as the result of prolonged diarrhoea.

Iron

There is relatively little iron in foodstuffs, and what little there is is rather badly absorbed, so that much of the world's population exists in chronic slight anaemia. This may even be true of Western populations, particularly infants and young children. Any serious loss of blood may tilt the balance unfavourably, so that women are more at risk than adult males, because of net iron loss in menstruation and pregnancy. At the other extreme, since iron metabolism is regulated by control of uptake, there is no mechanism for excreting excess iron, so that iron storage diseases, e.g. siderosis, with accompanying tissue necrosis, may easily occur.

The absorption of iron from the gut (see Fig. 17.2) is markedly affected by the presence of substances, such as phosphate and phytate, which complex it, particularly the ferric ion. For instance, ferric phosphate, which has been used as a dietary iron supplement, is almost completely unabsorbed. Any reducing substances that will help to keep the iron in the Fe^{2+} form, such as fructose or ascorbic acid, will promote absorption. Nevertheless, some free Fe^{3+} (probably mostly com-

plexed with amino acids or other ligands) is absorbed. The efficiency varies from 1 to 10%. The iron in meat is more efficiently absorbed than that in vegetable products. This is partly because most of the iron in meat, other than liver, is in haem (see below), but in addition, meat or fish promote the absorption of iron from vegetables eaten simultaneously.

The iron in haem is absorbed into the mucosal cells by a separate mechanism from non-haem iron. The haemoproteins are denatured in the stomach (probably here the iron spontaneously oxidizes to the Fe^{3+} form), and the haem is absorbed as such. Inside the cells the tetrapyrrole ring is broken up, for which the enzyme xanthine oxidase is said to be essential, and the iron is released. The efficiency of absorption of haem is about 20%.

Mucosal cells contain some ferritin (see p. 273), the iron in which turns over rather slowly. There is little doubt that an increased demand by the body for iron increases the efficiency of absorption, and vice versa, but the molecular mechanism, and the nature of the controlling signal, are still not known.

Iron is transported from the intestinal mucosa to other sites of metabolism by an α_2-globulin called *transferrin*, which usually has considerable excess capacity. Transferrin binds Fe^{3+}. In vitro experiments suggest that transfer of Fe^{3+} to reticulocytes (which happens without the free iron appearing in the medium) is less efficient if transferrin is lightly loaded with iron. A saturation of 16% or below is generally taken as evidence of iron deficiency. Radio-immunoassay shows that small amounts of ferritin circulate in blood and that the concentration varies with iron status. Although this is diagnostically useful, there is no suggestion that ferritin is important in iron transport, as opposed to storage.

When red cells are destroyed in the reticulo-endo-

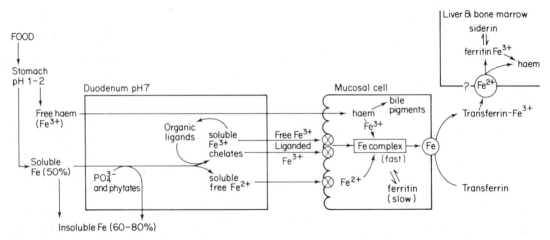

Fig. 17.2   Mechanism of absorption of iron in the duodenum, together with transfer to ferritin stores in liver and bone marrow.

thelial system at the end of their life-span, the haem is very efficiently conserved and re-used. It is stored in a storage protein called *ferritin*, found in most tissues, but chiefly in the liver. Ferritin consists of a nexus of Fe^{3+} atoms surrounded by a protein called apoferritin; the whole complex is brown. Apoferritin is synthesized *de novo* in response to iron loading. If the loading is excessive, molecules of ferritin come together to form a larger complex called *siderin*, which is also brown. It is at this stage that *siderosis*, with accompanying necrosis, may occur. It is very unusual to find siderin in the tissues of infants or adolescents, since the demands of growth keep them on the verge of iron deficiency.

Copper deficiency gives rise to anaemia, even if iron intake is normal. It has been proposed that iron atoms must be reduced to the Fe^{2+} form before they can be released from ferritin, and must be re-oxidized to Fe^{3+} before they can be taken up by transferrin. The liver synthesizes a blue copper-containing protein called *caeruloplasmin*, which also circulates in plasma. This enzyme oxidizes Fe^{2+} to Fe^{3+}, with concomitant reduction of O_2 to H_2O, and it is suggested that copper deficiency produces its effects by preventing the transfer of stored iron from ferritin to transferrin, so starving the haemopoietic tissues of iron.

Liver is not a very important haemopoietic tissue in the adult, but there is nevertheless considerable haem synthesis in hepatocytes. About 85% of this goes into cytochrome P_{450}, part of the mixed function oxidase system (p. 168). This is turned over rather rapidly, so that the consequences of iron deficiency may affect detoxication of drugs and other xenobiotics rather profoundly.

As a result of the efficient conservation of iron, daily requirements are rather small. The minimum requirements have been estimated to be 12 mg, most of which, as outlined above, will be lost in faeces.

There is no relation at all between the rate of Fe absorption and the serum iron level. Normally, 30–40% of the total Fe capacity of the plasma transferrin is occupied. The half-life of plasma Fe is about 90 minutes.

Trace elements

Copper. This is required by several enzymes, e.g. lysyl oxidase, ferroxidase (caeruloplasmin), cytochrome oxidase. Deficiency in humans is very rare. In Wilson's disease, copper export from liver by excretion in bile and by transfer to plasma as caeruloplasmin are defective, so that Cu accumulates in liver, and in other tissues, causing cirrhosis. Treatment is by administering a chelating agent, usually penicillamine.

Zinc. A hereditary defect in Zn absorption leads to *acrodermatitis enteropathica*, characterized by loss of skin and ulceration round anus and mouth, and severe diarrhoea. Similar symptoms are shown by animals on Zn-deficient diets, together with anorexia and cessation of growth. This has led to a realization of a high daily requirement for Zn in humans, about 24 mg/day, which is important, because there appears to be no store of Zn in the body. A protein in liver called *metallothionein* accumulates Zn, or in unfavourable circumstances Cd or Hg, but does not seem to be a storage protein.

The rapid appearance of symptoms related to the digestive tract and to skin, together with lowered immunocompetence, suggests interference with rapidly-dividing cells (cf. B vitamins, p. 275). It appears relevant that many enzymes concerned both in RNA and DNA synthesis, and in protein synthesis, are Zn-dependent.

Molybdenum and *chromium* deficiencies are not known in humans. The former element is required for xanthine oxidase; chromium is a component of the 'glucose tolerance factor', which has so far not been identified in humans.

Vanadium as vanadate ($H_2VO_4^-$) strongly inhibits the Na^+/K^+ ATP-ase (sodium pump) at concentrations (40 nM) which are similar to those found in tissues. Moreover, V does not accumulate in the body, so there must be some homeostatic control of its concentration. As would be expected for an inhibitor of the sodium pump, vanadate is very powerful diuretic in experimental animals.

Nothing is yet known of the physiological role of vanadium in humans.

Selenium is an essential component of glutathione peroxidase (see p. 140). Excess is toxic for farm animals, but there is no serious dysfunction in humans.

Iodine, usually as iodide, is required for the synthesis of thyroglobulin and the circulating thyroid hormones. The requirement is about 1 mg/day. Deficiency is found in several regions of the world where iodide is deficient in soil, and thus in plants, and it leads to endemic goitre and cretinism.

Fluoride. It was observed that in certain areas of the U.S.A. and Great Britain the teeth of the inhabitants became mottled with an unsightly yellow, but that these teeth were extraordinarily resistant to decay. Mottling was traced to the presence, in the drinking water supply, of an unusually high concentration of fluoride. This ion can substitute for OH^- in the hydroxylapatite which is the bone salt. It has been shown that there are concentrations of F^- in drinking water (1–2 p.p.m.) at which resistance to decay remains, but mottling does not occur. It has been suggested that suitable amounts of fluoride should be added to drinking water where it is not normally found. A good deal of controversy still centres round this proposal. In higher concentrations

fluoride is undoubtedly toxic. Fluoridosis is not known in man, but can cause distress, or even death, in farm animals.

Vitamins

These are chemical compounds of varying complexity which cannot be made by the organism and have to be supplied in small quantities in the diet. Sometimes an organism is able to make the substance in small amounts, not enough for its optimum requirements. Organisms differ in their synthetic abilities, so that what is a vitamin for one species is not for another. In this chapter, only those vitamins at present known to be necessary for human health are dealt with in any detail.

Definition. A definition which will exclude minerals, essential fatty acids, and amino acids is not easy to frame. The following has been suggested. A vitamin must be a substance:

1. chemically different from the three main nutrients: fats, carbohydrates, proteins;
2. necessary only in minute quantities in the diet;
3. whose absence must cause a specific deficiency disease.

Many vitamins, particularly those of the B group, are known to be coenzymes, but it does not appear that all vitamins have a coenzyme function.

Vitamin A

Vitamin A is a highly unsaturated alcohol with the following structure:

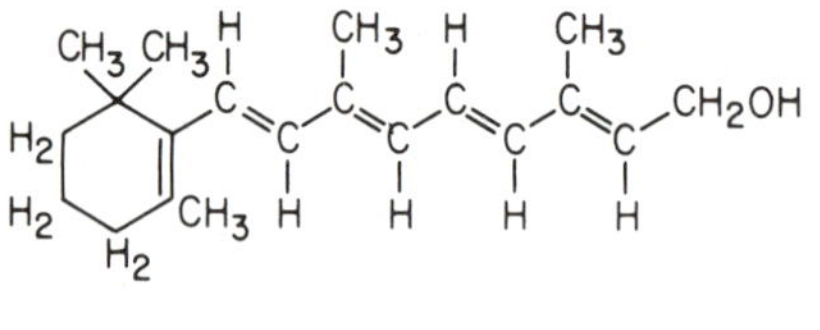

All–trans Vitamin A (retinol)

It is found free or esterified in milk, butter, eggs, and fish liver oils, but equally important sources are certain carotenes found in many plants (see Fig. 17.3). The conversion to vitamin A involves an oxidative fission in the centre of the chain. In order that at least one of the fission products shall be the vitamin, the structure of that terminal ring must be correct, and although there are a great many carotenes, most of them yellow pigments, only four of them are *provitamins A*. They are found in carrots, tomatoes, and to some extent in grass.

Hence the occurrence of provitamins A in milk. Infants and young children can carry out this transformation to a limited extent only; it is almost entirely inhibited by hypothyroidism.

Vitamin A deficiency in animals has several consequences: cessation of growth, failure of bone remodelling, abnormal deposition of keratin in skin and other epithelial tissues, reproductive disorders in both males and females, and night blindness. All these defects except the last two can be cured by vitamin A acid (*retinoic acid*), which suggests that the underlying molecular mechanisms are different for the first three.

The supposition that vitamin A derivatives are necessary for normal development of epithelial cells has led to their being tested in high doses as anti-tumour agents. Some success has been claimed against epitheliomas, but the treatment is still very speculative. Retinoic acid, or analogues of it, have been the most successful to date.

In man, vitamin A deficiency diseases are almost limited to the eyes. Night blindness (see below) is a mild defect, but keratomalacia, giving rise to irreversible corneal opacity, is perhaps the most serious vitamin deficiency disease at the present time, being responsible for thousands of cases of blindness in the tropics. There is a good deal of experimental evidence that vitamin A is necessary to prevent keratinization of mucoid

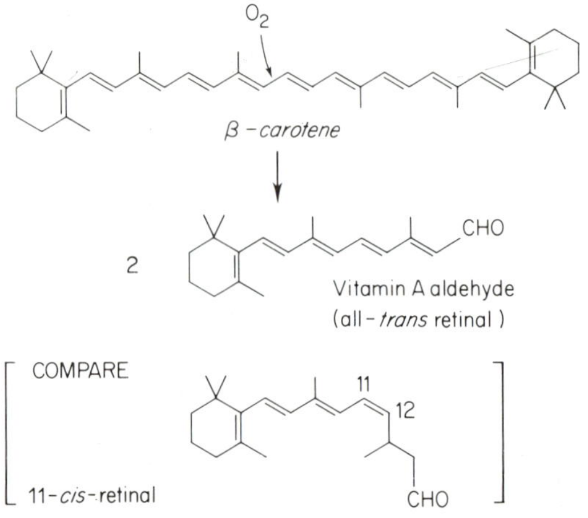

Fig. 17.3 Relationships between β-carotene and vitamin A aldehyde (retinal).

274

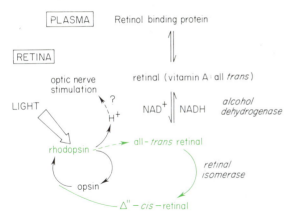

Fig. 17.4 Cyclic transformation of retinal in vision.

epithelium. Requirement: 5000–7000 international units (2 mg) per day.

Vitamin A and vision. The role of vitamin A in visual processes can largely be explained at the molecular level by Fig. 17.4.

Retinal is the aldehyde corresponding to vitamin A. It can exist in two configurations: all-*trans* (this is the main configuration taken up by vitamin A itself), and 11-*cis* (see Fig. 17.3).

Opsin is a protein which covalently binds 11-*cis* retinal through a Schiff's base link with the terminal —NH_2 of a lysyl residue, thus

$$RCHO + H_2N—lysyl + H^+ \rightleftharpoons$$
$$R \cdot CH = \overset{+}{N}H—lysyl + H_2O$$

The complex is called, because of its colour, *rhodopsin*. It is found only in rods, but it is probable that a series of similar protein–retinal complexes are responsible for colour vision in the cones of the retina.

When light of wavelengths centred on 500 nm falls on the rods, the opsin-bound retinal isomerizes to the all-*trans* form, which is only weakly bound. The

—$CH = \overset{+}{N}H$— bond breaks, and a proton is released. In subsequent stages a nerve impulse is generated by a mechanism which is not yet understood.

The all-*trans* retinal set free is then isomerized by 11-*cis retinal isomerase*. In the dark 11-*cis* retinal and opsin recombine to reform rhodopsin. Retinal is formed by oxidation of retinol by alcohol dehydrogenase using NAD^+. The reaction is reversible, so that all-*trans* retinol can slowly leak out into the plasma; in vitamin A deficiency, the concentrations of retinal and rhodopsin slowly fall, leading to night blindness.

For the metabolism of vitamin A, see Fig. 17.5.

Several months' supply of vitamin A can be stored in the liver. The substance is toxic in excess, which is usually the result of self-medication with vitamin supplements.

The vitamin B group

This was originally the 'water-soluble vitamin', in distinction to the 'fat-soluble vitamin A', but in fact a large number of different compounds share the functions of the original entity. A number of these compounds are coenzymes concerned with energy release in cellular oxidations, or with protein synthesis. It is therefore not surprising that B vitamin deficiencies should often manifest themselves as disorders of tissues in which rapid growth normally takes place even in the adult, namely skin and mucous epithelium, digestive glands, and bone marrow. It follows, too, that tissues with high rates of metabolism, including plant seeds, are good sources of these vitamins.

The known coenzyme functions of the B vitamins are shown in Table 17.1.

Vitamin B_1 (thiamine, aneurin)

Thiamine (aneurin, B_1). Found in cereals, in the germ, or in the husk of rice, also in lean meat, liver, kidney, eggs, and seeds. The most serious deficiency disease is *beri-beri*, still endemic in the East. It is characterized by weakness, lassitude, anorexia, and bradycardia, with death usually caused by heart failure. Less serious is generalized *peripheral neuritis*, not uncommon in chronic alcoholics, because alcoholic drinks are a source of energy, but a poor source of vitamins. Since thiamine is concerned with carbohydrate metabolism, the requirement is related to the carbohydrate intake; it is normally about 2 mg per day. A marked increase in blood pyruvate has been shown to occur in beri-beri, falling to normal on injection of the vitamin. Thiamine pyrophosphate is the prosthetic group of pyruvate decarboxylase, and in its absence pyruvate cannot be

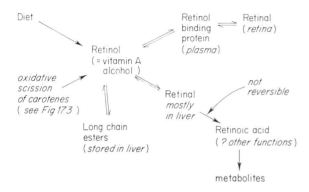

Fig. 17.5 The metabolism of vitamin A and its derivatives.

Table 17.1 **Known functions of the B vitamins**

Name	No.	Enzyme system	See page
Thiamine	B_1	Pyruvate oxidase	74
		α-Oxoglutarate oxidase	
		Branched-chain oxo-acid oxidases	144
		Transketolase	78
Riboflavin	B_2	Flavin mononucleotide (FMN)	36
		Flavin adenine dinucleotide (FAD) (Prosthetic groups of many oxidizing enzymes)	
Nicotinamide		Nicotinamide-adenine di-nucleotides (NAD and NADP)	34
Pyridoxine	B_6	Amino acid transaminases	127
		Amino acid decarboxylases	
Pantothenic acid		Coenzyme A	36
Folic acid		Transformylation	142
		Transhydroxymethylation	
		Transmethylation	
Biotin		Carboxylation of:	
		pyruvate	38
		acetyl-CoA	99
		propionyl-CoA	
Cobalamin	B_{12}	Rearrangement of methylmalonyl-CoA	38
		Methylation of homo-cysteine	106

oxidized. It is also the prosthetic group of α-oxoglutarate dehydrogenase, but reduction in the activity of this enzyme has not been recorded.

$$CH_2-(CHOH)_3-CH_2OH$$

Vitamin B_2 (riboflavin)

Riboflavin (B_2). Found in cereal germ, lean meat, liver, kidney, and in milk. It is unstable to acid and alkali and is also destroyed by light. Although riboflavin deficiency leads to complete cessation of growth in young rats, in humans the known symptoms of deficiency are minor: *cheilitis* is the best known. As riboflavin is a coenzyme or prosthetic group in a large number of systems carrying out cellular oxidations, the effects of deficiency might be expected to be more serious than they are. The reason for the resistance to deficiency in humans is unknown. Requirement: about 3 mg per day.

Nicotinic acid (*nicotinamide, niacin*). This vitamin is required for the synthesis of the oxidative coenzymes NAD and NADP. This is accomplished by the transfer of ribose phosphate (from PRPP, p. 55) to nicotinic acid, and the nucleotide is then amidated. Strictly speaking, therefore, nicotinic acid is the vitamin, but nicotinamide is the form which is obtained from foodstuffs. Hydrolysis of the amide group on the free pyridine derivative is so widespread that the distinction is not important. The most serious deficiency disease is *pellagra* (literally, rough skin), which is characterized by

Nicotinic acid   Nicotinamide

an exfoliative dermatitis particularly on areas of skin exposed to light, anorexia and digestive disturbances, and in its later stages by madness and death. It is a disease particularly of maize-eating areas, partly because most nicotinamide in maize is bound to the seed coat, and therefore unavailable, and partly because a certain amount of nicotinic acid (not enough for all the body's requirements) can be made from tryptophan (p. 146). Maize proteins are particularly deficient in tryptophan.

Clinical observation and nutritional research have shown that pellagra is not often due to a simple deficiency of nicotinamide; other vitamins, notably riboflavin and pyridoxine, are likely to be deficient in the pellagric diet, and to play some part in the development of the syndrome. The role of tryptophan is also not completely understood.

N-methyl nicotinamide is a major urinary metabolite of the vitamin. The parent acid is a very stable compound, resistant both to heat and to acids and alkalis. Requirement: about 20 mg per day.

Pyridoxine (B_6). This has not been shown to cure any specific deficiency disease in man. It is quite probable that the antituberculosis drug isonicotinic acid hydrazide (INH) acts by interfering with the function of pyridoxine in the bacillus. There are many enzymes requiring pyridoxine, often as pyridoxal(dehyde) phosphate. These enzymes are concerned with transaminations and oxidative decarboxylations of amino acids. Like riboflavin, pyridoxine is a prosthetic group rather than a coenzyme; it is fairly tightly bound to protein, and this may explain why deficiency rarely occurs.

Pyridoxine   Pyridoxal

Vitamin B_6

Pantothenic acid. This compound again has not been shown to cure any specific deficiency disease in man. It is very widely distributed in foods, which probably makes a deficiency unlikely to occur. Pantothenic acid

$$\underset{\displaystyle \text{CH}_3}{\overset{\displaystyle \text{CH}_3 \ \text{OH}}{\text{HO—CH}_2\text{—C—CH—CO—NH—CH}_2\text{—CH}_2\text{—COOH}}}$$

Pantothenic acid

is part of coenzyme A, the acyl carrier coenzyme necessary in fat metabolism and in the tricarboxylic acid cycle (see p. 36).

Biotin. Does not cure any specific deficiency disease

Biotin

in man, although a deficiency is sometimes produced by the administration of *avidin*, a biotin-binding protein

found in raw egg-white (it is destroyed by cooking). Biotin is the prosthetic group of some of the carboxylases. It is attached to the apo-enzymes by an amide link between its carboxyl and the ϵ-amino group of a lysine residue.

Folic acid. This is found in green leaves and in yeast. Deficiency chiefly shows itself in the development of certain types of anaemia, particularly in unweaned

Folic acid (pteroylglutamic acid)
(not reduced; mono-glutamate form)

children. A reduced form of the vitamin, tetrahydrofolic acid (formula on p. 35), is a prosthetic group or co-enzyme in transfers of one-carbon units, including the formation of thymidylic acid, a constituent of DNA (chapter 5). It is likely that folic acid anaemia is a deficiency in haemoglobin or erythrocyte synthesis resulting from diminished DNA or RNA synthesis. In bacteria, folic acid deficiency causes a more widespread

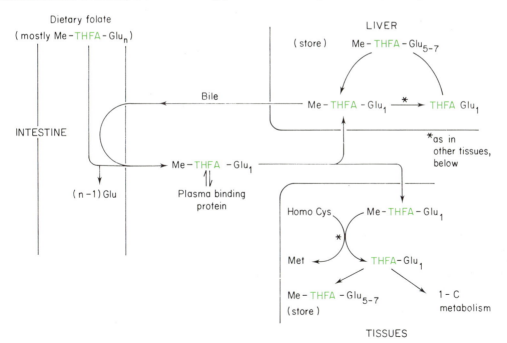

Fig. 17.6 Transport and storage of folate derivatives. Dietary folate is mostly a poly-glutamyl derivative of methyl-tetrahydrofolic acid (THFA), and this is also the major storage form in liver and other tissues. However, the vitamin is transported in blood only as the mono-glutamyl derivative, and this is also the form of the coenzyme (see chapter 8). Moreover, stored folate has to be demethylated before it can enter into one-carbon unit metabolism.

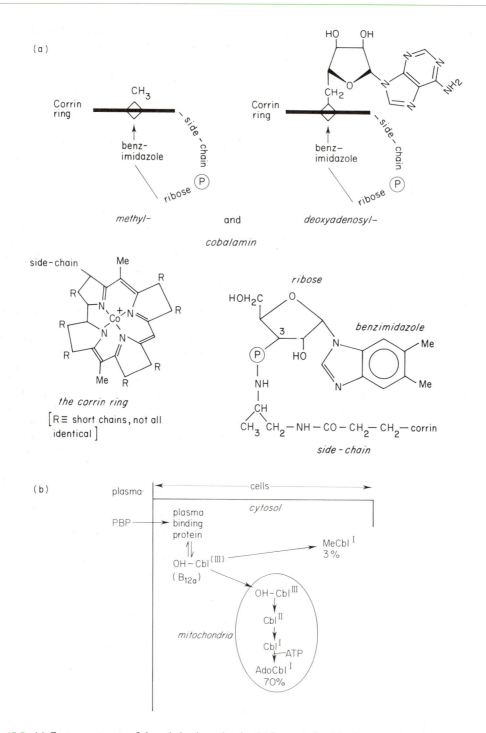

Fig. 17.7 (a) Component parts of the cobalamin molecule. (b) Interrelationships between the cobalamin coenzymes methyl- and deoxyadenosyl-cobalamin. Note that the cobalt atom has to be reduced to Co^+ before the methyl or deoxyadenosyl groups can be attached. Circulating cobalamin has an OH^- ion coordinated to the cobalt.

cessation of growth. The sulphonamide drugs produce this deficiency by interfering with the incorporation of *p-aminobenzoic acid* into the folic acid molecule. There are several active forms of folic acid: in the intestinal mucosa, the vitamin is reduced to THFA and methylated at N^5. It is this form that is transported to cells through the blood. Extra glutamyl groups may be added intracellularly. These processes are summarized in Fig. 17.6. Requirement: about 5 mg per day.

Cobalamin (B_{12}). This is not found in plants at all. It can be obtained from liver or lean meat, or from bacteria and moulds. The requirements for the normal adult are very small (about 1 μg per day). Deficiency shows itself in an anaemia, together with degeneration of the spinal cord, which may be fatal in long-standing cases. This is rare; it occurs occasionally after total gastrectomy and in *pernicious anaemia*. In this latter disease the stomach loses, in middle life, the capacity to produce an *intrinsic factor* essential for the absorption of cobalamin (the *extrinsic factor*). The most usual form of the vitamin contains one atom of trivalent unionized cobalt, which gives it a deep red colour, and a non-ionized cyanide radical. In the known coenzyme forms of the vitamin, this is replaced by covalently linked organic groups and the cobalt is reduced to Co^+ (see Fig. 17.7).

The underlying causes of pernicious anaemia are not at present understood. There is evidence that abnormal branched- and straight-chain fatty acids, which could have arisen from methylmalonyl or propionyl residues, accumulate in the neural lipids of B_{12}-deficient animals, but a connection between this and degeneration of the spinal cord has not yet been proved.

Requirements for pernicious anaemics: about 0.01 mg per day.

Vitamin C (ascorbic acid)

This is one of the water-soluble vitamins. In its chemical properties it is quite strongly acid, owing to ionization of the two hydroxyl groups separated by the

$$
\begin{array}{l}
\text{O=C}\!\!\rceil \\
\quad| \\
\text{HO—C} \\
\quad\|\ \ \text{O} \\
\text{HO—C} \\
\quad| \\
\text{H—C}\!\!\rfloor \\
\quad| \\
\text{HO—C—H} \\
\quad| \\
\text{CH}_2\text{OH}
\end{array}
$$

Vitamin C (L-ascorbic acid)

double bond. It is also oxidized with extraordinary readiness, even by dissolved oxygen at room temperature,

particularly if Cu^{2+} ions are present. The first oxidation product is *dehydroascorbic acid*:

$$
\begin{array}{ll}
\quad| & \quad| \\
\text{HO—C} & \text{O=C} \\
\quad\| & \rightarrow\ \ \quad| \ +2\text{H} \\
\text{HO—C} & \text{O=C} \\
\quad| & \quad|
\end{array}
$$

The disease resulting from vitamin C deficiency is *scurvy*. In its extreme form it is characterized by weakness, loss of teeth and shrinkage of the gums, peripheral haemorrhages, failure of wounds and fractures to heal, and often death. Subclinical scurvy is perhaps not uncommon; symptoms are lassitude, back pains, and hyperkeratosis of, and haemorrhage around, the hair follicles. Almost all the symptoms of ascorbic acid deficiency can be ascribed to lack of collagen, both in scar tissue and in the connective tissue supports of arterioles. Pre-collagen, the precursor of the fibrous form, is not secreted by fibroblasts in the absence of ascorbic acid, and as a result catabolism gradually depletes the connective tissue of its supporting protein. It is not known precisely how ascorbic acid affects pre-collagen formation. It is in some way connected with the hydroxylation of proline residues in procollagen, but is not the oxidizable co-substrate (chapter 9). It is even possible that it is required for the attachment of carbohydrate chains to hydroxyproline residues, a prerequisite of proper secretion of collagen from fibroblasts. Ascorbic acid is the oxidizable co-substrate for the hydroxylation of dihydroxyphenylethylamine to form noradrenaline, but this cannot be connected with its anti-scorbutic function, nor can its presence in adrenal cortex, whence it disappears on stimulation with ACTH. Ascorbic and dehydroascorbic acids are not hydrogen carriers (e.g. like NAD).

Ascorbic acid is found in citrus fruits, strawberries, blackcurrants, rosehips, potatoes, and green vegetables. The vitamin is readily destroyed by boiling. Daily requirements are 10–30 mg but symptoms may only develop after deprivation for 4–6 weeks because the vitamin is stored in various types of cell. In the U.S.A., daily requirements are put at 100 mg, and amounts of 1–5 g/day have been recommended as prophylactic against colds, but this is strongly disputed elsewhere. Ascorbic acid is partially metabolized to oxalate, which can form urinary stones, so that high intakes could be inadvisable.

Vitamin D

This is a fat-soluble vitamin. Two substances are active when taken by mouth: D_3 (cholecalciferol), a derivative of cholesterol, and D_2 (calciferol), a synthetic derivative of a plant sterol. The vitamins are not strictly sterols since ring B has been opened. The double bonds have the transoid configuration shown. The immediate

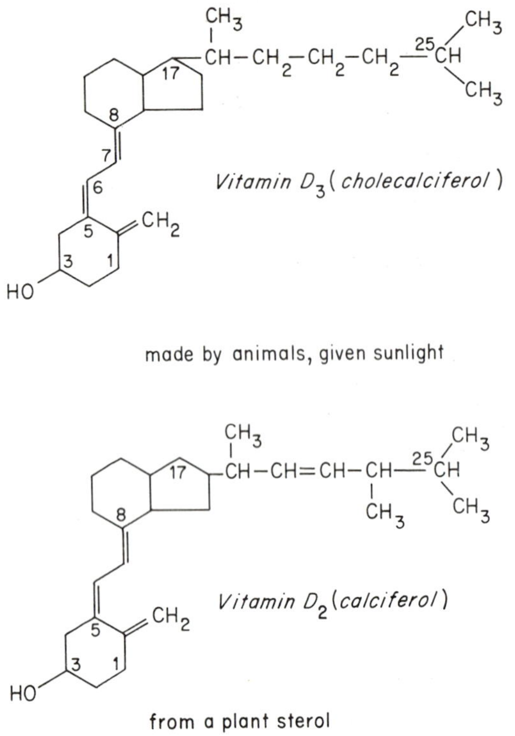

Vitamin D₃ (cholecalciferol)

$$CH_3 \quad CH_3$$

—CH—CH₂—CH₂—CH₂—²⁵CH

8, 17, 7, 6, 5, 3, 1, CH₂

HO

Vitamin D₃ (cholecalciferol)

made by animals, given sunlight

Vitamin D₂ (calciferol)

—CH—CH=CH—CH—²⁵CH

CH₃, CH₃

8, 17, 5, 3, 1, CH₂

HO

Vitamin D₂ (calciferol)

from a plant sterol

Fig. 17.8 Vitamins D₃ (cholecalciferol) and D₂.

precursor of D₃ is 7-dehydrocholesterol, which is found in the skin and sebum, and the agent promoting the change is ultra-violet light. If the skin is exposed sufficiently to direct sunlight, therefore, sufficient vitamin will be formed to make its intake in food unnecessary. There is indeed evidence that endogenously produced cholecalciferol derivatives are retained more efficiently in the body than those formed from ingested vitamin D. This evidence is based on measurement of plasma levels of 25-OH-cholecalciferol, where it is associated with a specific vitamin-D-binding protein. Prolonged exposure of the whole body to ultra-violet light does not cause the 25-OH-D₃ levels to rise to more than double the normal level, whereas high doses of oral vitamin D may cause it to rise 10-fold. As 25-OH-D₃ is the metabolite most closely associated with toxic effects of hypervitaminosis D, it is now being suggested that supplementation of foodstuffs should not be regarded as an adequate prophylactic, by comparison with UV irradiation of the skin.

As the preceding paragraph implies, cholecalciferol itself, although already notably modified from the parent compound cholesterol, is not the biologically active substance that modulates Ca (and P) metabolism, and so prevents rickets. The most important of the active substances, so far as present knowledge goes, is dihydroxy-cholecalciferol, which, as in Fig. 17.10

shows, is formed from calciferol by two successive hydroxylations. The first of these takes place in liver, and the second in kidney. Thus cholecalciferol is taken up by liver from plasma, although it is not stored there, and 25-OH-D₃ is re-secreted into plasma. The product of the second hydroxylation, 1,25-di-OH-D₃, has a rather short half-life, and it is this that suggests that maintenance of an effective concentration of the dihydroxy derivative in plasma depends on relatively slow, but steady, release of 7-dehydrocholesterol from irradiated skin, rather than on modification of cholecalciferol absorbed from the gut at irregular intervals.

It is now clear that production of 1,25-di-OH-D₃ is under hormonal control, as indicated in Fig. 17.9. It is found that its concentration in plasma rises when the body's demand for Ca is high, as in pregnancy or lactation. It appears that control is executed through parathyroid hormone action on renal cells, and not through direct feedback by plasma Ca^{2+}, or by calcitonin. Since dihydroxy-D₃ exerts a major effect on tissues other than kidney, particularly on intestinal mucosa (see below), this gives it the status of a hormone, with the kidney as an endocrine organ.

Absence of vitamin D causes *rickets*, a disease of infants. The bones become soft, the legs unable to support the weight, the flat bones deformed. The rib cage may become so deformed that death from respiratory failure occurs. It is not necessarily a disease of starvation. The skeleton of the newborn child is always deficient in Ca, and any derangement of Ca metabolism quickly makes the bones soft. In adults vitamin D deficiency causes *osteomalacia*, but the calcium reservoir in the adult is so large that the

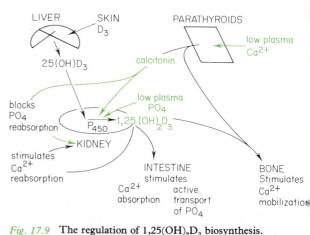

Fig. 17.9 The regulation of 1,25(OH)₂D₃ biosynthesis. The diagram shows that the parathyroids respond to low plasma $[Ca^{2+}]$ by secreting calcitonin, one of whose functions is to stimulate the hydroxylation of 25(OH)-D₃ in kidney mitochondria. Since 1,25(OH)₂D₃ must travel through the bloodstream before it can act on intestine and bone, this gives it the formal status of a hormone, with the kidney as an endocrine organ.

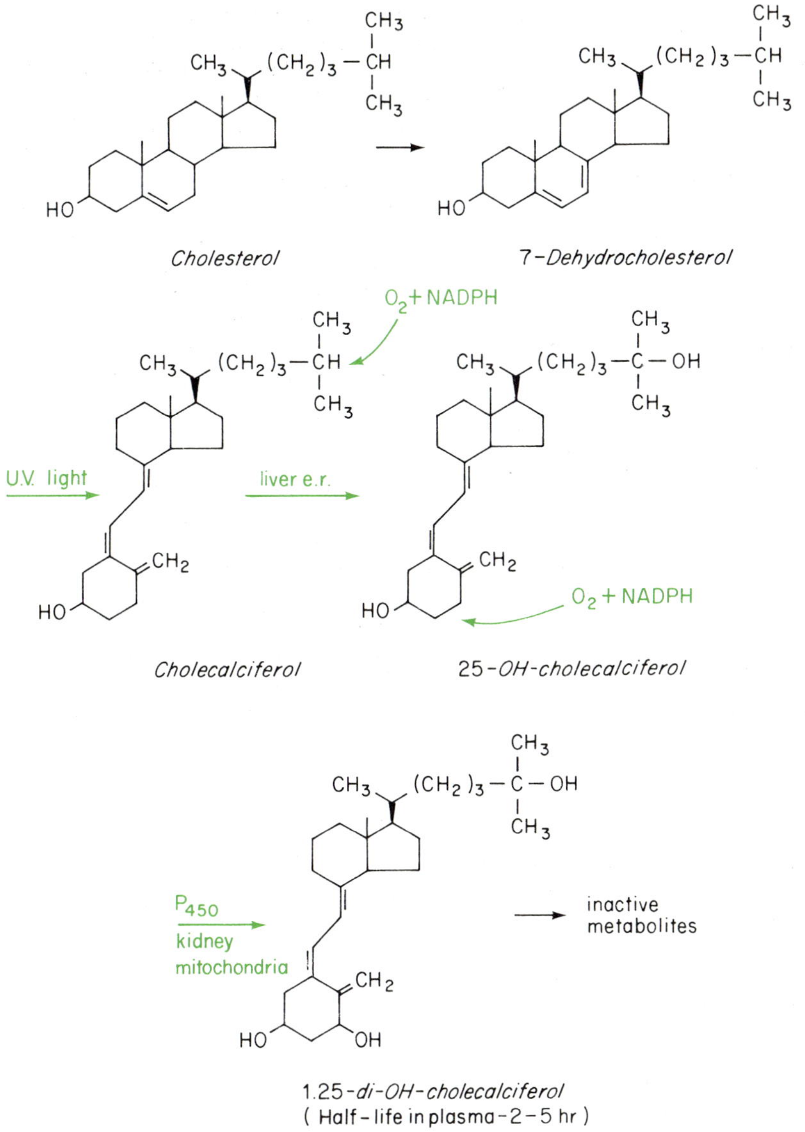

Fig. 17.10 The transformation of cholesterol into the major active metabolite 1,25-di-OH-cholecalciferol.

deficiency must be present for several years before softening of the bones is apparent (it is usually only seen associated with pregnancy and lactation).

Vitamin D has three effects: (1) to promote Ca absorption from the gut, (2) to promote mineral deposition in the skeleton, and (3) to inhibit phosphate and calcium excretion in the urine. None of these three effects is clearly established at the molecular level.

1. 1,25-di-OH-D$_3$ increases the absorptive capacity of the mucosal cells of duodenum and jejunum for calcium and phosphate. The hormone induces synthesis of an intracellular Ca-binding protein, but

it has been established that this occurs later than increased uptake from the lumen. A number of changes in the brush border are known to occur, but none is yet specifically linked to increased calcium uptake.

2. Bone matrix continues to be formed in the absence of vitamin D and epiphyseal growth zones are in fact greatly stimulated, but the matrix does not mineralize. Mineralization can be induced if the plasma phosphate level is brought up to normal by injection (cf. Table 17.2), but it is diffuse and not equivalent to normal mineralization from nucleation centres (cf. p. 271). It has however proved to be very difficult to

define a specific action of vitamin D on bone, other than on reabsorption. Possibly 24,25-di-OH-D$_3$ plays a more important part in bone formation.

3. 1,25-di-OH-D$_3$ stimulates tubular reabsorption of phosphate more than that of calcium. The mechanisms are unknown.

The best sources of vitamin D are fish liver oils and margarine. The requirement is about 500 IU per day. Unlike vitamin A, it is not stored in the liver and serious overdosage is toxic; it may cause calcification of the renal tubules. Several deaths, mostly of children, from vitamin D overdosage have been reported in the last few years.

Owing to the regulatory action of the parathyroid gland, the serum Ca level in rickets is rarely low, even when the rickets is directly due to Ca deficiency in the diet. Instead, the inorganic phosphate level in serum is usually diminished. This is made clear in Table 17.2, which shows mean values of serum Ca and P in normal subjects, in rickets, tetany, and hyperparathyroidism.

Vitamin E (tocopherol)

This vitamin is not known to be required by man. In animals its absence is said to cause many defects, ranging from sterility in rats to muscular dystrophy in rabbits. Tocopherol is fat-soluble and easily oxidized and reduced. It is known to protect unsaturated fats, and

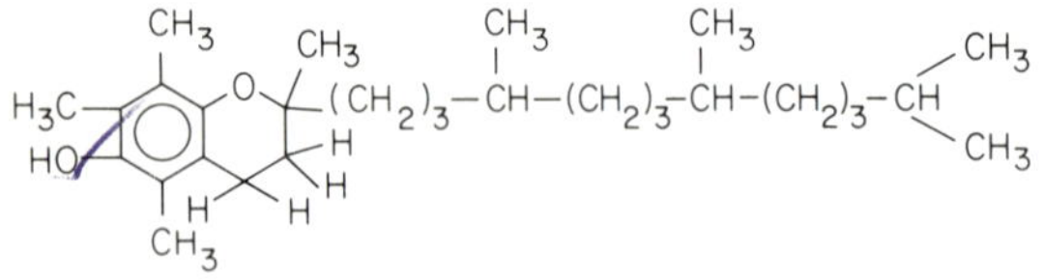

Vitamin E (α-tocopherol)

Several other tocopherols have vitamin E activity

particularly vitamin A, from oxidation. This it may do by inhibiting the formation of lipid peroxides. It is now widely added to artificially prepared dairy products, e.g. margarine. The human diet contains about 500 mg per day from many plant sources.

In animals, the functions of tocopherol and of selenium, especially in reproduction, are closely associated; a deficiency of one increases the need for the other. This is because both of them prevent peroxidation of lipids in membranes (selenium because it is a component of glutathione peroxidase (p. 140)).

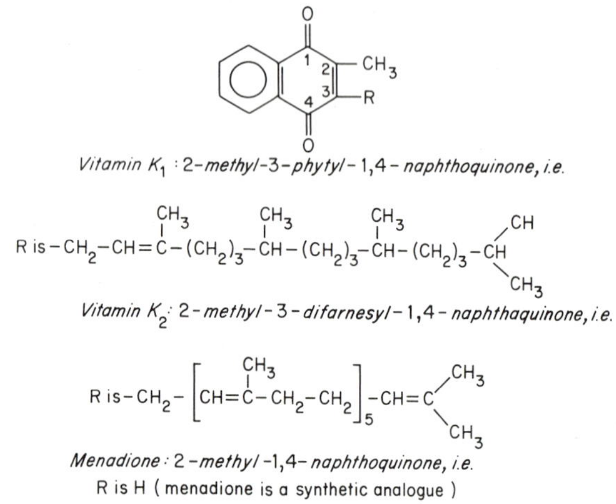

Table 17.2   **Serum Ca and P values**

	Average values, mg/100 ml (mmol)	
	Ca	P
Normal*	10.3 (2.6)	4.0 (1.3)
Rickets	9.0 (2.2)	3.0 (1.0)
Tetany	7.9 (2.0)	5.0 (1.6)
Hyperparathyroidism	15.9 (4.0)	2.2 (0.7)

* Adults. Inorganic P in children is normally much higher (about 5.6 mg/100 ml).

Vitamin K$_1$: 2-methyl-3-phytyl-1,4-naphthoquinone, i.e.

Vitamin K$_2$: 2-methyl-3-difarnesyl-1,4-naphthaquinone, i.e.

Menadione : 2-methyl-1,4-naphthoquinone, i.e.
R is H (menadione is a synthetic analogue)

Fig. 17.11   The K vitamins.

Vitamin K

A fat-soluble vitamin existing in two natural forms (see Fig. 17.11). Vitamin K$_1$ is found in green leafy plants. Vitamin K$_2$ has a different side chain (farnesyl instead of phytyl) and is synthesized by bacteria. Various synthetic compounds without either of these side chains are also potent vitamins. Lack of vitamin K gives rise to haemorrhage, because it is involved in the synthesis of carboxyglutamic acid, an amino acid residue which binds Ca^{2+}, and which is found only in prothrombin. The required daily amounts are so small that deficiency rarely arises, except in intestinal ailments or jaundice, which hinder its absorption, or during sulphonamide therapy, since intestinal bacteria may supply the requirement. It is often given as a routine precaution before operations or childbirth, but it is entirely ineffective as an immediate antidote to haemorrhage.

General Considerations on Vitamin Requirements

It is not easy to give more than rough estimates of vitamin needs for several reasons:

1. The requirements may vary with metabolic rate. This is particularly true of the B vitamins concerned with energy-supplying reactions. These must be supplied in greater amounts as metabolic activity increases.

2. The requirement may depend on the type of foodstuff. This is particularly true of thiamine; the requirement of this vitamin is often expressed as 1 mg per 4000 kJ carbohydrate oxidized.

3. On subsistence diets, an excess of one vitamin may precipitate a deficiency of another. This is again most true of the B vitamins, for obvious reasons, and it is considered unwise to supplement a diet with, say, nicotinamide or thiamine alone.

4. The rate of destruction of vitamins increases in fevers, so that the vitamin intake is often put up in convalescence. There is little evidence, however, that vitamins *prevent*, or increase resistance to, infections.

5. The requirement depends on the previous dietary history. Not all vitamins are stored to any extent; the most important stores are those of vitamins A, B_{12} and C.

6. An important source of supply of many vitamins is not the food as such, but bacteria living in the alimentary canal. It is difficult to make a quantitative estimate of this supply, but treatment of intestinal infections with insoluble sulphonamides, or other anti-bacterial drugs, has often been found to give rise to more or less serious avitaminoses. All the B vitamins, but particularly nicotinamide and B_{12}, and also vitamin K, are known to be absorbed from autolysing bacteria. Vitamins A, C, and D, however, are not synthesized by intestinal micro-organisms.

In certain circumstances bacteria in the intestine may themselves cause avitaminoses either by competing for the vitamins in food, or by causing conditions unfavourable to absorption (e.g. enteritis).

Index

$$\bar{x} = {}^{798}\!/_{10} \Rightarrow 79.8$$

$$\bar{y} = 819/10 \Rightarrow 81.9$$

$$S_x^2 = 64722/10 - (79.8)^2 = 104.16.$$

$$S_{xy} = 66045/10 - 79.8(81.9) \Rightarrow 68.88.$$